Urban Drainage

Fourth Edition

David Butler
Christopher Digman
Christos Makropoulos
John W. Davies

CRC Press
Taylor & Francis Group
Boca Raton London New York

CRC Press is an imprint of the
Taylor & Francis Group, an **informa** business

CRC Press
Taylor & Francis Group
6000 Broken Sound Parkway NW, Suite 300
Boca Raton, FL 33487-2742

© 2018 by Taylor & Francis Group, LLC
CRC Press is an imprint of Taylor & Francis Group, an Informa business

No claim to original U.S. Government works

Printed on acid-free paper by Ashford Colour Press Ltd

International Standard Book Number-13: 978-1-4987-5058-5 (Paperback)
978-1-4987-5293-0 (Hardback)

Library of Congress Cataloging-in-Publication Data

Names: Butler, David, 1959- author. | Digman, Christopher, author. | Makropoulos, Christos, author. | Davies, John W., author.
Title: Urban drainage / Professor David Butler, Dr. Christopher Digman, Dr. Christos Makropoulos and Professor John W. Davies.
Description: Boca Raton : Taylor & Francis, CRC Press, 2018. | Includes bibliographical references and index.
Identifiers: LCCN 2017043602 (print) | LCCN 2017045904 (ebook) | ISBN 9781498750592 (PDF) | ISBN 9781498750608 (VitalBook) | ISBN 9781498750615 (ePub) | ISBN 9781498750585 (pbk. : acid-free paper) | ISBN 9781498752930 (hardback : acid-free paper)
Subjects: LCSH: Urban runoff.
Classification: LCC TD657 (ebook) | LCC TD657 .B88 2018 (print) | DDC 628/.21--dc23
LC record available at https://lccn.loc.gov/2017043602

Visit the Taylor & Francis Web site at
http://www.taylorandfrancis.com

and the CRC Press Web site at
http://www.crcpress.com

Urban Drainage

Fourth Edition

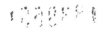

Contents

3 Wastewater   43

15   Structural design and construction   315

22 Smart systems

23 Global issues

Readership

In this book, we cover engineering and environmental aspects of the drainage of rainwater and wastewater from areas of human development. We present basic principles and engineering best practice. The principles are essentially universal but, in this book, are mainly illustrated by UK practice. We have also included introductions to current developments and recent research.

The book is primarily intended as a text for students on undergraduate and postgraduate courses in Civil or Environmental Engineering and researchers in related fields. We believe engineering aspects are treated with sufficient rigour and thoroughness to be of value to practising engineers as well as students, though the book does not take the place of an engineering manual.

The basic principles of drainage include wider environmental and societal issues, and these are of significance not only to engineers, but also to all with a serious interest in the urban environment, such as students, researchers and practitioners in environmental science, technology, policy and planning, geography and health studies. These wider issues are covered in particular parts of the book, deliberately written for a wide readership (indicated in the table overleaf). The material makes up a significant portion of the book, and if these sections are read together, they should provide a coherent and substantial insight into a fascinating and important environmental topic.

The book is divided into twenty-four chapters, with numerical examples throughout, and problems at the end of each chapter. Comprehensive reference lists that point the way to further, more detailed information, support the text. Our aim has been to produce a book that is both comprehensive and accessible, and to share our conviction with all our readers that urban drainage is a subject of extraordinary variety and interest.

Chapter	Coverage of wider issues
1	All
2	2.5, 2.6, 2.7
11	11.1
12	12.1, 12.2, 12.3
16	16.1, 16.2
17	17.1, 17.2
18	18.1, 18.2
19	19.1, 19.2, 19.3
21	21.1, 21.2, 21.4, 21.5
23	All
24	All

Acknowledgements

We remain hugely grateful to those who gave us support in writing the first through third editions of this book and to all our colleagues around the world.

For help with the fourth edition, we particularly thank Dr. Arturo Casal-Campos, Dr. Albert Chen, Dr. Peter Melville-Shreeve, and Dr. Fanlin Meng, University of Exeter; Panayiotis Kossieris, NTUA; Professor Richard Ashley, Ecofutures; Trevor Hillas, WSP/Parsons Brinckerhoff; Richard Kellagher, HR Wallingford; Steve Kenney, United Utilities; Peter Bridgens, Yorkshire Water; Andrew Arnison, Jon Boon, Dorotka Browne, Gerrard Buffham, Ian Cranshaw, Mark Futter, Adrian Johnson, Martin Long, Matthew Mullarkey, Steve Parkinson, Stephen Riisneas, and the late Chris Barker, MWH.

Notation

a	constant	$C1\text{-}C4$	empirical coefficients
a_{50}	effective surface area for infiltration	d	depth of flow
A	catchment area		depression storage
	cross-sectional area	d'	sediment particle size
	plan area	d_c	critical depth
A_b	area of base	d_m	hydraulic mean depth
A_D	impermeable area from which runoff received	d_n	normal depth
		d_u	upstream depth
A_{gr}	sediment mobility parameter	d_1	depth upstream of hydraulic jump
A_i	impervious area	d_2	depth downstream of hydraulic jump
A_o	area of orifice	d_{50}	sediment particle size larger than 50% of all particles
A_p	gully pot cross-sectional area		
$API5$	FSR 5-day antecedent precipitation index	D	internal pipe diameter
			rainfall duration
ARF	FSR rainfall areal reduction factor		wave diffusion coefficient
b	width of weir		longitudinal dispersion coefficient
	sediment removal constant	D_o	orifice diameter
	constant	D_{gr}	sediment dimensionless grain size
b_p	width of Preissman slot	D_P	gully pot diameter
b_r	sediment removal constant (runoff)	DWF	dry weather flow
b_s	sediment removal constant (sweeping)	e	voids ratio
B	flow width		sediment accumulation rate in gully
B_c	outside diameter of pipe	E	specific energy
B_d	downstream chamber width (high side weir)		gully hydraulic capture efficiency
			industrial effluent flow-rate
	width of trench at top of pipe	$EBOD$	Effective BOD_5
B_u	upstream chamber width (high side weir)	f	soil infiltration rate
			potency factor
c	concentration	f_c	soil infiltration capacity
	channel criterion	f_o	soil initial infiltration rate
	design number of appliances	f_s	number of sweeps per week
	wave speed	f_t	soil infiltration rate at time t
c_0	dissolved oxygen concentration	F_m	bedding factor
c_{0s}	saturation dissolved oxygen concentration	F_r	Froude number
c_v	volumetric sediment concentration	Fs	specific force
c_w	weir height	F_{se}	factor of safety
C	runoff coefficient	g	acceleration due to gravity
	consequence of an occurrence	G	water consumption per person
C_d	coefficient of discharge	G'	wastewater generated per person
C_v	volumetric runoff coefficient	h	head
C_R	dimensionless routing coefficient	h_a	acceleration head

h_f	head loss due to friction	m	Weibull's event rank number
h_L	total head loss		reservoir outflow exponent
h_{local}	local head loss	M	mass
h_{max}	depth of water		empirical coefficient
	gully pot trap depth	M_s	mass of pollutant on surface
H	total head	$MT\text{-}D$	FSR rainfall depth of duration D with
	difference in water level		a return period T
	height of water surface above weir	n	number
	crest		Manning's roughness coefficient
	depth of cover to crown of pipe		porosity
H_c	height of culvert (internal vertical	n_{DU}	number of discharge units
	dimension)	N	total number
H_{min}	minimum difference in water level for	$NAPI$	new antecedent precipitation index
	non-drowned orifice	O	outflow rate
i	rainfall intensity	p	pressure
i_e	effective rainfall intensity		probability of appliance discharge
i_n	net rainfall intensity		BOD test sample dilution
I	inflow rate		projection ratio
	pipe infiltration rate	P	wetted perimeter
	rainfall depth		perimeter of infiltration device
IF	effective impervious area factor		population
$IMKP$	maximum rainfall density over 5		power
	minutes		probability
j	time		height of weir crest above channel
J	housing density	P_d	downstream weir height (high side
	criterion of satisfactory service		weir)
	empirical coefficient	P_F	peak factor
k	constant	PF	porosity factor
k_b	effective roughness value of sediment	PI	precipitation index
	dunes	P_s	surcharge pressure
k_{DU}	dimensionless frequency factor	P_u	upstream weir height (high side weir)
k_L	local head loss constant	$PIMP$	FSR percentage imperviousness
k_s	pipe roughness	PR	WP percentage runoff
k_T	constant at T °C	q	flow per unit width
k_1	depression storage constant		appliance flow-rate
k_2	Horton's decay constant	Q	flow-rate
k_3	unit hydrograph exponential decay	Q_{av}	average flow-rate
	constant	Q_b	gully bypass flow-rate
k_4	pollutant washoff constant	Q_c	gully capacity
k_5	amended pollutant washoff constant	Q_d	continuation flow-rate (high side weir)
k_{20}	constant at 20°C	Q_f	pipe-full flow-rate
K	routing constant	Q_{max}	maximum flow
	constant in CSO design (Figure 12.14)	Q_{min}	minimum flow
	Rankine's coefficient	Q_o	wastewater baseflow
	empirical coefficient	Q_p	peak flow-rate
K_{LA}	volumetric reaeration coefficient	Q_r	runoff flow-rate
K_{PBOD}	BOD$_5$ potency factor	Q_u	inflow (high side weir)
K_{Pn}	pollutant n potency factor	$\bar{Q}$	gully approach flow
L	length	$\bar{Q}_L$	limiting gully approach flow
	load-rate	r	probability that an event will equal or
	gully spacing		exceed the design storm at least once in
L_E	equivalent pipe length for local losses		N years
L_I	initial gully spacing		number of appliances discharging
L_w	weir length		simultaneously

	FSR ratio of 60 min to 2 day 5 year return period rainfall	v_t	threshold velocity required to initiate movement
r_b	oxygen consumption rate in the biofilm	V	volume
		V_f	volume of first flush
r_s	oxygen consumption rate in the sediment	V_I	inflow volume
		V_o	outflow volume
r_{sd}	settlement deflection ratio		baseflow volume in approach time
r_w	oxygen consumption rate in the bulk water	V_t	basic treatment volume
		w	channel bottom width
R	hydraulic radius		pollutant-specific exponent
	ratio of drained area to infiltration area	W	width of drainage area
	total risk		pollutant washoff rate
R_e	Reynolds number	W_b	sediment bed width
Res	urban drainage resilience index	W_c	soil load per unit length of pipe
$RMED$	FEH median of annual rainfall maxima	W_{csu}	concentrated surcharge load per unit length of pipe
s	ground slope	W_e	effective sediment bed width
S	storage volume		external load per unit length of pipe
	soil moisture storage depth	W_s	settling velocity
S_c	critical slope	W_t	crushing strength per unit length of pipe
S_d	sediment dry density		
S_f	hydraulic gradient or friction slope	W_w	liquid load per unit length of pipe
S_G	specific gravity	x	longitudinal distance
S_o	pipe, or channel bed, slope		return factor
$SAAR$	FSR standard average annual rainfall	X	chemical compound
SMD	FSR soil moisture deficit	y	depth
$SOIL$	FSR soil index	Y	chemical element
SPR	standard percentage runoff	Y_d	downstream water depth (high side weir)
t	time		
	pipe wall thickness	Y_u	upstream water depth (high side weir)
t'	duration of appliance discharge	z	potential head
t_c	time of concentration		side slope
t_e	time of entry	Z	reflectivity
t_f	time of flow	Z	index of hydrogen sulphide generation
t_F	mean duration of flooding	Z'	modified Z formula index
T_p	time for hydrograph to reach peak value	Z_1, Z_2	pollutant specific constant
t_p	time to peak		FSR growth factor
T	rainfall event return period	$\bar{z}$	depth to centroid of flow cross-section
	wastewater temperature	α	channel side slope angle to horizontal
	pump cycle (time between starts)		number of reservoirs
T'	mean interval between appliance use		turbulence correction factor
T_a	approach time		empirical coefficient
T_c	time between gully pot cleans	β	empirical coefficient
u	unit hydrograph ordinate	γ	empirical coefficient
U^*	shear velocity	ε	empirical coefficient
$UCWI$	FSR urban catchment wetness index		gully pot sediment retention efficiency
v	mean velocity	ε'	gully pot cleaning efficiency
v_c	critical velocity	ζ	sediment washoff rate
v_f	pipe-full flow velocity	η	sediment transport parameter
V_F	total flood volume		pump efficiency
v_{GS}	gross solid velocity	θ	transition coefficient for particle Reynold's number
v_L	limiting velocity without deposition		
v_{max}	maximum flow velocity		angle subtended by water surface at centre of pipe
v_{min}	minimum flow velocity		

	Arrhenius temperature correction factor	ν	kinematic viscosity
Φ	vertical slope angle	ρ	density
κ	sediment supply rate	τ_b	critical bed shear stress
λ	friction factor	τ_o	boundary shear stress
λ_b	friction factor corresponding to the sediment bed	γ	unit weight
			temperature correction factor
λ_c	friction factor corresponding to the pipe and sediment bed	χ	surface sediment load
		χ_u	ultimate (equilibrium) surface sediment load
λ_g	friction factor corresponding to the grain shear factor		
		ω	counter
μ	coefficient of friction	ψ	shape correction factor for part-full pipe
μ'	coefficient of sliding friction		

Units are not specifically included in this notation list, but have been included in the text.

Abbreviations

AA	Annual allowance		CSO	Combined sewer overflow
AMP	Asset management planning		DCLG	Department for Communities and Local Government
AOD	Above ordnance datum			
ARF	Areal reduction factor		DD	Directional drilling
ASCE	American Society of Civil Engineers		DDF	Depth-duration-frequency
ASR	Aquifer storage recovery		DDT	Dichlorodiphenyltrichloroethane
ATU	Allylthiourea		DEFRA	Department for environment, food and rural affairs
AWWA	American Waterworks Association			
BAU	Business as usual		DEM	Digital elevation model
BeST	Benefits of SuDS Tool		DG5	OFWAT performance indicator
BHRA	British Hydrodynamics Research Association		DO	Dissolved oxygen
			DoE	Department of the Environment
BMP	Best management practice		DOT	Department of Transport
BOD	Biochemical oxygen demand		DN	Nominal diameter
BRE	Building Research Establishment		DSD	Drop size distribution
BS	British Standard		DTM	Digital terrain model
CA	Cellular automata		DU	Discharge unit
CAD	Computer aided drawing/design		EA	Environment Agency
CAPEX	Capital expenditure		EC	Escherichia coli
CARP	Comparative acceptable river pollution procedure		ECOsan	Ecological sanitation
			EEA	European Environment Agency
CBOD	Carbonaceous biochemical oxygen demand		EEC	European Economic Community
			EFRA	Exceedance flood risk assessment
CCD	Charge coupling device		EGL	Energy grade line
CCF	Climate change factor		EMC	Event mean concentration
CCTV	Closed-circuit television		EM-DAT	Emergency Event Database
CCW	Consumer Council for Water		EN	European Standard
CDM	Construction (Design and Management) Regulations		EPA	Environmental Protection Agency (US)
			EPS	Ensemble prediction systems
CEC	Council of European Communities		EQO	Environmental quality objectives
CEN	European Committee for Standardisation		EQS	Environmental quality standards
			EWPCA	European Water Pollution Control Association
CESMM4	Civil engineering standard method of measurement, fourth edition		EWS	Early warning systems
			EU	European Union
CFD	Computational fluid dynamics		FC	Faecal coliform
CIWEM	Chartered Institution of Water and Environmental Management		FEH	Flood Estimation Handbook
			FEH99	Flood estimation handbook model (1999 version)
CIH	Critical input hyetograph			
CIRIA	Construction Industry Research and Information Association		FEH13	Flood estimation handbook model (2013 version)
COD	Chemical oxygen demand			

FFC	Flood Forecasting Centre		MAFF	Ministry of Agriculture, Fisheries and Food
FFG	Flash flood guidance		MC	Monte Carlo
FGS	Flood guidance statement		MDG	Millennium development goal
FIO	Faecal indicator organism		MDPE	Medium density polyethylene
FOG	Fats, oils and grease		MH	Manhole
FORGEX	FEH focused rainfall growth curve extension method		MPC	Model-based Predictive Control
FS	Faecal streptococci		MPN	Most probable number
FSR	Flood Studies Report		NBOD	Nitrogenous biochemical oxygen demand
FWR	Foundation for Water Research		NERC	Natural Environment Research Council
GHG	Greenhouse gas			
GIS	Geographical information system		NGO	Non-governmental organisation
GL	Ground level		NOAA	National Ocean and Atmospheric Administration
GLUE	Generalised likelihood uncertainty estimation		NOD	Nitrogenous oxygen demand
GMT	Greenwich Mean Time		NRA	National Rivers Authority
GRA	Global resilience analysis		NWC	National Water Council
GRP	Glass reinforced plastic		NWP	Numerical weather predictions
HF	High frequency		NWS	National Weather Service (US)
HDPE	High density polyethylene		Ofwat	Water Services Regulation Authority
HGL	Hydraulic grade line			
HMSO	Her Majesty's Stationery Office		OD	Outside diameter
HR	Hydraulics Research		OGC	Open geospatial consortium
HRS	Hydraulics Research Station		Open MI	Open Modelling Interface and Environment
HSE	Health and Safety Executive			
IAHR	International Association of Hydraulic Engineering and Research		OPEX	Operational expenditure
IAWPRC	International Association on Water Pollution Research and Control		OPP	Orangi pilot project
			OS	Ordnance Survey
IAWQ	International Association on Water Quality		PAH	Polyaromatic hydrocarbons
			PCB	Polychlorinated biphenyl
ICBM	Integrated component-based model		PE	Polyethylene
ICE	Institution of Civil Engineers		PFA	Pulverised fuel ash
ICP	Inductively coupled plasma		PID	proportional-integral-derivative
IDF	Intensity - duration - frequency		PLC	Programmable logic controller
IE	Intestinal enterococci		POST	Parliamentary Office of Science and Technology
IL	Invert level			
IPCC	Intergovernmental Panel on Climate Change		PPE	Personal protective equipment
			PVC-U	Unplasticised polyvinylchloride
IoH	Institute of Hydrology		PU	Polyurethane
IUD	Integrated urban drainage		QUALSOC	Quality impacts of storm overflows: consent procedure
IUDM	Integrated urban drainage model			
IUWCM	Integrated urban water cycle model		RBF	Radial basis functions
IUWSM	Integrated urban water system model		RCP	Representative concentration pathway
IWA	International Water Association			
IWEM	Institution of Water and Environmental Management		ReFH	Revitalised flood hydrograph model
			RRL	Road Research Laboratory
LC50	Lethal concentration to 50% of sample organisms		RTC	Real-time control
			RTU	Remote terminal units
LiDAR	Light detection and ranging		RWH	Rainwater harvesting
LOD	Limit of deposition		SAR	Synthetic aperture radar
LSTD	Limiting solids transport distance		SAAR	Standard annual average rainfall
MAC	Maximum allowable concentration			

SCADA	Supervisory Control and Data Acquisition	TWL	Top water level
SCM	Stormwater control measure	UCWI	Urban catchment wetness index
SDD	Scottish Development Department	UHF	Ultra high frequency
SDI	Sustainable Development Indicators	UK	United Kingdom
SDG	Sustainable Development Goal	UKCIP	UK climate impacts programme
SEPA	Scottish Environmental Protection Agency	UKWIR	United Kingdom Water Industry Research
SOD	Sediment oxygen demand	ULFT	Ultra low flush toilet
SG	Specific gravity	UPM	Urban pollution management
SM	Sewerage Management Plan	UWOT	Urban water optioneering tool
SRM	Sewerage Rehabilitation Manual	UWWTD	Urban Wastewater Treatment Directive
SRM	Sewerage risk management [note now two definitions of SRM]	VHF	Very high frequency
SS	Suspended solids	VIP	Ventilated improved pit latrine
SRES	Special report on emissions scenarios	VSAT	Very small aperture terminal
STC	Standing Technical Committee	WAA	Water Authorities Association
STEPS	Short-term ensemble prediction system	WaSSP	Wallingford Storm Sewer Package
SUDS	Sustainable (urban) drainage systems	WaPUG	Wastewater Planning User Group
SuDS	Sustainable drainage system	WC	Water closet (toilet)
SWMP	Surface water management plan	WEF	Water Environment Federation (US)
SWMM	Stormwater management model	WFD	Water Framework Directive
SWO	Stormwater outfall	WMO	World Meteorological Organisation (include?)
TBC	Toxicity-based consents	WMO	World Meteorological Organisation
TISCIT	Totally Integrated Sonar and CCTV Inspection Technique	WP	Wallingford Procedure
TKN	Total Kjeldahl nitrogen	WPCF	Water Pollution Control Federation (US)
TNO	The Netherlands Organisation for applied scientific research	WSA	Water Services Association
TOC	Total organic carbon	WSUD	Water sensitive urban design
TOTEX	Total expenditure	WTP	Wastewater treatment plant
TRRL	Transport & Road Research Laboratory	WO	Welsh Office
TWL	Top water level	WRAP	Winter rain acceptance potential
		WRc	Water Research Centre

Authors

David Butler, FREng, is professor of Water Engineering and co-director of the Centre for Water Systems at the University of Exeter, United Kingdom.

Christopher Digman is technical director for urban drainage at Stantec and a visiting professor at the University of Sheffield, United Kingdom.

Christos Makropoulos is an associate professor in the School of Civil Engineering at the National Technical University of Athens, Greece.

John W. Davies is a lecturer in Civil Engineering at the University of Plymouth and emeritus professor of Civil Engineering at Coventry University, United Kingdom.

Chapter 1

Introduction

In this chapter, which forms an introduction and scene-setter for the whole book, we introduce exactly what urban drainage is (Section 1.1) and its relationship with urbanisation (1.2). The next three sections (1.3 through 1.5) introduce in turn the particular need for this infrastructure system, the history of its implementation, and then some diverse geographical examples. The various types of urban drainage system are introduced in Section 1.6 and placed in the wider context of urban water systems in Section 1.7. The chapter concludes with a discussion of the changing context and drivers affecting the development of urban drainage (1.8).

1.1 WHAT IS URBAN DRAINAGE?

Drainage systems are needed in developed urban areas because of the interaction between human activity and the natural water cycle. This interaction has two main forms: the abstraction of water from the natural cycle to provide a water supply for human life, and the covering of land with impermeable surfaces that divert rainwater away from the local natural system of drainage. These two types of interaction give rise to two types of water that require drainage.

The first type, *wastewater*, is water that has been supplied to support life, maintain a standard of living, and satisfy the needs of industry. After use, if not drained properly, it could cause pollution and create health risks. Wastewater contains dissolved material, fine solids, and larger solids, originating from WCs, from washing of various sorts, from industry, and from other water uses.

The second type of water requiring drainage, *stormwater*, is rainwater (or water resulting from any form of precipitation) that has fallen on a built-up area. If stormwater were not drained properly, it would cause inconvenience, damage, flooding, and further health risks. It contains some pollutants, originating from rain, the air, or the catchment surface.

Urban drainage systems handle these two types of water with the aim of managing the impact on human life and the environment. Thus, urban drainage has two major interfaces: with the public and with the environment (Figure 1.1). The public is usually on the transmitting rather than receiving end of services from urban drainage ("flush and forget"), and this may partly explain the lack of public awareness and appreciation of a vital urban service.

In many urban areas, drainage is based on a completely artificial system of sewers: pipes and structures that collect and dispose of this water. In contrast, isolated or low-income communities normally have no main drainage. Wastewater is treated locally (or not at all), and stormwater is drained naturally into the ground. These sorts of arrangements have generally existed when the extent of urbanisation has been limited. However, as discussed

Figure 1.1 Interfaces with the public and the environment.

later in the book, current practice encourages the use of more natural and non-pipe-based drainage arrangements wherever possible.

Urban drainage presents a classic set of modern environmental challenges: the need for cost-effective and socially acceptable technical improvements in existing systems, the need for assessment of the impact of those systems, and the need to search for sustainable and resilient solutions. As in all other areas of environmental concern, these challenges cannot be considered to be the responsibility of one profession alone. Policy-makers, engineers, planners, builders, environment specialists, together with all citizens, have a role. And these roles must be played in partnership. Engineers must understand the wider issues, while those who seek to influence policy must have some understanding of the technical problems. This is the reasoning behind the format of this book, as explained in the Preface. It is intended as a source of information for all those with a serious interest in the urban environment.

1.2 EFFECTS OF URBANISATION

Let us consider further the effects of human development on the passage of rainwater. Urban drainage replaces one part of the natural water cycle and, as with any artificial system that takes the place of a natural one, it is important that the full effects are understood.

In nature, when rainwater falls on a natural surface, some water returns to the atmosphere through evaporation, or transpiration by plants; some infiltrates the surface and becomes groundwater; and some runs off the surface (Figure 1.2a). The relative proportions depend on the nature of the surface, and vary with time during the storm. (Surface runoff tends to increase as the ground becomes saturated.) Both groundwater and surface runoff are likely to find their way to a river, but surface runoff arrives much faster. The groundwater will

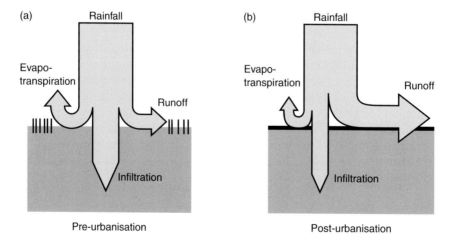

Figure 1.2 Effect of urbanisation on rate of rainfall.

become a contribution to the river's general baseflow rather than being part of the increase in flow due to any particular rainfall.

Development of an urban area, involving covering the ground with artificial surfaces, has a significant effect on these processes. The artificial surfaces increase the amount of surface runoff in relation to infiltration, and therefore increase the total volume of water reaching the river during or soon after the rain (Figure 1.2b). Surface runoff travels quicker over hard surfaces and through sewers than it does over natural surfaces and along natural streams. This means that the flow will both arrive and die away faster; therefore, the peak flow will be greater (see Figure 1.3). (In addition, reduced infiltration means poorer recharge of groundwater reserves.)

This obviously increases the danger of sudden flooding of the river. It also has strong implications for water quality. The rapid runoff of stormwater is likely to cause pollutants and sediments to be washed off the surface or scoured by the river. In an artificial environment, there are likely to be more pollutants on the catchment surface and in the air than there would be in a natural environment. Also, drainage systems in which there is mixing of wastewater and stormwater may allow pollutants from the wastewater to enter the river.

The existence of wastewater in significant quantities is itself a consequence of urbanisation. Much of this water has not been made particularly "dirty" by its use. Just as it is a standard convenience in a developed country to turn on a tap to fill a basin, it is a standard convenience to pull the plug to let the water "disappear." Water is also used as the principal medium for disposal of bodily waste, and varying amounts of bathroom litter, via WCs.

In a developed system, much of the material that is added to the water while it is being turned into wastewater is removed at a wastewater treatment plant prior to its return to

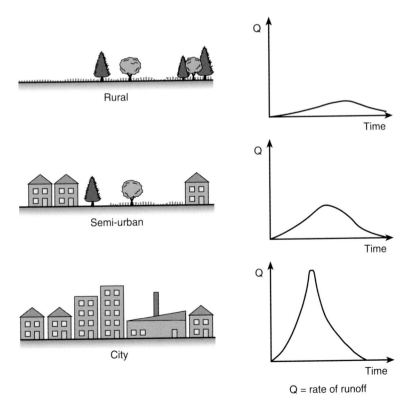

Figure 1.3 Effect of urbanisation on peak rate of runoff.

the urban water cycle. Nature itself would be capable of treating some types of material, bodily waste for example, but not in the quantities created by urbanisation. The proportion of material that needs to be removed will depend in part on the capacity of the river to assimilate what remains.

So the general effects of urbanisation on drainage, or the effects of replacing natural drainage by urban drainage, are to produce higher and more sudden peaks in river flow, to introduce pollutants, and to create the need for artificial wastewater treatment. While to some extent impersonating nature, urban drainage may also impose heavily upon it.

1.3 URBAN DRAINAGE PRIORITIES

1.3.1 Public health

In human terms, the most valuable benefit of an effective urban drainage system is the maintenance of public health. This particular objective is often overlooked in modern practice and yet is of extreme importance, particularly in protection against the spread of diseases.

Despite the fact that some vague association between disease and water had been known for centuries, it was in 1855 that a precise link was demonstrated. This came about as a result of the classic studies of John Snow in London concerning the cholera epidemic sweeping the city at the time. That diseases such as cholera are almost unknown in the industrialised world today is in major part due to the provision of centralised urban drainage (along with the provision of a microbiologically safe, potable supply of water).

Urban drainage has a number of major roles in maintaining public health and safety. Human excreta (particularly faeces) are the principal vector for the transmission of many communicable diseases. Urban drainage has a direct role in effectively removing excreta from the immediate vicinity of habitation. However, there are further potential problems in large river basins in which the downstream discharges of one settlement may become the upstream abstraction of another. In the United Kingdom (UK), some 30% of water supplies are so affected. This clearly indicates the vital importance of disinfection of water supplies as a public health measure.

Also, of particular importance in tropical countries, standing water after rainfall can be largely avoided by effective drainage. This reduces the mosquito habitat and hence the spread of malaria and other diseases.

While many of these problems have apparently been solved, it is essential that in industrialised countries, as we look for ever more innovative sanitation techniques, we do not lose ground in controlling serious diseases. Sadly, while we may know much about waterborne and water-related diseases, some rank among the largest killers in societies where poverty and malnutrition are widespread. Millions of people around the world still lack any hygienic and acceptable method of excreta disposal. The global issues associated with urban drainage in low-income communities are returned to in more detail in Chapter 23.

1.3.2 Minimising adverse impacts

It has already been stated that the basic function of urban drainage is to collect and convey wastewater and stormwater. In the UK and other developed countries, this has generally been taken to cover all wastewater, and all it contains (subject to legislation about hazardous chemicals and industrial effluents). For stormwater, the aim has been to remove rainwater (for storms up to a particular severity) with the minimum of inconvenience to activities on the surface.

Most people would see the efficient removal of stormwater as part of "progress." In a developing country, they might imagine a heavy rainstorm slowing down the movement of people and goods in a sea of mud, whereas in a city in a developed country they would probably consider that it should take more than mere rainfall to stop transport systems and businesses from running smoothly. Nowadays, however, as with other aspects of the environment, the nature of progress in relation to urban drainage, its consequences, desirability, and limits, are being closely and frequently reassessed.

The traditional aim in providing storm drainage has been to remove water from surfaces, especially roads, as quickly as possible. It is then disposed of, usually via a pipe system, to the nearest watercourse. This, as we have discovered in Section 1.2, can cause damage to the environment and increase the risk of flooding elsewhere. So, drainage systems should protect people and property from stormwater, and should also limit the impact of the drained flow on the receiving water. For this reason, and to achieve other environmental and social benefits, interest in more natural methods of disposing of stormwater is increasing. These include infiltration, storage, and vegetated systems (discussed in full in Chapter 21), and the general intention is to attempt to reverse the trend illustrated in Figure 1.3: to decrease the peak flow of runoff and increase the time it takes to reach the watercourse.

Another way in which attempts are being made to reverse the effects of urbanisation on drainage described in Section 1.2 is to reduce the non-biodegradable content in wastewater. Public campaigns have been mounted to persuade people not to treat the WC as a rubbish bin.

These tendencies towards reducing the dependence on "hard" engineering solutions to solve the problems created by urbanisation, and the philosophy that goes with them, are associated with the word *sustainability* and are further considered in Chapter 24.

The imperative of climate change affects urban drainage directly and indirectly. Clearly our approaches to urban drainage must adapt to climate change because of the predicted changes in rainfall patterns. We consider this in Chapter 4. Also, drainage engineers must play their part in minimising carbon emissions (see Chapter 14 for further discussion).

1.4 HISTORY

1.4.1 Ancient civilisations

There is plenty of archaeological evidence that ancient civilisations demonstrated no less skill in drainage provision than in other aspects of urban infrastructure. Here we look briefly at some interesting examples of success in creating effective drainage systems, without modern scientific knowledge.

The Minoans on Crete (between about 3000 and 1000 BC) had some well-developed drainage systems, for stormwater and wastewater. At the palace of Knossos there is evidence of separate stone conduits for stormwater and for wastewater, and even an early type of flushed toilet (De Feo et al., 2014).

At a similar time, in parts of what is now Afghanistan, Pakistan, and north-west India, the civilisation of the Indus Valley excelled in urban planning and infrastructure, including provision of drainage. Typically, drainage channels were built into the centres of streets, and these carried both stormwater and wastewater. Wastewater did not flow directly into the channel. From each house it flowed first via a pipe to a small pit where solids settled out; at a certain depth liquids overflowed into the channel. In some cases the channels were covered with stones or bricks (Burian and Edwards, 2002).

Many towns and cities in ancient Greece (around 500–100 BC) had covered drainage channels built into the streets. An interesting detail of drainage in ancient Greece is that most

Greek theatres, being large, impermeable, outdoor spaces, had well-thought-out arrangements for draining stormwater. Existing parts of the theatre's structure, including stairs and corridors, led the runoff towards the lower part of the theatre, the circular "orchestra" or main performing space. A duct, semi-circular in plan, sometimes covered, at the edge of the orchestra, collected the runoff and carried it away to an outfall (De Feo et al., 2014).

The Romans (particularly during the period from 500 BC to 300 AD) are well known for their public health engineering feats, particularly the impressive aqueducts bringing water into their cities; but they also excelled in systems to take water away. And as well as removal of water, they placed importance on systems for reuse of rainwater. Urban drainage was achieved by sophisticated systems of open channels and buried sewers ("cloacae"). In Rome these provided drainage for a population of up to 1 million, including 900 public baths. The largest and most well-known of the cloacae is the Cloaca Maxima, which drained the lowest-lying parts of Rome, including the Forum, to the River Tiber. Part of its function was to provide land drainage for marshy areas, and in doing so reduced the presence of mosquitoes and therefore the risk of malaria. Construction of the Cloaca Maxima started in the sixth century BC, and it is still in use today.

1.4.2 Ancient to modern

After the Romans, urban drainage practices did not develop significantly for many centuries. In medieval Europe, drainage did not go far beyond the use of ditches that followed natural land drainage patterns. The English word *sewer* is derived from an Old French word, *esseveur*, meaning "to drain off," related to the Latin *ex-* (out) and *aqua* (water). The *Oxford English Dictionary* gives the earliest meaning as "an artificial water-course for draining marshy land and carrying off surface runoff into a river or sea."

It was not until the rapid growth of cities in the eighteenth and nineteenth centuries that things had to change. This was typically in response to serious problems with public health, leading to an urgent search for solutions and to developments in urban drainage practice. We consider the example of London.

1.4.3 London

In London, before the eighteenth century, sewers had the meaning given above and their alignment was loosely based on the natural network of streams and ditches that preceded them. In a quite unconnected arrangement, bodily waste was generally disposed of into cesspits (under the residence floor), which were periodically emptied. Flush toilets (discharging to cesspits) became quite common around 1770–1780, but it remained illegal until 1815 to connect the overflow from cesspits to the sewers. This was a time of rapid population growth and, by 1817, when the population of London exceeded one million, the only solution to the problem of under-capacity was to allow cesspit overflow to be connected to the sewers. Even then, the cesspits continued to be a serious health problem in poor areas, and, in 1847, 200,000 of them were eliminated completely by requiring houses to be connected directly to the sewers.

This moved the problem elsewhere – namely, the River Thames. By the 1850s, the river was filthy and stinking (Box 1.1) and directly implicated in the spread of deadly cholera.

There were cholera epidemics in 1848–1849, 1854, and 1867, killing tens of thousands of Londoners. The Victorian sanitary reformer Edwin Chadwick passionately argued for a dual system of drainage, one for human waste and one for rainwater: "the rain to the river and the sewage to the soil." He also argued for small-bore, inexpensive, self-cleansing sewer pipes in preference to the large brick-lined tunnels of the day. However, the complexity and

BOX 1.1   MICHAEL FARADAY'S ABRIDGED
LETTER TO *THE TIMES* OF 7 JULY, 1855

I traversed this day by steamboat the space between London and Hungerford Bridges [*on the River Thames*], between half-past one and two o'clock. The appearance and smell of water forced themselves on my attention. The whole of the river was an opaque pale brown fluid. The smell was very bad, and common to the whole of the water. The whole river was for the time a real sewer.

If there be sufficient authority to remove a putrescent pond from the neighbourhood of a few simple dwellings, surely the river which flows for so many miles through London ought not be allowed to become a fermenting sewer. If we neglect this subject, we cannot expect to do so with impunity; nor ought we to be surprised if, ere many years are over, a season give us sad proof of the folly of our carelessness.

cost of engineering two separate systems prevented his ideas from being put into practice. The solution was eventually found in a plan by Joseph Bazalgette to construct a number of "combined" interceptor sewers on the north and the south of the river to carry the contents of the sewers to the east of London. The scheme, an engineering marvel (Figure 1.4), was mostly constructed by 1875, and much of it is still in use today (Halliday, 2001).

Again, though, the problem had simply been moved elsewhere. This time, it was the Thames estuary, which received huge discharges of wastewater. Storage was provided to allow release on the ebb tide only, but there was no treatment. Downstream of the outfalls, the estuary and its banks were disgustingly polluted. By 1890, some separation of solids was carried out at works on the north and south banks, with the sludge dumped at sea. Biological treatment was introduced in the 1920s, and further improvements followed.

Figure 1.4 Construction of Bazalgette's sewers in London. (From *The Illustrated* London News, 27 August 1859, reproduced with permission of The Illustrated London News Picture Library.)

Table 1.1 Percentage of population connected to main sewers in selected European countries, 2009 figures

Country	Percentage (%) population connected to sewer
Greece	88
Netherlands	98
Norway	82
Poland	64
Romania	42
United Kingdom	96

Source: European Environment Agency 2013. Urban Waste Water Treatment. www.eea.europa.eu/data-and-maps/indicators/urban-waste-water-treatment/urban-waste-water-treatment-assessment-3

However, it was not until the 1970s that the quality of the Thames was such that salmon were commonplace and porpoises could be seen under Blackfriars Bridge.

1.5 GEOGRAPHY

The main factors that determine the extent and nature of urban drainage provision in a particular region are wealth; the climate and other natural characteristics; the intensity of urbanisation; and history, legislation, and politics. The greatest differences are the result of differences in wealth. Most of this book concentrates on urban drainage practices in countries that can afford fully engineered systems. The differences in countries that cannot will be apparent from Chapter 23 where we take a more global view.

Countries in which rainfall tends to be occasional and heavy have naturally adopted different practices from those in which it is frequent and generally light. For example, it is common in Australia to provide "minor" (underground, piped) systems to cope with low quantities of stormwater, together with "major" (overground) systems for larger quantities. We return to this concept in Chapter 11. Other natural characteristics have a significant effect. Sewers in the Netherlands, for example, must often be laid in flat, low-lying areas and, therefore, must be designed to run frequently in a pressurised condition.

Intensity of urbanisation and history of development have a strong influence on the percentage of the population connected to a main sewer system. Table 1.1 gives percentages in a number of European countries.

Historical and political factors determine the age of the system (which is likely to have been constructed during a period of significant development and industrialisation), characteristics of operation such as whether or not the water/wastewater industry is publicly or privately financed, and strictness of statutory requirements for pollution control and the manner in which they are enforced. Countries in the European Union (EU) are subject to common requirements, described throughout this book.

Boxes 1.2 through 1.4 present a selection of examples to give an idea of the wide range of different urban drainage issues throughout the world.

1.6 TYPES OF SYSTEM

Piped systems consist of drains carrying flow from individual properties, and sewers carrying flow from groups of properties or larger areas. The word *sewerage* refers to the whole

BOX 1.2 ORANGI PILOT PROJECT, PAKISTAN

Orangi, a "katchi abadi," or squatter community, in Karachi, has a population of over one million. Until the 1980s there were no sewers: people had to empty bucket latrines into the narrow alleys. In a special self-help programme, quite different from government-sponsored improvement schemes, the community built its own sewers, with no outside contractors. A small septic tank was placed between the toilet and the sewer to reduce the entry of solids into the pipe. The system had a simplified design and was built up alley-by-alley, as the people made a commitment to the improvements. This was a great success for community action and created major improvements in the immediate environment. In Orangi about 100,000 households developed their own sanitation systems, and in 11 other towns in Pakistan at least 40,000 households are known to have used the same approach (Hasan, 2006).

Growing from the work on sewers, the Orangi Pilot Project (OPP) has gone on to coordinate activity in a number of areas including housing, education, water supply, social development, and enterprise support (OPP-RTI, 2012). The OPP "has supported one of the world's largest programmes for improved provision for sanitation in low income areas – in Orangi and in many other cities and small urban centres – as well as supporting improvements in other forms of infrastructure and in services" (UN-Habitat, 2006, p67). "The growing influence of OPP can be seen not only in neighbourhood solutions but in developing city-wide policies and priorities" (UN-Habitat, 2003, p237).

BOX 1.3 BELO HORIZONTE, BRAZIL

Belo Horizonte, capital of the state of Minas Gerais, was the first modern "planned" city in Brazil. Streets are set out on a broadly spaced grid with ample green spaces. There is a high standard of drinking water provision with connections to virtually all residents. But although sewerage systems reach 92% of the population, there are serious issues with pollution and flooding of watercourses. This is partly because some of the main interceptor sewers needed to connect the sewerage system to the wastewater treatment plants have never been constructed. There has also been excessive development on flood-prone areas, and there is a serious problem with illicit interconnections between the wastewater and stormwater systems.

In the 1970s and 1980s, during rapid population growth, the emphasis was on heavily engineered river and drainage works. More recently water management processes have been democratised, with community participation in planning, in development of infrastructure, and even in budgeting. Starting in 2006, Belo Horizonte was also a demonstration city for the SWITCH project, an EU-funded initiative for the practical application of research on sustainable urban water management (Knauer et al., 2011). The main areas of activity have been rainwater harvesting, providing infiltration and detention facilities, and developing wetlands. Sites have been developed to demonstrate the effectiveness of the approaches and to engage stakeholders (including the local community, local officials, and engineers). Many of the sites have deliberately been located at schools to maximise the learning impact. Belo Horizonte is judged to be achieving genuine transition to integrated urban water management, which is bringing many benefits.

BOX 1.4   WASTEWATER HEAT RECOVERY IN VANCOUVER, CANADA

Olympic Village, Vancouver, is a mostly residential neighbourhood originally built as accommodation for athletes at the 2010 winter games. Approximately 70% of its heat energy needs are satisfied by heat recovered from wastewater. Heat that would otherwise be wasted is extracted from sewer flows using heat exchangers and transferred to a hot water distribution system (Baber, 2010). There is no contact between the wastewater itself and the external hot water system (Figure 1.5).

"It's very similar to geothermal energy," Chris Baber, project manager for the city of Vancouver's Neighbourhood Energy Utility, said of the sewer-heat system. Much as geothermal systems use heat exchangers to extract heat from the soil, the sewer-heat system uses exchangers to extract the otherwise waste heat from the city's sewers. The heat can then be used to warm up buildings and provide hot water.

At present the system provides hot water to buildings with a total floor area of nearly 400 000 m². The City of Vancouver plans to increase this to over 700 000 m² as part of its aim of achieving 100% renewable energy for Vancouver before 2050 (City of Vancouver, 2015).

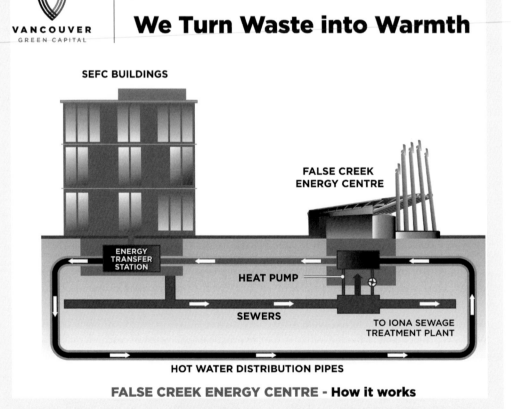

Figure 1.5 Sewer-heat recovery system in Vancouver. (Courtesy of the City of Vancouver.)

centralised infrastructure system: pipes, manholes, structures, pumping stations, and so on. There are basically two types of conventional sewerage: a combined system in which wastewater and stormwater flow together in the same pipe, and a separate system in which wastewater and stormwater are kept in separate pipes.

Non-pipe systems (called SuDS or sustainable drainage systems in the UK) manage *stormwater* flows closer to the source at which they are generated (decentralised), using the infiltration and storage properties of semi-natural features. Some schemes for reducing dependence on main drainage also involve more localised collection and treatment of *wastewater*. However, movements in this direction, while of great significance, are still only in their relatively early stages (as described in Chapter 24).

1.6.1 Combined systems

In the UK, most of the older sewerage is combined, and this accounts for about 70% by total length. Many other countries have a significant proportion of combined sewers: in France and Germany, for example, the figure is also around 70%, and in Denmark it is 45%.

A sewer network is a complex branching system, and Figure 1.6 presents a simplification of a typical arrangement, showing a very small proportion of the branches. The figure is a plan of a town located beside a natural water system of some sort: a river or estuary, for example. The combined sewers carry both wastewater and stormwater together in the same pipe, and the ultimate destination is the wastewater treatment plant (WTP), located, in this case, a short distance out of the town.

In dry weather, the system carries wastewater flow. During rainfall, the flow in the sewers increases as a result of the addition of stormwater. Even in quite light rainfall, the stormwater flows will predominate, and in heavy falls the stormwater could be 50 or even 100 times the average wastewater flow.

It is simply not economically feasible to provide capacity for this flow along the full length of the sewers – which would, by implication, carry only a tiny proportion of the capacity most of the time. At the WTP, it would also be unfeasible to provide this capacity in the treatment processes. The solution was to provide structures in the sewer system that, during

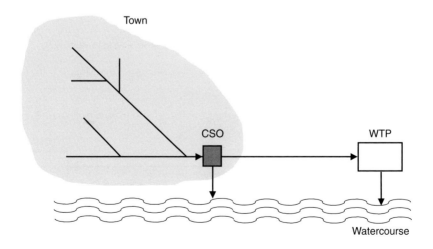

Figure 1.6 Combined system. (Schematic plan.)

medium or heavy rainfall, divert flows above a certain level out of the sewer system and into a natural watercourse. These structures are called combined sewer overflows (CSOs). A typically located CSO is included in Figure 1.6.

The key drawback of combined sewer overflows and by extension combined sewers in general is that large volumes of minimally treated wastewater are discharged during rainfall to the environment when they operate. To put it simply, CSOs cause pollution. Chapter 12 considers the design and operation of combined sewers and CSOs in more detail.

1.6.2 Separate systems

Most sewerage systems constructed in the UK since 1945 are separate (about 30%, by total length). Figure 1.7 is a sketch plan of the same town as shown in Figure 1.6, but this time sewered using the separate system. Wastewater and stormwater are carried in separate pipes, usually laid side by side. Wastewater flows vary during the day, but the pipes are designed to carry the maximum flow all the way to the WTP. The stormwater is not mixed with wastewater and can be discharged to the watercourse at a convenient point. The first obvious advantage of the separate system is that CSOs, and the pollution associated with them, are avoided.

An obvious disadvantage might be cost. It is true that the pipework in separate systems is more expensive to construct, but constructing two pipes instead of one does not cost twice as much. The pipes are usually constructed together in the same excavation. The stormwater pipe (the larger of the two) may be about the same size as the equivalent combined sewer, and the wastewater pipe will be smaller. So the additional costs are due to a slightly wider excavation and an additional, relatively small pipe.

Separate systems do have drawbacks of their own. These relate to the fact that perfect separation is effectively impossible to achieve. First, it is difficult to ensure that polluted flow is carried only in the wastewater pipe. Stormwater can be polluted for many reasons, including the washing-off of pollutants from the catchment surface. Second, it is very hard to ensure that no rainwater finds its way into the wastewater pipe. These points are considered in more detail in Chapter 5.

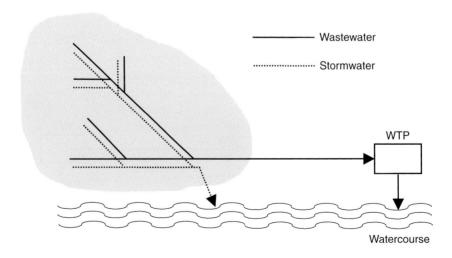

Figure 1.7 Separate system. (Schematic plan.)

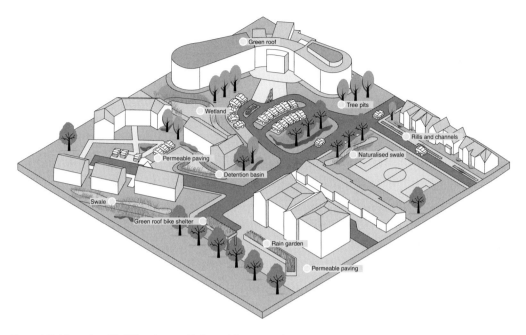

Figure 1.8 Non-pipe (SuDS) scheme. (Adapted from graphic on www.susdrain.org.)

1.6.3 Hybrid and partially separate systems

Some towns have hybrid systems, for example, "partially separate" systems, in which waste-water is mixed with some stormwater, while the remaining stormwater is conveyed by a separate pipe. Many other towns have hybrid systems for more accidental reasons: for example, because a new town drained by a separate system includes a small old part drained by a combined system, or because misconnections resulting from ignorance, malpractice, or malfunctioning dual manholes have caused unintended mixing of the two types of flow.

1.6.4 Non-pipe systems

Non-pipe systems manage stormwater closer to its source of generation, typically on or near the ground surface, using a variety of devices (see Figure 1.8). This has clear advantages and disadvantages. Advantages include the ability to limit flows downstream and hence contrib-ute to flood risk management, the ability to partially treat stormwater prior to discharge, and the opportunity to provide local amenity and biodiversity gains. The disadvantages involve the need for extra space, which is often at a premium in urban areas, and more practical matters concerning ownership and the maintenance of long-term performance. Stormwater management is considered in more detail in Chapter 21.

An initial comparison of the various systems is given in Table 1.2. We develop these ideas further throughout the remainder of this book.

1.7 URBAN WATER SYSTEM

In this section we look at how urban drainage fits within the whole urban water system. Figure 1.9 is a diagrammatic representation of the urban water system with a general

Table 1.2 Combined sewers, separate sewers and SuDS compared

	Combined sewers	Separate sewers	SuDS[a]
CSOs	Required part of system	Not required	Not required, and SuDs limit stormwater entering sewers, so reducing CSO operation
WTPs	Larger works inlets needed and stormwater storage required	Smaller works	Smaller works
Pumping	Higher if pumping of flow to treatment is necessary	Stormwater rarely pumped	Rarely required
Sediment	Deposition in pipes can lead to blockage and higher maintenance needs	More feasible to design for self-cleansing	Limits downstream discharge of sediment but requires on-site maintenance
Flooding (storm conditions)	If flooding occurs, foul conditions will be caused	Any flooding will be by stormwater only	Any flooding will be by stormwater only
Pipework	House drainage simple	More house drains, with risk of misconnections and difficult to identify	Reduced pipework overall
Construction costs	Lower than separate system	Some extra cost	Similar to or lower than separate system
Maintenance costs	Highest	Lower than combined	Different maintenance regime required
Space requirements	Minimum underground space	Additional underground space due to two pipes	Higher above-ground space normally needed
Flow rate	Attenuation in pipes and tanks	Attenuation in pipes and tanks	Attenuation at source
Flow volume	No reduction	No reduction	Significant reduction
Treatment of stormwater	No treatment	Minimal treatment	Significant treatment
Long-term performance	Systems in operation for over a century	Systems in operation for many decades	Systems in operation for years
Amenity	No contribution	No contribution	Significant contribution if carefully designed
Biodiversity	No contribution	No contribution	Significant contribution if carefully designed

[a] A piped system for wastewater would also be required for this case.

representation of the flow paths and the interrelationship of the main elements. Solid arrows represent intentional flows and dotted arrows unintentional ones. Heavy-bordered boxes indicate "sources," and dashed, heavy-bordered boxes show "sinks." This figure shows a separate sewer system, but a similar diagram can be produced for a combined system, and non-pipe systems can also be represented.

There are two main inflows. The first is rainfall that falls onto catchment surfaces such as "impervious" roofs and paved areas and "pervious" vegetation and soil. It is at this point that the quality of the flow is degraded as pollutants on the catchment surfaces are washed off. This is a highly variable input that can only be properly described in statistical terms (as considered in Chapter 4). The resulting runoff retains similar statistical properties to rainfall (Chapter 5). There is also the associated outflow of evaporation, whereby some water

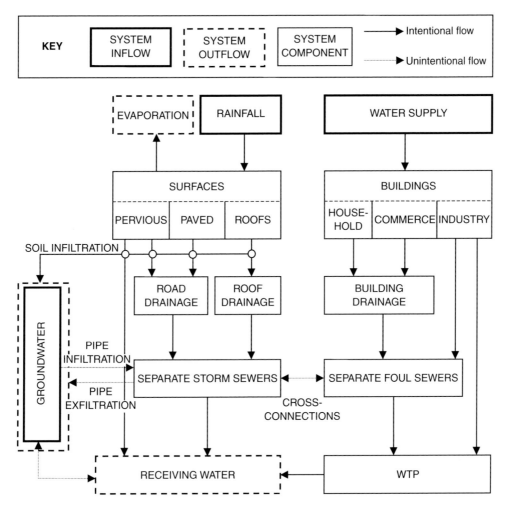

Figure 1.9 Urban drainage (separate sewers) and the urban water system.

is removed from the system. Rainfall that does not run off will find its way into the ground and eventually the receiving water or infiltrate into the drainage system. The component that runs off is conveyed by the roof and highway drainage as stormwater directly into the storm sewer. Discharges from the storm sewer to the environment are intermittent and are statistically related to the rainfall inputs.

The second inflow is water supply. Water consumption is more regular than rainfall, although even here there is some variability (Chapter 3). The resulting wastewater is closely related in timing and magnitude to the water supply. The wastewater is conveyed by the building drainage directly to the foul sewer. An exception is where an industry treats its own waste separately and then discharges treated effluent directly to the receiving water. The quality of the water (originally potable) deteriorates during usage. The foul sewer conveys the wastewater to the WTP with patterns related to the water consumption. Unintentional flow may leave the pipes via exfiltration to the ground (Chapter 3). At other locations, groundwater may act as a source and add water into the system via pipe infiltration (Chapter 3). Misconnections can cause unintentional mixing of the wastewater with stormwater in either pipe (Chapter 5). The treatment plant, in turn, discharges to the receiving water.

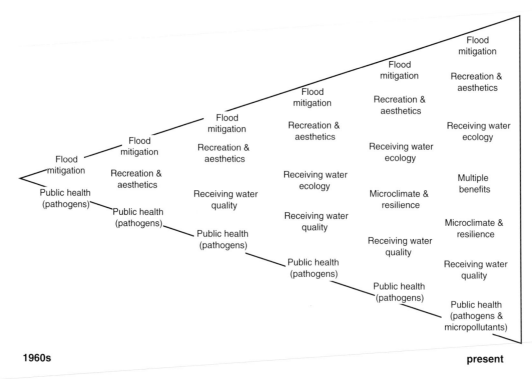

Flood
mitigation

Flood
mitigation

Flood
mitigation

Flood
mitigation

Flood
mitigation

Flood
mitigation

Flood
mitigation

Recreation &
aesthetics

Recreation &
aesthetics

Recreation &
aesthetics

Recreation &
aesthetics

Recreation &
aesthetics

Receiving water
ecology

Flood
mitigation

Public health
(pathogens)

Receiving water
quality

Receiving water
ecology

Receiving water
ecology

Multiple
benefits

Public health
(pathogens)

Public health
(pathogens)

Receiving water
quality

Microclimate &
resilience

Microclimate &
resilience

Public health
(pathogens)

Receiving water
quality

Receiving water
quality

Receiving water
quality

Public health
(pathogens)

Public health
(pathogens)

Public health
(pathogens &
micropollutants)

1960s   **present**

Figure 1.10 Changing context and drivers of urban drainage over time. (Adapted from Fletcher, T.D. et al. 2015. *Urban Water Journal*, 12(7), 525–542.)

Non-pipe solutions are effectively included in the "SURFACES" components of Figure 1.9. Alterations to paved and roof surfaces such as porous pavements, rain gardens, and rainwater harvesting will reduce runoff rates into downstream combined sewers, local receiving waters, or groundwater. This will, in effect, mitigate flooding and reduce watercourse pollution. For separate systems, similar interventions can help eliminate the need for some storm sewers entirely.

1.8 CHANGING CONTEXT

Urban drainage has come a long way in a relatively short time. Figure 1.10 illustrates the changing context and drivers over the last 50 years or so. It is increasingly being realised that urban drainage sits within a wider context sometimes characterised as water cycle management, integrated urban water management, or sustainable water management (see Chapter 24 for more details). Furthermore, these areas also sit within wider urban planning and improvements. Such concepts recognise the "joined-up" nature of water systems and the fact that there is only one water cycle, however we might try and dissect it. In addition, it is increasingly being recognised that stormwater and wastewater are not nuisances or threats that should be immediately dealt with but are, in fact, resources that should be managed.

PROBLEMS

1.1   Do you think urban drainage is taken for granted by most people in developed countries? Why? Is this a good or bad thing?

1.2 How does urbanisation affect the natural water cycle?

1.3 Some claim that urban drainage engineers, throughout history, have saved more lives than doctors and nurses. Can that be justified, nationally and internationally?

1.4 What have been the main influences on urban drainage engineers since the start of their profession?

1.5 "Mixing of wastewater and stormwater (in combined sewer systems) is fundamentally irrational. It is the consequence of historical accident, and remains a cause of significant damage to the water environment." Explain and discuss this statement.

1.6 Explain the characteristics of piped and non-piped systems. Discuss the advantages and disadvantages of each.

1.7 Describe how urban drainage systems interact with the overall urban *water* system. (Use diagrams.) How would you represent rainwater harvesting on the diagram?

1.8 In what way does urban drainage form a part of integrated urban water management?

KEY SOURCE

Tyson, J.M., Evans, T.D., and Orman, N. 2013. *Urban Drainage and the Water Environment: A Sustainable Future? A Review of Current Knowledge.* 2nd Edn., Foundation for Water Research, FR/R0011.

REFERENCES

Baber C. 2010. Tapping into waste heat. *Water Environment and Technology*, 22(12), 40–45.

Burian, S.J. and Edwards, F.G. 2002. Historical perspectives of urban drainage. *Global Solutions for Urban Drainage: Proceedings of the 9th International Conference on Urban Drainage*, Portland, Oregon, September, on CD-ROM.

City of Vancouver. 2015 Renewable City Strategy. www.vancouver.ca/files/cov/renewable-city-strategy-2015.pdf.

De Feo, G., Antoniou, G., Fardin, H.F., El-Gohary, F., Zheng, X.Y., Reklaityte, I. et al. 2014. The historical development of sewers worldwide. *Sustainability*, 6, 3936–3974.

European Environment Agency. 2013. Urban Waste Water Treatment. www.eea.europa.eu/data-and-maps/indicators/urban-waste-water-treatment/urban-waste-water-treatment-assessment-3.

Fletcher, T.D., Shuster, W., Hunt, W.F., Ashley, R., Butler, D., Arthur, S. et al. 2015. SUDS, LID, BMPs, WSUD and more – The evolution and application of terminology surrounding urban drainage. *Urban Water Journal*, 12(7), 525–542.

Halliday, S. 2001. *The Great Stink of London: Sir Joseph Balzalgette and the Cleansing of the Victorian Metropolis*, Sutton Publishing.

Hasan, A. 2006. Orangi Pilot Project: The expansion of work beyond Orangi and the mapping of informal settlements and infrastructure. *Environment and Urbanization*, 18(2), 451–480.

Knauer, S., Nascimento, N., de Oliveira, Butterworth, J., Smits, S., and Lobina, E. 2011. Towards Integrated Urban Water Management in Belo Horizonte, Brazil: A Review of the SWITCH Project. www.switchurbanwater.eu/outputs/pdfs/W6-2_CBEL_RPT_SWITCH_City_Paper_-_Belo_Horizonte.pdf.

OPP-RTI. 2012. Orangi Pilot Project (OPP) – Institutions and Programs. www.oppinstitutions.org.

UN-Habitat. 2003. *Water and Sanitation in the World's Cities: Local Action for Global Goals*, Earthscan.

UN-Habitat. 2006. *Meeting Development Goals in Small Urban Areas*, Earthscan.

Chapter 2

Water quality

2.1 INTRODUCTION

The chemical and biological quality aspects of wastewater and stormwater conveyed in urban drainage systems are important for several reasons:

- Significant water quality changes can occur throughout the drainage system.
- Decisions made in the sewer system have significant effects on the wastewater treatment plant (WTP) performance.
- Direct discharges from drainage systems (e.g., combined sewer overflows, stormwater outfalls) can have a serious pollutional impact on receiving waters.

This chapter looks at the basic approaches to characterising wastewater and stormwater including outlines of the chemical basics (Section 2.2), the main water quality parameters and tests used in practice (2.3), and common processes (2.4). Typical test data are given in Chapters 5 and 7. It also considers the water quality impacts of discharges from urban drainage systems (Section 2.5), discusses relevant legislation and water quality standards (2.6), and provides an introduction to the urban pollution management procedure (2.7).

2.2 BASICS

2.2.1 Strength

Water has been called the *universal solvent* because of its ability to dissolve numerous substances. The term *water quality* relates to all the constituents of water, including both dissolved substances and any other substances carried by the water.

The *strength* of polluted liquid containing a constituent of mass M in water of volume V is its *concentration* given by $c = M/V$, usually expressed in mg/L. This is numerically equivalent to parts per million (ppm) assuming the density of the mixture is equal to the density of water (1000 kg/m^3). The plot of concentration c as a function of time t is known as a *pollutograph* (see Figure 13.9 for an example). Pollutant mass-flow or flux is given by its *load rate* $L = M/t = cQ$, where Q is the liquid flow rate.

EXAMPLE 2.1

A laboratory test has determined the mass of constituent in a 2 L wastewater sample to be 0.75 g. What is its concentration (c) in mg/L and ppm? If the wastewater discharges at a rate of 600 L/s, what is the pollutant load rate (L)?

Solution

$$c = \frac{M}{V} = \frac{750}{2} = 375\,\text{mg/L} = 375\,\text{ppm}$$

$$L = cQ = 0.375 = 600 = 225\,\text{g/s}$$

In order to calculate the average concentration, either of wastewater during the day or of stormwater during a rain event, the *event mean concentration* (EMC) can be calculated as a flow weighted concentration c_{av}:

$$c_{av} = \frac{\sum Q_i c_i}{Q_{av}} \tag{2.1}$$

where c_i is the concentration of each sample i (mg/L), Q_i is the flow rate at the time the sample was taken (L/s), and Q_{av} is the average flow-rate (L/s).

2.2.2 Equivalent concentrations

It is a common practice when dealing with a pollutant (X) that is a compound to express its concentration in relation to the parent element (Y). This can be done as follows:

*Concentration of compound X as **element Y** =*

$$concentration\ of\ compound\ X \times \frac{atomic\ weight\ of\ element\ Y}{molecular\ weight\ of\ compound\ X} \tag{2.2}$$

The conversion of concentrations is based on the gram molecular weight of the compound and the gram atomic weight of the element. Atomic weights for common elements are given in standard texts (e.g., Droste, 1997).

Expressing substances in this way allows easier comparison between different compounds of the same element, and more straightforward calculation of totals. Of course, it also means care needs to be taken in noting in which form compounds are reported (see Example 2.2).

EXAMPLE 2.2

A laboratory test has determined the mass of orthophosphate (PO_4^{3-}) in a 1 L stormwater sample to be 56 mg. Express this in terms of phosphorus (P).

Solution

Gram atomic weight of P is 31 g
Gram atomic weight of O is 16 g
Gram molecular weight of orthophosphate is $31 + (4 \times 16) = 95$ g

Hence, from Equation 2.2,

$$56\,\text{mg}\ PO_4^{3-}\,/\,L = 56 \times \frac{31\,\text{gP}}{95\,\text{g}\,PO_4^{3-}} = 18.3\ PO_4^{3-} - P\,/\,L$$

2.3 PARAMETERS

There is a wide range of quality parameters used to characterise wastewater, and these are described in the following section. Further details on these and many other water quality parameters and their methods of measurement can be found elsewhere (e.g., AWWA, 1999; DoE various). Specific information on the range of concentrations and loads encountered in practice is given in Chapters 3 (wastewater) and 5 (stormwater).

2.3.1 Sampling and analysis

There are three main methods of sampling: grab, composite, and continuous. *Grab* samples are simply discrete samples of fixed volume taken to represent local conditions in the flow. They may be taken manually or extracted by an automatic sampler. A *composite* sample consists of a mixture of a number of grab samples taken over a period of time or at specific locations, taken to more fully represent the composition of the flow. *Continuous* sampling consists of diverting a small fraction of the flow over a period of time. This is useful for instruments that give almost instantaneous measurements, for example, pH and temperature. The Environment Agency provides further details (EA, 2014).

In sewers, where flow may be stratified, samples need to be taken throughout the depth of flow if a true representation is required. Mean concentrations can then be calculated by weighting with respect to the local velocity and area of flow.

In all of the tests available to characterise wastewater and stormwater, it is necessary to distinguish between precision and accuracy. In the context of laboratory measurements, *precision* is the term used to describe how well the analytical procedure produces the same result on the same sample when the test is repeated. *Accuracy* refers to how well the test reproduces the actual value. It is possible, for example, for a test to be precise, but inaccurate with all values closely grouped, but around the wrong value! Techniques that are both precise and accurate are required.

2.3.2 Solids

Solid types of concern in wastewater and stormwater can broadly be categorised into four classes: gross, grit, suspended, and dissolved (see Table 2.1). Gross and suspended solids may be further subdivided according to their origin as wastewater and stormwater.

2.3.2.1 Gross solids

There is no standard test for the gross solids found in wastewater and stormwater, but they are usually defined as solids (specific gravity [SG] = 0.9–1.2) captured by a 6 mm mesh screen (i.e., solids >6 mm in two dimensions). Gross sanitary solids (also variously known as aesthetic, refractory, or intractable solids) include faecal stools, toilet paper, and "sanitary

Table 2.1 Basic classification of solids

Solid type	Size (μm)	SG (–)
Gross	>6000	0.9–1.2
Grit	>150	2.6
Suspended	≥0.45	1.4–2.0
Dissolved	<0.45	—

refuse" such as women's sanitary protection, condoms, bathroom litter, and so on. Faecal solids and toilet paper break up readily and may not travel far in the system as gross solids. Gross stormwater solids consist of debris such as bricks, wood, cans, paper, and so on.

The particular concern about these solids is their "aesthetic impact" when they are discharged to the aqueous environment and find their way onto riverbanks and beaches. They can also cause maintenance problems by deposition and blockage (in pipes and also at pumping stations), and can cause blinding of screens at WTPs, particularly during storm flows.

2.3.2.2 Grit

Again, there is no standard test for determination of grit, but it may be defined as the inert, granular material (SG ≈2.6) retained on a 150 μm sieve. Grit forms the bulk of what is termed *sewer sediment*, and the nature and problems associated with this material are revisited in Chapter 16.

2.3.2.3 Suspended solids

The suspended solids (SSs) content is the solid matter (both organic and inorganic) maintained in suspension, and retained when a sample is filtered (0.45 μm pore size). In the SS test, the residue is washed, dried, and weighed under standard conditions and expressed as a concentration. The accuracy of the SS test is approximately ±15%.

The finer fractions of suspended solids (<63 μm) are extremely efficient carriers of pollutants, carrying greater than their proportionate share (see Sections 5.4.2 and 16.5.2). High concentrations may have a number of adverse effects on the receiving water, including increased turbidity, reduced light penetration, blanketing of the bed, and interference with many types of fish and aquatic invertebrates. Even after deposition, the pollutants attached to these enriched sediments still present a risk, since they can cause a "delayed" sediment oxygen demand (see Section 2.4.3), or may be resuspended at high flows.

By definition, solids not in suspension (i.e., with a diameter <0.45 μm) are dissolved (see Example 2.3).

2.3.2.4 Volatile solids

The solids retained during the SS test can be ignited at 550°C in a muffle furnace. The residue is known as non-volatile or fixed material. The volatilised fraction (the volatile solids) gives an indication of the organic content of the SS.

2.3.3 Oxygen

A key to understanding the reactions occurring anywhere within the urban drainage system is the measurement and prediction of the oxygen levels in the aqueous phase. Dissolved oxygen (DO) levels depend on physical, chemical, and biochemical activities in the system.

2.3.3.1 Dissolved oxygen

Oxygen in water is only sparingly soluble. In equilibrium with air, the solubility of dissolved oxygen (DO) in water is referred to as its *saturation* value, and it decreases with the increase of both temperature and purity (salinity, solids content) and with the decrease in atmospheric pressure (Table 2.2). Hence, warm water (even with no impurities) is, in effect, a pollutant source.

Table 2.2 Dissolved oxygen concentration (under standard conditions) in water as function of temperature

Temperature (°C)	DO (mg/L)
0	14.62
5	12.80
10	11.33
15	10.15
20	9.17
25	8.38
30	7.63

Dissolved oxygen (DO) can be measured analytically using the Winkler titration method. Titration is a laboratory technique where measured volumes of a reagent (the titrant) are incrementally added to a sample up to the equivalent amount of the constituent being analysed. Membrane electrodes are now more commonly and conveniently used both in the laboratory and for *in situ* measurements.

EXAMPLE 2.3

In a standard laboratory test, a crucible and filter pad are dried and their combined mass measured at 64.592 g. A 250 mL wastewater sample is drawn through the filter under vacuum. The filter and residue are then placed on the crucible in an oven at 104°C for drying. The new combined mass is 64.673 g. The crucible and its contents are next placed in a muffle furnace at 550°C. After cooling, the combined mass is measured as 64.631 g. Determine (1) the suspended solids concentration of the sample and (2) the volatile fraction of the suspended solids.

Solution

Mass of suspended solids removed:

> Crucible + filter + solids = 64.673 g
> Crucible + filter = 64.592 g
> Mass of suspended solids = 0.081 g

Concentration of SS:

> 81 (mg)/0.250 (L) = 324 mg/L

Mass of volatile suspended solids removed:

> Initial crucible + filter + solids = 64.673 g
> Final crucible + filter + solids = 64.631 g
> Mass of volatile solids = 0.042 g

Volatile fraction of suspended solids:

> 42 (mg)/81 (mg) = 0.52

The concentration of DO is an excellent indicator of the "health" of a receiving water. All the higher forms of river life require oxygen. Coarse fish, for example, require in excess of 3 mg/L (see Table 2.3). In the absence of toxic impurities, there is a close correlation between DO and biodiversity.

Table 2.3 Oxygen requirements of fish species

Characteristic species	Minimum DO concentration (mg/L)	Minimum saturation (%)	Comment
Trout, bullhead	7–8	100	Fish require much oxygen
Perch, minnow	6–7	<100	Need more oxygen for active life
Roach, pike, chub	3	60–80	Can live for long periods at this level
Carp, tench, bream	<1	30–40	Can live for short periods at this level

Source: Adapted from Gray, N.F. 1999. *Water Technology. An Introduction for Scientists and Engineers.* Arnold.

2.3.4 Organic compounds

Wastewater and stormwater contain significant quantities of organic matter in both particulate and soluble forms. Organic compounds in water are unstable and are readily oxidised either biologically or chemically to stable, relatively inert, end products such as carbon dioxide, nitrates, sulphates, and water. There are three main categories of biodegradable organics present:

- Carbohydrates such as sugars, starch, and cellulose
- Proteins, which are complex molecules built up of amino acids and urea
- Lipids and fats

Decomposition of organic matter by micro-organisms consumes DO. In urban drainage systems the main implication is oxygen depletion in

- Sewers, resulting in an anaerobic environment (considered in Chapter 17)
- Receiving waters (considered later in this chapter)

An indirect indication of the amount of organic material in a wastewater can be derived from one of two tests: the biochemical oxygen demand (BOD) or the chemical oxygen demand (COD). A third option is the total organic carbon (TOC) test, which gives a more direct measure of the carbon content of the sample under test.

2.3.4.1 Biochemical oxygen demand (BOD₅)

This test is a laboratory simulation of the microbial processes occurring in water contaminated with organic compounds. It measures the DO consumed in a sample diluted in a 300 mL bottle during a specified incubation period (usually 5 days at a temperature of 20°C in darkness). The DO is used by micro-organisms as they break down organic material and certain inorganic compounds. Thus,

$$BOD_5 = \frac{(c_{DOI} - c_{DOF})}{p} \tag{2.3}$$

where P is the sample dilution (volume of sample/volume of bottle), c_{DOI} is the initial dissolved oxygen concentration (mg/L), and c_{DOF} is the final dissolved oxygen concentration (mg/L)

Measured amounts of the wastewater sample are diluted with prepared water containing nutrients and DO. Seed micro-organisms are added if insufficient amounts are available in the sample. Equation 2.3 assumes the dilution water has negligible BOD₅ (see Example 2.4).

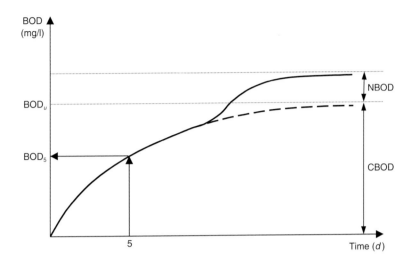

Figure 2.1 Development of biochemical oxygen demand (BOD) with time.

The test may also measure the oxygen used to oxidise reduced forms of nitrogen (nitrogenous demand – NBOD) unless an inhibitor (e.g., allylthiourea [ATU]) is used. The evolution of BOD with time is shown in Figure 2.1.

The test does not give a measure of the *total* oxidisable organic matter because of the presence of considerable quantities of carbonaceous matter resistant to biological oxidation. Over 5 days, only the readily biodegradable fraction of organic material present in the water will be broken down.

The test can be extended up to 10–20 days to reach the ultimate carbonaceous CBOD ≈1.5 BOD_5. BOD tests are subject to inhibition if the wastewater contains any toxic components (e.g., trace metals in runoff), and they should be seen as an indication rather than as an accurate determination.

BOD_5 is a common parameter used in the control of treated wastewater effluent quality through the setting and monitoring of discharge consent standards (e.g., 25 mg/L) (CEC, 1991a). Rivers are considered to be polluted if their BOD_5 exceeds 5 mg/L, and 3 mg/L is required for salmonid rivers (CEC, 1978).

EXAMPLE 2.4

A laboratory test for BOD_5 (ATU) is carried out by mixing a 5 mL sample with distilled water into a 300 mL bottle. Prior to the test the DO concentration of the mixture was 7.45 mg/L and after 5 days it had reduced to 1.40 mg/L. What is the BOD_5 concentration of the sample?

Solution

Dilution, $p = 5/300 = 0.0167$

Equation 2.3:

$BOD_5 = (7.45 – 1.40)/0.0167 = 363$ mg/L

2.3.4.2 Chemical oxygen demand (COD)

The COD test measures the oxygen equivalent of the organic matter that can be oxidised by a strong chemical oxidising agent (potassium dichromate) in an acidic medium. The standard test lasts for 3 hours using either a titrimetric or colorimetric method. Colorimetric test methods rely on measuring the intensity of light from colour changes in the reaction. Almost all organic compounds are oxidised. Some inorganics are also oxidised, but ammonium and ammonia are not. Thus, this measurement is a good estimation of the total content of organic matter.

One of the main limitations of the COD test is its inability to differentiate between biodegradable and biologically inert organic matter. However, if sufficient data are available, it is often possible to empirically relate the COD with BOD values, such as

$$c_{BOD} \approx a \times c_{COD} \tag{2.4}$$

where c_{BOD} is the biochemical oxygen demand concentration (mg/L), c_{COD} is the chemical oxygen concentration (mg/L), and a is the 0.4–0.8 (–).

It should be stressed, however, that "a" will vary between wastewaters, and no universal relationship has been found between the two parameters. Nevertheless, it is a good indicator of the wastewater treatability.

The COD of a sample can be further differentiated into several forms. The first major category is inert (suspended or soluble) material that is non-biodegradable within the timescale associated with an urban drainage system. The second is biodegradable matter, which in turn can be divided into readily and slowly degradable material. The former refers to material that can be immediately oxidised by micro-organisms and the latter to matter, which is degraded more slowly. The approximate relationship between the COD fractions and BOD is summarised in Figure 2.2. The methods for characterising the various COD fractions are still under development and are not yet standardised (Henze, 1992).

2.3.4.3 Total organic carbon (TOC)

Unlike the BOD and COD tests, the TOC test directly measures the TOC content of a sample. The test is based on the fact that carbon dioxide (CO_2) is a product of combustion.

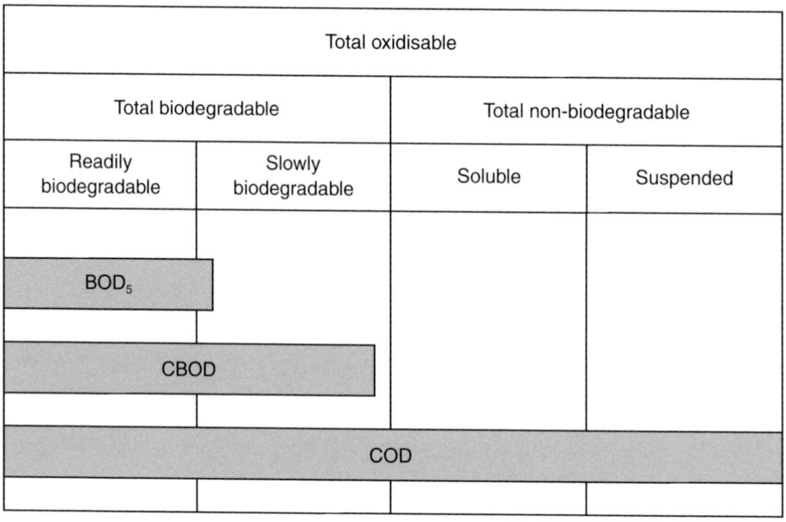

Total oxidisable			
Total biodegradable		Total non-biodegradable	
Readily biodegradable	Slowly biodegradable	Soluble	Suspended
BOD$_5$			
CBOD			
COD			

Figure 2.2 Relationship between BOD and COD for organic matter.

It is carried out in an instrument containing a small furnace at 950°C in the presence of a catalyst (having first removed the *inorganic* carbon). After this, the CO_2 released is measured for the known volume of sample. It can be performed very rapidly using a single instrument and is especially applicable in the analysis of small concentrations of organic matter.

COD is approximately related to TOC as follows (see Example 2.5):

$$COD \approx 2.5 \times TOC \qquad (2.5)$$

EXAMPLE 2.5

If organic material can be represented by the chemical formula $C_6H_{12}O_6$ (glucose), calculate the *theoretical* relationship between COD and TOC.

Solution

In the COD and TOC tests, organic material is oxidised to carbon dioxide and water:

$$C_6H_{12}O_6 + 6O_2 \rightarrow 6CO_2 + 6H_2O$$

From the above equation, each mole of organic material (molecular weight = 180) requires 6 moles of oxygen (molecular weight 32) for oxidation. As the weight in grams of a substance is the number of moles × molecular weight:

$$COD = \frac{6\,moles\,O_2}{1\,mole\,C_6H_{12}O_6} = \frac{6 \times 32}{1 \times 180} = 1.067\,gO_2/gC_6H_{12}O_6$$

Also, each mole of organic material contains 6 atoms of carbon, so

$$TOC = \frac{6\,atoms\,C}{1\,mole\,C_6H_{12}O_6} = \frac{6 \times 12}{1 \times 180} = 0.4\,gC/gC_6H_{12}O_6$$

$$\therefore \; COD = \frac{1.067}{0.4} \times TOC = 2.67\,TOC$$

2.3.5 Nitrogen

Nitrogen exists in four main forms: organic (in the protein that makes up much matter), ammonia (or ammonia salts), nitrite, and nitrate. Total nitrogen is the sum of all forms, although in wastewater and stormwater, organic and ammonia nitrogen make up most of the total. The concentration of nitrogen in domestic wastewater is usually related to the BOD_5.

Excessive levels of nitrogen discharged to receiving waters can promote the growth of undesirable aquatic plants such as algae and floating macrophytes. In severe cases, the receiving water can experience *eutrophic* symptoms such as water discoloration, odours, and depressed oxygen levels (considered in more detail later in this chapter).

2.3.5.1 Organic nitrogen (org.N)

Organic nitrogen includes such natural materials as proteins and peptides, nucleic acids and urea, and numerous synthetic organic materials, although not *all* organic nitrogen compounds.

Analytically, organic nitrogen and ammonia are determined together using the Kjeldahl method. In this test, the aqueous sample is boiled to remove any pre-existing ammonia and then digested, during which the organic nitrogen is converted to ammonia. The amount of ammonia produced is then determined as detailed in the next section.

2.3.5.2 Ammonia nitrogen (NH$_3$–N)

Ammonia nitrogen exists in solution in two forms, as the ammonium ion (NH$_4^+$) and as ammonia gas (NH$_3$) depending on the pH and temperature of the wastewater:

$$NH_3 + H^+ \rightleftharpoons NH_4^+ \tag{2.6}$$

At values of pH ≤ 7, virtually all the ammonia is present as ammonium. At pH 9, for example, 35% is present as NH$_3$.

Ammonia nitrogen is determined analytically by raising the pH, and using a distillation process. The final measurement is then made by titration or by colorimetry. Sometimes organic nitrogen and ammonia nitrogen are determined together in the total Kjeldahl nitrogen (TKN) test, which is similar to the basic Kjeldahl method except any pre-existing ammonia is *not* removed.

Any unionised ammonia (NH$_3$) present in wastewater discharged to the environment is particularly toxic to fish, depending on the dissolved oxygen of the receiving water. Ammonia also exerts an oxygen demand during its conversion to nitrite and subsequently to nitrate.

2.3.5.3 Nitrite and nitrate nitrogen (NO$_2^-$ – N, NO$_3^-$ – N)

Nitrite is the intermediate oxidation state of nitrogen. It is relatively unstable, easily oxidised, and its presence shows that oxidation of nitrogenous matter is taking place. Nitrate forms the most highly oxidised state of nitrogen found in wastewater. Determination is usually by colorimetric methods.

2.3.6 Phosphorus

Phosphorus can be expressed as total, organic, or inorganic (ortho- and poly-) phosphorus. Organic phosphate is a minor constituent of wastewater and stormwater, with most phosphorus being in the inorganic form. Polyphosphates consist of combinations of phosphorus, oxygen, and hydrogen atoms. Orthophosphates (e.g., PO$_4^{3-}$, HPO$_4^{2-}$, H$_2$PO$_4^-$, H$_3$PO$_4$) are simpler compounds and may be in solution or attached to particles. Orthophosphates can be determined directly, but poly- and organic phosphate must first be converted to orthophosphates before determination.

Phosphorus-containing compounds are also implicated in receiving water eutrophication. Generally, phosphorus is the controlling nutrient in urban freshwater systems. Salmonid rivers have upper limits of 0.065 mg P/L (CEC, 1978).

2.3.7 Sulphur

Sulphurous compounds are found mainly in wastewater in the form of organic compounds and sulphates (SO$_4^{2-}$). Under anaerobic conditions, these are reduced to sulphides (S−), mercaptans, and certain other compounds. The principal product, hydrogen sulphide (H2S), is formed mainly by biofilms on the walls of sewers and in sediment deposits.

Hydrogen sulphide is a flammable and very poisonous gas and, when escaping into the atmosphere, can cause serious odour nuisance. It is acutely toxic to aquatic organisms and could be a factor in fish kills near CSOs. Hydrogen sulphide in damp conditions can be oxidised biologically to sulphuric acid (H$_2$SO$_4$), which may cause serious damage to sewer materials, especially concrete (see Chapter 17).

2.3.8 Hydrocarbons

Hydrocarbons are organic compounds containing only carbon and hydrogen. They are classified into four groups based on molecular structure: aliphatic or straight chained, branch chained, aromatic (based on the benzene ring), and alicyclic. In this text, we are mainly concerned with the petroleum-derived group commonly found in stormwater, which includes petrol, lubricating, and road oils. They are among the more stable organic compounds and are not easily biodegraded. Most have a strong affinity for suspended particulate matter. They are determined by extraction with carbon tetrachloride.

Hydrocarbons are lighter than water, and virtually insoluble, causing films and emulsions on the water surface and reducing atmospheric reaeration. Those in accumulated sediments can persist for long periods and exert a chronic impact on bottom-dwelling organisms, as well as can be remobilised by subsequent storm events.

2.3.9 FOG

FOG is a general term used to include the fats, oils, greases, and waxes of plant- or food-based origin present in wastewater. These lipids are complex organic molecules with building blocks of three, two, or one chain of fatty acids known as tri-, di-, and monoglycerides, respectively. These are very stable compounds and do not easily biodegrade naturally. They do not fully mix with water, and readily accumulate and float on water surfaces. However, in the presence of hot water and detergents, separation takes place at low rates. They are determined gravimetrically by extraction with trichlorotrifluroethane (Freon).

Fats in sewer systems can cause blockages (see Chapter 17). When discharged to the environment they cause films and sheens on the water surface, inhibiting oxygen transfer.

2.3.10 Heavy metals and synthetic compounds

A considerable number of heavy metals and synthetic organic and inorganic chemicals can be found in wastewater and stormwater. Among the many constituents of concern are metal species such as arsenic, cyanide, lead, cadmium, iron, copper, zinc, and mercury. Metals can exist in particulate, colloidal, and dissolved (labile) phases depending mainly on the prevailing redox and pH conditions. The concentration of individual metals is often determined by atomic absorption spectrophotometry, although newer, multi-element equipment (e.g., ICP) is becoming more widespread.

Metals in stormwater are predominantly in the particulate phase. This is important because the environmental mobility and bioavailability (and hence toxicity) of metals are highly related to their concentration in solution. Many (particularly the more soluble forms of zinc and copper) are known to have toxic effects on aquatic life and can inhibit biological processes at the WTP.

Herbicides and pesticides can be toxic to a variety of aquatic life at very low concentrations. Some of the more toxic varieties (e.g., chlorinated organics, DDT and PCBs) are no longer used, but their residues can still be found in the environment.

2.3.11 Micro-organisms

Direct determination of the presence of pathogenic micro-organisms (e.g., salmonella, enteroviruses) in wastewater and stormwater is not normally carried out. Indeed, even microbiological indicator tests are not routinely undertaken. However, faecal indicator organisms (FIOs), such as total coliforms and faecal coliforms (*Escherichia coli*), and faecal streptococci (FS) are known to occur in large numbers in both wastewater and stormwater.

One of the major objectives of urban drainage systems (as discussed in Chapter 1) is the protection of public health, particularly reducing the risk associated with human contact with excreta. Wastewater, even when treated at WTPs, is not routinely disinfected before discharge to a receiving water unless it is classified as bathing water (see discussion following). Discharges from CSOs and SWOs are also not routinely disinfected, and bacterial standards for recreational activity can be violated during even modest storm events.

2.3.12 Priority substances

The Water Framework Directive (Section 2.6.1.3) defines a European priority list of 33 substances, extended to 45 substances in 2013, posing a threat to (or via) the aquatic environment as given in Table 2.4. These are defined as "substances or groups of substances that are toxic, persistent and liable to bioaccumulate, and other substances or groups of substances which give rise to an equivalent level of concern." The substances that are thought to pose the greatest threat are further identified as "priority" hazardous substances. Environmental quality standards for the concentrations of the priority substances in surface water bodies were also set.

2.4 PROCESSES

2.4.1 Hydrolysis

Hydrolysis is an important precursor to both aerobic and anaerobic transformations. It consists of the natural reaction of large organic molecules with water in the presence of enzymes

Table 2.4 EU hazardous and priority hazardous substances

Hazardous substance	Priority hazardous substance
251-835-4	Anthracene
Aclonifen	Brominated diphenylethers
Alachlor	Cadmium and its compounds
Atrazine	C_{10-13}-chloroalkanes
Benzene	DEHP – Di(2-ethylhexyl)phthalate
Bifenox	Dicofol
Chlorfenvinphos	Dioxins and dioxin-like compounds
Chlorpyrifos	Endosulfan
Cybutryne	Heptachlor and heptachlor epoxide
Cypermethrin	Hexabromocyclododecane
1,2-Dichloroethane	Hexachlorobenzene
Dichloromethane	Hexachlorobutadiene
Dichlorvos	Hexachlorocyclohexane
Diuron	Mercury and its compounds
Fluoranthene	Nonylphenols
Lead and its compounds	Pentachlorobenzene
Naphthalene	Perfluorooctane sulfonic acid
Nickel and its compounds	Polyaromatic hydrocarbons
Octylphenols	Quinoxyfen
Pentachlorophenol	Tributyltin compounds
Simazine	Trifluralin
Terbutryn	
Trichlorobenzenes	
Trichloromethane	

Source: Council of the European Communities (CEC). 2013a. *Amending Directives 2000/60/EC and 2008/105/EC as regards Priority Substances in the Field of Water Policy,* Directive 2013/39/EU.

to produce smaller molecules that are potentially available for utilisation by bacteria. It is temperature dependent.

Suspended organic particles are hydrolysed, bringing them into solution. For example, insoluble cellulose is slowly hydrolysed in two stages to form dextrose (Inman, 1979):

$$(C_6H_{10}O_5)n + nH_2O \rightarrow nC_6H_{12}O_6$$

Urea represents 80% of the nitrogen content of fresh wastewater. It is relatively rapidly hydrolysed to ammonia-nitrogen (by the enzyme *urease*) at a rate of 3 mg.N/L per hour at 12°C in stored samples (Painter, 1958). Polyphosphates will slowly hydrolyse to the ortho-phosphate form.

2.4.2 Aerobic degradation

Aerobic processes are those carried out by aerobic bacteria in the presence of free oxygen, whereby larger, but soluble, organic molecules are degraded to simple and stable end products. Such micro-organisms may be freely suspended individually or in flocs in the flow, or be attached to pipe walls or sediment beds as biofilms.

The particular class of organism that carries out this reaction is aerobic *heterotrophic* bacteria. These take organic material as a "food" source to provide energy or to synthesise new bacterial cells. End products of this reaction are CO_2, H_2O and other oxidised forms such as nitrate, phosphate, and sulphate. This can be simply represented as

$$C,H,O,N,P,S + O_2 \rightarrow H_2O + CO_2 + NO_3- + PO_4^{3-} + SO_4^{2-+} \text{newcells} + \text{energy}$$

Unless it is replaced, the oxygen in water can quickly be used up producing an environment hostile to aerobic micro-organisms. Conditions in the bulk flow in gravity sewers are likely to be aerobic.

2.4.2.1 Nitrification

In an aerobic environment with low levels of organic material, heterotrophic micro-organisms will no longer be able to thrive. However, another group of organisms known as *autotrophs* (nitrosomas and nitrobacter) can thrive, provided there is a sufficient source of oxygen. These utilise inorganic nutrients as an energy source (carbon dioxide and oxidised forms of nitrogen, phosphorus, and sulphur). Nitrification is the biological process by which ammonia (an inorganic nutrient) is converted first to nitrite and then to nitrate. This can be summarised as

$$NH_4^+ + 2O_2 \rightarrow NO_3- + 2H^+ + H_2O + \text{newcells} + \text{energy}$$

The nitrification reaction consumes large quantities of oxygen and alkalinity.

It is unusual for nitrification to occur naturally in urban drainage networks, but it is possible towards the end of long, well-aerated outfall sewers, particularly in warmer climates.

2.4.3 Denitrification

In the denitrification process, nitrate is reduced to nitrogen gas by the same heterotrophic bacteria responsible for carbonaceous oxidation. Reduction occurs when dissolved oxygen levels are at or near zero. A carbon source must be available to the bacteria.

Denitrification can be stimulated in anaerobic sewer environments by adding a source of nitrate (see Chapter 17).

2.4.4 Anaerobic degradation

Anaerobic processes are those carried out by anaerobic bacteria in the absence of oxygen, whereby large organic molecules are degraded to simple organic gases as end products, resulting in a partial breakdown of the substrate.

The class of organisms known as anaerobic heterotrophic bacteria carries out this reaction. These take organic material as a food source to provide energy or to synthesise new bacterial cells. They must take their oxygen from dissolved inorganic salts; therefore, they produce reduced forms of end products together with methane and carbon dioxide. Although this is a three or more stage process, it can be simply represented as

$$C,H,O,N,P,S + H_2O \rightarrow CO_2 + CH_4 + NH_3 + H_2S + newcells + energy$$

The products of anaerobic processes are more objectionable and sometimes more dangerous than those of aerobic processes. For example, hydrogen sulphide is malodorous, and methane is combustible. Anaerobic conditions are likely to occur when sediment beds form in sewers, and are common in pressurised rising mains.

2.5 RECEIVING WATER IMPACTS

All receiving waters can assimilate wastes to some extent, depending on their natural self-purification capacity. Problems arise when pollutant loads exceed this capacity, thus harming the aquatic ecology and restricting the potential use of the water (e.g., water supply, recreation, fisheries). Discharges in urban areas can be continuous or intermittent, depending on their source. The aim of the urban drainage designer (in pollution terms) is, therefore, to balance the effects of these discharges against the assimilation capacity of the receiving water including background levels of pollution from other sources, so as to optimise water quality and minimise treatment costs.

2.5.1 Emissions

Urban drainage emissions can be categorised as direct and indirect, listed as below.

Direct – from the sewer system

- Intermittent discharges from CSOs consisting of a mixture of stormwater, domestic, commercial, and industrial wastewater with groundwater and sewer deposits
- Intermittent discharges from separate storm sewer outfalls or direct stormwater discharges consisting mainly of runoff from urban surfaces

Indirect – via the treatment plant

- Continuous low-level inputs from normally functioning WTPs
- Intermittent shock loads (of suspended solids and/or ammonia) from WTPs disturbed by wet weather transient loads

Intermittent discharges are particularly difficult to quantify and regulate because of their nature. Their acute (immediate) impact can only be measured during a spill event, and their chronic (long-term) effects are often difficult to isolate from background pollution. Therefore, design standards and performance criteria specifically tailored to intermittent discharges are needed. These are discussed in more detail in the following section.

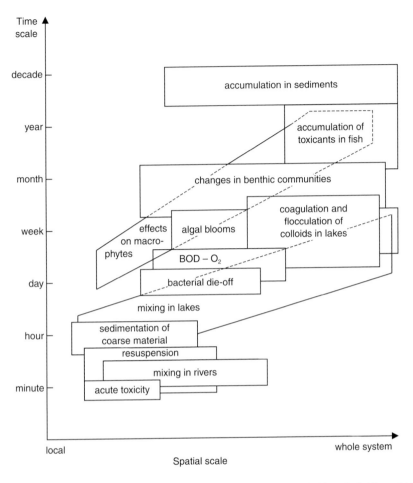

Figure 2.3 Time and spatial scales for receiving water impacts. (Based on Aalderink, R.H., and Lijklema, L. 1985. Water quality effects in surface waters receiving stormwater discharges, in *Water in Urban Areas, TNO Committee on Hydrological Research. Proceedings and Information,* 33, 143–159, with permission of the authors.)

2.5.2 Processes

The processes occurring in receiving waters subject to discharges from urban drainage systems include (House et al., 1993)

- Physical: Transport, mixing, dilution, flocculation, erosion, sedimentation, thermal effects, and reaeration
- Biochemical: Aerobic and anaerobic oxidation, nitrification, and adsorption and desorption of metals and other toxic compounds
- Microbiological: Growth and die-off and toxicant accumulation

The extent and importance of individual processes will depend on the temporal and spatial scales as shown in Figure 2.3. For example, a shock load into flowing water in a river will transport downstream relatively quickly, interacting with the water column as it progresses. Thus, a significant length of water will be exposed to contamination for a short period. A load discharged to stagnant water in a lake will disperse more slowly and will generally persist for a longer period. We can identify three relevant timescales:

- Short term (acute)
- Medium term (delayed)
- Long term (chronic, cumulative)

2.5.3 Impacts

Impacts can be divided into direct water quality effects (DO depletion, eutrophication, toxics), public health issues, and aesthetic influences. These are summarised for various determinands and receiving water types in Table 2.5.

2.5.3.1 DO depletion

The most important phenomena caused by intermittent discharges (particularly CSOs) are the following:

- Mixing of low DO spills with the receiving water.
- Degradation of discharged (dissolved and particulate) organic matter that exerts an *immediate* oxygen demand on the receiving water. In the case of a river, these occur in a plug moving downstream.
- *Delayed* sediment oxygen demand (SOD) caused by the deposited sediment and the scouring effect of discharges after the polluted plug has passed. Typical undisturbed SOD levels are 0.15–2.75 g/m².d, elevated to 240–1500 g/m².d during storm flow conditions (House et al., 1993). As the rate of decomposition is low, the area affected can be extensive. Sensitive benthic organisms are rapidly eliminated but are soon replaced by high-population densities of a few species tolerant of the low oxygen silty conditions.

The relative magnitude of these effects will depend on the specific circumstances of the discharge and the recipient. Hvitved-Jacobsen (1986) indicates that, in larger rivers,

Table 2.5 Qualitative assessment of receiving water impacts of urban discharges

Receiving water	Water quality			Public health		Aesthetics	
	Dissolved oxygen	Nutrients	Sediments	Toxics	Microbials	Clarity	Sanitary debris
Streams							
Steep	–	–	–	x	xx	–	xx
Slack	x	–	x	x	xx	–	xx
Rivers							
Small	xx	–	x	x	xx	–	xx
Large	x	–	x	x	xx	x	xx
Estuaries							
Small	x	x	x	x	xx	x	xx
Large	–	–	x	–	xx	x	xx
Lakes							
Shallow	x	xx	x	x	xx	x	xx
Deep	x	x	x	x	xx	x	xx

Source: House, M.A. et al. 1993. *Water Science and Technology*, 27(12), 117–158.

xx Probable; x Possible; – Unlikely.

immediate consumption dominates, whereas small rivers with flows <0.5 m³/s often show depletion due to delayed consumption.

The most apparent consequences of reduced DO levels are fish kills. Additionally, odour problems may be experienced due to putrefaction.

2.5.3.2 Eutrophication

If large quantities of nutrients such as nitrogen or phosphorus are discharged into receiving waters, excessive growth of aquatic weeds and algae may occur. This can lead to

- Oxygen depletion
- Anaerobic conditions in bottom muds
- Fish kills
- Aesthetic problems

This is a long-term problem usually associated with shallow, stagnant waters such as lakes, estuaries, and the coastal zone, but rivers may also be affected. Intermittent discharges are usually a relatively small constituent of the total nutrient load.

2.5.3.3 Toxics

Intermittent discharges are a significant source of elevated levels of ammonia (toxic to fish), chlorides, metals, hydrocarbons, and trace organics, which cause toxic impacts. These may be either acute or chronic depending on the specific circumstances. The effect on the receiving water biota is to rapidly reduce species diversity and abundance leading to complete elimination at excessive concentrations. Downstream recovery is generally slower than the loss rate, with tolerant species returning, often at population densities greater than initial levels, due to lack of competition.

The difficulties in assessing toxic impacts based on single parameter values have led to the development of toxicity-based consents (TBCs). Typically, these are derived from laboratory and field ecotoxicological studies on fish and invertebrates. Results are expressed as LC50 values that indicate short-term lethal concentrations of a particular pollutant resulting in 50% mortality.

2.5.3.4 Public health

Relatively high concentrations of a variety of pathogens may be expected from both combined sewer overflow (CSO) and stormwater outfall (SWO) discharges. Bacterial contamination is a relatively short-term problem as die-off usually occurs within several days, although this is a longer period than the discharge itself. In addition, bacteria tend to adhere to suspended solids. As many of these particles will settle, bacteria can become established in the receiving waterbed, considerably extending their survival times.

The risk to public health depends on the degree of potential human exposure, and this will be greatest if the receiving water is used for contact recreational purposes. Swimmers are therefore at greatest risk. Studies have now demonstrated the relationship between gastroenteritis and FS levels (Wyer et al., 1995).

2.5.3.5 Aesthetics

In addition to chemical and biological impacts, public perception of water quality is also important. Research has shown that the public has a good idea of what might be considered a polluted river but is less certain as to what might be considered a clean river. The public tends to misperceive as polluted even rivers of high chemical and biological quality. However, solids of obvious sanitary origin near to receiving waters are considered to be offensive.

2.6 RECEIVING WATER STANDARDS

2.6.1 Legislation and regulatory regime

2.6.1.1 Urban Waste Water Treatment Directive

The EU Urban Waste Water Treatment Directive (CEC, 1991a) aims to protect the environment from the adverse effects of wastewater discharges from urban areas and biodegradable industrial wastewater by specifying uniform performance standards for wastewater treatment and collection across the European Union. The standards are more prescriptive for WTPs and less so for collection systems. Compliance dates for various aspects of the directive ranged from 31 December 1998, to 31 December 2005, with extended deadlines for newer member states. However, by the end of 2009/2010 the degree of compliance was still variable across the Union. For example, in terms of collecting systems, compliance averaged at 94% across all EU member states, but varied from as high as 100% to as low as 30%. Compliance with required secondary treatment had reached 82% and tertiary treatment 77% (CEC, 2013b).

The directive requires member states to ensure all urban areas with a population equivalent of 2,000 or more are provided with collection systems. Further, the design, construction, and maintenance of collecting systems needs to be undertaken in accordance with best technical knowledge not entailing excessive costs with respect to

- Volume and characteristics of urban wastewater
- Prevention of leaks
- Limitation of pollution of receiving waters due to overflows

The directive acknowledges the difficulties that arise during periods of unusually heavy rainfall, and allows member states to decide on their own measures to limit pollution from overflows. These could be based on one of the following:

- Dilution rates
- Collection system capacity in relation to dry weather flow
- A certain number of overflows per year

2.6.1.2 Bathing Water Directive

The EU Bathing Water Directive (CEC, 2006) regulates the bacteriological water quality of coastal and inland bathing waters based on two faecal indicator parameters: intestinal *enterococci* and *Escherichia coli*. Three levels or standards are set for inland waters and for coastal and transitional waters: excellent, good, and sufficient (see Table 2.6). There is also a "poor" classification for waters that do not reach these standards. All EU bathing waters were due to be classified by the end of the 2015 bathing season at which time all were to be at least "sufficient." In 2014, 95% of bathing waters were found to meet minimum water quality standards with more than 83% meeting the "excellent" standard (EEA, 2015). Failure to comply with the standards, when it does happen, typically comes about by a combination of continuous wastewater discharges and intermittent CSO discharges.

Table 2.6 Bathing water quality standards

Waters	Parameter	Excellent quality	Good quality	Sufficient
Inland	Intestinal enterococci (cfu/100 mL)	200[a]	400[a]	330[b]
	Escherichia coli (cfu/100 mL)	500[a]	1000[a]	900[b]
Coastal and transitional	Intestinal enterococci (cfu/100 mL)	100[a]	200[a]	185[b]
	Escherichia coli (cfu/100 mL)	250[a]	500[a]	500[b]

Source: CEC. 2006. *Directive Concerning the Quality of Bathing Water*, 2006/7/EC.

[a] Based on a 95-percentile evaluation.
[b] Based on a 90-percentile evaluation.

Hence, control of overflow discharges becomes important where they discharge directly into bathing waters or to estuarine or coastal waters that lead to bathing waters.

2.6.1.3 *Water Framework Directive*

The Water Framework Directive (WFD) incorporates the main requirements for water management in Europe into one single, holistic system based on river basins (CEC, 2000). New or reorganised river basin district authorities are to be formed, each with a management plan and a "programme of measures" aimed at achieving the goals of the directive. The key guiding goal is to achieve "good status" of groundwaters and surface waters; "good" meaning that water meets the standards established in existing water directives and, in addition, new ecological quality standards. A surface water is defined as of good ecological quality if there is only a slight departure from the biological community that would be expected in conditions of minimal anthropogenic impact. *Good chemical status* is defined in terms of compliance with all of the quality standards established for chemical substances at the European level. A new mechanism for controlling the discharge of dangerous *priority hazardous substances* (Section 2.3.12) is provided. A combined approach to setting standards is to be used where both emission limit values and environmental quality standards are to be legally binding. Derogations from good status are allowed in unforeseen or exceptional circumstances (e.g., droughts, floods). Environmental quality standards are specified as a maximum allowable concentration (MAC-EQS) to protect against short-term, acute exposure, and/or an Annual allowance (AA-EQS) to provide protection against long-term, chronic effects (Lundy et al., 2012)

The WFD is designed to provide an "umbrella" to existing directives as well as new standards being incorporated. River basin authorities will designate specific protection zones within their area (i.e., bathing, shellfish, drinking water, or protected natural areas) where the standards in the respective existing EU directives apply, but zones with higher objectives may also be established where more stringent standards must be met. Good ecological and chemical status is the minimum for all waters. The Urban Waste Water Treatment Directive and the Nitrates Directive (CEC, 1991b) are considered as tools to achieve the objectives of the river basin management plans.

The directive also sets rules for groundwater. All direct discharges to groundwater are prohibited and a requirement is introduced to monitor groundwater bodies so as to detect changes in composition due to non-point pollution and take measures to reverse them.

The initial period of implementation of the directive was 15 years to 2015 (9 years to prepare plans and 6 years more to achieve specific targets) plus an additional deferment of up to two periods of 6 years if justified on technical or economic grounds (Kallis and Butler, 2001). By 2009, 43% of surface water bodies had achieved good status, and this was projected to increase to 53% by 2015 on the basis of planned measures (CEC, 2012).

2.6.1.4 Marine Strategy Framework Directive

The Marine Strategy Framework Directive required member states to develop strategies to protect their marine waters with the aim of achieving "good environmental status" by 2020. This is being implemented by protecting the marine environment, preventing its deterioration, and restoring degraded marine ecosystems (CEC, 2008). The directive is complementary to the Water Framework Directive.

2.6.2 Permitting intermittent discharges

It is common that continuous discharges to receiving waters (e.g., WTPs) are regulated with discharge permits or licences and are routinely monitored. The degree of legal control over intermittent discharges (such as from CSOs) is much more variable throughout the EU (Zabel et al., 2001). There are two main approaches: emission standards and environmental quality standards (EQSs), with the former being the most common approach. The continuing use of emission standards such as spill frequency, duration, or volume (despite their convenience) is surprising given the poor correlation between reducing CSO spill frequency or volume and improving receiving water quality (Lau et al., 2002). However, many overflows in the United Kingdom are or will be routinely or continuously monitored by 2020 (see Chapter 12).

Environmental quality standards will be discussed in the next section, and emission standards based on surrogate indicators will be explained in Chapter 12 together with CSO design criteria.

2.6.3 Environmental quality standards

Environmental quality standards are routinely used to control continuous discharges. They are based on a receiving water's ability to assimilate discharged pollutants without detriment to legitimate uses of the water – based on environmental quality objectives (EQOs). These standards are monitored by checking the compliance of routinely taken samples against water quality criteria (usually 90 or 95 percentiles). However, discharges from drainage systems tend to be infrequent and of short duration, although they can be of high pollutant concentration, resulting in a disproportionately high impact on receiving water "users." In addition, widespread routine sampling is not feasible for intermittent discharges.

In response to this, intermittent standards have been derived, which take into account the particular characteristics of the discharges and their impact (FWR, 2012). They provide acceptable concentrations of river quality determinands for short- and long-term exposure and the recovery period in between based on aquatic life, shellfish, bathing, and amenity protection.

2.6.3.1 Aquatic life standards

Intermittent standards to protect river aquatic life, based on the LC50 values mentioned earlier, are given in the *UPM Manual* (FWR, 2012). The standards consist of a relationship between three variables:

- Pollutant concentration
- Return period of an event in which that concentration is exceeded
- Duration of the event

Table 2.7 Intermittent standards for dissolved oxygen and ammonia concentration/ duration thresholds for sustaining cyprinid fisheries

Return period (months)	DO concentration (mg/L)[a]		
	1 hour	6 hours	24 hours
1	4.0	5.0	5.5
3	3.5	4.5	5.0
12	3.0	4.0	4.5
	NH_3–N concentration (mg/L)[b]		
1	0.150	0.075	0.030
3	0.225	0.125	0.050
12	0.250	0.150	0.065

Source: FWR. 2012. *Urban Pollution Management Manual*, 3rd edn, Foundation for Water Research, http://www.fwr.org/UPM3.

[a] Applicable when NH_3–N < 0.02 mg/L.
[b] Applicable when DO > 5 mg/L, pH > 7 and T > 5°C.

Table 2.7 shows this three-way relationship for dissolved oxygen and unionised ammonia based on sustaining cyprinid fisheries. Thus, minimum river DO levels of 3.0–5.5 mg/L are allowed, depending on the duration and frequency of the storm event. Fish kills due to ammonia poisoning can be avoided if NH_3–N levels are limited to 0.03–0.25 mg/L.

The manual also describes how higher percentile water quality criteria (e.g., 99 percentile) can be set as an alternative means of protecting receiving water ecosystems.

2.6.3.2 Shellfish standards

Shellfish water EQSs are based on the impact of all discharges (both intermittent and continuous) and are designed to achieve a water quality standard of 1,500 faecal coliforms per 100 mL for at least 97% of the time over the long term (FWR, 2012).

2.6.3.3 Bathing standards

Bathing water EQSs are based on limiting the concentration of intestinal *enterococci* and *Escherichia coli* to the levels given in Table 2.6 for at least 98.2% of the bathing season (May to September) as judged over an average period of about 10 years.

2.6.3.4 Amenity standards

Receiving waters can be classified, in terms of their amenity value, by the amount of public contact:

- High amenity – water used for bathing and water-contact sports, watercourses through parks and picnic sites, shellfish waters
- Moderate amenity – water used for boating, watercourses near popular footpaths or through housing developments or town centres
- Low amenity – limited public interest

Amenity guidelines are currently based on emission standards and provision of good engineering design. These are discussed further in Chapter 12.

2.7 URBAN POLLUTION MANAGEMENT (UPM)

The *UPM Manual* (FWR, 2012) sets out a methodical procedure for dealing with wet weather discharges from sewer systems. There are three recurring themes in the guidance. The first is that analysis should be *holistic*, covering all elements in the system that determine the pollution impact of a sewer system: rainfall, the sewer system itself, the WTP, and the receiving river. The second is that the level of detail of any study, and in particular of the models used, should be appropriate and that, in the right circumstances, a holistic approach may also be *simple*. Third, the approach should be underpinned by relevant water quality standards (of appropriate complexity and rigour) with models able to demonstrate compliance with those standards.

Experience in using the UPM procedure and progress made in applying it in the context of improving unsatisfactory intermittent discharges was summarised by Clifforde and Crabtree (2005).

PROBLEMS

2.1   Sediment transported in a sewer as bed-load has been measured at a concentration of 20 ppm (by volume). What is the concentration in mg/L if the specific gravity of the sediment is 2.65? [53 mg/L]

2.2   Plot the following hydrograph and pollutograph (concentration and load rate).

Time (hours)	Flow (L/s)	COD (mg/L)
0.5	80	50
1	170	160
1.5	320	380
2	610	400
2.5	670	230
3	590	130
3.5	380	70
4	220	40
4.5	100	20
5	50	0

Compute the average and flow-weighted (event mean) concentration of COD. [148 mg/L, 208 mg/L]

2.3   A wastewater sample has an organic nitrogen content of 15 mg org.N-N/L and an ammonium concentration of 35 mg NH_4^+/L. If the nitrite and nitrate concentrations are negligible, what is the total nitrogen concentration of the sample? [42 mg N/L].

2.4   Define the main types of solids found in urban drainage systems, and discuss their importance.

2.5   Compare and contrast the main methods for determining the organic content of a wastewater sample.

2.6   Describe the main forms of nitrogen found in wastewater. Why are they of interest?

2.7   What are priority and hazardous substances, and why are they important?

2.8   What type of emissions can be expected from an urban drainage system? Explain how their impact may be acute, delayed, or chronic.

2.9   Discuss the main types of receiving water impact caused by intermittent discharges.

2.10 What are the implications of the Water Framework Directive for urban drainage discharges?

2.11 Explain the difference between Environmental Quality Objectives and Environmental Quality Standards.

2.12 How do intermittent standards differ from ordinary water quality standards? How are they derived?

2.13 The following table shows the number of times that dissolved oxygen fell below 4 mg/L for 6 hours or more in a river. Is this in compliance with the 4 mg/L–6 hour–1 year standard? [No]

Year	1	2	3	4	5	6	7	8	9	10
Number	0	1	1	3	0	0	2	1	0	3

KEY SOURCES

Achleitner, S., De Toffol, S., Englehard, C., and Rauch, W. 2005. The implementation of the European Water Framework Directive – Upcoming steps and future challenges for sanitary engineering practice. *Environmental Management*, 35(4), 517–525.

Chave, P. 2001. *The EU Water Framework Directive. An Introduction*, IWA Publishing.

Ellis, J.B. (ed.) 1985. *Urban Drainage and Receiving Water Impacts*, Pergamon Press.

FWR 2012. *Urban Pollution Management Manual*, 3rd edn, Foundation for Water Research, http://www.fwr.org/UPM3.

Hvitved-Jacobsen, T. 1986. Conventional pollutant impacts on receiving waters, in *Urban Runoff Pollution* (eds H.C. Torno, J. Marsalek, and M. Desbordes), Springer-Verlag, 345–378.

Kallis, G., and Butler, D. 2001. The EU water framework directive: measures and implications. *Water Policy*, 3, 125–142.

Marsalek, J. 1998. Challenges in urban drainage: Environmental impacts, impact mitigation, methods of analysis and institutional issues, in *Hydrodynamic Tools for Planning, Design, Operation and Rehabilitation of Sewer Systems* (eds J. Marsalek, C. Maksimovic, E. Zeman, and R. Price), NATO ASI Series 2. Environment, 44, Kluwer Academic, 1–23.

Marsalek, J., Jimenez-Cineros, B., Karamouz, M., Malmqvist, P.-A., Goldenfum, J., and Cocat, B. 2008. *Urban Water Cycle Processes and Interactions*. Urban Water Series Volume 2, UNESCO Publishing—Taylor and Francis.

REFERENCES

Aalderink, R.H. and Lijklema, L. 1985. Water quality effects in surface waters receiving stormwater discharges, in *Water in Urban Areas, TNO Committee on Hydrological Research*, Proceedings and Information No. 33, 143–159.

American Water Works Association (AWWA). 1999. *Standard Methods for the Examination of Water and Wastewater*, 20th edn, AWWA.

Council of the European Communities (CEC). 1978. *Directive Concerning Quality of Freshwater Supporting Fish*, 78/659/EEC.

Council of the European Communities (CEC). 1991a. *Directive Concerning Urban Waste Water Treatment*, 91/271/EEC.

Council of the European Communities (CEC). 1991b. *Directive Concerning Protection of Water against Pollution by Nitrates from Agriculture*, 91/276/EEC.

Council of the European Communities Council of the European Communities (CEC). 2000. *Directive Establishing a Framework for Community Action in the Field of Water Policy*, 2000/60/EC.

Council of the European Communities (CEC). 2006. *Directive Concerning the Quality of Bathing Water*, 2006/7/EC.

Council of the European Communities (CEC). 2008. *Directive Establishing a Framework for Community Action in the Field of Marine Environmental Policy (Marine Strategy Framework Directive)*, 2008/56/EC.

Council of the European Communities (CEC). 2012. *Report on the Implementation of the Water Framework Directive (2000/60/EC) River Basin Management Plans*, COM/2012/0670 final.

Council of the European Communities (CEC). 2013a. *Amending Directives 2000/60/EC and 2008/105/EC as regards Priority Substances in the Field of Water Policy*, Directive 2013/39/EU.

Council of the European Communities (CEC). 2013b. *Seventh Report on the Implementation of the Urban Waste Water Treatment Directive (91/271/EEC)*, COM/2013/0574 final.

Clifforde, I.T. and Crabtree, R.W. 2005. 10 years experience of integrated wastewater management and planning in the United Kingdom. *Proceedings of the 10th International Conference on Urban Drainage*, Copenhagen, August, on CD-ROM.

DoE (various). *Methods for the Examination of Waters and Associated Materials* (Separate booklets for individual determinations), HMSO.

Droste, R.L. 1997. *Theory and Practice of Water and Wastewater Treatment*, John Wiley.

Environment Agency (EA). 2014. *Monitoring of Discharges to Water and Sewer*, Technical Guidance Note (Monitoring) M18, Environment Agency, version 4.

European Environment Agency (EEA). 2015. *European Bathing Water Quality in 2014*, Report No. 1/2015.

Gray, N.F. 1999. *Water Technology. An Introduction for Scientists and Engineers*. Arnold.

Henze, M. 1992. Characterisation of wastewater for modelling of activated sludge processes. *Water Science and Technology*, 25(6), 1–15.

House, M.A., Ellis, J.B., Herricks, E.E., Hvitved-Jacobsen, T., Seager, J., Lijklema, L. et al., 1993. Urban drainage – impacts on receiving water quality. *Water Science and Technology*, 27(12), 117–158.

Inman, J.B. 1979. Sewage and its pretreatment in sewers. Chap. 7. in *Developments in Sewerage – 1* (ed. R.E. Bartlett), Applied Science.

Lau, K.T., Butler, D., and Schütze, M. 2002. Is combined sewer overflow spill frequency/volume a good indicator of receiving water quality impact? *Urban Water*, 4(2), 181–189.

Lundy, L., Ellis, J.B., and Revitt, D.M. 2012. Risk prioritisation of stormwater pollutant sources. *Water Research*, 46, 6589–6600.

Painter, H.A. 1958. Some characteristics of a domestic sewage. *Journal of Biochemical and Microbiological Technology and Engineering*, 1(2), 143–162.

Wyer, M.D., Fleisher, J.M., Gough, J., Kay, D., and Merrett, H. 1995. An investigation into parametric relationships between enterovirus and faecal indicator organisms in the coastal waters of England and Wales. *Water Research*, 29(8), 1863–1868.

Zabel, T., Milne, I., and McKay, G. 2001. Approaches adopted by the European Union and selected Member States for the control of urban pollution. *Urban Water*, 3(1–2), 25–32.

Chapter 3

Wastewater

3.1 INTRODUCTION

Wastewater, or sewage, is one of the two major urban water-based flows that form the basis of concern for the drainage engineer. The other, stormwater, is described in Chapter 5. Wastewater is the main liquid waste of the community. Safe and efficient drainage of wastewater is particularly important to maintain public health (because of the high levels of potentially disease-forming micro-organisms in wastewater) and to protect the receiving water environment (due to large amounts of oxygen-consuming organic material and other pollutants in wastewater). This chapter provides background information and summary data on wastewater. The quantification for design purposes is dealt with in Chapter 8.

The basic sources of wastewater are summarised in Figure 3.1 and consist of

- Domestic
- Non-domestic (commercial and industrial)
- Infiltration/inflow

In practice, the relative importance of the components varies with a number of factors, including

- Location (climatic conditions, the availability of water and its characteristics, and individual domestic water consumption)
- Diet of the population
- Presence of industrial and trade effluents
- Type of collection system (i.e., separate or combined)
- Condition of the collection system

This chapter is concerned with the generation and characteristics of wastewater. It collates quantity and quality information on the various sources of wastewater and discusses their relative importance. Section 3.2 covers domestic sources of wastewater, Section 3.3 non-domestic, and Section 3.4 the contribution of infiltration and inflow. Wastewater quality issues are dealt with in Section 3.5.

3.2 DOMESTIC

In many (but not all) networks, the domestic component of wastewater is the most important. Domestic wastewater is generated primarily from residential properties but also includes contributions from institutions (e.g., schools, hospitals) and recreational facilities

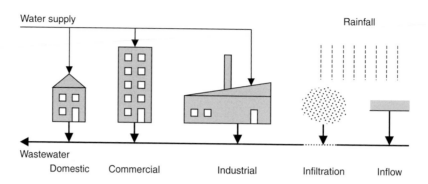

Figure 3.1 Sources of wastewater.

(such as leisure centres). In terms of flow quantity, the defining variable is domestic water *consumption*, which is linked to human behaviour and habits. In fact, very little water is actually consumed, or lost from the system. Instead, it is used intermittently (degrading its quality) and then discharged as wastewater. Hence, in this section we look at the links between water usage and wastewater discharge and, in particular, how these vary with time.

3.2.1 Water use

Important factors affecting the magnitude of *per capita* water demand include the following.

3.2.1.1 Climate

Climatic effects such as temperature and rainfall can significantly affect water demand. Water use tends to be greatest when it is hot and dry, due largely to increased garden watering/sprinkling and landscape irrigation. The impact on wastewater is less pronounced, as this additional water will probably not find its way into the sewer.

3.2.1.2 Demography

It has been demonstrated that household occupancy levels are important, with larger families tending to have lower *per capita* demand (Willis et al., 2013). While, at the other end of the scale, retired people have been shown to use more water than the rest of the population (Russac et al., 1991).

3.2.1.3 Socio-economic factors

The greater the affluence or economic capabilities of a community, the greater the water use tends to be. Work in the UK (Russac et al., 1991) and Australia (Willis et al., 2013) has confirmed the link between water demand and economic indicators such as dwelling type, dwelling rateable value, and household income. This is probably due to greater ownership and use of water-using domestic appliances such as washing machines, dishwashers, and power showers.

3.2.1.4 Development type

Dwelling type is important. In particular, dwellings with gardens may use more water than flats or apartments.

3.2.1.5 Extent of metering and water conservation measures

Water undertakers with metered supplies usually charge their customers based on the quantity of water used in a given period. Systems with unmetered services charge a flat rate for unlimited water use. In theory at least, metered supplies should prevent waste of water by users, reduce actual water use, and therefore reduce wastewater flows. Water is not universally metered in the UK with only about 50% of houses metered, albeit with a rising trend and wide regional variations (CCW, 2016). It has been shown by Ornaghi and Tonin (2015) in their study of the widespread rollout of meters in Southeast England (January 2011–September 2014) to result in a water saving of 16.5%. Variable tariffs allied to smart meters could increase these savings further. Water conservation measures such as low-flow taps/showers, low-flush toilets, and recycling/reuse systems also reduce water demand. The potential impacts on the urban drainage system of more widespread use of such measures are discussed further in Chapters 9 and 24.

3.2.1.6 Quantification

Water consumption per head of population can be extremely varied, as shown in Figure 3.2. However, average domestic water usage in England and Wales stood at 140 L/hd.d in 2015/2016 (CCW, 2016). Usage ranges from 95 to 224 L/hd.d in the 11 countries cited by Friedler et al. (2013).

Water is used in three main areas in the home. Approximately one third of the water is used for WC flushing; one third for personal washing via the wash basin, bath, and shower; and the final third for other uses such as washing up, laundry, and food/drink preparation (see Table 3.1). It is notable that only a very small percentage of this potable standard water is actually drunk (1–2 L/hd.d).

3.2.2 Water–wastewater relationship

As mentioned earlier, there is a strong link between water usage and wastewater disposal, with relatively little supplied water being "consumed" or taken out of the system. On a daily basis we can simply say

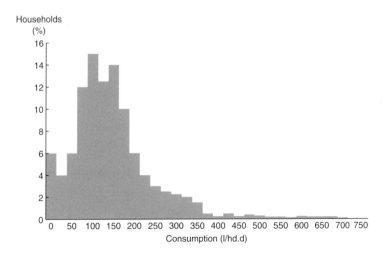

Figure 3.2 Variation of *per capita* water consumption. (Based on Russac, D.A.V., et al. 1991. *Journal of the Institution of Water and Environmental Management*, 5(3), June, 342–351. With permission of the Chartered Institution of Water and Environmental Management, London.)

Table 3.1 Percentage of water consumed for various purposes

Component	Water consumed (%)		
	Household	Commercial	Industrial and agricultural
WC flushing	27	35	5
Showering/bathing	28	26	1
Urinal flushing	—	15	2
Food preparation/drinking	15	9	13
Laundry (washing machine)	16	8	—
Washing-up (including dishwasher)	10	2	—
Car washing/garden use	5	4	17
Other	1	1	62

Source: Adapted from Friedler, E. et al. 2013. Wastewater composition. Chapter 17 in *Source Separation and Decentralization for Wastewater Management* (eds. T.A. Larsen, K.M. Udert, and J. Lienert), IWA Publishing.

$$G' = xG \qquad (3.1)$$

where G is the water consumption per person (L/hd.d), G' is the wastewater generated per person (L/hd.d), and x is the return factor, given in Table 3.2 (–).

It is estimated that, in the UK, about 95% of water used is returned to the sewer network. The other 5% is made up of water used externally (watering the garden and washing the car, for example) and to miscellaneous losses within the household. In hotter climates with low rainfall, this proportion can be up to 40%.

Figure 3.3 shows a comparison made throughout the day between water use and wastewater flow in a catchment. In general, water use exceeds wastewater flow, especially in the early evening when gardens are being watered. At night this situation is reversed due to sewer infiltration flows.

3.2.3 Temporal variability

It is emphasised that both wastewater quantity and quality vary widely from the very long term to the short term. Hence, any particular reported value should be related to the timescale over which it was measured.

Table 3.2 Percentage of water discharged as wastewater

Country	x (%)
United Kingdom	95
Middle East	
Poor housing	85
Good housing	75
United States	60

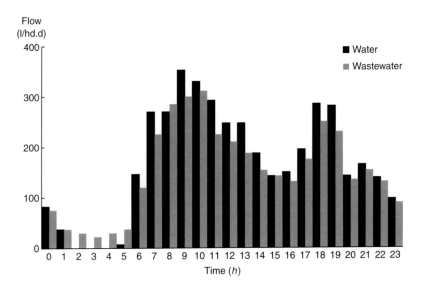

Figure 3.3 Typical diurnal plot of water consumption and wastewater flow.

3.2.3.1 Long term

Until recently the long-term trend has been a steady increase in *per capita* consumption on an annual basis, reflecting a number of factors such as increased ownership of water-using domestic appliances. However, that trend has now been reversed downwards in the UK due to an increased emphasis on household water efficiency.

3.2.3.2 Annual

Variations within the year due to seasonal effects can be observed in water demand. Evidence (Thackray et al., 1978) suggests that WC flushing decreases in summer (probably due to increased rate of body evaporation) and that bathing/showering increases. Outside water use increases significantly from gardening, and this can dominate the demand during summer months. For example, during the dry summer of 1995, increases in average demand of 50% or higher were observed in some areas. Figure 3.4 shows the monthly trends in the Anglian region for 3 years where the average consumption in July was up to 25% greater than in one of the winter months. The effect on wastewater flows is less clearly defined, but typically, summer dry weather flow discharges normally exceed winter flows by 10%–20%.

3.2.3.3 Weekly

Variations in water demand and wastewater production can occur within the week, from day to day. Butler (1991) and Parker and Wilby (2013) found increased water consumption at weekends, probably due to increased WC flushing and bathing. This may also be due to a transfer of location rather than an increase *per se*.

3.2.3.4 Diurnal

A basic diurnal pattern showing variation from hour to hour of wastewater is given in Figure 3.3. Minimum flows occur during the early morning hours when activity is at its

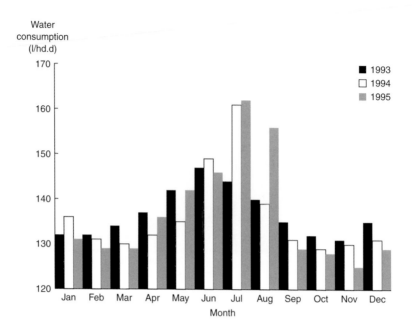

Figure 3.4 Comparison of the effects on water consumption in a dry summer (Anglian region).

lowest. The first peak generally occurs during the morning, the exact timing of which is dependent on the social activities of the community, but in this example, it is between 09:00 and 10:00. A second flow peak occurs in the early evening between 18:00 and 19:00, and then a third can also be distinguished between 21:00 and 22:00, but this is less clearly defined in magnitude and timing. Detailed timing within the diurnal cycle is also affected by the day of the week, with some differences noted at weekends.

3.2.4 Appliances

Wastewater production is strongly linked to the widespread ownership and use of a wide range of domestic appliances, such as those in Table 3.3. The contribution of each individual appliance depends on both the volume of flow discharged after each operation and the frequency with which it is used.

Table 3.3 shows typical discharge volumes of six different domestic appliances. Particularly large volumes are discharged by washing machines and during bathing, while relatively little is used during each use of the wash basin.

Figure 3.5 illustrates how the discharges from the individual appliances go to make up the general wastewater diurnal pattern. The most important contributor overall is the WC, which although only of modest volume is used very frequently throughout the day, and particularly at peak periods. Further discussion of the implications of the diurnal wastewater pattern is given in Chapter 8.

3.3 NON-DOMESTIC

3.3.1 Commercial

This category includes businesses such as shops, offices, and light industrial units, and commercial establishments such as restaurants, laundries, public houses, and hotels.

Table 3.3 Average extant domestic appliance discharge volumes

Appliance	Volume (l)*
WC (per flush)	9.5
Bath (per use)	80
Shower (per use)	35
Washing machine (per cycle)	80
Dishwasher (per cycle)	35

Source: After Lallana, C. et al. 2001. *Sustainable Water Use in Europe, Part 2: Demand Management*, Environmental Issue Report 19, European Environment Agency.

*   Introduction of low flow variants of these appliances will gradually reduce average volumes over time.

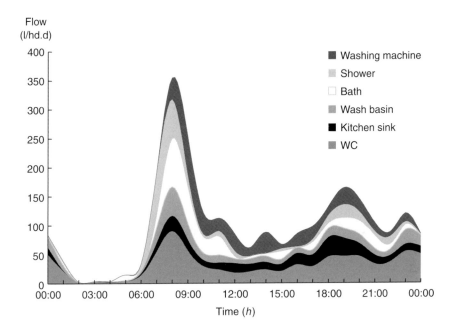

Figure 3.5 Appliance diurnal discharge patterns.

Demand is generated by drinking, washing, and sanitary facilities, but patterns of use are inevitably different to those generated by domestic usage. For example, Table 3.1 shows how toilet/urinal usage is an even more dominant component of water use (50%) than in the domestic environment. Much less detailed information is available on commercial usage than on domestic usage.

3.3.2 Industrial

The component of wastewater generated by industrial processes can be important in specific situations but is more difficult to characterise in general because of the large variety of industries. Table 3.1 shows that many of the most important components of usage found in domestic and commercial premises are much less important in industry and agriculture.

In most cases, effluents result from the following water uses:

- Sanitary (e.g., washing, drinking, personal hygiene)
- Processing (e.g., manufacture, waste and by-product removal, transportation)
- Cleaning
- Cooling

The detailed rate of discharge will vary from industry to industry and will depend significantly on the actual processes used. Water consumption is often expressed in terms of volume used per mass of product. So, for example, papermaking consumes 50–150 m³/t and dairy products 3–35 m³/t.

Industrial effluents can be highly variable (in both quantity and quality) as a consequence of batch discharges, operation start-ups and shut-downs, working hours, and other factors. These may change significantly at weekends. Depending on the relative magnitude of the flows, industrial discharges can completely alter the normal diurnal patterns of flow. There may also be significant seasonal changes in demand, for example, due to agro-industrial practices responding to the needs of food production.

Other important factors include the size and structure of the organisation, the extent of process water recycling, and the availability and cost of water (Shang et al., 2016).

3.4 INFILTRATION AND INFLOW

Unlike the other sources of wastewater, infiltration and inflow are not deliberate discharges but occur as a consequence of the existence of a piped network. Infiltration and inflow have been briefly introduced in Chapter 1 and are defined as water that enters the sewer system, including laterals and private sewers, through indirect and direct means, respectively.

Infiltration is flow that enters the sewer system through defective drains and sewers (cracks and fissures), pipe joints, couplings, and manholes arising from

- Extraneous groundwater
- Spring water
- Seawater
- Water from other leaking pipes

Inflow is actually stormwater that enters separate foul sewers from illegal or misconnected yard gullies, through roof downpipes, or through manhole covers (see Chapter 5 for further discussion on misconnections). This is extremely common, and a survey of one separate system (Inman, 1975) found that 40% of all houses had some arrangement whereby stormwater could enter the wastewater sewer. Inflow can also arise from adjacent storm sewers, land drainage, streams, and other watercourses.

3.4.1 Problems

The presence of excessive amounts of infiltration may cause one or more of the following problems:

- Reduced effective sewer capacity leading to possible surcharging and/or flooding preventing the flushing of toilets, and so on, or pollution for long periods of time
- Overloading of pumping stations and wastewater treatment works

- Higher frequency of combined sewer overflow operation, possibly in dry weather during periods of high groundwater levels
- Increased entry of sediment (soil), resulting in higher maintenance requirements and possible surface subsidence

3.4.2 Quantification

The extent of infiltration is site specific, but when excessive, is usually a result of poor design and construction and will generally deteriorate as the system physically degrades. Influencing factors include the following (Bishop et al., 1998; Karpf and Krebs, 2013):

- Age of the system
- Standard of materials and methods
- Standard of workmanship in laying pipes
- Settlement due to ground movement
- Height of groundwater level (varies seasonally)
- Type of soil (hydraulic conductivity)
- Properties of the sewer trench
- Aggressive chemicals in the ground
- Extent of the network – total length of sewer (including house connections); type of pipe joint, number of joints and pipe size; number and size of manholes and inspection chambers
- Proximity of other drainage networks to transfer flow (e.g., laid in the same trench)
- Frequency of surcharge

The amount of infiltration may range widely and can reach serious proportions in old systems. In the UK, rates up to 50% of average dry weather flow have been measured. The proportion of total infiltration arising from house connections may be up to 50% in places, but this is difficult to predict with any accuracy (Kohout et al., 2010). More complete details of the causes, costs, and control of infiltration can be found in White et al. (1997).

3.4.3 Exfiltration

Exfiltration is the opposite of infiltration. Under certain circumstances, wastewater (or stormwater) is able to leak out of the sewer into the surrounding soil and groundwater. This creates the potential for infiltration into another system and groundwater contamination, which could be critical in areas where an aquifer is used for drinking water supply.

Rutsch et al. (2008) argue that there are contradictory viewpoints on whether or not exfiltration is a serious problem. Values of exfiltration (as with infiltration) are variable, with published rates as low as 1% and as high as 13% of dry weather flow in UK case studies (Rueedi et al., 2009; Wakida and Lerner, 2005), with even higher rates noted in unpublished studies. A study in Canada (Guérineau et al., 2014) found exfiltration to vary between 0.6 and 15.7 m^3/d per kilometre during dry weather, and 1.1 and 19.5 m^3/d per kilometre during wet weather. This variability is a result of the inherent differences across systems and of the many alternative methods of quantification used, many of which are indirect.

Factors affecting the likelihood of exfiltration are similar to those discussed for infiltration. More complete details of the causes, costs, control, and implications of exfiltration can be found in Anderson et al. (1996) and Reynolds and Barrett (2003).

3.5 WASTEWATER QUALITY

Wastewater contains a complex mixture of natural organic and inorganic material present in various forms, from coarse grits, through fine suspended solids, to colloidal and soluble matter. Much is in the form of highly putrescible compounds. In addition, a small proportion of man-made substances, derived from commercial and industrial practices, will be present.

In fact, wastewater is 99.9% water although the remaining 0.1% is significant, particularly if it is allowed to enter the environment. Fresh domestic wastewater is typically cloudy grey in colour with some recognisable solids and has a musty/soapy odour. With time (2–6 hours depending on ambient conditions), the waste "ages" and gradually changes in character as a result of physical and biochemical processes. Stale wastewater is dark grey/black with smaller and fewer recognisable solids, and "older" flows can have a pungent "rotten eggs" odour due to the presence of hydrogen sulphide.

Wastewater quality is variable in respect to both location and time. In addition, the techniques commonly used for sampling and analysis are subject to error (see Chapter 2). Therefore, caution is needed in interpreting standard or typical values. Such data should never be assumed to accurately represent the wastewater from a particular community – this can only be properly confirmed by a (possibly extensive) testing programme or access to historic data.

3.5.1 Pollutant sources

Wastewater quality is influenced by the contaminants discharged into it derived mainly from human, household, and industrial activities. The quality of the carriage water (the original drinking water) or infiltrating groundwater can also be influential.

3.5.1.1 Human excreta

Human excreta are responsible for a large proportion of the pollutants in wastewater. Adults produce on average 200–300 g of faeces and 1–1.5 kg of urine per day. Faeces account for 10–15 g/hd.d of biochemical oxygen demand (BOD) and urine 6 g/hd.d of BOD, but together only contribute a small proportion of wastewater fats (Friedler et al., 2013).

Excreta are also an important source of nutrients. The bulk (94%) of the organic nitrogen in wastewater is derived from excreta. Of this percentage, 50% derives from urine (urea), which is most abundant in fresh wastewater as it is rapidly converted to ammonia under both aerobic and anaerobic conditions (see Chapter 2). Approximately 50% of the phosphorus discharged to sewer (1.5 g/hd.d) is derived from excreta. Excreta also contain about 1 g/hd.d of sulphur (Gilmour et al., 2008).

The bulk of the micro-organisms in wastewater originate in faeces; urine is relatively microbe free.

3.5.1.2 Toilet/WC

A wide range of large (gross) solids is discharged, either deliberately or accidentally, via the toilet such as condoms, sanitary towels, panty liners, tampons, disposable diapers, toilet paper, paper towels, wet wipes, and cotton buds. Toilet paper is used in large quantities. Although this typically disintegrates in a matter of hours in the turbulent flow in sewers (Eren and Karadagli, 2012), it is only slowly biodegradable due to the presence of cellulose fibres. Spence et al. (2016) found by in-sewer sampling that 12–23 g/hd.d is disposed of influenced particularly by the demographics of the sampled catchments. Tests have shown that coloured papers contribute some 15% of the wastewater chemical oxygen demand (COD). In total, some 0.15 sanitary items/hd are disposed of each day (Friedler et al., 1996).

A number of cleaning, disinfecting, and descaling chemicals are also routinely discharged into the system via the toilet.

3.5.1.3 Food

Digested food is the source of many of the excreta-related pollutants mentioned above. However, undigested food is a major contributor of fats, oils, and grease including butter, margarine, cooking oils, vegetable fats, meats, cereals, and nuts. Food residues are also a source of some organic nitrogen and phosphorus and of salt (NaCl).

Food waste disposers are not widely used in the UK but are common in Australia, New Zealand, and the United States (Evans, 2012). Clearly, food waste load entering the sewer will be greater if this appliance is attached to the kitchen sink. Significant benefits (e.g., increased biogas production) are claimed on the basis of trials in Sweden, with very few drawbacks (e.g., increased water consumption, sewer deposition) in evidence (Evans et al., 2010; Mattsson et al., 2014).

3.5.1.4 Washing/laundry

Washing and laundry activities add soaps and detergents to the sewer (e.g., washing machine and dishwater detergents). The polyphosphate builders used in synthetic detergents contribute approximately 50% of the phosphorus load. Phosphorus concentrations have diminished significantly in countries where legislation has imposed significant reductions in the amounts of phosphorus used by manufacturers of detergents (Morse et al., 1993). Values for total phosphorus are approximately 0.3 g/hd.d from laundry activity and 0.2 g/hd.d from dishwashers (Gilmour et al., 2008).

3.5.1.5 Industry

The characteristics of industrial wastewaters, or trade effluents as they are often called, are similar to those of domestic wastewater in that they are likely to contain a very high proportion of water, and the impurities may be present as suspended, colloidal, or dissolved matter. But in addition, a very large variety of pollutant types can be generated and industrial wastewater may contain

- Extremes of organic content
- A deficiency of nutrients
- Inhibiting chemicals (acids, toxins, bactericides)
- Resistant organic compounds
- Heavy metals and accumulative persistent organics

Processing liquors from the main industrial processes tend to be relatively strong, while wastewaters from rinsing, washing, and condensing are comparatively weak. Discharges may be seasonal and vary considerably from day to day both in volume and strength.

3.5.1.6 Carriage water and groundwater

The sulphate present in wastewater is derived principally from the mineral content of the municipal water supply or from saline groundwater infiltration (see Chapter 2).

In hard water areas, the use of softeners can result in significant increases in the wastewater chloride concentrations. Infiltration of saltwater (if present) can contribute similarly.

3.5.2 Pollutant levels

Typical values and ranges of pollutant levels in UK wastewater are given in Table 3.4.

Table 3.4 Pollutant concentrations and unit loads for wastewater

Parameter type	Parameter		Unit load (g/hd.d)	Concentration (mg/L) mean (range)
Physical	Suspended solids			
		Volatile	48	240
		Fixed	12	60
	Total		60	300 (180–450)
	Gross (sanitary) solids			
		Sanitary refuse	0.15*	
		Toilet paper	7	
	Temperature			18 (15–20) °C: summer
				10°C: winter
Chemical	BOD$_5$			
		Soluble	20	100
		Particulate	40	200
	Total		60	300 (200–400)
	COD			
		Soluble	35	175
		Particulate	75	375
	Total		110	550 (350–750)
	TOC		40	200 (100–300)
	Nitrogen			
		Organic N	4	20
	Ammonia		8	40
	Nitrites			0
	Nitrates			<1
	Total		12	60 (30–85)
	Phosphorus			
		Organic	1	5
		Inorganic	2	10
	Total		3	15
	pH			7.2 (6.7–7.5): hard water
				7.8 (7.6–8.2): soft water
	Sulphates		20	100: dependent on water supply
	FOG			100
Microbiological	Total coliforms			10^7–10^8 MPN/100 mL
	Faecal coliforms			10^6–10^7 MPN/100 mL
	Viruses			10^2–10^3 infectious units/100 mL

Source: Adapted from Ainger, C.M. et al. 1997. *Dry Weather Flow in Sewers*, CIRIA R177.

* Items/hd.d.

PROBLEMS

3.1 Classify the major sources of wastewater, and discuss the factors affecting their prevalence in practice.

3.2 Explain the quantitative link between water demand and wastewater generation. What are the major factors influencing domestic water use?

3.3 What are the main differences between domestic, commercial, and industrial water demand?

3.4 Describe how wastewater varies at various timescales, and explain the significance of this.

3.5 Compare and contrast the mechanisms, amounts, and implications of infiltration and exfiltration.

3.6 What are the main sources of pollutants in wastewater, and what is their importance?

KEY SOURCES

Ainger, C.M., Armstrong, R.J., and Butler, D. 1997. *Dry Weather Flow in Sewers*, CIRIA R177.

Friedler, E., Butler, D., and Alfiya, Y. 2013. Wastewater composition. Chapter 17 in *Source Separation and Decentralization for Wastewater Management* (eds. T.A. Larsen, K.M. Udert, and J. Lienert), IWA Publishing.

REFERENCES

Anderson, G., Bishop, B., Misstear, B., and White, M. 1996. *Reliability of Sewers in Environmentally Sensitive Areas*, CIRIA PR44.

Bishop, P.K., Misstear, B.D., White, M., and Harding, N.J. 1998. Impacts of sewers on groundwater quality. *Journal of the Chartered Institution of Water and Environmental Management*, 12(June), 216–223.

Butler, D. 1991. The influence of dwelling occupancy and day of the week on domestic appliance wastewater discharge. *Building and Environment*, 28(1), 73–79.

Consumer Council for Water (CCW). 2016. *Delving into water 2016: Performance of the water companies in England and Wales 2011–12–2015–16*, Birmingham, UK.

Eren, B. and Karadagli, F. 2012. Physical disintegration of toilet papers in wastewater systems: Experimental analysis and mathematical modeling. *Environmental Science and Technology*, 46(5), 2870–2876.

Evans, T.D. 2012. Domestic food waste: The carbon and financial costs of the options. *Municipal Engineer*, 165(ME1), 3–10.

Evans, T.D., Andersson, P., Wievegg, A., and Carlsson, I. 2010. Surahammar: A case study of the impacts of installing food waste disposers in 50% of households. *Water and Environment Journal*, 24(4), 309–319.

Friedler, E., Brown, D.M., and Butler, D. 1996. A study of WC derived sewer solids. *Water Science and Technology*, 33(9), 17–24.

Gilmour, D., Blackwood, D., Comber, S., and Thornell, A. 2008. Identifying human waste contribution of phosphorus loads to domestic wastewater. *Proceedings of the 11th International Conference on Urban Drainage*, Edinburgh, September, on CD-ROM.

Guérineau, H., Dorner, S., Carrière, A., McQuaid, N., Sauvé, S. Aboulfadl, K. et al., 2014. Source tracking of leaky sewers: A novel approach combining fecal indicators in water and sediments. *Water Research*, 58(1), 50–61.

Inman, J. 1975. Civil engineering aspects of sewage treatment works design. *Proceedings of the Institution of Civil Engineers, Part 1*, 58(May), 195–204, discussion, 669–672.

Karpf, C. and Krebs, P. 2013. Modelling of groundwater infiltration into sewer systems. *Urban Water Journal*, 10(4), 221–229.

Kohout, D., Princ, I., and Pollert, J. 2010. Measuring and scale-up of infiltration and exfiltration in house connections. In *Assessing Infiltration and Exfiltration on the Performance of Urban Sewer Systems (APUSS)* (eds. B. Ellis and J.L. Bertrand-Krajewski), IWA Publishing.

Lallana, C., Krinner, W., Estrela, T., Nixon, S., Leonard, J., and Berland, J.M. 2001. *Sustainable Water Use in Europe, Part 2: Demand Management*, Environmental Issue Report 19, European Environment Agency, Copenhagen, Denmark.

Mattsson, J., Hedstrom, A., and Viklander, M. 2014. Long-term impacts on sewers following food waste disposer installation in housing areas. *Environmental Technology*, 35(21), 2643–2651.

Morse, G.K., Lester, J.N., and Perry, R. 1993. *The Economic and Environmental Impact of Phosphorus Removal from Wastewater in the European Community*, Selper Publications.

Ornaghi, C. and Tonin, M. 2015. *The Effect of Metering on Water Consumption: Policy Note*, University of Southhampton, UK.

Parker, J.M. and Wilby, R.L. 2013. Quantifying household water demand: A review of theory and practice in the UK. *Water Resources Management*, 27(4), 981–1011.

Reynolds, J.H. and Barrett, M.H. 2003. A review of the effects of sewer leakage on groundwater quality. *Journal of the Chartered Institution of Water and Environmental Management*, 17(1), March, 34–39.

Rueedi, J., Cronin, A.A., and Morris, B.L. 2009. Estimation of sewer leakage to urban groundwater using depth-specific hydrochemistry. *Water and Environment Journal*, 23, 134–144.

Russac, D.A.V., Rushton, K.R., and Simpson, R.J. 1991. Insights into domestic demand from a metering trial. *Journal of the Institution of Water and Environmental Management*, 5(3), June, 342–351.

Rutsch, M., Rieckermann, J., Cullmann, J., Ellis, J.B., Vollertsen, J., and Krebs, P. 2008. Towards a better understanding of sewer exfiltration. *Water Research*, 42, 10–11, 2385–2394.

Shang, Y., Wang, J., Ye, Y., Lei, X., Gong, J., and Shi, H. 2017. An analysis of the factors that influence industrial water use in Tianjin, China. *International Journal of Water Resources Development*, 33(1). doi:10.1080/07900627.2016.1164672.

Spence, K.J., Digman, C., Balmforth, D., Houldsworth, J., Saul, A., and Meadowcroft, J. 2016. Gross solids from combined sewers in dry weather and storms, elucidating production, storage and social factors. *Urban Water Journal*, 13(8), 773–789.

Thackray, J.E., Cocker, V., and Archibald, G. 1978. The Malvern and Mansfield studies of water usage. *Proceedings of the Institution of Civil Engineers, Part 1*, 64, 37–61.

Wakida, F.T. and Lerner, D.N. 2005. Non-agricultural sources of groundwater nitrate: A review and case study. *Water Research*, 39, 3–16.

White, M., Johnson, H., Anderson, G., and Misstear, B. 1997. *Control of Infiltration to Sewers*, CIRIA R175.

Willis, R.M., Stewart, R.A., Giurco, D.P., Talebpour, M.R., and Mousavinejad, A. 2013. End use water consumption in households: Impact of socio-demographic factors and efficient devices. *Journal of Cleaner Production*, 60(1), 107–115.

Chapter 4

Rainfall

4.1 INTRODUCTION

As already noted, urban drainage systems deal with both wastewater and stormwater. Most stormwater is the result of rainfall. Other forms of precipitation – snow, for example – are also contributors, but rainfall is the most significant in most places. Methods of representing and predicting rainfall are therefore crucial in the design, analysis, and operation of drainage systems.

The detailed study of rainfall is the work of meteorologists and hydrologists, and their work primarily entails interpreting and predicting nature – always a difficult task. They work primarily using observation, which is the origin of all our knowledge about rainfall. The more we observe, the more we learn.

Observation provides historical records and allows derivation of relationships between rainfall event properties (particularly intensity, duration, and frequency). Often, urban drainage engineers and modellers require long periods of rainfall (including different storms and the dry periods between) to input to models in order to see how a drainage system behaves. These may be real historical records, but in that case, it is hard to know exactly how representative a particular portion of history actually is. For this reason, it may be more appropriate to use specially created synthetic sets of data that represent the properties of actual rainfall.

This chapter describes the main methods of rainfall measurement and considers rainfall data requirements for different applications (Section 4.2). Historical rainfall data are analyzed in Section 4.3 and the data are represented as single events (4.4) or multiple events (4.5) and described. Section 4.6 discusses climate change and its important implications for urban drainage. Data and techniques presented in this chapter are used in subsequent chapters on design and analysis.

4.2 MEASUREMENT

4.2.1 Rain gauges

Rain gauges are the most common device for measuring rainfall. A standard nonrecording gauge (Figure 4.1a) collects rain falling on a standard area (a 127 mm diameter funnel with the rim placed 300 mm above ground level) over a known period of time. The volume of the stored rainfall is measured manually and, if necessary, converted to rainfall intensity (depth/time) by dividing by the collection area. Collection periods range from 6 hours to 1 month, but 1 day is typical. Since urban drainage systems can respond in less than 6 hours, the data from nonrecording gauges is of limited value in this application.

 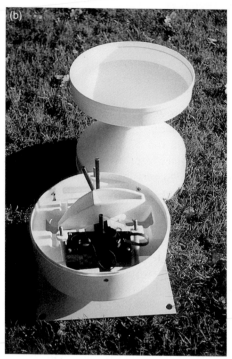

Figure 4.1 Rain gauges: (a) standard; (b) tipping bucket. (Courtesy of Hydrokit, Poole.)

Recording gauges are able to provide a continuous record of rainfall. The tipping-bucket rain gauge collects rainfall over short periods of time in a balanced reservoir consisting of two miniature compartments. Rainwater enters the first compartment until the weight of the water makes it tilt. Water begins to enter the second compartment while the first empties (Figure 4.1b). Thus, the gauge produces a series of tips with a changing frequency depending on the rainfall intensity. The number of tips per unit time is therefore related to the rainfall intensity. A record is made either of the number of tips in a set time interval or the time of each tip. Typically, this is recorded electronically and stored in the memory of a data logger on site or transmitted over telephone lines or mobile-phone networks to a central station. The data can be downloaded to a computer at convenient time intervals for processing. The range of rainfall depth resolution is 0.1–0.5 mm/tip.

A number of other types are also available including weighing, acoustic, and optical gauges. Their function as well as their pros and cons are described in the publication by the World Meteorological Organization (WMO, 2008).

4.2.1.1 Siting

The siting of gauges must be carefully planned in order to obtain representative data for the catchment. There are general rules (WMO, 2008; WRc, 1987) about the distances from obstacles, the level of the gauge orifice above ground level, and so on, but these rules cannot always be fully adhered to in the urban environment. In addition, in urban areas, some rain falls on roofs, so the siting of gauges on roofs is acceptable if the number of gauges on the different elevation levels corresponds approximately to the proportions of the different types of urban surface.

If more than one gauge is in use, careful synchronisation of records is crucial. This can be achieved manually via accurate clock setting at each gauge or remotely via a grid of sensors. Settings should be regularly checked. As a rule, rain gauges should be visited and checked at least once a week.

4.2.2 Radar

Unlike the ground rain gauges that provide point estimates of rainfall values, radars can survey large areas, capturing the spatial and temporal variability of rainfall that is of high importance in the design and modelling of urban drainage systems. Radar stations sweep out short pulses (typically 2 microseconds) of electromagnetic waves, in a narrow 1° bandwidth around a vertical axis, at falling raindrops and measure the time and strength of the return signal to infer the location and the intensity of rainfall accordingly (Collier, 1996). The sweep is repeated at various angles of elevation, usually between 0.5° and 4°, to take account of local topography and buildings that clutter the image.

Three main types of weather radar are available depending on the wavelength of the transmitted signal (from larger to shorter length): S-band, C-band, and X-band. The large wavelength radars, such as the S-band, are able to cover larger areas, given that they are not easily attenuated; however, there are problems associated with the interaction of the beam with the ground and the vertical variation of reflectivity. S-band radars are the biggest and most expensive given that larger dishes are needed to achieve a small beam width. On the contrary, C- and X-band radars require less power and possess smaller dishes, but sensitivity to attenuation is a critical issue. C-band radars balance power, size, and cost and are widely used in European countries, along with S-band units. X-band radars are more sensitive and can detect tiny droplets and, therefore, are often used to provide very short-term weather observations and forecasts that are of interest to urban hydrology applications such as early warning systems.

In the British Isles there are currently 18 weather (C-band) radar stations (Met Office, 2013). These radars complete a series of scans every 5 minutes, providing detailed spatial information on rainfall intensities (approximately one measurement per 1 or 2 km) to a 75 km radius and the overall picture of the extent of precipitation (5 km resolution) to a range of 255 km.

The most commonly used equation to convert the power of the reflected signal, or equally reflectivity Z, to rainfall intensity i is (WMO, 2008)

$$Z = ai^b \tag{4.1}$$

where a and b are constants linked with drop size distribution (DSD).

Alternative values can be assigned to the two constants for different types of storms (e.g., Collier, 1996), while the most typical values for stratiform rainfall are $a = 200$ and $b = 1.60$ (Marshall et al., 1947).

Since radars provide an indirect measurement of rainfall, multiple sources of errors and uncertainties are inserted in the estimation (Harrison et al., 2000; Met Office, 2013):

- Permanent echoes caused by obstacles or hills near the radar (ground clutter).
- Spurious echoes caused by aircraft and interference from other radars.
- Attenuation of the signal by precipitation, especially for C- and X-band radars.
- The radar beam may not detect lower-level clouds at long distances, under-predicting rainfall intensity.

- Evaporation of rainfall at lower levels beneath the beams may result in the over-prediction of rainfall intensity.
- Increase of rainfall intensity at lower levels upon hills where medium-level frontal clouds accumulate other small droplets as they pass through moist, cloudy layers at low levels. This may cause under-estimation of rainfall values.
- False hypothesis for DSD that may result in incorrect transformation of reflectivity to intensity. In general, radars under-estimate rainfall from frontal clouds with small drops and over-estimate rainfall falling from convective shower clouds.
- Miscalibration.

The typical pre-processing of the radar picture to correct some of the above errors comprises (Golz et al., 2005) (1) removal of ground clutter and spurious echoes, (2) correction of radial anomalies, and (3) correction of attenuation effects. Indeed, not all errors can be corrected, some can only be detected (Einfalt et al., 2002). Special attention has been given by the scientific and engineering community to the correction of errors inserted by the transformation of reflectivity into rainfall amounts (Einfalt et al., 2005; Goudenhoofdt and Delobbe, 2013; Smith et al., 2009). In general, the higher are the rainfall intensity and variability, the wider the error bandwidths in the reflectivity to rainfall conversion function become due to difficulty in accurately determining DSD.

Considerable reduction of some of these errors, as well as more accurate identification of DSD have been achieved with the deployment of Polarimetric Doppler Weather Radar, also referred to as *dual-polarization radar* (Bringi and Chandrasekar, 2001). This type of radar transmits and receives signals around both the horizontal and vertical axes, enabling a more accurate sizing of droplets and estimation of rainfall intensities (Bringi et al., 2011).

To improve the accuracy of radar estimates at local spatial scales, useful in urban hydrology, ground measurements from rain gauge networks are often required. This process, referred to as *adjustment*, *combination*, or *merging*, dynamically corrects the error between ground observations and radar estimates. A great number of adjustment procedures can be found in the literature, which, in general, can be classified into two main categories (Wang et al., 2013): (1) mean bias reduction techniques (e.g., Seo et al., 1999; Smith et al., 2007) and (2) error variance minimisation techniques (e.g., Gerstner and Heinemann, 2008; Todini, 2001). In the context of urban drainage applications, the dynamic geostatistical adjustment methods seem more suitable to reproduce the high variability of rainfall at fine spatio-temporal scales (Nielsen et al., 2014; Sinclair and Pegram, 2005; Wang et al., 2015), in contrast to mean bias reduction techniques that have been widely applied at coarser scales.

Radar data are of particular value when radars are continuously operating and feed data online for flood warning (see Section 23.3), real-time control (see Section 23.2), weather forecasting, and water quality prediction (Dale and Stidson, 2007). Data can also be used offline for long-term model simulations (Chapter 19), and rainfall event assessment (e.g., rainfall extremes, high spatial variation) and can provide rapid and extensive urban water management information (Neale, 2008).

4.2.3 Satellites

Satellite imagery has also been used, usually by indirectly relating cloud-top temperature to rainfall (Rosenfeld and Collier, 1998). Petrovic and Elgy (1994) used satellite infrared data to estimate the areal distribution of rain. Despite the fact that rainfall estimates from satellites are less accurate than ground gauges and radars, they can scan much larger areas, sampling also oceans and remote regions. Two different techniques exist to estimate precipitation from space (WMO, 2008). The first method, named *cloud indexing*, analyses the structure of the clouds (i.e., the amount and type of cloudiness) to establish a relationship between a precipitation index and a function of the cloud surface area. The main limitation of this method is the bias from the presence of non-precipitating clouds, such as cirrus. The second method, called *life history* uses consecutive images from geostationary satellites, sampling every half hour, to take into consideration the stage of development of the cloud and estimate precipitation. Various techniques and methods have been developed to reduce the error in rainfall estimation from satellites, using observations from radars and rain gauges (Ebert et al., 2007; Krajewski et al., 2000).

The unique advantage of satellites to cover much larger areas and provide weather forecasts with great lead times is of high importance in online applications of urban drainage such as early warning systems and real-time control (Bajracharya et al., 2017; Chiang et al., 2007; Fan et al., 2016; see also Chapter 22).

4.2.4 Data requirements

The appropriate level of detail in rainfall measurement depends on how the data will be used. Three broad categories can be identified.

In *design and planning*, the task is to produce the overall dimensions of the system. Examples include determining the peak flow rate in storm sewers (see Chapter 10) or the total volume of storm detention tanks (Chapters 12 and 13).

In *checking and evaluation*, the performance of the designed system is assessed under extreme or onerous conditions. This usually requires more effort than design and, consequently, more detailed rain data.

The third task, *analysis and operation*, is concerned with the evaluation of systems that already exist, and examples include the verification of a flow simulation model with real flow data (Chapter 19) or operation of a system in real time (Chapter 22). This latter task has the most stringent requirements for rainfall data.

Table 4.1 lists the rainfall data requirements for each of the three main engineering tasks. The *rainfall record duration* is the length of historical data available for analysis, measured

Table 4.1 Requirements for rainfall data in urban drainage applications

Engineering task	Rainfall record duration (years)	Rain gauge location (relative to catchment)	Temporal resolution (minutes)	Spatial resolution (km^2/gauge)	Synchronisation error (minutes)
Design/planning					
Sewers	>10	Near	Block rain	Homogeneous	≤30
CSO volumes	>5	Vicinity	≤15	Homogeneous	≤30
Checking/evaluation					
Sewers	>20	Adjacent	1	Homogeneous	≤10
CSO volumes	>10	Adjacent	5	≤5	≤5
Analysis/operation					
Calibration	Several	Within	2	2	0.25
Verification	Events	Within			0.25
Real-time control	Online	Within	2[a]	2[a]	0.25

Source: Adapted from Schilling, W. 1991. *Atmospheric Research*, 27, 5–21.

[a] No less than three in total.

in years. This should be significantly longer than the return period (defined in the next section) of the storm event used in system design. The *gauge location* is ideally within the catchment, but this is less important in design and more important in analysis. *Temporal resolution* is the desired time period between rainfall measurements, and *spatial resolution* indicates the desired distance between rain gauges. It is preferable to have several gauges in all catchments to provide data checks and detect spatial variations, including storm movement. Minimising *synchronisation errors* becomes important when multiple gauges are used.

4.3 ANALYSIS

4.3.1 Basics

Rain data measured at an individual rain gauge are most commonly expressed either as *depth* in millimeters or *intensity* in millimeters per hour (mm/hour). This type of *point* rainfall data is therefore representative of one particular location in the catchment. Such data are of greater value if they can be related statistically to two other important rainfall variables: duration and frequency (or probability).

The rainfall *duration* refers to the time period D over which the rainfall falls. However, duration is not necessarily the time period for the whole storm, as any event can be subdivided and analysed for a range of durations. It is common to represent the *frequency* or *probability* of the rainfall as a *return period*. An annual maximum rainfall event has a return period of T years if it is equalled or exceeded in magnitude once, on average, every T years. Thus, a rainfall event that occurs on average 20 times in 100 years has a probability of being equalled or exceeded of 0.2, and a return period of 5 years. Annual maximum storm events are often used to determine return period because it is assumed that the largest event in 1 year is statistically independent of the largest event in any other year.

4.3.2 IDF relationships

4.3.2.1 Definition

The most commonly used tool for the design and planning of drainage systems is the intensity-duration-frequency (IDF) relationship of maximum rainfall amounts. A typical set of IDF curves is given in Figure 4.2 where it can be seen that (for a particular return period) rainfall intensity and duration are inversely related. As the duration increases, the intensity reduces. This confirms the commonsense observation that heavy storms only last a short time, but drizzle can go on for long periods. Also, frequency and intensity are related, as rarer events (greater return periods) tend to have higher intensities (for a given duration).

4.3.2.2 Derivation

Several semi-empirical as well as statistically consistent methods have been developed to construct IDF curves (Chow et al., 1988; Shaw et al., 2010). A generalised IDF formula, which encompasses all the empirical relationships in the literature, is as in Equation 4.2 (Koutsoyiannis et al., 1998):

$$i = \frac{a}{(D^\alpha + b^\beta)} \quad (4.2)$$

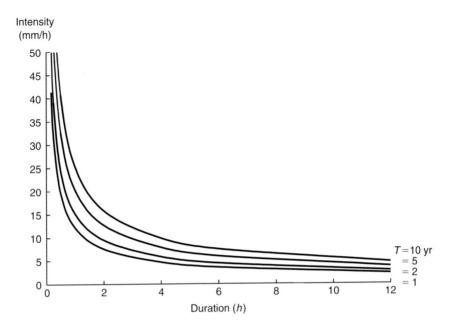

Figure 4.2 Typical intensity-duration-frequency curves.

where a, b, α and β are non-negative coefficients with α, $\beta \leq 1$.

Estimation of the parameters is based on the frequency analysis of maximum rainfall depths and comprises the following steps:

- For each duration, the series of annual maximum values is retrieved (i.e., from 1 minute to 48 hours) and is fitted to a suitable statistical distribution (e.g., log-normal, Gumbel, Pareto).
- For each duration, the rainfall intensities are estimated from the fitted distribution, for a set of return periods (e.g., 5, 10, 20, 50, 100 years).
- For each return period, the intensities of the previous step are inserted in a numerical procedure to establish a relationship between i and D—that is, to estimate IDF parameters.

The last step results in different values of IDF parameters for each T. Alternative approaches enable the estimation of unique values for the parameters of the denominator, allowing only parameter a to vary with T (see Koutsoyiannis et al., 1998).

EXAMPLE 4.1

Using the data presented in Figure 4.2, determine the intensity of a 1-year return period 2-hour duration rainfall event. For a similar duration event of a 10-years return period, find the appropriate rainfall depth.

Solution

For $T = 1$ year, $D = 2$ hours $\therefore i = 7.5$ mm/hours

For $T = 10$ years, $D = 2$ hours $\therefore i = 16$ mm/hours $\therefore d = 16 \times 2 = 32$ mm

---

4.3.2.3 IDFs in practice

In most situations, it is not necessary to derive such a set of curves, since maps with IDF information are available for many countries. For example, the *NOAA Atlas 14* provides IDF curves as well as other information for rainfall extremes for the United States (hdsc. nws.noaa.gov). The *Atlas* is divided into volumes based on geographic sections of the country, while the data are available for operational purposes via the *Precipitation Frequency Data Server* web application. Environment Canada provides short duration IDF curves for each province and territory across Canada (Environment Canada, 2016).

The *Flood Studies Report* (*FSR*; Natural Environment Research Council [NERC], 1975) gave point rainfall depth-duration-frequency data for the whole of the UK for durations from 1 minute to 48 hours originally available as a set of paper maps. The *FSR* IDF model was updated by the *Flood Estimation Handbook* (*FEH*; Institute of Hydrology, 1999), a revised version of which forms current best practice for estimation of rainfall-runoff at river catchment scale (>0.5 km²). However, both methods are still in use with the *FEH* recommended for the 193-year return period (to be used when estimating the 150-year flood event), the higher of the two methods for the 1000-year return period, and the *FSR* only for the 10,000-year return period (Stewart et al., 2013).

4.3.3 Wallingford Procedure

The *Wallingford Procedure* is a set of documents (DoE/NWC, 1981) that explain the basis for and development of an early sewer modelling package (WaSSP). Although the model is no longer in use, the procedure still provides a useful reference document and set of maps where *FSR*-based methods and rainfall can be abstracted for any location in the UK and used for urban drainage design and analysis. Aspects of the procedure are returned to in Chapters 5 and 10, but the rainfall estimation approach is explained here.

The *FSR* and the *Wallingford Procedure* both use a standard notation when specifying rainfall information. Thus, *MT-D* represents the depth of rainfall (in millimeters) occurring for duration D with a return period T years. Durations specified in minutes start at any minute in the hour, those in hours start "on the hour," and those in days begin at 9 a.m. GMT.

The method is based on working from standard M5–60 minutes rainfall and the ratio (r) of M5–60 minutes/M5–2 day rainfall depth, both of which are mapped for the UK and given in Figures 4.3 and 4.4, respectively. The M5–60 minutes rainfall effectively denotes the quantity of rainfall in an area, and the ratio r reflects the "type" of rainfall. Low values of r (<0.2) represent rain that mostly falls as drizzle, whereas values >0.4 indicate the prevalence of much higher-intensity storms.

By means of coefficients (or *growth factors*) Z_1 (Figure 4.5) and Z_2 (Table 4.2), the standard M5–60 minutes rainfall can be related to

- The 5-year rain depth for the required duration (M5–D)
- The depth of rain for the required duration and return period (MT-D)

Example 4.2 shows how this approach can be used to produce an IDF relationship.

IDF relationships of point rainfall are widely used in urban drainage applications. In particular, they are essential in the application of the Rational Method (Chapter 10).

4.3.4 Areal extent

Point rainfall is not necessarily representative of rainfall over a larger area, because average rainfall intensity decreases with increasing area. In order to deal with this problem, and

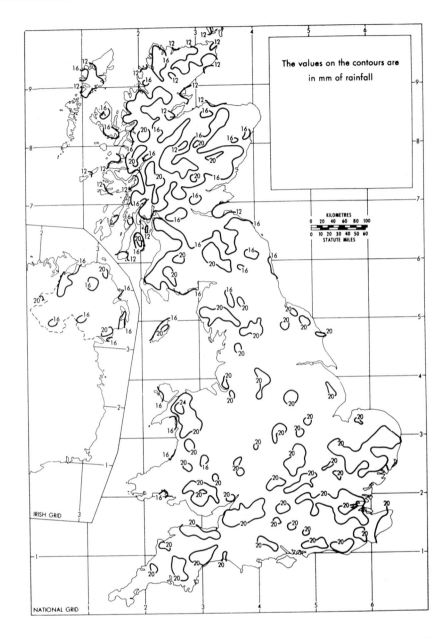

The values on the contours are in mm of rainfall

Figure 4.3 Rainfall depths of 5-year return period and 60-minutes duration: M5–60 minutes. (Reproduced from "The Wallingford Procedure" with permission of HR Wallingford Ltd.)

avoid overestimating flows from larger catchments, *areal reduction factors* (ARFs) have been developed based on the comparison of point and areal data from areas where several gauges exist.

In the *Wallingford Procedure*, the ARF is calculated from

$$\text{ARF} = 1 - f_1 D^{-f_2} \tag{4.3}$$

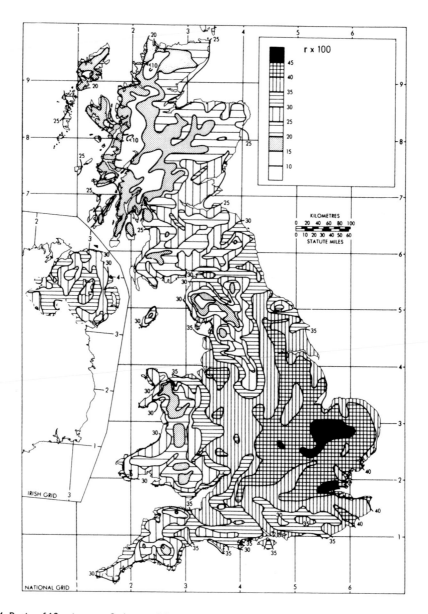

Figure 4.4 Ratio of 60-minute to 2-day rainfalls of 5-year return period: *r*. (Reproduced from "The Wallingford Procedure" with permission of HR Wallingford Ltd.)

where f_1 is $0.0394A^{0.354}$, f_2 is $0.040 - 0.0208 \ln (4.6 - \ln A)$, A is the catchment area (km²).

The expression is valid for UK catchment areas <20 km² and storm durations of 5 minutes to 48 hours (see Example 4.3).

The ARF is lower the larger the catchment area and the shorter the event duration. For urban drainage applications, some specifications set a maximum size of catchment for ARF calculation such that ARF will exceed 0.9 for all event durations (CIWEM, 2016).

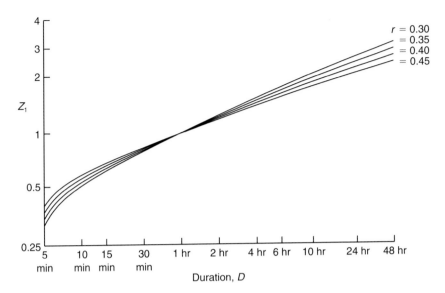

Figure 4.5 Relationship between Z_1 and D for different values of r $(0.30 \leq r \leq 0.45)$. (Based on "The Wallingford Procedure" with permission of HR Wallingford Ltd.)

4.3.5 Flood Estimation Handbook

The *FEH* was originally published by the Institute of Hydrology (1999) as a major revision to supersede the *FSR* and its supplementary reports. The current handbook is further supported by software packages (i.e., WINFAP-FEH, ReFH for flood estimation) and a web application (fehweb.ceh.ac.uk) for point data abstraction. *FEH99* (as it is now known) provides design rainfall depths by a depth-duration-frequency (DDF) model that covers a wide range of durations, from 1 hour up to 8 days, and return periods extended up to 1000 years or longer (Faulkner, 1999).

The *FEH99* DDF model was subsequently revisited after concerns about the apparent high design rainfall depths that are provided for higher return periods (from 100 to

Table 4.2 Ratio Z_2—relationship between rainfall of return period T(MT) and M5 for England and Wales

M5 (mm)	M1	M2	M3	M4	M5	M10	M20	M50
5	0.62	0.79	0.89	0.97	1.02	1.19	1.36	1.56
10	0.61	0.79	0.90	0.97	1.03	1.22	1.41	1.65
15	0.62	0.80	0.90	0.97	1.03	1.24	1.44	1.70
20	0.64	0.81	0.90	0.97	1.03	1.24	1.45	1.73
25	0.66	0.82	0.91	0.97	1.03	1.24	1.44	1.72
30	0.68	0.83	0.91	0.97	1.03	1.22	1.42	1.70
40	0.70	0.84	0.92	0.97	1.02	1.19	1.38	1.64
50	0.72	0.85	0.93	0.98	1.02	1.17	1.34	1.58
75	0.76	0.87	0.93	0.98	1.02	1.14	1.28	1.47
100	0.78	0.88	0.94	0.98	1.02	1.13	1.25	1.40

Source: After DoE/NWC. 1981. *Design and Analysis of Urban Storm Drainage. The Wallingford Procedure. Volume 1: Principles, Methods and Practice*, Department of the Environment, Standing Technical Committee Report No. 28.

10,000 years). Although a new model, referred to also as the *FEH13* DDF rainfall model (Stewart et al., 2013), was developed to allow a better estimation of design rainfall for reservoir flood risk assessment, the analysis showed that it could be applied over the full range of return periods from 2 to over 10,000 years. In general, the *FEH13* DDF model provides lower short-duration design rainfall than *FEH* for most areas of England, in contrast to Wales and Scotland where a considerable increase in the rainfall amounts is apparent.

For rainfall durations shorter than 1 hour, the use of rainfall statistics from the *FSR* is currently recommended. However, a recent study with data from 19 rain gauges across England and Wales showed that the *FEH* model provides less biased results than the *FSR* model, at sub-hourly durations (Prosdocimi et al., 2014).

EXAMPLE 4.2

Determine the relationship between intensity and duration for 10-year return period storms in the London area.

Solution

Calculate the range of M10–D rainfall intensities.

Read from Figure 4.3: M5–60 minutes = 20 mm
Read from Figure 4.4: M5–60 minutes/M5–2 day, $r = 0.45$
Read from Figure 4.5: Z_1 values for various Ds
Read from Table 4.2: Z_2 values for various Ms

(1) Storm duration (D)		(2) M5–60 minutes rainfall total (mm)	(3) Z_1	(4) M5–D rainfall total (mm) (2) × (3)	(5) Z_2	(6) M10–D rainfall total (mm) (4) × (5)	(7) Intensity (mm/hour) (6) ÷ (1)
Hours	Minutes						
	5	20	0.39	7.8	1.21	9.4	112.8
	10	20	0.55	11.0	1.22	13.4	80.4
	15	20	0.63	12.6	1.23	15.5	62.0
	30	20	0.77	15.4	1.24	19.1	38.2
1		20	1.0	20	1.24	24.8	24.8
2		20	1.2	24	1.24	29.8	14.9
4		20	1.4	28	1.22	34.2	8.6
6		20	1.5	30	1.22	36.6	6.1
10		20	1.65	33	1.21	39.9	4.0
24		20	2.0	40	1.19	47.6	2.0
48		20	2.3	46	1.17	53.8	1.1

EXAMPLE 4.3

Adjust the point rainfall intensity of 25 mm/hour for a 15-minute storm falling over a 200 ha urban catchment.

Solution

$D = 0.25$ hour, $A = 2$ km² $\therefore$ Equation 4.3 is valid.

$$f_1 = 0.0394 \times 2^{0.354} = 0.050$$
$$f_2 = 0.040 - 0.0208 \ln (4.6 - \ln 2) = 0.012$$
$$\text{ARF} = 1 - 0.050 \times 0.25^{-0.012} = 0.95$$

Areal intensity = 24 mm/hour

4.4 SINGLE EVENTS

So far we have considered rainfall to consist of just a fixed rainfall depth for a given duration. Clearly, this is unrealistic as rainfall intensity varies with time throughout the storm. This is represented as a plot of rainfall intensity against time called an *hyetograph* (or "storm profile").

4.4.1 Synthetic design storms

A design storm is an idealised storm profile to which a statistically based return period has been attached. The defined pattern in time is designed to reproduce (albeit imperfectly) the "shape" of observed storms. The shape depends mainly on the type of event: a frontal storm usually has the highest intensities near the middle, and a convective storm has the highest intensities near the beginning.

The simplest (and least realistic) form of design storm is *block rainfall*, which may be simply derived from an IDF curve. Block rainfall has the same intensity over its duration and therefore has a rectangular time distribution. It is widely used in the Rational Method (given in Chapter 10) and has the advantage of being simple, quick to use, and easily understandable.

In order to facilitate more accurate design solutions, profiles that better represent observed rainfall profiles have been produced. This has become more important with the advent of more sophisticated surface routing methods and flow simulation models. A number of shapes have been proposed over the years, based successively on more comprehensive data sets.

Information on storm profiles can be found in the *FSR* based on the analysis of a wide range of storm events. A family of standard, symmetrical profiles was produced, with maximum rainfall intensity at the centre of the storm and varied in amplitude. The *peakedness* of a profile is defined as the ratio of maximum to mean intensity, and the *percentile peakedness* is the percentage of storms that are equally or less peaked. The profile shape was not found to vary significantly with storm duration, return period, or geographical region. However, on average, summer storms were found to be more peaked than winter ones.

The *Wallingford Procedure* recommends the 50 percentile summer profile (i.e., the storm that is more peaked than 50% of all summer storms) for design of drainage systems (see Figure 4.6). A rainfall profile can be estimated by distributing the mean intensity over the storm duration as shown in Example 4.4. The *Procedure* also recommends a method of smoothing the point rainfall profile to allow for areal extent.

EXAMPLE 4.4

Determine the intensity of a 50 percentile summer storm of mean intensity 25 mm/hour and duration 15 minutes at its one-third and one-half points.

Solution

1/3 point = 33.3% duration. From Figure 4.6, % of mean intensity = 80
$i_{33} = 0.8 \times 25 = 21$ mm/hour @ $t = 5$ minutes

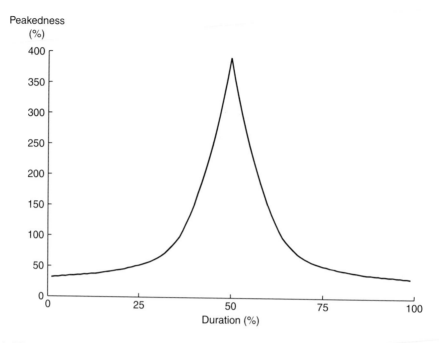

Figure 4.6 FSR 50 percentile summer storm profile.

> 1/2 point $= 50\%$ duration. From Figure 4.6:
> $i_{50} = 3.9 \times 25 = 98$ mm/hour @ $t = 7.5$ minutes

4.4.2 Historical single events

Historical single events are hyetographs of point rainfall constructed from measured data. Unlike design storms, they are not idealised and do not have an attached return period. Recorded data should be at intervals of 5 minutes, and preferably 1 minute. Their main use is in the verification of flow simulation models with measured hyetographs and simultaneous observations of flow. A model so verified is then assumed to give an accurate picture of catchment response (as considered in Chapter 19).

If the model allows consideration of spatially varying rainfall, the hyetographs used should adequately reflect the patterns of rainfall in time *and* space. The direction and movement (tracking) of a storm can be important in certain catchments (especially large ones) and can be a source of error if neglected (Ngirane-Katashaya and Wheater, 1985). Storms moving longitudinally downstream through the catchment have the greatest influence.

4.4.3 Critical input hyetograph (Superstorm)

To fully assess the performance of a sewer system, design storms of different return periods should be applied for a wide range of durations. Then, a *critical duration* can be determined by defining the event that has the greatest impact on the sewer system in terms of specific criteria, for example, risk of failure or cost of damages.

To alleviate the high computational burden imposed by time-consuming simulation models that need to run time and time again to find this critical event, the Critical Input Hyetograph (CIH), or Superstorm, method can be applied. This method substitutes a given set of rainfall hyetographs, of specific return period, with a long-duration single synthetic

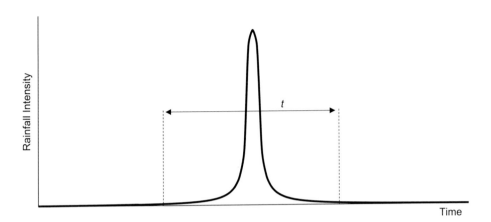

Figure 4.7 Typical critical input hyetograph.

hyetograph that encompasses the critical intensity and/or critical volume. In this case, only one simulation per return period is needed to characterise the behaviour of the sewer system under extreme conditions (CIWEM, 2016).

The CIH is composed so that, for any given duration, the middle section has the same average intensity as the critical, observed or synthetic, hyetograph with the same duration. A typical CIH is given in Figure 4.7, where t determines the section of the graph that has the same average intensity as the critical hyetograph with t duration, used to generate the CIH.

This method, also known as the Chicago Method, was studied first by Keifer and Chu (1957), while other CIH generators, based either on observed rainfall events or on synthetic *FEH* or *FSR* hyetographs, are also available (e.g., Newton et al., 2013).

One drawback of the CIH method is the computational effort that is required to analyse the observed rainfall series so as to determine the critical rainfall hyetographs of different durations. A further drawback is that the method tends to generate storm events with longer durations than what are likely to be the critical durations for some parts of the drainage network.

4.5 MULTIPLE EVENTS

4.5.1 Historical time series

An historical series of rainfall events is the full set of all measured point rainfall for a particular time period (which would include all single historical events and the intervening dry periods) at a particular location. These are used in conjunction with pre-calibrated, continuous simulation models, for long-term analyses. In effect, the use of time-series rainfall shifts the frequency analysis from the rainfall stage to the simulated runoff stage, since the complete series of historical events is run through the model. The return period of any characteristic of interest (e.g., peak flow, combined sewer overflow [CSO] operation) is obtained by the conventional ranking procedure and the use of a plotting formula, as described in Section 4.3.2. A typical time series is shown in Figure 4.8.

The advantage of time series is that they are almost certain to contain the conditions that are critical for the catchment being studied. Their main disadvantage is that large amounts of data from recording rain gauges are required, and it is unlikely that this will be available

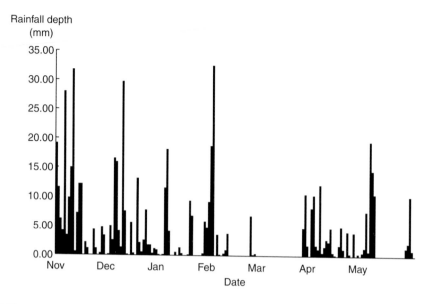

Figure 4.8 Time-series rainfall (6 months of daily data).

for the particular site under consideration (although rainfall data from radar can be used). However, the recent employment of weather radars to measure rainfall supports a wider spatial and temporal coverage. Further to that, the limited availability of data can also be overcome to a limited extent by the use of regional annual time series.

4.5.1.1 Annual time series

An annual rainfall time series is a sequence of historic rainfall events that is statistically representative of the annual pattern at a given location. Three time series have been derived in the UK by selecting typical months from a 40-year rainfall record and assembling them to form a typical year (Henderson, 1986). Their regions of applicability and basic hydrological properties are given in Table 4.3.

The time series recommended for use is the one with the closest rainfall characteristics, not necessarily the closest geographically. Nevertheless, regionalisation procedures are still required to account for differences in hydrological factors between the catchment being studied and the region. Garside (1991) describes the procedures in detail.

Annual time series may be used either in chronological order or ranked with the highest intensity/depth event first and the lowest last. The three series mentioned above are available

Table 4.3 Hydrological characteristics of WRc annual time series

Region	M5–60 (mm)	r	SAAR[a] (mm)
South-east England	20	0.40	630
Central eastern England	19	0.40	590
South-west England[b]	19	0.33	930

[a] Standard annual average rainfall.
[b] Can also be used for northwest England and Wales.

as 99-event sequences, with each event associated with a particular catchment wetness (UCWI) and duration of antecedent dry weather period. The series can be sampled (e.g., by selecting the first five storms and every fifth thereafter) to reduce computing time requirements. Their main use is to represent the significant events in a typical year for situations where only short return periods are significant, such as day-to-day hydraulic performance of existing and rehabilitated systems, including overflow spill events. The results of most interest generally are the total annual spill volume and the spill frequency from an overflow in particular, and the sewer system in general.

Unfortunately, the annual time series suffers from several faults that make it difficult to use, including its limitation to just 1 year, the assumptions in its derivation, and the need to rely on regionalisation. These concerns have led to the development of synthetic time series for rainfall.

4.5.2 Synthetic time series

4.5.2.1 Synthetic series

A synthetic series simply overcomes the problem of using large numbers of different events in the annual time series by representing their pattern with just a small number of synthetic storms. The synthetic storms can be based on conventional rainfall parameters such as storm depth, catchment wetness, and storm peakedness.

Rainey and Osborne (1991) have derived two synthetic series based on the annual time series described previously. Their regions of applicability and basic hydrological characteristics are given in Table 4.4.

4.5.2.2 Stochastic rainfall generation

An alternative approach is the development of statistically based rainfall models. Output from these models is a continuous rainfall time series, which is statistically similar to the historical data for the catchment.

Current stochastic rainfall generation models are based on Poisson-cluster models, first developed by Rodriguez-Iturbe et al. (1987). This type of model is able to reproduce the characteristics of rainfall at multiple fine time scales, via a small number of parameters. These models have a plausible physical basis that assumes any rainfall event is triggered by arriving *storm origins* from which a sequence of *rain cells* is generated. It is assumed that

- The storm origins arrive based on a Poisson process.
- Each storm origin generates a random number of rain cells.
- The intensity of each cell is constant throughout the duration.
- The intensity and duration of each cell follow assumed statistical distributions.
- The total intensity at any point in time is the sum of the intensities of the rain cells.

Clustering of the rain cells is accomplished using Bartlett-Lewis or Neyman-Scott models (Onof et al., 2000). Their structure (illustrated in Figure 4.9) is characterised by a Poisson

Table 4.4 HR synthetic rainfall series

Region	R	SAAR (mm)
East	0.40	610
West	0.33	930

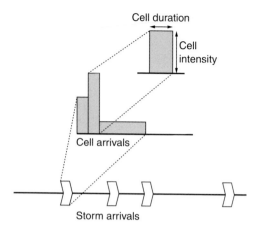

Figure 4.9 Poisson cluster rectangular pulse model schematic. (Adapted from Wheater, H.S., et al. 2005. *Stochastic Environmental Research and Risk Assessment*, 19, 403–416.)

process of cluster (or "storm") arrivals. In the Bartlett-Lewis process, it is assumed that a cell is located at the storm origin. The storm arrival is followed by a "Poisson process" of cell arrivals over a random duration (typically chosen as exponentially distributed, so that the number of cells has a geometric distribution). In the alternative Neyman-Scott process, cells arrive so that the times elapsing from the storm arrival (which is notional and not observed) are independently exponentially distributed, and the number of cells per cluster is randomly distributed (typically Poisson or geometrical). In both cases, the rectangular pulse is assumed to have a random intensity and duration. This defines models with five (if the random intensity is exponentially distributed) or, more generally, six parameters (Wheater et al., 2005).

The two basic models have been further modified to enable a better reproduction of different profiles of rainfall and their temporal properties, such as the proportion of dry intervals. Rodriguez-Iturbe et al. (1988) and Entekhabi et al. (1989) extended the Bartlett-Lewis and Neyman-Scott models, respectively, so that the parameter that specifies the duration of cells is randomly varied between storms according to a gamma distribution. The parameterisation of the first model entails also the variation of parameters of cell origin and storm duration, proportionally to the cell duration parameter. In the case of exponential distribution for cell intensity, the randomised models have six parameters.

Greater diversity of rainfall types can also be achieved either by allowing different cell types within any storm (Cowpertwait, 1994) or by superposing multiple independent processes for different storm types (Cowpertwait, 2004; Cowpertwait et al., 2007). In the former case, Cowpertwait (1994) formulated a two-cell Neyman-Scott model that requires eight parameters for exponentially distributed cell intensities, with the additional three being mean cell intensity and cell duration for the second cell type, and the probability of a cell being of this type. If two parameter distributions are used for the cell intensities, 10 parameters in all are needed. In the latter case, Cowpertwait (2004) superposed two independent Neyman-Scott processes to simulate the convective and stratified rainfall at a site in New Zealand. It is worth mentioning that despite the considerable flexibility that the above techniques provide in the simulation of different types of rainfall, the large number of parameters poses extra difficulties in the fitting of the models.

The fitting for all of the Poisson-cluster models, mentioned in this section, is typically handled as an optimisation problem that minimises the departure between the statistical properties of the process, expressed as a function of the model parameters, and the corresponding observed values. The typical properties are the mean, variance, autocorrelation function, and probability dry, while analytical expressions about some temporal characteristics of storms can also be used (e.g., Onof and Wheater, 1994; Onof et al., 1994).

Further to these properties, the performance of the models regarding the distribution of extremes is of high importance for urban drainage applications. Empirical analyses have shown that the use of longer-tailed distribution for cell intensity (i.e., Gamma, Weibull, or Pareto), along with the use of the third- or higher-order moment in the fitting procedure, enables a better reproduction of extreme values, especially for lower time scales and high return periods (Cowpertwait, 1998; Onof and Wheater, 1994; Verhoest et al., 2010; Wheater et al., 2005).

The rectangular intensity of the rain cell does not allow the original models to reproduce the high variability of sub-hourly rainfall (e.g., 5 minute), which is important in the design and evaluation of stormwater systems. To address this issue two different approaches have been proposed. The first concerns the replacement of the constant rectangular cell intensities by a cluster of instantaneous pulses that occur following a third Poisson process. Cowpertwait et al. (2007) applied this modification to the original Bartlett-Lewis model and Kaczmarska et al. (2014) to the randomised model version. The superposition of the two modified models to provide even greater flexibility has also been examined. The second approach concerns the re-parameterisation of the randomised Bartlett-Lewis model so as to introduce dependence between cell intensity and duration (Kaczmarska et al., 2014). Specifically, the cell intensity parameter varies between storms, proportionally to cell duration parameter. Both modifications enabled the reproduction of the main rainfall characteristics from 5-minute scale up to daily scale.

4.5.2.3 Stochastic disaggregation models

Observed data at fine time scales are often limited, and in some cases completely unavailable, in contrast to daily values, which have been routinely collected for at least a few decades. Taking advantage of the available coarser rainfall records, disaggregation techniques can be employed to produce possible, statistically consistent, realisations of rainfall events at finer time scales. Two "disaggregators" for sub-hourly time scales are "RainSim" and "TSRsim," while another software, "HyetosMinute," which incorporates the modified versions of the Bartett-Lewis for sub-hourly scales, has also been developed.

The first disaggregator is the RainSim family of models. In its initial version (Cowpertwait, 1994), the model simulated rainfall time series either at a single location or distributed across a region of up to 200 km in diameter. RainSim version 2 was developed to include third moment properties, important for the modelling of extreme rainfall, providing point simulation with a single rain cell type. Version 3 includes improved model calibration and provides a spatial–temporal modelling capability (Burton et al., 2008).

The second approach, TSRsim, is coupled to a Bartlett-Lewis rectangular pulse model, BaLeReP (UKWIR, 2003). This utilises a multi-scaling random cascade (Lovejoy and Schertzer, 1995) that downscales the coarse-scale rainfall intensity in stages defined by a halving of the timescale. At each stage, the intensity is multiplied by identically distributed and independent random variables to obtain the finer-scale intensities.

Both methods are widely used and perform well, although there is some evidence that short-duration storm intensities are under-predicted by TSRsim and over-predicted by RainSim, and the converse occurs for long durations (Kellagher, 2005; UKWIR, 2003; Kellagher, R. 2008, personal communication).

HyetosMinute (Kossieris et al., 2016) combines the use of the Bartlett-Lewis model, as rainfall generator, along with accurate adjusting procedures (Koutsoyiannis and Onof, 2001) to disaggregate daily rainfall into any sub-hourly timescale, down to 1 minute.

4.6 CLIMATE CHANGE

4.6.1 Causes

The design of urban drainage systems that are able to last for a long time necessitates an understanding of phenomena and processes that could alter the characteristics of the various input parameters (such as precipitation, temperature, and evaporation) we used to design the systems in the first place. Having gone through the process of accurately reproducing historic rainfall time series to use in the design of future systems, using the methods described above, it would be unfortunate if future rainfall did not actually follow historic patterns. This is where it becomes important for the drainage engineer to understand and assess the impact of global climate change on the various input design variables of the drainage system. The evidence for global climate change is compelling, with records showing that the global-average surface air temperature has risen by around 0.85°C over the period 1880–2012 (IPCC, 2014; Figure 4.10). In Central England, the temperature has risen by about a degree Celsius since 1980 (Jenkins et al., 2008), with 2014 being the warmest fourth wettest year since 1910.

The main clue to the cause of these rises is the significant increase in the concentration of greenhouse gases (GHGs) observed over the past 200 years, attributed to the earth's climate system variability and emissions from human activities (mostly through the combustion of fossil fuel and industrial activities). In 2010, the anthropogenic GHG emissions (e.g., carbon dioxide, methane, nitrous oxide) were around 49 $GtCO_2$/year, increasing by a rate of 2.2% per year since 2000. This intensifies the natural greenhouse effect, trapping more energy in the lower atmosphere. Other human activities generate pollutants that actually cool the climate (sulphur dioxide that transforms into aerosols).

Unravelling the impact of various forces in the climate system and making predictions into the future require the combined use of different types of climate models, such as the

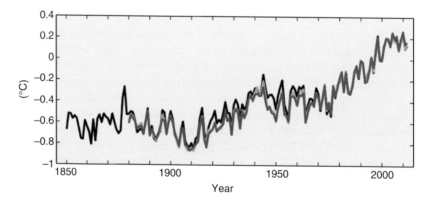

Figure 4.10 Annually and globally averaged combined land and ocean surface temperature anomalies relative to the average over the period 1986 to 2005. Colors indicate different data sets. (Adapted from IPCC. 2014. *Climate Change 2014 Synthesis Report Summary Chapter for Policymakers.* https://www.ipcc.ch/report/ar5/syr/.)

Atmosphere-Ocean General Circulation Models, Earth System Models, Earth System Models of Intermediate Complexity, and Regional Complexity models (see Flato et al., 2013). A great variety of such models have been developed from leading climate centres, such as the HadCM3 model developed at the Hadley Centre in the UK. The Coupled Model Intercomparison Projects (CMIP3 and CMIP5) (Meehl et al., 2007; Taylor et al., 2012) ensembles the output from different models to provide multi-model projections, used in the assessments of the Intergovernmental Panel on Climate Change (IPCC).

The CMIP3 and CMIP5 projections showed that only by incorporating natural *and* human factors could the model fit the data adequately, especially the warming since the 1970s (see Figure 4.11). This and other evidence have led the IPCC to conclude that "most of the warming observed over the last 50 years is *likely* to have been due to increasing concentrations of greenhouse gases" (IPCC, 2001). In the fourth assessment (IPCC, 2007), the wording was revised to "... is *very* likely...," while in the last assessment (IPCC, 2014) it is stated that "It is *extremely* likely that more than half of the observed... anthropogenic increase in GHG concentrations" (our emphasis).

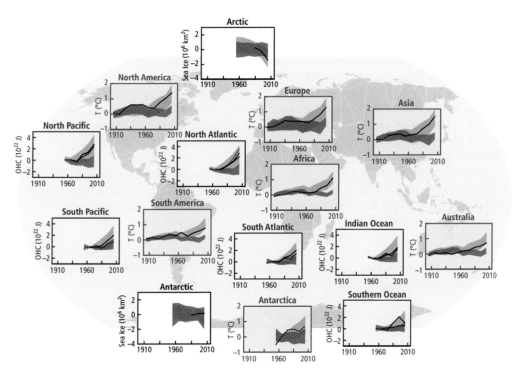

Figure 4.11 Comparison of observed and simulated change in continental surface temperatures on land, Arctic and Antarctic September sea ice extent, and upper ocean heat content in the major ocean basins. Global average changes are also given. Anomalies are given relative to 1880–1919 for surface temperatures, to 1960–1980 for ocean heat content, and to 1979–1999 for sea ice. All time series are decadal averages, plotted at the centre of the decade. For temperature panels, observations are dashed lines if the spatial coverage of areas being examined is below 50%. For ocean heat content and sea ice panels, the solid lines are where the coverage of data is good and higher in quality, and the dashed lines are where the data coverage is only adequate, and, thus, uncertainty is larger. Model results shown are CMIP5 multimodel ensemble ranges, with shaded bands indicating the 5%–95% confidence intervals. (Adapted from IPCC. 2014. *Climate Change 2014 Synthesis Report Summary Chapter for Policymakers* (p. 49). https://www.ipcc.ch/report/ar5/syr/.)

4.6.2 Future trends

To obtain climate change projections, under uncertain driving forces of these changes, different emission scenarios have been developed and examined by the IPCC. The first two assessments (IPCC, 2001, 2007) have been based on the IPCC Special Report on Emissions Scenarios (SRES, 2000) that categorises the scenarios into four broader families (A1, A2, B1, B2) covering a wide range of demographic, economic, and technological key factors and resulting GHG emissions. These scenarios do not include any adaptation policies to climate change, and no attempt is made to prioritise these or attach a probability of occurrence. The scenarios used in the fifth assessment (IPCC, 2014) are called Representative Concentration Pathways (RCPs) and also cover cases with climate adaptation policies.

The SRES scenarios have been used as a basis for the UKCIP02 (Hulme et al., 2002) and the UKCP09 (Murphy et al., 2009) climate projections for the UK. In the most recent UKCP09, three different emission scenarios are examined: high (A1FI), medium (A1B), and low (B1); in contrast with the UKCIP02 that includes four (A2 and B2 scenarios for medium emissions, instead of A1B). The UKCP09 scenarios are summarised in Table 4.5.

Although the HadCM3 model was used in both UKCIP02 and UKCP09 projections, there are fundamental methodological differences and improvements between the two systems. The UKCIP02 system provides deterministic projections of the climate variables (single value at a location) with a spatial resolution of 50 km². In contrast, UKCP09 projections are probabilistic, compiled by sampling several HadCM3 model variants, along with other international global models, to account for uncertainties in the inherent climate variability and the model structural errors. For most variables, the UKCP09 projections are given at three temporal scales (month, season, and year) with a temporal resolution of 25 km² (Jenkins et al., 2009). Projections at finer time (daily and hourly) and temporal scales (5 km²) can be obtained by the UKCP09 weather generator that disaggregates the higher-level series (Jones et al., 2009).

In general, all areas of the UK are expected to become warmer, so by the 2080s the highest change in mean summer temperature, at the 50% probability level, is predicted to be between 3.1°C (for the low emissions scenario) and 5.3°C (for the high emissions scenario). The south-east of England is expected to warm more than the north-west, with more frequent high summer temperatures and fewer very cold winters.

The projections of total annual precipitation show very little change, but there is significant seasonal variability across the country. The highest change in the mean winter precipitation, under central estimates, is seen along the western side of the country and is

Table 4.5 UKCP09 Projections and IPCC Scenarios

Scenario	IPCC	Description
High emissions	A1FI	Rapid economic growth, global population that peaks in mid-century and declines thereafter, and the rapid introduction of new and more efficient technologies. Convergence among regions, capacity building, and increased cultural and social interactions, with a substantial reduction in regional differences in per capita income. Fossil-intensive energy resources.
Medium	A1B	The same as A1FI with a balance exploitation model across resources (fossil and non-fossil).
Low	B1	Convergent world with the same global population as in A1FI, but with rapid changes in economic structures towards a service and information economy, with reductions in material intensity, and the introduction of clean and resource-efficient technologies. Emphasis on global solutions to economic, social, and environmental sustainability, including improved equity, but without additional climate initiatives.

predicted to be between +30 mm (for the low-emissions scenario) and +47 mm (for the high-emissions scenario). In contrast, summer precipitation is projected to decrease for the whole country, with the biggest changes in southern England.

Precipitation on the wettest day, in the medium-emissions scenario, is expected to increase by up to 40%–50% in winter for a few southern areas, but it will decrease by up to 10% in summer. For the same scenario and under central estimates, the heavy rain days (rainfall greater than 25 mm) will increase by a factor of between 2 and 3.5 in winter, and 1 to 2 in summer.

The analysis of changes in the intensity and frequency of extreme events, based on UKCP09 projections, showed clearly extreme events in winter become more frequent (lower return period), but it was not possible to obtain a clear signal for similar changes in the summer events (Fowler and Wilby, 2010; Sanderson, 2010). According to Fowler and Ekstrom (2009), the winter, spring, and autumn extreme precipitation will increase from 5% to 30% by the 2080s, depending on region and season. The spatial and temporal resolution of UKCP09 projections do not allow clear conclusions to be drawn about changes in short-duration convective events that typically cause most of the sewer flood events in the summer.

Figure 4.12 shows an example of the predicted changes to the magnitude and spatial extent of extreme rainfall depth in the UK. This was produced by comparing output generated by the Hadley Centre HadRM3 regional climate model based on the medium-high scenario for the 2080s with equivalent data generated from the *FEH* (Section 4.3.5) and is presented as anomalies (i.e., ratios) between the two. Results show that most areas will experience heavier 6-hour events (up to 20% "uplift"), although the Midlands is predicted to have smaller rainfall amounts, especially at higher return periods. An analysis of mean rainfall indicates less rainfall in summer months and a shorter period of the year over which there is increased rainfall (UKWIR, 2003).

4.6.3 Design rainfall under climate change

Regarding the urban drainage applications, the impact of climate change in the intensity and frequency of extreme rainfall events is of great importance, since a possible increase in these characteristics may result in more frequent and severe sewer overflows and flood events. The incorporation of UKCP09 projections in an urban modelling framework is the subject of several projects, such as UKWIR Climate Change Modelling for Sewerage Networks (UKWIR, 2010) and Modifying Existing Rainfall Design Sets for Climate Change (WRc, 2010).

Typically, changes in the intensity and frequency of extreme rainfall events are modelled via an "uplift" to the existing *FEH* or *FSR* design rainfall, where the rainfall intensities at each time step are multiplied by the climate change factor (CCF). For instance, in the case of an increase in rainfall intensities of 20%, a CCF equal to 1.20 should be applied. CCF factors are provided in

- UKCP09 projections for changes in wettest winter day (Murphy et al., 2009)
- EA/Defra guidance (EA, 2011)
- UKWIR CL10/A205 Report (UKWIR, 2015)

4.6.4 Implications

There are many studies, based either on global or regional models, that support the hypothesis of an increase of extreme precipitations in a future climate over many regions

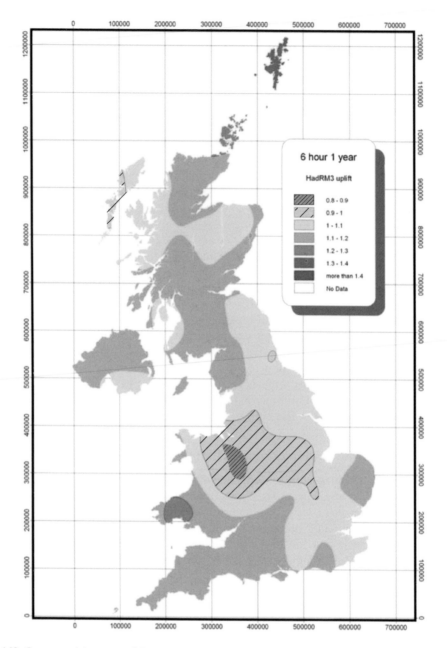

Figure 4.12 One-year 6-hours rainfall event anomalies. (Reproduced with permission from UKWIR. 2003. *Climate Change and the Hydraulic Design of Sewerage Systems*. Summary report no. 03/CL/10/0. UK Water Industry Research.)

(see Mailhot and Duchesne, 2010). The potential implications are as follows (Ofwat, 2011):

- Increased volume and flow rate that may exceed the capacity of existing sewer systems leading to more frequent surcharging, surface flooding, and property damage (e.g., Butler et al., 2007)

- Greater deterioration of sewers due to more frequent surcharging
- More frequent CSO spills
- Greater buildup and mobilisation of surface pollutants in summer
- Poorer water quality in rivers due to extra stormwater outfall and CSO spills and reduced base flows in summer
- Increased flows of dilute wastewater at wastewater treatment plants due to higher rainfall and infiltration, potentially leading to poorer treatment by biological processes

Increased temperature is also associated with more rapid chemical and biological reactions. This may lead to a greater number of sulphide-related problems (e.g., corrosion, odour—see Chapter 17) but may actually reduce wastewater organic concentrations due to increased degradation rates.

4.6.5 Solutions

It is unlikely that major upgrades of the existing sewer network will be carried out in response to *potential* climate change. Increases in runoff will need to be dealt with in a number of ways, such as

- Increased application of SuDS: infiltration devices, above-ground storage (see Chapter 21)
- More widespread capture and reuse of rainwater (see Chapter 24)
- Planned exceedance flow routing (major–minor systems) (see Chapter 11)
- Increased application of real-time control (see Chapter 22).

PROBLEMS

4.1 Describe the main types of rain gauges and assess their relative merits in urban drainage applications.

4.2 Calculate the average rainfall intensity for 5-year and 50-year return period storms of 30 minutes duration using the generalised IDF formula with parameters $v = 1$, $\theta = 0.180$, $\eta = 0.90$, and $\omega(T) = 22.184\ T^{0.31}$, where T is the return period in years. [51.70, 105.55 mm/hour]

4.3 Derive a series of IDF curves (1, 2, and 5 years) for a location in Coventry.

4.4 Construct the 50 percentile rainfall intensity profile for a 20 mm/hour, 20-minute duration summer storm.

4.5 What is a synthetic design storm, and how can it be represented? What are the differences from an historical event?

4.6 What is the critical input hyetograph, and what are its main advantages?

4.7 Explain what you understand by an historical series of rainfall. What is an annual time series?

4.8 What are the benefits of using rainfall synthetic time series?

4.9 Describe how stochastic rainfall generators work. What is the main advantage of Poisson-cluster models against simple Poisson models?

4.10 Describe the modifications that allow Poisson-cluster models to capture the high variability of sub-hourly rainfall.

4.11 Explain the main purpose of disaggregation techniques.

4.12 What are the main impacts of climate change, and how might urban drainage systems be adapted to take account of them?

KEY SOURCES

CIWEM 2016. *Rainfall Modelling Guide 2016*. The Chartered Institution of Water and Environmental Management, London.

Collier, C.G. 1996. *Applications of Weather Radar Systems: A Guide to Uses of Radar Data in Meteorology and Hydrology*, 2nd edn, Praxis Publishers, John Wiley and Sons, Chichester/London.

DoE/NWC. 1981. *Design and Analysis of Urban Storm Drainage. The Wallingford Procedure. Volume 1: Principles, Methods and Practice*, Department of the Environment, Standing Technical Committee Report No. 28.

Hulme, M., Jenkins, G.J., Lu, X., Turnpenny, J.R., Mitchell, T.D., Jones, R.G. et al., 2002. *Climate Change Scenarios for the United Kingdom: The UKCIP02 Scientific Report*. Tyndall Centre for Climate Change Research, University of East Anglia, Norwich.

IPCC. 2014. Climate Change 2014 Synthesis Report Summary Chapter for Policymakers. https://www.ipcc.ch/report/ar5/syr/

Onof, C., Chandler, R.E., Kakou, A., Northrop, P., Wheater, H.S., and Isham, V.S. 2000. Rainfall modelling using Poisson-cluster processes: A review of developments. *Stochastic Environmental Research and Risk Assessment*, 14, 384–411.

Shaw, E.M., Beven, K.J., Chappell, N.A., and Lamb, R. 2010. *Hydrology in Practice*, 4th edn, CRC Press, Boca Raton, FL.

REFERENCES

Bajracharya, S.R., Shrestha, M.S., and Shrestha, A.B. 2017. Assessment of high-resolution satellite rainfall estimation products in a streamflow model for flood prediction in the Bagmati basin, Nepal. *Journal of Flood Risk Management*, 10(1), 5–16.

Bringi, V.N. and Chandrasekar, V. 2001. *Polarimetric Doppler Weather Radar. Principles and Applications*. Cambridge University Press, New York.

Bringi, V.N., Rico-Ramirez, M.A., and Thurai, M. 2011. Rainfall Estimation with an Operational Polarimetric C-Band Radar in the United Kingdom: Comparison with a Gauge Network and Error Analysis. *Journal of Hydrometeorology*, 12, 935–954.

Burton, A., Kilsby, C.G., Fowler, H.G., Cowpertwait, P.S.P., and O'Connell, P.E. 2008. RainSim: A spatial–temporal stochastic rainfall modelling system. *Environmental Modelling and Software*, 23, 1356–1369.

Butler, D., McEntee, B., Onof, C., and Hagger, A. 2007. Sewer storage tank performance under climate change. *Water Science and Technology*, 56(12), 29–35.

Chiang, Y.-M., Hsu, K.-L., Chang, F.-J., Hong, Y., and Sorooshian, S. 2007. Merging multiple precipitation sources for flash flood forecasting. *Journal of Hydrology*, 340(3–4), 183–196.

Chow, V.T., Maidment, D.R., and Mays, L.W. 1988. *Applied Hydrology*, McGraw-Hill, New York.

Collier, C.G. 1996. *Applications of Weather Radar Systems: A Guide to Uses of Radar Data in Meteorology and Hydrology*, 2nd edn, Praxis Publishers, John Wiley and Sons, Chichester/London.

Cowpertwait, P., Isham, V., and Onof, C. 2007. Point process models of rainfall: Developments for fine-scale structure. *Proceedings of the Royal Society: Mathematical Physical and Engineering Sciences*, 463, 2569–2587.

Cowpertwait, P.S.P. 1994. A generalized point process model for rainfall. *Proceedings of the Royal Society: Mathematical Physical and Engineering Sciences*, A447, 23–37.

Cowpertwait, P.S.P. 1998. A Poisson-cluster model of rainfall: High-order moments and extreme values. *Proceedings of the Royal Society: Mathematical Physical and Engineering Sciences*, 454, 885–898.

Cowpertwait, P.S.P. 2004. Mixed rectangular pulses models of rainfall. *Hydrology and Earth System Sciences*, 8, 993–1000.

Dale, M. and Stidson, R. 2007. Weather radar for predicting beach bathing water quality. *WaPUG Autumn Conference*, Blackpool.

EA. 2011. *Adapting to Climate Change: Advice for Flood and Coastal Erosion Risk Management Authorities.* Environment Agency, Bristol, UK.

Ebert, E.E., Janowiak, J.E., Kidd, C. 2007. Comparison of near-real-time precipitation estimates from satellite observations and numerical models. *Bulletin of the American Meteorological Society*, 88, 47–64.

Einfalt, T., Arnbjerg-Nielsen, K., and Spies, S. 2002. An enquiry into rainfall data measurement and processing for model use in urban hydrology. *Water Science and Technology*, 45(2), 147–152.

Einfalt, T., Jessen, M., and Mehlig, B. 2005. Comparison of radar and rain gauge measurements during heavy rainfall. *Water Science and Technology*, 51, 195–201.

Entekhabi, D., Rodriguez-Iturbe, I., and Eagleson, P.S. 1989. Probabilistic representation of the temporal rainfall process by a modified Neyman-Scott Rectangular Pulses Model: Parameter estimation and validation. *Water Resources Research*, 25, 295–302.

Environment Canada. 2016. *Engineering Climate Datasets.* http://climate.weather.gc.ca/prods_servs/engineering_e.html.

Fan, F.M., Collischonn, W., Quiroz, K.J., Sorribas, M.V., Buarque, D.C., and Siqueira, V.A. 2016. Flood forecasting on the Tocantins River using ensemble rainfall forecasts and real-time satellite rainfall estimates. *Journal of Flood Risk Management*, 9, 278–288.

Faulkner, D. 1999. *Rainfall Frequency Estimation. Volume 2 of the Flood Estimation Handbook*, Institute of Hydrology, Wallingford.

Flato, G., Marotzke, J., Abiodun, B., Braconnot, P., Chou, S.C., Collins, W. et al., 2013. *Evaluation of Climate Models. Climate Change 2013: The Physical Science Basis.* Contribution of Working Group I to the Fifth Assessment Report of the Intergovernmental Panel on Climate Change.

Fowler, H.J. and Ekstrom, M. 2009. Multi-model ensemble estimates of climate change impacts on UK seasonal precipitation extremes. *International Journal of Climatology*, 29(3), 385–416.

Fowler, H.J. and Wilby, R.L. 2010. Detecting changes in seasonal precipitation extremes using regional climate model projections: Implications for managing fluvial flood risk. *Water Resources Research*, 46, 1–17.

Garside, I.G. 1991. *Using Annual Rainfall Time Series*, WaPUG User Note No. 23.

Gerstner, E.M. and Heinemann, G. 2008. Real-time areal precipitation determination from radar by means of statistical objective analysis. *Journal of Hydrology*, 352, 296–308.

Golz, C., Einfalt, T., Gabella, M., and Germann, U. 2005. Quality control algorithms for rainfall measurements. *Atmospheric Research*, 77, 247–255.

Goudenhoofdt, E. and Delobbe, L. 2013. Statistical characteristics of convective storms in Belgium derived from volumetric weather radar observations. *Journal of Applied Meteorology and Climatology*, 52, 918–934.

Harrison, D.L., Driscoll, S.J., and Kitchen, M. 2000. Improving precipitation estimates from weather radar using quality control and correction techniques. *Meteorological Applications*, 7, 135–144.

Henderson, R.J. 1986. *Rainfall Time Series for Sewer System Modelling*, WRc Report No. ER195E.

Institute of Hydrology. 1999. *Flood Estimation Handbook*, 5 volumes, Institute of Hydrology, Wallingford.

Intergovernmental Panel on Climate Change (IPCC). 2001. *Climate Change 2001: Synthesis Report.* Contribution of Working Groups I, II and III to the Third Assessment Report. www.grida.no/climate.

Intergovernmental Panel on Climate Change (IPCC). 2007. *Climate Change 2007: Synthesis Report.* Contribution of Working Groups I, II and III to the Fourth Assessment Report. www.ipcc.ch/pdf/assessment-report/ar4/syr/ar4_syr.pdf.

Intergovernmental Panel on Climate Change (IPCC). 2014. Climate Change 2014 Synthesis Report Summary Chapter for Policymakers. https://www.ipcc.ch/report/ar5/syr/.

Jenkins, G.J., Murphy, J.M., Sexton, D.M.H., Lowe, J.A., Jones, P., and Kilsby, C.G. 2009. *UK Climate Projections: Briefing Report*, Met Office Hadley Centre, Exeter.

Jenkins, G.J., Perry, M.C., and Prior, M.J. 2008. *The Climate of the United Kingdom and Recent Trends.* Met Office Hadley Centre, Exeter.

Jones, P., Harpham, C., Kilsby, C., Glenis, V., and Burton, A. 2009. *Projections of Future Daily Climate for the UK from the Weather Generator*. UK Climate Projections.

Kaczmarska, J., Isham, V., and Onof, C. 2014. Point process models for fine-resolution rainfall. *Hydrological Sciences Journal*, 59, 1972–1991.

Keifer, C.J. and Chu, H.H. 1957. Synthetic storm pattern for drainage design. *ASCE Journal of the Hydraulics Division*, 83, 1–25.

Kellagher, R. 2005. Comparison and assessment of 2 stochastic extreme time series rainfall generators. *WaPUG Autumn Conference*, Blackpool.

Kossieris, P., Makropoulos, C., Onof, C., and Koutsoyiannis, D. 2016. A rainfall disaggregation scheme for sub-hourly time scales: Coupling a Bartlett-Lewis based model with adjusting procedures. *Journal of Hydrology*, doi:10.1016/j.jhydrol.2016.07.015.

Koutsoyiannis, D., Kozonis, D., and Manetas, A. 1998. A mathematical framework for studying rainfall intensity-duration-frequency relationships. *Journal of Hydrology*, 206, 118–135.

Koutsoyiannis, D. and Onof, C. 2001. Rainfall disaggregation using adjusting procedures on a Poisson cluster model. *Journal of Hydrology*, 246, 109–122.

Krajewski, W.F., Ciach, G.J., McCollum, J.R., and Bacotiu, C. 2000. Initial validation of the Global Precipitation Climatology Project monthly rainfall over the United States. *Journal of Applied Meteorology*, 39, 1071–1086.

Lovejoy, S. and Schertzer, D. 1995. *Multifractals and Rain. New Uncertainty Concepts in Hydrology and Hydrological Modelling*. Cambridge University Press, Cambridge, 62–103.

Mailhot, A. and Duchesne, S. 2010. Design criteria of urban drainage infrastructures under climate change. *Journal of Water Resources Planning and Management.*, 136, 201–208.

Marshall, J.S., Langille, R.C., and Palmer, W.M.K. 1947. Measurement of rainfall by radar. *Journal of Meteorology*, 4(6), 186–192.

Meehl, G.A., Covey, C., Delworth, T., Latif, M., McAvaney, B., Mitchell, J.F.B. et al. 2007. The WCRP CMIP3 multimodel dataset: A new era in climatic change research. *Bulletin of the American Meteorological Society*, 88, 1383–1394.

Meteorological Office. 2013. National Meteorological Library and Archive Fact sheet 15—Weather radar 22.

Murphy, J.M., Sexton, D.M.H., Jenkins, G.J., Boorman, P.M., Booth, B.B.B., Brown, C.C. et al., 2009. *UK Climate Projections Science Report: Climate Change Projections*. Met Office Hadley Centre, Exeter.

Natural Environment Research Council (NERC). 1975. *Flood Studies Report*, 5 vol., Institute of Hydrology, Wallingford.

Neale, W. 2008. Practical use of radar rainfall data at Thames Water. *WaPUG Autumn Conference*, Blackpool.

Newton, C., Jarman, D., Memon, F.A., Andoh, R., and Butler, D. 2013. Implementation and assessment of a critical input hyetograph generation for use in a decision support tool for the design of flood attenuation systems. *International Conference on Flood Resilience: Experiences in Asia and Europe*, Exeter, September.

Ngirane-Katashaya, G. and Wheater, H.S. 1985. Hydrograph sensitivity to storm kinematics. *Water Resources Research*, 2(3), 337–345.

Nielsen, J.E., Thorndahl, S., and Rasmussen, M.R. 2014. A numerical method to generate high temporal resolution precipitation time series by combining weather radar measurements with a nowcast model. *Atmospheric Research*, 138, 1–12.

Ofwat. 2011. Future Impacts on Sewer Systems in England and Wales: Summary of a hydraulic modelling exercise reviewing the impact of climate change, population and growth in impermeable areas up to around 2040, Birmingham, UK.

Onof, C. and Wheater, H.S. 1994. Improvements to the modelling of British rainfall using a modified random parameter Bartlett-Lewis Rectangular Pulse Model. *Journal of Hydrology*, 157, 177–195.

Onof, C., Wheater, H.S., and Isham, V. 1994. Note on the analytical expression of the inter-event time characteristics for Bartlett-Lewis type rainfall models. *Journal of Hydrology*, 157, 197–210.

Petrovic, J. and Elgy, J. 1994. The use of Meteostat infrared imagery for enhancement of areal rainfall estimates. *Remote Sensing and GIS in Urban Waters*, UDT '94 (eds. C. Maksimovic, J. Elgy, and V. Dragalov), Moscow, 29–40.

Prosdocimi, I., Stewart, E.J., Svensson, C., and Vesuviano, G. 2014. *Depth-Duration-Frequency Analysis for Short-Duration Rainfall Events*. Report SC090031/R, Environment Agency.

Rainey, C.M.L. and Osborne, M.P. 1991. Design for storage using synthetic rainfall. *WaPUG Autumn Meeting*. www.ciwem.org/groups/wapug.

Rodriguez-Iturbe, I., Cox, D.R., and Isham, V.S. 1987. Some models for rainfall based on stochastic point processes. *Proceedings of the Royal Society: Mathematical Physical and Engineering Sciences*, A410, 269–288.

Rodriguez-Iturbe, I., Cox, D.R., and Isham, V.S. 1988. A point process model for rainfall: Further developments. *Proceedings of the Royal Society: Mathematical Physical and Engineering Sciences*, 417, 283–298.

Rosenfeld, D. and Collier, C.G. 1998. Estimating surface precipitation. Chapter 14.1, in *Global Energy and Water Cycles* (eds. K.A. Browning and R. Gurney), Cambridge University Press, 124–133.

Sanderson, M. 2010. Changes in the frequency of extreme rainfall events for selected towns and cities, Ofwat, Birmingham, UK.

Schilling, W. 1991. Rainfall data for urban hydrology: What do we need? *Atmospheric Research*, 27, 5–21.

Seo, D.-J., Breidenbach, J., and Johnson, E. 1999. Real-time estimation of mean field bias in radar rainfall data. *Journal of Hydrology*, 223, 131–147.

Sinclair, S. and Pegram, G. 2005. Combining radar and rain gauge rainfall estimates using conditional merging. *Atmospheric Science Letters*, 6, 19–22.

Smith, J.A., Baeck, M.L., Meierdiercks, K.L., Miller, A.J., and Krajewski, W.F. 2007. Radar rainfall estimation for flash flood forecasting in small urban watersheds. *Advances in Water Resources*, 30, 2087–2097.

Smith, J.A., Hui, E., Steiner, M., Baeck, M.L., Krajewski, W.F., and Ntelekos, A.A. 2009. Variability of rainfall rate and raindrop size distributions in heavy rain. *Water Resources Research*, 45, 1–12.

Special Report on Emissions Scenarios (SRES). 2000. A Special Report of Working Group III of the IPCC. https://www.ipcc.ch/pdf/special-reports/emissions_scenarios.pdf.

Stewart, E.J., Jones, D.A., Svensson, C., Morris, D.G., Dempsey, P., Dent, J.E. et al., 2013. *Reservoir Safety—Long Return Period Rainfall*. R&D Technical Report WS 194/2/39/TR (two volumes), DEFRA/EA.

Taylor, K.E., Stouffer, R.J., and Meehl, G.A. 2012. An overview of CMIP5 and the experiment design. *Bulletin of the American Meteorological Society*, 93, 485–498.

Todini, E. 2001. A Bayesian technique for conditioning radar precipitation estimates to rain-gauge measurements. *Hydrology and Earth System Sciences*, 5, 187–199.

UK Water Industry Research (UKWIR). 2003. *Climate Change and the Hydraulic Design of Sewerage Systems*. Summary report no. 03/CL/10/0.

UK Water Industry Research (UKWIR). 2010. *Climate Change Modelling for Sewerage Networks*. Report no. 10/CL/10/15.

UK Water Industry Research (UKWIR). 2015. *Rainfall Intensity for Sewer Design*. Report no. 15/CL/10/16.

Verhoest, N.E.C., Vandenberghe, S., Cabus, P., Onof, C., Meca-Figueras, T., and Jameleddine, S. 2010. Are stochastic point rainfall models able to preserve extreme flood statistics? *Hydrological Processes*, 24, 3439–3445.

Wang, L.P., Ochoa-Rodriguez, S., Onof, C., and Willems, P. 2015. Singularity-sensitive gauge-based radar rainfall adjustment methods for urban hydrological applications. *Hydrology and Earth System Sciences*, 19, 4001–4021.

Wang, L.P., Ochoa-Rodríguez, S., Simões, N.E., Onof, C., and Maksimović, Č. 2013. Radar-raingauge data combination techniques: A revision and analysis of their suitability for urban hydrology. *Water Science and Technology*, 68, 737–747.

Wheater, H.S., Chandler, R.E., Onof, C.J., Isham, V.S., Bellone, E., Yang, C. et al. 2005. Spatial-temporal rainfall modelling for flood risk estimation. *Stochastic Environmental Research and Risk Assessment*, 19, 403–416.

World Meteorological Organization (WMO). 2008. *Guide to Meteorological Instruments and Methods of Observation*, 7th edn, Geneva.

Water Research Centre (WRc). 1987. *A Guide to Short Term Flow Surveys of Sewer Systems*, Water Authorities Association.

Water Research Centre (WRc). 2010. *Modifying Existing Rainfall Design Sets for Climate Change*, Collaborative Project CP: 383, Report no. P8100.05.

Chapter 5

Stormwater

5.1 INTRODUCTION

Stormwater (or surface water runoff) is the second major urban flow of concern to the drainage engineer. Safe and efficient drainage of stormwater is particularly important to maintain public health and safety (due to the potential impact of flooding on life and property) and to protect the receiving water environment. Reliable data on the quantity and quality of existing and projected stormwater flows are a prerequisite for cost-effective urban drainage design and analysis.

Stormwater is generated by precipitation, typically rainfall, and consists of that proportion that runs off from urban surfaces (see Figure 5.1). Hence, the properties of stormwater, in terms of quantity and quality, are intrinsically linked to the nature and characteristics of both the rainfall and the catchment. This chapter focuses on the generation and characteristics of stormwater, with many of the concepts underlying the design and analysis techniques described in later chapters. Section 5.2 explains the mechanisms of runoff generation, while Section 5.3 discusses overland flow processes. Stormwater quality issues are dealt with in Section 5.4.

5.2 RUNOFF GENERATION

The transformation of a rainfall hyetograph into a surface runoff hydrograph involves two principal parts. First, *losses* due to interception, depression storage, infiltration, and evapo-transpiration are deducted from the rainfall. Second, the resulting *effective rainfall* is transformed by *surface routing* into an *overland flow* hydrograph.

For most urban drainage applications, the runoff processes are at least as important as the pipe flow processes (discussed later in Chapter 7) and of equal importance to the rainfall processes.

Much of the rainfall that reaches the ground does not, in fact, run off. It is "lost" immediately or as it runs overland. The water may be completely lost from the catchment surface by processes such as evapo-transpiration, it may be temporarily retained in local depression storage, or it may eventually find its way to the drainage system via groundwater.

5.2.1 Initial losses

5.2.1.1 Interception and wetting losses

Interception consists of the collection and retention of rainfall by vegetation cover. There is an initial retention period, after which excess rain falls through the foliage or flows to the soil over the stems. The interception rate then rapidly approaches zero. The interception loss for impervious areas is small in magnitude (<1 mm) and is normally neglected or combined with depression storage.

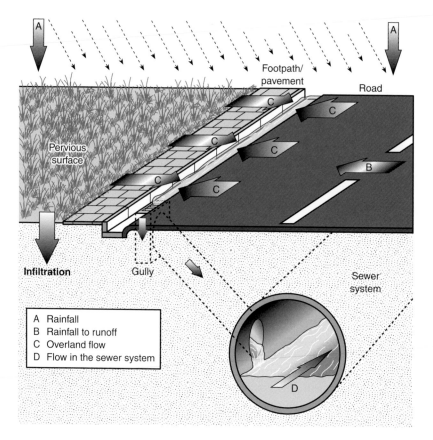

Figure 5.1 Stormwater runoff generation processes.

5.2.1.2 Depression storage

Depression storage accounts for rainwater that has become trapped in small depressions on the catchment surface, preventing the water from running off. Infiltration, evaporation, or leakage will eventually remove the water that has been retained. Factors affecting the magnitude of depression storage are surface type, slope, and rainfall return period (Kidd and Lowring, 1979). Depression storage d (mm) can be represented as

$$d = \frac{k_1}{\sqrt{s}} \tag{5.1}$$

where k_1 is a coefficient depending on surface type (0.07 for impervious surfaces and 0.28 for pervious surfaces) (mm), and s is the ground slope (−).

Typical values for d are 0.5–2 mm for impervious areas, 2.5–7.5 mm for flat roofs, and up to 10 mm for gardens.

5.2.1.3 Representation

For intense summer storms in urban areas, the initial losses are not important, but for less severe storms or for less urbanised catchments, they should not be neglected. For modelling purposes, the combined initial losses are usually subtracted from the rainfall at the beginning of the storm to leave the *net* rainfall. This is illustrated in Example 5.1.

5.2.2 Continuing losses

5.2.2.1 Evapo-transpiration

Evapo-transpiration is the vaporisation of water from plants and open water bodies and therefore its removal from surface runoff. Although it is a continuing, constant loss, its effect during short-duration rainfall events is negligible. For example, the average daily value of potential evaporation in the United Kingdom during the summer months is 2–3 mm. Consequently, it is normally neglected in most models or considered to be lumped into the initial losses.

5.2.2.2 Infiltration

Infiltration represents the process of rainfall passing through the ground surface into the pores of the soil. The infiltration capacity of a soil is defined as the rate at which water infiltrates into it. The magnitude depends on factors including soil type, structure and compaction, initial moisture content, surface cover, and the depth of water on the soil. The infiltration rate tends to be high initially but decreases exponentially to a final quasi-steady rate when the upper soil zone becomes saturated.

EXAMPLE 5.1

For an urban catchment of average slope 1% with an estimated interception loss of 0.5 mm, calculate the net rainfall profile (based on initial losses only) of the following storm:

Time (minutes)	0–10	10–20	20–30	30–40
Rainfall intensity (mm/h)	6	12	18	6

Take $k_1 = 0.1$ mm.

Solution

Interception loss $= 0.5$ mm
Depression storage loss from Equation 5.1: $d = 0.1/\sqrt{0.01} = 1$ mm

Time (minutes)	0–10	10–20	20–30	30–40
Rainfall intensity (mm/h)	6	12	18	6
Rainfall depth (mm)	1	2	3	1
Net rainfall depth (mm)	0	1.5	3	1
Net rainfall intensity (mm/h)	0	9	18	6

Initial losses are deducted from rainfall *depth* at the beginning of the storm.

A common empirical relationship used to represent infiltration is Horton's (1940) equation:

$$f_t = f_c + (f_o - f_c)e^{-k_2 t} \tag{5.2}$$

where f_t is the infiltration rate at time t (mm/h), f_c is the final (steady state) infiltration rate or capacity (mm/h), f_o is the initial rate (mm/h), and k_2 is the decay constant (h^{-1}).

The equation is valid when $i > f_c$. These parameters depend primarily on soil/surface type and initial moisture content of the soil. The range of values encountered for f_c, f_o, and k are given in Table 5.1. Careful adaptation of the equation is required to render it suitable for application in continuous simulation models.

Other, more physically based approaches have been formulated, such as Green and Ampt's (1911) equation and Richard's (1933) equation. These are less widely implemented and used in urban drainage models.

5.2.2.3 Representation

Continuing losses are always important in urban catchments, but are of most prominence in areas with relatively large open spaces. A simplified, but common, approach to representing them is by a constant proportional loss model applied after initial losses have been deducted to produce the *effective* rainfall:

$$i_e = C\, i_n \tag{5.3}$$

where i_e is the effective rainfall intensity (mm/h), C is the dimensionless runoff coefficient (–), and i_n is the net rainfall intensity (mm/h).

The runoff coefficient C depends primarily on land use, soil and vegetation type, and slope. It is also influenced by rainfall characteristics (e.g., intensity, duration) and antecedent conditions. Values of C range from 0.70 to 0.95 for impervious surfaces such as pavements and roofs, and from 0.05 to 0.35 for pervious surfaces. A more comprehensive listing of coefficients is given in Table 10.3. This model also forms the basis of the Rational Method used for estimating stormwater peak flow rates. This important method is described in more detail in Chapter 10.

5.2.3 Fixed runoff equation

For urban catchments in the UK, the dimensionless runoff coefficient can be estimated from Equation 5.4 (where $C = PR/100$) originally prepared as part of the *Wallingford Procedure* (DoE/NWC, 1981). This is a regression equation derived from data obtained from 17 catchments and 510 (summer) events:

$$\begin{aligned} PR &= 0.829\, PIMP + 25.0\, SOIL + 0.078\, UCWI - 20.7 \quad [PR > 0.4\, PIMP] \\ PR &= 0.4\, PIMP \qquad\qquad\qquad\qquad\qquad\qquad\quad\;\; [PR \le 0.4\, PIMP] \end{aligned} \tag{5.4}$$

Table 5.1 Typical Horton parameters for various surface types

Surface type	f_o (mm/h)	f_c (mm/h)	k_2 (h^{-1})
Coarse textured soils	250	25	2
Medium textured soils	200	12	2
Fine textured soils	125	6	2
Clays/paved areas	75	3	2

where PR is the percentage runoff (%), PIMP is the percentage impervious area of the catchment (25–100), SOIL is a soil index (0.15–0.50), UCWI is the urban catchment (antecedent) wetness index (30–300).

This equation is reasonably reliable provided it is used with variables that are within the range of those upon which it is based (shown in brackets). Since its development, it has been used successfully to represent many hundreds of catchments (see Example 5.2). The principal variables are described in further detail in the following sections.

5.2.3.1 PIMP

The percentage imperviousness represents the degree of urban development of the catchment and is defined as

$$PIMP = \frac{A_i}{A} \times 100 \tag{5.5}$$

where A_i is the impervious (roofs and paved areas) area (ha), and A is the total catchment area (ha).

EXAMPLE 5.2

Calculate the effective rainfall profile for the storm specified in Example 5.1. The rain falls on a catchment that is 78% impervious, has a soil type index of 0.25, and has a standard average annual rainfall (SAAR) of 540 mm.

Solution

$SAAR = 540$ mm
Read from Figure 5.2: $UCWI = 40$
Equation 5.4:

$$PR = 0.829 \times 78 + 25.0 \times 0.25 + 0.078 \times 40 - 20.7 = 53\%$$

Equation is valid as $[PR = 53] > [0.4 \times 78 = 31]$

Total net rainfall depth = 5.5 mm (from Example 5.1)
Runoff rainfall depth $\quad = 0.53 \times 5.5 = 2.9$ mm
Runoff loss $\qquad\qquad = 5.5\text{–}2.9 = 2.6$ mm
Continuing loss $\qquad\quad = 2.6/0.5 = 5.2$ mm/h (over 30 minutes)

The profile is therefore as follows:

Time (minutes)	0–10	10–20	20–30	30–40
Net rainfall profile (mm/h)	0	9	18	6
Effective rainfall profile (mm/h)	0	3.8	12.8	0.8

5.2.3.2 SOIL

The SOIL index is based on the winter rain acceptance parameter in the *Flood Studies Report* (NERC, 1975) and is a measure of infiltration potential of the soil. It can be obtained from maps in the *Flood Studies Report* or the *Wallingford Procedure* (DoE/NWC, 1981).

5.2.3.3 UCWI

The urban catchment wetness index (UCWI) represents the degree of wetness of the catchment at the start of a storm event. As UCWI increases, so does the PR value reflecting the increased runoff expected from a wetter catchment. It can be estimated for design purposes from its relationship with the SAAR given in Figure 5.2. A map of average annual rainfall is given in the *Wallingford Procedure*.

When simulating historical events,

$$UCWI = 125 + 8\ API5 - SMD \qquad\qquad (5.6)$$

where API5 is the 5-day antecedent precipitation index, and SMD is the soil moisture deficit.

API5 is calculated according to a methodology described in the Wallingford Procedure based on rainfall depths in the 5 days prior to the event. SMD is a measure of the amount of water that can be retained within the soil matrix, values of which are available for UK locations from the Meteorological Office.

5.2.3.4 Limitations

In specific circumstances, the PR equation has been found to have limitations:

- For catchments with relatively low PIMP (that is, with a large proportion of pervious surface), particularly those with light soils in dry conditions, the equation tends to under-predict the runoff volume.

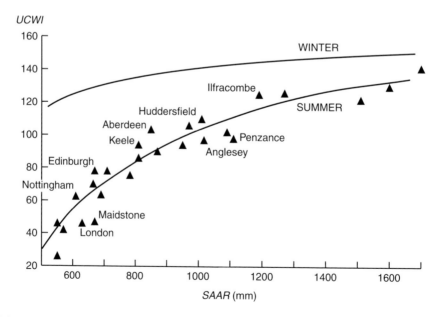

Figure 5.2 Relationship between UCWI and SAAR. (Based on Packman, J.C. 1986. Runoff estimation in urbanising and mixed urban/rural catchments. *IPHE Seminar on Sewers for Adoption*, Imperial College, London, with permission of the Chartered Institution of Water and Environmental Management, London.)

- During long-duration storms, catchment surfaces can be significantly wetted, increasing the proportion of runoff. This expected increase in runoff is not properly represented.
- The equation was developed for use with discrete rainfall events and is not directly applicable for continuous simulation using rainfall time series now commonly applied (see Section 4.5).

5.2.4 Variable runoff equation

The variable ("new") runoff equation (Packman, 1990) was developed to try to overcome some of the limitations mentioned in the previous section. It was based on data from 11 catchments and 112 events and has two major differences. The first is that runoff is calculated separately for impervious and pervious areas (not combined, as in the fixed equation). The second difference lies in the way API is allowed to vary during the storm rather than being a fixed value.

The model has three components: initial losses (see Section 5.2.1), runoff from impervious areas, and runoff from pervious areas.

It has the following form:

$$PR = IF.PIMP + (100 - IF.PIMP)\ NAPI/PF \qquad (5.7)$$

where IF is the effective impervious area factor, $NAPI$ is the new antecedent precipitation index (mm), and PF is the porosity factor (mm).

5.2.4.1 Impervious area runoff

The model deals with continuous losses following deduction of the initial losses (depression storage $d = 0.5$ for paved surfaces). Impervious areas are dealt with, simply, in two parts:

- A proportion of the surface is assumed to be directly connected to the drainage network and to generate 100% runoff ($IF.PIMP$). The effective impervious area factor can be estimated from Table 5.2.
- The rest of the surface is assumed to be less effectively connected and to behave hydrologically as if it were pervious. This part is therefore added to the pervious area total (remainder of the right-hand side of Equation 5.7).

5.2.4.2 Pervious area runoff

The pervious area and the less effectively connected impervious area are taken together, and the runoff is calculated using a soil moisture storage model, applied progressively throughout a storm (rather than once, before the storm).

Table 5.2 Effective impervious area factor

Surface type	IF
Normal urban paved surfaces	0.6
Roof surfaces	0.8
Well-drained roads	0.8
Very high-quality roads	1.0

Source: After Osborne, M.P. 2009. A New Runoff Volume Model, WaPUG User Note No. 28, version 3.

The new antecedent precipitation index $NAPI$ in Equation 5.7 is defined as a 30-day API with evapo-transpiration and initial losses subtracted from the rainfall. This formulation takes account of evaporation, better represents the rate of drying out of different soil types, and can be continuously updated during the storm. Details of its calculation are given by Osborne (2009). The default value for the porosity factor PF, representing soil moisture storage depth S, is 200 mm, which notionally represents the soil depth that is wetting and drying. A detailed analysis of the performance of this model is given by Balmforth et al. (2006).

5.2.5 UK Water Industry Research runoff equation

The UK Water Industry Research (UKWIR) runoff equation (2014) is a further development of the variable runoff equation with the intention of overcoming some of the weaknesses of previous models.

$$PR = \sum_{n=1}^{N}\left(IF_n \cdot PIMP_n + (1 - IF_n) \cdot PIMP_n \cdot \frac{PI_{pv}^{\beta}}{PF_{pv}} \right)$$
$$+ \left((1 - PIMP_{TOTAL}) \cdot \frac{(NAPI_s + PI_s)^{\alpha} \cdot SPR}{PF_s} \right) \tag{5.8}$$

where n is the number of surface types, 1 to N, PI_{pv} is the precipitation index for paved surfaces, β is the empirical coefficient for paved surfaces, PF_{pv} is the porosity factor for paved surfaces, $NAPI_s$ is the $NAPI$ for pervious surfaces, PI_s is the PI for pervious surfaces, α is the empirical coefficient for pervious surfaces, SPR is the standard percentage runoff, and PF_s is the porosity factor for pervious surfaces.

Further details of the model and an explanation of the various parameters and their values can be found in the SuDS Manual (Woods Ballard et al., 2016).

5.3 OVERLAND FLOW

Once the losses from the catchment have been accounted for, the effective rainfall hyetograph can be transformed into a surface runoff hydrograph—a process known as *overland flow* or *surface routing*. In this process, the runoff moves across the surface of the sub-catchment to the nearest entry point to the sewerage system.

There are two general approaches currently used for routing overland flow. The most common utilises the *unit hydrograph* method, although this is actually implemented in a number of different ways. The second, more physically based approach, usually utilises a *kinematic wave* model.

5.3.1 Unit hydrographs

The unit hydrograph is a widely used concept in hydrology that has also found application in urban hydrology. It is based on the premise that a unique and time-invariant hydrograph results from effective rain falling over a particular catchment. Formally, it represents the outflow hydrograph resulting from a unit depth (generally 10 mm) of effective rain falling uniformly over a catchment at a constant rate for a unit duration D: the D-h unit hydrograph is shown in Figure 5.3. The ordinates of the D-h unit hydrograph are given as $u(D, t)$, at any time t. D is typically 1 hour for natural catchments but could, in principle, be any time period.

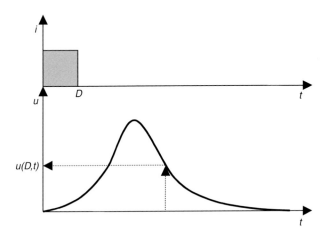

Figure 5.3 The unit hydrograph.

Once derived, the unit hydrograph can be used to construct the hydrograph response to any rainfall event based on three guiding principles:

- Constancy: The time base of the unit hydrograph is constant, regardless of the intensity of the rain.
- Proportionality: The ordinates of the runoff hydrograph are directly proportional to the volume of effective rain – doubling the rainfall intensity doubles the runoff flow rates.
- Superposition: The response to successive blocks of effective rainfall, each starting at particular times, may be obtained by summing the individual runoff hydrographs starting at the corresponding times.

This approximate, linear approach (known as *convolution*) is stated succinctly in Equation 5.9. If a rainfall event has n blocks of rainfall of duration D, the runoff $Q(t)$ at time t is

$$Q(t) = \sum_{\omega=1}^{N} u(D, j) I_\omega \qquad (5.9)$$

$$Q(t) = u(D,t)I_1 + u(D,t - D)I_2 \ldots + (D,t - (N - 1)D)I_n$$

where $Q(t)$ is the runoff hydrograph ordinate at time t (m³/s), $u(D,j)$ is the D-h unit hydrograph ordinate at time j (m³/s), I_ω is the rainfall depth in the Wth of N blocks of duration D (m), and j is the $t - (\omega - 1)D$ (s).

Further detail and examples on using unit hydrographs are given in Shaw et al. (2010).

In order to use this concept in ungauged urban catchments for design purposes, some way of predicting unit hydrographs is required, based on catchment characteristics. Three methods of doing this are in current use: synthetic unit hydrographs, the time–area method, or reservoir models.

5.3.2 Synthetic unit hydrographs

The detailed shape of the unit hydrograph reflects the characteristics of the catchment from which it has been derived. When converted into dimensionless form, it is found that similar shapes are observed in catchments in the same region.

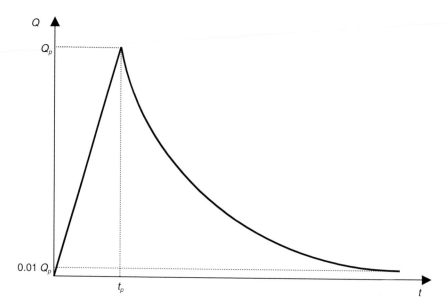

Figure 5.4 Synthetic unit hydrograph.

Harms and Verworn (1984) derived a dimensionless unit hydrograph suitable for urban areas, as shown in Figure 5.4. This has a linear rise up to the peak flow, an exponential recession, and an endpoint at 1% of peak flow:

$$Q = \frac{t}{t_p} Q_p \quad 0 < t < t_p \tag{5.10}$$

$$Q = Q_p e^{-(t - t_p / k_3)} \quad t \geq t_p \tag{5.11}$$

where Q is the flow rate (m³/s), Q_p is the peak flow rate (m³/s), t is the time (s), t_p is the time to peak (s), and k_3 is the exponential decay constant (s⁻¹).

The three parameters, Q_p, t_p, and k can be related to catchment characteristics.

This is the most direct application of the unit hydrograph approach, but it is the least common in practice.

5.3.3 Time–area diagrams

An alternative approach is to derive a time–area diagram, which is a special case of a unit hydrograph. To do this, lines of equal flow "travel time" to the catchment outfall are delineated, called *isochrones*. The maximum travel time represents the time of concentration (t_c) of the catchment (considered in Chapter 10). The time–area diagram that is constructed by summing the areas between the isochrones (see Figure 5.5) defines the response of the catchment.

When combined with rainfall in depth increments of $I1, I2, \ldots I_N$, flow at any time $Q(t)$ is

$$Q(t) = \sum_{\omega=1}^{N} \frac{dA(j)}{dt} I_\omega \tag{5.12}$$

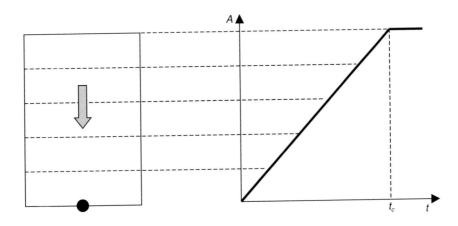

Figure 5.5 Linear time–area diagram.

where $dA(j)/dt$ is the slope of the time–area diagram at time j.

The time–area diagram can be used in design as an extension to the Rational Method (Chapter 10 has a description and examples) and is also implemented in some simulation models using standard time–area diagram profiles and empirical equations to estimate *tc*.

5.3.4 Reservoir models

The third approach is to propose the analogy that the catchment surface acts on the flow generated by an effective rainfall profile as one or more reservoirs connected in series. Each reservoir then experiences inflows of rainfall (and/or inflows from upstream reservoirs) and outflows of runoff. The model is based on the two equations of continuity and storage:

$$\frac{dS}{dt} = I - O \tag{5.13}$$

$$S = KO^m \tag{5.14}$$

where I is the inflow rate (m³/s), O is the outflow rate (m³/s), S is the storage volume (m³), K is the reservoir time constant (s), and m is the exponent (–).

These equations are returned to in Chapter 8, in the context of reservoir routing.

If m is taken to be equal to unity (a physical impossibility, but a conceptual convenience), then the reservoir is referred to as "linear." Nash (1957) proposed that the overland flow process could be represented as a series of identical linear reservoirs, where the output from one reservoir is considered as the input to a second, and so on. Assuming an instantaneous inflow of unit volume, the unit hydrograph time to peak t_p and peak flow Q_p are

$$t_p = (\alpha - 1)K \tag{5.15}$$

and

$$Q_p = \frac{1}{K(\alpha - 1)!}\left(\frac{\alpha - 1}{e}\right)^{n-1} \tag{5.16}$$

This approach can be used by specifying the number of reservoirs (a) and the time constant K, where a and K can be related to catchment characteristics. Alternatively, a and K can be used as calibration constants in models. A linear reservoir cascade is also a special case of the unit hydrograph approach.

The reservoirs are termed linear because, in Equation 5.14, the storage S is linearly related to outflow O, so from Equation 5.13 outflow must be linearly related to inflow. However, other non-unity values of m (for example, 0.67) can be used, resulting in a non-linear response from a *single* conceptual reservoir. The parameters K and m become calibration constants.

5.3.5 Kinematic wave

A more physically based approach is to simplify and solve the equations of motion to give

$$\frac{\partial d}{\partial t} + \frac{\partial q}{\partial x} = i_e \tag{5.17}$$

$$q = \frac{1}{n} d^{5/3} s^{1/2} \tag{5.18}$$

where d is the depth of flow (m), q is the flow per unit width $= Q/b$ (m²/s), i_e is the effective rainfall intensity (m/s), t is the time (s), x is the longitudinal distance (m), n is the Manning's roughness coefficient (m$^{-(1/3)}$/s), and s is the catchment slope (−).

Equation 5.17 is a continuity equation, and Equation 5.18 is a simplified momentum equation based on Manning's equation. The kinematic wave approximation is explained further in Chapter 19 and Manning's equation in Chapter 7. Manning's n for urban surfaces is typically 0.05–0.15, an order of magnitude greater than pipe roughness.

5.4 STORMWATER QUALITY

Urban stormwater contains a complex mixture of natural organic and inorganic materials, with a small proportion of man-made substances derived from transport, commercial, and industrial practices. These materials find their way into the drainage system from atmospheric sources and as a result of being washed off or eroded from urban surfaces. In certain respects, stormwater can be as polluting as wastewater.

It was stressed in Chapter 3 that wastewater is variable in character. The quality of stormwater is even more variable from place to place and from time to time. As with wastewater, care should be taken in interpreting "standard" or "typical" values.

5.4.1 Pollutant sources

Stormwater quality is influenced by rainfall and, especially, by the catchment. The major catchment sources include vehicle emissions, corrosion, and abrasion; building and road corrosion and erosion; bird and animal faeces; street litter deposition, fallen leaves and grass residues; and spills.

5.4.1.1 Atmospheric pollution

Pollutants in the urban atmosphere result mainly from human activities: heating, vehicular traffic, industry or waste incineration, for example. They may either be absorbed and dissolved by precipitation (known as wet fallout), to be carried directly into the drainage system with the stormwater; or they may settle on land surfaces (as dry fallout), and subsequently be washed off. Dry fallout particles can be transported by winds over long distances.

Although atmospheric sources are accepted as a major contributor to stormwater pollution, the importance of dry and wet fallout appears to be dependent on the site and the pollutant. In Gothenburg, for example, wet fallout has been identified as the dominant form of atmospheric pollutant (at 60%) for nitrogen, phosphorus, lead, zinc, and cadmium. Even higher percentages have been noted in some locations.

Dry fallout is thought to be of more importance in urban areas or areas with significant sources of solids. In Sweden, it was estimated that 20% of the organic matter, 25% of phosphorus, and 70% of the total nitrogen in stormwater can be attributed to atmospheric fallout (Malmqvist, 1979). Granier et al. (1990) found that approximately half the total loads of lead and chromium come from the atmosphere, with a significantly lower proportion of zinc.

5.4.1.2 Vehicles

Vehicle emissions include volatile solids and polyaromatic hydrocarbons (PAHs) derived from unburned fuel, exhaust gases and vapours, lead compounds (from petrol additives), and hydrocarbon losses from fuels, lubrication, and hydraulic systems.

Pollutants are generated by the everyday passage of traffic. Tyre wear releases zinc and hydrocarbons. Vehicle corrosion releases pollutants such as iron, chromium, lead, and zinc. Other pollutants include metal particles, especially copper and nickel, released by wear of clutch and brake linings. Most metals are predominantly associated with the particulate phase.

Wear of the paved surface will release various substances: bitumen and aromatic hydrocarbons, tar and emulsifiers, carbonates, metals, and fine sediments, depending on the road construction technique and materials used.

5.4.1.3 Buildings and roads

Urban erosion produces particles of brick, concrete, asphalt, and glass. These particles can form a significant constituent of sediment in stormwater. The extent of pollution depends on the condition of the buildings/roads. Roofs, gutters, and exterior paint can release varying amounts of particles, again depending on condition. Metallic structures, such as street furniture (e.g., fences, benches) corrode, releasing toxic substances such as chromium. Roads and pavements degrade over time, releasing particles of various sizes.

5.4.1.4 Animals

Urine and faeces deposited on roads and pavements by animals (pets or wild) are a source of bacterial pollution in the form of faecal coliforms and faecal streptococci. They are also a source of high oxygen demand.

5.4.1.5 De-icing

The most commonly used de-icing agent is salt (sodium chloride). Salt applications to roads cause the annual chloride loads in stormwater to be (on average) 50–500 times higher than would occur naturally (Stotz, 1987). Rock salt contains other impurities, including an insoluble fraction shown to contribute 25% of the winter suspended solids load in a motorway study (Colwill et al., 1984). The presence of salt also accelerates corrosion of vehicles and metal structures.

5.4.1.6 Urban debris

Urban surfaces can contain large amounts of street debris, litter, and organic materials such as dead and decaying vegetation. Litter will generally result in elevated levels of solids and greater consequential oxygen demand. Fallen leaves and grass cuttings may lie on urban surfaces, particularly in road gutters, and decompose, or may be washed into gullies.

5.4.1.7 Spills/leaks

Household cleaners, motor fluids/lubricants, and bin washings are sometimes illegally discarded or spilled into gutters and gullies. The range and amounts of these pollutants vary considerably, depending on land use and public behaviour. However, domestic sources of chemical pollutants are usually minor when compared with industrial spills or illicit toxic waste disposal.

5.4.2 Surface pollutants

The bulk of the pollutants, derived from the sources mentioned, are attached to particles of sediment that deposit temporarily on the catchment surface. Analysis of this particulate material shows a large range of sizes from below 1 μm to above 10 mm. Larger, denser sediment causes particular problems in the drainage system itself, and this is addressed further in Chapter 16. Despite typically comprising less than 5% of the material present, particulates less than 50 μm in size have most of the pollutants associated with them: 25% of the COD, up to 50% of the nutrients, and 15% of total coliforms (Ellis, 1986). These, of course, are the particles most that are readily washed off by the stormwater.

The situation with pollutants washed off from the surface is somewhat different. The median particle size (d_{50}) is much smaller at around 50 μm (see Figure 5.6), with d_{50} positively correlated with the runoff rate (Figure 5.7).

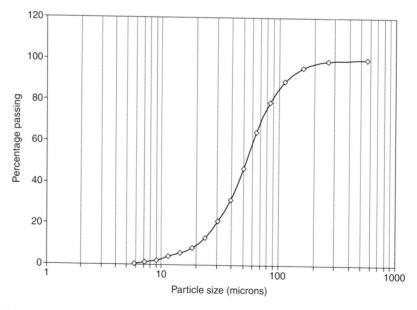

Figure 5.6 Particle size analysis for solids in urban road runoff. (Based on Memon, F.A. and Butler, D. 2005. *Urban Water Journal*, 2(3), 171–182.)

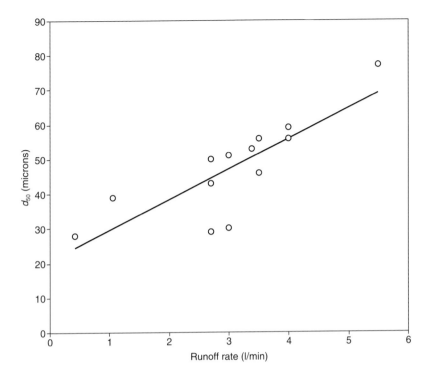

Figure 5.7 Relationship between median particle size (d_{50}) and runoff flow rate. (Based on Memon, F.A. and Butler, D. 2005. *Urban Water Journal*, 2(3), 171–182.)

5.4.3 Pollutant levels

Pollutants in stormwater include solids, oxygen-consuming materials, nutrients, hydrocarbons, heavy metals and trace organics, and bacteria. Typical values and ranges of pollutant discharges from stormwater systems are given in Table 5.3. Table 5.3 demonstrates the inherent variability of runoff quality. The average quality of "clean" and "dirty" catchments can vary by a factor of 10, and the variation in quality between stormwater events for any single catchment can vary by a factor of 3 (Ellis, 1986). Runoff quality will depend on a number of factors, including

- Geographical location
- Road and traffic characteristics
- Building and roofing types
- Weather, particularly rainfall

As a specific example, for two catchments in France, the total atmospheric contribution to the pollutants studied (SS, COD, 7 metals, and 4 pesticides) was mainly less than 20%. The biggest contribution (>50%) came from the catchment surface runoff (Becouze-Lareure et al., 2016).

5.4.4 Representation

The most common methods used to quantify stormwater quality are event mean concentrations, regression equations/rating curves, and buildup/washoff models. These are described in the following sections.

Table 5.3 Pollutant event mean concentrations (range and mean) and unit loads for urban stormwater

Quality parameter	EMC (mg/L)	Unit load (kg/imp ha.yr)[a]
Suspended solids (SS)	21–2582 (90)	347–2340 (487)
BOD$_5$	7–22 (9)	35–172 (59)
COD	20–365 (85)	22–703 (358)
Ammoniacal nitrogen	0.2–4.6 (0.56)	1.2–25.1 (1.76)
Total nitrogen	0.4–20.0 (3.2)	0.9–24.2 (9.9)
Total phosphorus	0.02–4.30 (0.34)	0.5–4.9 (1.8)
Total lead	0.01–3.1 (0.14)	0.09–1.91 (0.83)
Total zinc	0.01–3.68 (0.30)	0.21–2.68 (1.15)
Total hydrocarbons	0.04–25.9 (1.9)	0.01–43.3 (1.8)
Polyaromatic hydrocarbons (PAH)	0.01	0.02
Fecal coliforms (*Escherichia coli*)	400–50000 (MPN/100 mL)	0.9–3.8 (2.1) (110^9 counts/ha)

Source: After Ellis, J.B. and Mitchell, G. 2006. *Water and Environment Journal*, 20(1), 19–26.

[a] imp ha = impervious area measured in hectares.

5.4.4.1 Event mean concentrations

Event mean concentrations (EMCs), such as those given in Table 5.3, can be used in relation to environmental quality standards (EQSs) expressed in terms of maximum allowable concentrations (MACs) or annual allowances (AAs). These were discussed in Chapter 2. They can also be readily integrated into standard flow simulation models. Lundy et al. (2012) have demonstrated how such data can be used to estimate the risks associated with pollutants from a variety of different land use types and activities.

5.4.4.2 Regression equations

In this approach, the quality of stormwater is statistically regressed against a number of describing variables, for example, catchment characteristics or land use. This is the quality equivalent to the PR equation discussed earlier in the chapter. Regression equations can usually be relied on to give good representations on the catchment(s) on which they were based and perhaps similar ones. They will be less accurate on other catchments, but can often give a reasonable first approximation.

5.4.4.3 Buildup

The most common model-based approach to quality representation is by separately predicting pollutant buildup and washoff. In practice, the distinction between these two processes is not clearly defined. The factors affecting buildup of pollutants on impervious surfaces include

- Land use
- Population
- Traffic flow

- Effectiveness of street cleaning
- Season of the year
- Meteorological conditions
- Antecedent dry period
- Street surface type and condition

Buildup on the surface dM_s/dt can be assumed to be linear, so

$$\frac{dM_s}{dt} = aA \tag{5.19}$$

where M_s is the mass of pollutant on surface (kg), a is the surface accumulation rate constant (kg/ha.d), A is the catchment area (ha), and t is the time since the last rainfall event or road sweeping (d).

Accumulation rate a values for solids in residential areas are up to 5 kg/imp ha.d.

Detailed observation of sites in the United States (Sartor and Boyd, 1972) revealed that, despite there being no rainfall or street cleaning, the pollutant deposition often has a reducing rate of increase rather than a uniform linear increase. The first-order removal concept can be used to represent this, which implies that equilibrium is reached when the supply rate of pollutants matches their removal:

$$\frac{dM_s}{dt} = aA - bM_s \tag{5.20}$$

where b is the removal constant (d^{-1}).

The equilibrium mass on the catchment is therefore A (a/b). Novotny and Chesters (1981) reported values for b from a U.S. medium-density residential area of 0.2–0.4 d^{-1}. Studies in London (Ellis, 1986) suggest equilibrium is reached within 4–5 days where vehicle-induced re-suspension is dominant. The levelling-off phenomenon is most profound in areas where

- Adjacent pollution traps (pervious areas) are available.
- Vehicle-induced wind and vibration is high.

This will be the case on motorways and trunk roads, and in busy commercial/industrial areas.

Pollutants other than solids can be predicted using "potency" factors (defined in Chapter 20).

5.4.4.4 Washoff

Washoff occurs during rainfall/runoff by raindrop impact, erosion, or solution of the pollutants from the impervious surface. Important factors include

- Rainfall characteristics
- Topography
- Solid particle characteristics
- Street surface type and condition

This extensive range of influences makes pollutant washoff difficult to quantify with any certainty. The simplest approach adopted is to assume there is effectively an infinite store of pollutants always available on the surface to be washed off, and hence no buildup. Experimental evidence suggests this assumption may be valid in British conditions (Mance and Harman, 1978). Washoff can then be modelled as a function of rainfall intensity:

$$W = z_1 i^{z_2} \tag{5.21}$$

where W is the pollutant washoff rate (kg/h), i is the rainfall intensity (mm/h), and z_1, z_2 are the pollutant-specific constants.

The exponent z_2 usually has values between 1.5 and 3 for particulate pollutants and <1 for those in solution (Delleur, 1998). Price and Mance (1978) found z_2 to be 1.5 and z_1 0.02 on some British catchments. This is a convenient form for ready addition to flow models.

Alternatively, a first-order relationship, where the rate at which pollutants washoff is assumed to be directly proportional to the amount of pollutant remaining on the surface, can be used:

$$W = -\frac{dM_s}{dt} = k_4 i M_s(t) \tag{5.22}$$

where k_4 is the washoff constant (mm^{-1}).

Integrating Equation 5.21 gives

$$M_s(t) = M_s(0)e^{-k_4 it} \tag{5.23}$$

where $M_s(0)$ is the initial amount of pollutant on surface (kg), $M_s(t)$ is the amount of pollutant on surface after time t (kg), and $M_w(t)$ is the amount of pollutant washed off after time t (kg).

Since $M_s(t) = M_s(0) - M_w$,

$$M_w(t) = M_s(0)[1 - e^{-k_4 it}] \tag{5.24}$$

Typical values for k_4 are 0.1–0.2 mm^{-1}. As with the buildup parameters, this requires calibration for each individual catchment. Washoff concentration (c) can be obtained as

$$c = \frac{W}{Q} = \frac{k_4 M_s}{A_i} \tag{5.25}$$

where A_i is the catchment impervious area (ha).

EXAMPLE 5.3

A storm of duration 30 minutes and intensity 10 mm/h falls on a 1.5 ha impervious area urban catchment. If the initial pollutant mass on the surface is 12 kg/ha, calculate:

1. The mass of pollutant washed off during the storm ($k_4 = 0.19$ mm^{-1})
2. The average pollutant concentration

Solution

a. $M_s(0) = 12 \times 1.5 = 18$ kg
From Equation 5.24,

$$M_w(0.5) = 18[1 - e^{-0.19 \times 10 \times 0.5}] = 11.0 \text{ kg}$$

b. $c = \dfrac{M_w(0.5)}{Q} = \dfrac{11.0(\text{kg})}{0.01(\text{m/h}) \times 0.5(\text{h}) \times 15,000(\text{m}^2)} = 0.147 \text{kg/m}^3$

$$= 147 \text{ mg/L}$$

A disadvantage of this formulation is that pollutant concentration will only decrease with time as M_s decreases. This can be remedied by introducing an exponent w for i in Equation 5.22, where i is in the range 1.4–1.8:

$$W = k_5 i^w M_s \tag{5.26}$$

where k_5 is the amended washoff constant (mm^{-1}).

5.4.5 Sewer misconnections

The issue of sewer misconnection has been discussed in Chapter 3 in the context of inflow of stormwater into separate foul sewers. However, an arguably more serious situation arises when wastewater from buildings gets into separate stormwater sewers (via soil and waste drainage systems) with the potential to directly impact on the water quality and the amenity value of an area. The main pollutant loading contributions arise from misconnected toilets (BOD, NH_4-N, FIOs), kitchen sinks (BOD and PO_4-P), washing machines (PO_4-P and BOD), and, to a lesser extent, dishwashers (PO_4-P).

The scale of the issue is hard to establish with any accuracy, but Ellis and Butler (2015) report that the average misconnection rate in England and Wales is about 3% with a working range of 1%–5%. However, outlier rates as high as 30% or more can occur arising from specific local conditions. The organic and nutrient loads from misconnections at this level have been shown to have a significant impact on receiving water quality and the achievement of river quality standards (Revitt and Ellis, 2016). Methods to detect and eliminate illicit discharges include visual reconnaissance to document flowing stormwater outfalls in dry weather, sampling to identify possible sources, and testing for key chemical/microbiological determinands (Irvine et al., 2011). Chandler and Lerner (2015) have developed a technique based on the passive sampling of optical brighteners. These chemicals, commonly found in toilet paper, sanitary products, detergents, and cleaning products, do not occur naturally in the environment but are commonly associated with illicit discharges.

PROBLEMS

5.1 List and explain the initial rainfall losses. How are they represented mathematically?

5.2 List and explain the continuing rainfall losses. How are they represented mathematically?

5.3 What are the disadvantages of the fixed runoff equation? How are some of these overcome with the variable runoff equation?

5.4 Reassess the effective rainfall profile calculated in Example 5.2 using Horton's equation as the continuing loss model where $f_0 = 10$ mm/h, $f_c = 1$ mm/h, and $k_2 = 1$ h^{-1}
[0, 0.4, 10.6, 0 mm/h]

5.5   A 1 mm, 10-minute unit hydrograph for an urban catchment is given in the table. Derive the runoff hydrograph resulting from the effective rainfall hyetograph also given.   [Peak = 2500 L/s @ 40 minutes]

Time (minutes)	0–10	10–20	20–30	30–40	40–50	50–60	60–70
UH flow rate (L/s)	0	250	500	375	250	125	0
Rainfall intensity (mm)	1	2	3	0	1	0	0

5.6   Compare and contrast the three main approaches to predicting unit hydrographs in ungauged urban catchments.

5.7   A three-reservoir Nash cascade ($K = 12$ minutes) is used to represent the runoff response of a 10 ha urban catchment. Calculate the 10 mm 1 hour unit hydrograph peak flow and time to peak.   [278 L/s, 24 minutes]

5.8   What are the main sources of pollutants in stormwater, and what is their importance?

5.9   Explain the main ways in which stormwater quality is modelled. What are their relative merits?

5.10   For the conditions described in Example 5.3, reevaluate the stormwater pollutant concentration at 10-minute time intervals.   [166, 122, 88 mg/L]

5.11   What are the main problems associated with sewer misconnections?

KEY SOURCES

Luker, M. and Montague, K.N. 1994. *Control of Pollution from Highway Drainage Discharges*, CIRIA R142.

Marsalek, J., Maksimovic, C., Zeman, E., and Price, R. (eds.). 1998. *Hydroinformatic Tools for Planning, Operation, Design, Operation and Rehabilitation of Sewer Systems*, NATO ASI Series 2: Environment – Vol. 44, Kluwer Academic Press.

Shaw, E.M., Beven, K.J., Chappell, N.A., and Lamb, R. 2010. *Hydrology in Practice*, 4th edn, CRC Press.

Torno, H.C., Marsalek, J., and Desbordes, M. (eds.). 1986. *Urban Runoff Pollution*, NATO ASI Series G: Ecological Sciences – Vol. 10, Springer-Verlag.

REFERENCES

Balmforth, D., Digman, C., Kellagher, R., and Butler, D. 2006. *Designing for Exceedance in Urban Drainage—Good Practice*, CIRIA C635, l.

Becouze-Lareure, C., Dembélé, A., Coquery, M., Cren-Olivé, C., Barillon, B., and Bertrand-Krajewski, J.-L. 2016. Source characterization and loads of metals and pesticides in urban wet weather discharges, *Urban Water Journal*, 13(6), 600–617.

Chandler, D.M. and Lerner, D.N. 2015. A low cost method to detect polluted surface water outfalls and misconnected drainage, *Water and Environment Journal*, 29, 202–206.

Colwill, D.M. Peters, C.J., and Perry, R. 1984. *Water Quality in Motorway Runoff*. TRRL Supplementary Report 823, Transport and Road Research Laboratory.

Delleur, J.W. 1998. Modelling quality of urban runoff, in *Hydroinformatic Tools for Planning, Operation, Design, Operation and Rehabilitation of Sewer Systems* (eds. J. Marsalek, C. Maksimovic, E. Zeman, and R. Price), NATO ASI Series 2: Environment – Vol. 44, Kluwer Academic Press, 241–285.

DoE/NWC. 1981. *Design and Analysis of Urban Storm Drainage. The Wallingford Procedure. Volume 1: Principles, Methods and Practice*. Department of the Enviroment, Standing Technical Committee Report No. 28.

Ellis, J.B. 1986. Pollutional aspects of urban runoff, in *Urban Runoff Pollution* (eds. H.C. Torno, J. Marsalek, and M. Desbordes), NATO ASI Series G: Ecological Sciences – Vol. 10, Springer-Verlag, 1–38.

Ellis, J.B. and Butler, D. 2015. Surface water sewer misconnections in England and Wales: Pollution sources and impacts. *Science of the Total Environment*, 526, 98–109.

Ellis, J.B. and Mitchell, G. 2006. Urban diffuse pollution: Key data information approaches for the water framework directive. *Water and Environment Journal*, 20(1), 19–26.

Granier, L., Chevrevil, M., Carru, A., and Letolle, R. 1990. Urban runoff pollution by organochlorines (polychlorinated biphenyls and lindane) and heavy metals (lead, zinc and chromium). *Chemosphere*, 21(9), 1101–1107.

Green, W.H. and Ampt, G.A. 1911. Studies of soil physics, 1: The flow of air and water through soils. *Journal of Agricultural Science*, 4(1), 1–24.

Harms, R.W. and Verworn, H.-R. 1984. HYSTEM—ein hydrologisches Stadtent-waesser-ungsmodell. Teil I: Modellbeschreibung. *Korrespondenz Abwasser*, 31(2), 112–117.

Horton, R.E. 1940. An approach towards a physical interpretation of infiltration capacity. *Proceedings of Soil Science Society of America*, 5, 399–417.

Irvine, K., Rossi, M.C., Vermette, S., Bakert, J., and Kleinfelder, K. 2011. Illicit discharge detection and elimination: Low cost options for source identification and trackdown in stormwater systems. *Urban Water Journal*, 8(6), 379–395.

Kidd, C.H.R. and Lowring, M.J. 1979. *The Wallingford Urban Sub-catchment Model*. IoH Report No. 60, Institute of Hydrology.

Lundy, L., Ellis, J.B., and Revitt, D.M. 2012. Risk prioritisation of stormwater pollutant sources. *Water Research*, 46, 6589–6600.

Malmqvist, P.-A. 1979. Atmospheric fallout and street cleaning—Effect on urban storm water and snow. *Progress in Water Technology*, 10, 417–431.

Mance, G. and Harman, M. 1978. The quality of urban stormwater runoff, in *Urban Storm Drainage* (ed. P.R. Helliwell), Pentech Press, 603–617.

Memon, F.A. and Butler, D. 2005. Characterisation of pollutants washed off from road surfaces during wet weather. *Urban Water Journal*, 2(3), 171–182.

Nash, J.E. 1957. The form of the instantaneous unit hydrograph. *International Association of Science Hydrology Publication*, 45(3), 114–121.

Natural Environment Research Council. 1975. *Flood Studies Report*, 5 volumes, Institute of Hydrology.

Novotny, V. and Chesters, G. 1981. *Handbook of Nonpoint Pollution, Sources and Management*, Van Nostrand Reinhold.

Osborne, M.P. 2009. *A New Runoff Volume Model*, WaPUG User Note No. 28, version 3.

Packman, J.C. 1986. Runoff estimation in urbanising and mixed urban/rural catchments. *IPHE Seminar on Sewers for Adoption*, 2/1-2/12, Imperial College, London.

Packman, J.C. 1990. New hydrology model. *Proceedings of WaPUG Conference*, Birmingham, UK, June.

Price, R. and Mance, G. 1978. A suspended solids model for storm water runoff, in *Urban Storm Drainage* (ed. P.R. Helliwell), Pentech Press, 546–555.

Revitt, D.M. and Ellis, J.B. 2016. Urban surface water pollution problems arising from misconnections. *Science of the Total Environment*, 551–552, 163–174.

Richards, L.A. 1933. Capillary conduction of liquids through porous mediums. *Physics*, 1, 318–333.

Sartor, J.D. and Boyd, G.B. 1972. *Water Pollution Aspects of Street Surface Contaminants*. EPA Report No. EPA/R2/72/081.

Stotz, G. 1987. Investigations of the properties of the surface water runoff from federal highways in the FRG. *Science of the Total Environment*, 59, 329–337.

UK Water Industry Research (UKWIR). 2014. *Development of the UKWIR Runoff Model: Main Report*. Report 14/SW/01/6, London.

Woods Ballard, B., Wilson, S., Udale-Clark, H., Illman, S., Scott, T., Ashley, A., and Kellagher, R. 2016. *The SuDS Manual*, 5th edn, CIRIA C753.

Chapter 6

System components and layout

6.1 INTRODUCTION

This chapter gives an overview of the elements that make up any piped urban drainage system operating under gravity, including building drainage (Section 6.2) and other main system components (Section 6.3). The main stages in the design process are also described (Section 6.4). Gravity system "exceedance" components are described separately in Chapter 11. Piped systems operating under different pressure regimes are discussed in Chapter 14, and non-piped system components (SuDS) are covered in Chapter 21.

6.2 BUILDING DRAINAGE

Even though urban drainage engineers are not normally involved directly in the planning, design, and construction of building drainage, it is important that they are at least aware of the main components and layout of systems in and around buildings. This includes, in particular, an understanding of how building drains connect with the main sewer system. Building drainage systems are subject to BS EN 12056-2: 2000 and Approved Document H (2015), part of the 2010 Building Regulations.

6.2.1 Soil and waste drainage

6.2.1.1 Inside

A common arrangement for the soil (WC/toilet) and waste (other appliances) drainage of modern domestic properties is shown in Figure 6.1. This illustrates a 2-storey dwelling with appliances on both floors connected to a single vertical stack. Each appliance is protected by a trap (U-bend or S-bend) and water seal to prevent odours reaching the house from the downstream drainage system.

The stack (typically 100–150 mm in diameter) has a top open to atmosphere that should be at least 900 mm from the top of any adjacent opening into the property. The flow regime in a vertical stack is quite different than that in sloping pipes. Flow tends to adhere to the perimeter of the pipe, forming an annulus with a central air core. The pressure of the air in the core varies with height, depending on the appliances in use, and can be both positive and negative. Design rules and details have been devised to avoid the risk of water being siphoned from the traps.

The distance x from the lowest branch to the invert of the building drain typically exceeds 450 mm for dwellings up to 3 storeys high.

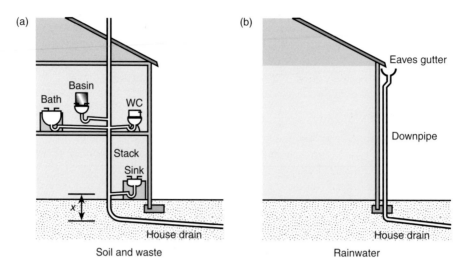

Figure 6.1 Typical building drainage arrangement in a 2-storey house.

6.2.1.2 Outside

The individual lengths of drains connecting each property to the public sewer tend to be short (usually <20 m) and small (diameter <150 mm). However, in terms of the total length of the whole piped system, they make up a surprisingly large fraction – perhaps as much as half.

6.2.1.3 Components

Outside building drainage systems have a number of common components, particularly associated with providing access for testing, inspection, and blockage-clearance from the surface. A *rodding eye* permits rodding along the drain from the surface. It consists of a vertical or inclined riser pipe with a sealed, removable cover. Access can also be gained using an *access chamber* over a pipe fitting with a sealed, removable cover. *Inspection chambers* are also used and consist of shallow access points on the drain, and also have a sealed, removable cover. Woolley (1988) gives comprehensive coverage of building drainage details.

Building drains are typically designed using procedures similar to small foul sewers (described in Chapter 9). Gradients tend to be quite steep (>1:80 for 100 mm diameter pipes), although field evidence suggests that very flat drains are no more likely to block than steep ones (Lillywhite and Webster, 1979); good quality construction is more influential in reducing blockage potential. Drains and private sewers are relatively shallow with a minimum cover of 0.75 m under gardens and 1.25 m under roads and paths. The height x, plus the length and gradient of the drain, determine the minimum feasible depth of the public sewer.

6.2.1.4 Layout

The main aim of the layout of external building drainage is to minimise the length of pipe-work and associated components, while ensuring that adequate accessibility is maintained. Generally, changes of direction should be minimised and appropriate access points provided where necessary.

Building drains carrying soil and waste should discharge only to a public foul or combined sewer. Some existing installations still feature an interceptor trap with water seal in the last inspection chamber before the sewer. These were provided to reduce the risk of odour release into the building drainage and to discourage the entry of rodents. However, they have tended to fall into disuse and disrepair and can be a source of blockage and odour problems in their own right. Today, they are not normally specified, and water service providers have often undertaken programmes to remove them in problem areas.

6.2.2 Roof drainage

A conventional arrangement for the roof drainage of domestic properties is given in Figure 6.1. This shows a 2-storey property with a pitched roof, drained by an eaves gutter connected to a single, vertical downpipe, positioned at one end. A typical eaves gutter is a 75 mm half-round channel with a nominal fall. Its capacity can be estimated using the theory of spatially varied flow, and also depends on the position and spacing of the outlets.

Flow in the rainwater downpipe is annular, just like in the soil and waste stack. The type of inlet dictates capacity. For single-family dwellings, downpipes are 75–100 mm in diameter.

The downpipes can discharge directly to a separate storm sewer but will need a water seal trap if connected to a combined sewer. Roof drainage should *not* be discharged to separate foul sewers.

The design of roof drainage systems is similar to that of small storm sewer networks (Chapter 10 gives details). In this situation, the catchments are very small (<60 m²), the time of concentration is low (1–2 minutes), and so short-duration, high-intensity rainfall events are critical for pipe capacity estimation. Approved Document H of The Building Regulations 2010 gives rainfall intensities appropriate for different parts of the United Kingdom, for example, 0.018–0.020 L/s/m² for Devon and Cornwall (equivalent to 65–72 mm/h). This figure is multiplied by roof area to determine runoff, with allowance for the slope of pitched roofs.

Readers are referred to Swaffield et al. (2015) and Wise and Swaffield (2002) for a fuller exposition of building drainage hydraulic design.

6.3 SYSTEM COMPONENTS

The design of sewer systems is covered in Chapters 9 and 10, but this section introduces the main "hardware" components of the sewerage.

6.3.1 Sewers

Sewer pipes have historically been made of a wide range of materials including vitrified clay, concrete, fibre cement, PVC-U and other polymers, pitch fibre, and brick. Information on current pipe materials and jointing methods is given in Chapter 15. Sewer cross-sectional shapes have also varied historically, and some of these are discussed in Chapter 7. Most new sewers are circular in cross section and range upward in diameter from 150 mm.

6.3.1.1 Vertical alignment

Figure 6.2 illustrates how the vertical position of a sewer is defined by its invert level (IL). The invert of a pipe refers to the lowest point on the inside of the pipe. The IL is the vertical distance of the invert above some fixed level or *datum* (e.g., in the UK, above ordnance datum [AOD]). Other important levels shown in Figure 6.2 are the *soffit level*, which is the highest point on the inside of the pipe, and the *crown*, which is the highest point

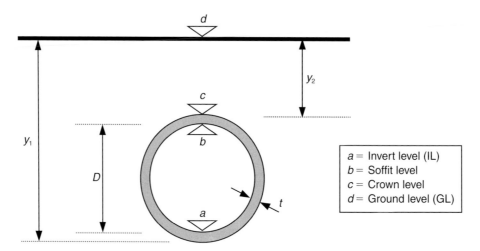

Figure 6.2 Level definitions associated with sewers.

on the outside of the pipe. Using the nomenclature defined in Figure 6.2, $b = a + D$ and $c = b + t = a + D + t$, where D is the internal diameter of the pipe (mm) and t is the pipe wall thickness (mm).

EXAMPLE 6.1

A 375 mm diameter pipe with 15 mm walls has an invert level of 52.665 m. If the ground level is 54.930, calculate the pipe: (a) soffit level, (b) depth, and (c) cover.

Solution
    a. Soffit level: $b = a + D = 52.665 + 0.375 = 53.040$ m
    b. Depth (Equation 6.1): $y_1 = d - a + t = 54.930 - 52.665 + 0.015 = 2.280$ m
    c. Cover (Equation 6.2): $y_2 = y_1 - D - 2t = 2.280 - 0.375 - 0.030 = 1.875$ m

---

The depth of the pipe (y_1) is, therefore,

$$y_1 = d - a + t \tag{6.1}$$

and the cover of the pipe (y_2) is

$$y_2 = d - c = y_1 - D - 2t \tag{6.2}$$

Figure 6.3 shows a typical sewer vertical alignment plotted on a *longitudinal profile*. The profile contains the main information required in the vertical plane to construct the pipeline.

Two invert levels are given at two of the manholes, since one refers to the exit and one to the entry level. At MH34, dissimilar diameter pipes meet, and good practice recommends (as shown) that soffit not invert levels are matched. "Chainage" refers to the plan (horizontal) distance along the pipe from a specific point. The top line of the profile box is fixed at a given level above datum (in this case 75 m). All other vertical levels can then be scaled from this line. The scale of the drawing is usually distorted to give more detail in the vertical plane.

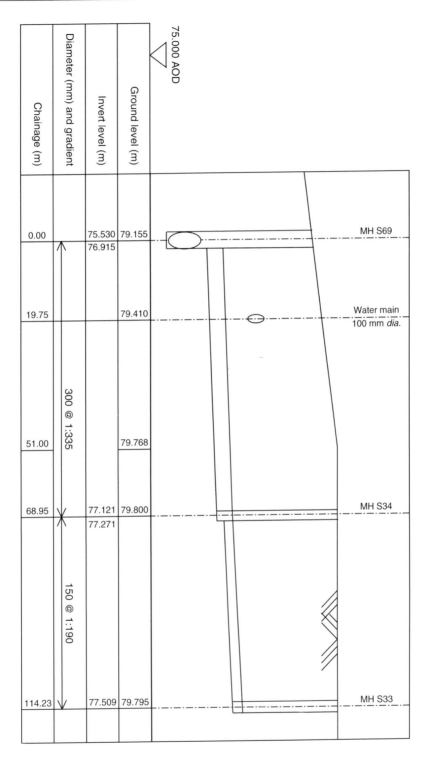

Figure 6.3 Longitudinal profile of a sewer.

Normal practice is to ensure that individual pipes between manholes have a constant gradient. Sewers are usually constructed under the highway with the storm sewer being on the centre line and the foul sewer being offset laterally and slightly lower. Should exfiltration occur, this will avoid pollution of the stormwater system. Sewers should be laid deep enough:

- To drain the lowest appliance in the premises served
- To withstand surface loads
- To prevent the contents from freezing

Typically, minimum cover for rigid pipes is 0.9 m under gardens and fields and 1.2 m under roads. More detailed aspects of the structural design of sewers are discussed in Chapter 15.

6.3.1.2 Horizontal alignment

Figure 6.4 shows a typical sewer horizontal alignment with two ways of numbering the system. Figure 6.4a numbers the *pipes* in the form (*x.y*), where *x* refers to the sewer branch

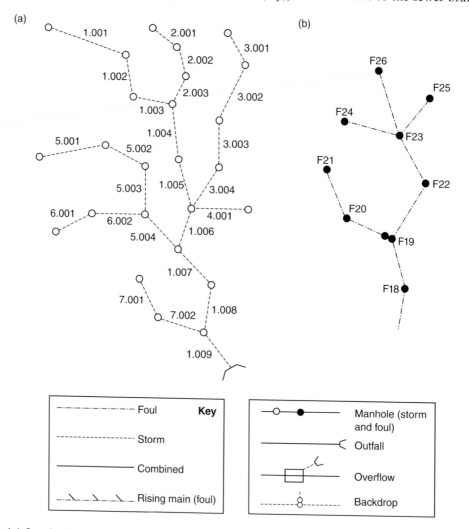

Figure 6.4 Standard sewer symbols and numbering systems.

and *y* refers to the individual pipe within the branch. Figure 6.4b numbers the manholes with a unique code. The horizontal positions of manholes may be identified by their grid reference. Standard symbols are given in the key.

Good engineering practice is to ensure that individual pipe runs (between manholes) are straight in plan. However, larger (man-entry) sewers can be built with slight curves, if necessary. Further aspects of good practice in horizontal and vertical alignment are discussed later in the chapter under layout design.

6.3.2 Manholes

As with building drainage systems, access points are required for testing, inspection, and cleaning. In sewer systems, access is usually by manholes that differ from inspection chambers in that they are deeper (>1 m) and can be entered if necessary. Manholes are provided at

- Changes in direction
- Heads of runs
- Changes in gradient
- Changes in size
- Major junctions with other sewers
- Every 90–100 m

In larger pipes, where man-access is possible (although undesirable), the spacing of manholes may be increased up to 200 m. In tunnels, the lengths increase further and are dictated by operational requirements. Manholes are commonly constructed of precast concrete rings as specified in BS 5911-3: 2010 + A1: 2014. Figure 6.5 shows a detail of a precast concrete ring manhole. Smaller manholes may have precast benching. The diameter of the manhole depends on the size of sewer and the orientation and number of inlets. Requirements for manhole covers and frames are given in BS EN 124-1: 2015. More details on manholes can be found in Woolley (1988) and WRc (2012).

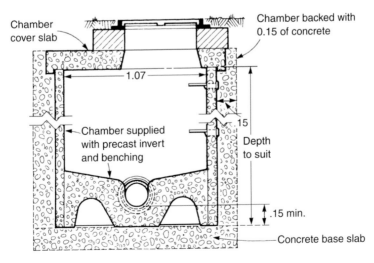

Figure 6.5 Precast concrete ring manhole. (Reproduced from Woolley, L. 1988. *Drainage Details*, 2nd edn, Taylor and Francis with permission of E&FN Spon.)

In situations where a high-level sewer is connected to one of significantly lower level, a backdrop manhole can be used. These are typically used to bring the flow from higher-level laterals into a manhole rather than lowering the length of the last sewer lengths. Drops may be external or internal to the manhole, or sloping ramps may be used, depending on the drop height and the diameter of the pipe. Figure 6.6 shows an externally placed vertical backdrop manhole. Drop manholes can require additional maintenance.

Alternatives to straightforward drop arrangements, including vortex drop shafts, are considered in Chapter 8.

Historically, some separate sewer systems were provided with dual manholes consisting of a single chamber but giving access to both foul and storm sewers. However, the rodding eye stoppers have often become lost or broken resulting in flow transfer and pollution, and these are no longer specified.

6.3.3 Gully inlets

Surface runoff (stormwater) is admitted from roads and other paved areas via inlets known as *gullies* or *catch basins*. Gullies consist of a grating and usually an underlying sump (a

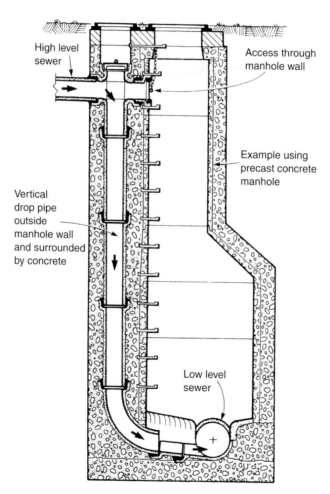

Figure 6.6 Backdrop manhole. (Reproduced from Woolley, L. 1988. *Drainage Details*, 2nd edn, Taylor and Francis with permission of E&FN Spon.)

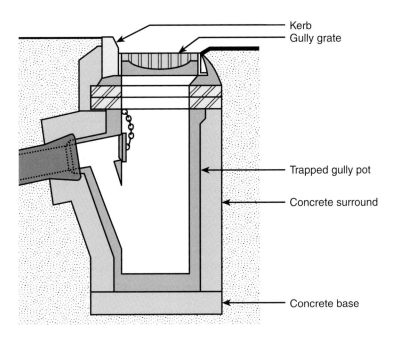

Figure 6.7 Trapped road gully.

gully pot) to collect heavy material in the flow. A water seal is incorporated to act as an odour trap for those gullies connected to combined sewers (see Figure 6.7). The gully is connected to the sewer by a lateral pipe. The relevant standard for gully gratings and frames is BS EN 124-1: 2015 and for precast concrete gully pots, BS 5911-6: 2004 + A1: 2010.

The size, number, and spacing of gullies determine the extent of surface ponding of runoff during storm events. Gullies are always placed at low points and, typically, are spaced along the road channel, adjacent to the curb. The simplest approach is to specify a standard of 50 m spacing or to require one gully per 200 m^2 of impervious area. More accurate (but more involved) gully spacing methods are explained in Chapter 8.

6.3.4 Ventilation

Ventilation is required in all urban drainage systems, but particularly in foul and combined sewers and storage tanks. It is needed to ensure that aerobic conditions are maintained within the pipe, and to avoid the possibility of buildup of toxic or explosive gases. The implications of anaerobic conditions and health and safety issues are discussed in Chapter 17.

Nearly all sewer systems are ventilated passively, without air extraction equipment. However, some major pumping stations and wastewater treatment plants are mechanically ventilated. In larger and older schemes, above-ground ventilation shafts have been used to ensure good circulation of air. Care is needed in siting these structures to avoid odour nuisance. Some schemes use ventilated manhole covers. More modern practice is to utilise the ventilation provided by the soil stacks on individual buildings (Figure 6.1) or individual vent pipes, such as on large storage chambers. Air is drawn through the system by the low pressure induced by the flow of air over the top of the stacks, and by the fall of wastewater. The water seals on domestic appliances avoid backup of sewer gases into the building interior.

6.4 DESIGN

6.4.1 Stages

A number of fundamental stages need to be followed to design a rational and cost-effective urban drainage system. These are illustrated in Figure 6.8 and are valid for any type of system. Further details on sewer sizing are given in Chapters 9–12.

The first stage is to define the *contributing area* (catchment area and population) and display it on a (digital) topographical map. Sewer software packages can often do this

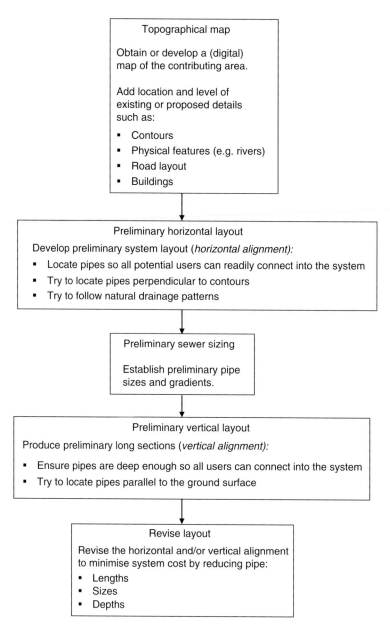

Figure 6.8 Urban drainage design procedure.

automatically. The map should already include contours, but other pertinent natural (e.g., rivers) and man-made (e.g., buildings, roads, services) features should also be marked up. Possible outfall or overflow points should be identified and investigations made as to the capacity of the receiving water body.

The next stage is to produce a preliminary *horizontal alignment* aiming to achieve a balance between the requirement to drain the whole contributing area and the need to minimise pipe run lengths. Least-cost designs tend to result when the pipe network broadly follows the natural drainage patterns and is branched, converging to a single major outfall.

Having located the pipes horizontally, the pipe sizes and gradients can now be calculated based on estimated flows from the contributing area as described in Chapters 9–12. Generally, sewers should follow the slope of the ground as far as possible to minimise excavation. However, gradients flatter than 1:500 should be avoided as they are difficult to construct accurately. A preliminary *vertical alignment* can then be produced, again bearing in mind the balance between coverage of the area and depths of pipes. The alignment can be plotted on longitudinal profiles as shown in Figure 6.3. The final stage involves revising both the horizontal and vertical alignment to minimise cost by reducing pipe lengths, sizes, and depths while meeting the hydraulic design criteria. Longer sewer runs may be cost-effective if shorter runs would require costlier excavation and/or pumping.

6.4.2 Sewers for adoption

When new residential areas are built, it is common for developers to construct the building drainage and some of the "communal" sewers. The aim of the developers and of the subsequent house owners is for the drainage network to be accepted by the local sewerage undertaker and be adopted as part of the network under its control. To aid this process, a set of design and construction guidelines, called *Sewers for Adoption* has been produced, which is generally agreed and accepted (WRc, 2012).

The main areas covered by this document are

- Design and construction advice
- Standard details
- Model civil engineering specification
- Form of agreement (Section 104 of the Water Industry Act 1991)

Appropriate reference to the recommendations in Sewers for Adoption is made in later chapters.

PROBLEMS

6.1 Describe the main components of building drainage. Compare and contrast them with those used in main sewer systems.

6.2 A 375 mm internal diameter, 1:258 gradient sewer connects two manholes A and B. The upstream manhole A has coordinates E 274.698, N 842.393, and the soffit level of the sewer leaving it is 16.438 m. Assuming negligible fall across manhole B (E 342.812, N 864.844), what is the invert level of the 450 mm diameter exiting pipe? If the cover level at manhole B is 18.590 m, what is its depth?

[15.710 m, 2.880 m]

6.3 Explain how manholes differ from inspection chambers. Why and where would you locate manholes? Where are backdrop manholes used and why?

6.4 Explain the function of road gullies. Why, where, and how would you locate them?

6.5 How are sewer systems ventilated?

6.6 List the main stages in urban drainage design. Assess the main factors affecting the horizontal and vertical alignment of the system.

KEY SOURCES

Swaffield, J., Gormley, M., Wright, G., and McDougall, I. 2015. *Transient Free Surface Flows in Building Drainage Systems*, Routledge, Oxon.

Wise, A.F.E. and Swaffield, J.A. 2002. *Water, Sanitary and Waste Services for Buildings*, 5th edn, Taylor and Francis.

Woolley, L. 1988. *Drainage Details*, 2nd edn, Taylor and Francis.

Water Research Centre (WRc). 2012. *Sewers for Adoption: A Design and Construction Guide for Developers*, 7th edn, Water UK.

REFERENCES

Approved Document H. 2015. *Drainage and Waste Disposal*, The Building Regulations 2010, HM Government.

BS 5911-3: 2010 + A1: 2014. Concrete Pipes and Ancillary Concrete Products. Specification for Unreinforced and Reinforced Concrete Manholes and Soakaways.

BS 5911-6: 2004 + A1: 2010. Concrete Pipes and Ancillary Concrete Products. Specification for Road Gullies and Gully Cover Slabs.

BS EN 124-1: 2015. Gully Tops and Manhole Tops for Vehicular and Pedestrian Areas. Definitions, Classification, General Principles of Design, Performance Requirements and Test Methods.

BS EN 12056-2: 2000. Gravity drainage systems inside buildings. Sanitary pipework, layout and calculation.

Lillywhite, M.S.T. and Webster, C.J.D. 1979. Investigations of drain blockages and their implications for design. *The Public Health Engineer*, 7(2), 53–60.

Chapter 7

Hydraulics

7.1 INTRODUCTION

An understanding of hydraulics is needed in the design of new drainage systems in order to specify the appropriate size of system components, especially pipes, channels, and tanks. It is also needed in the analysis and modelling of existing systems in order to predict the relationship between flow rate and depth for varying inflows and conditions.

Study of civil engineering hydraulics tends to concentrate on two main types of flow. The first is *pipe flow* in which a liquid flows in a pipe under pressure. The liquid always fills the whole cross section, and the pipe may be horizontal, or inclined up or down in the direction of flow. The second is *open-channel flow*, in which a liquid flows in a channel by gravity, with a free surface at atmospheric pressure. The liquid only fills the channel when the flow rate equals or exceeds the designed capacity, and the bed of the channel slopes down in the direction of flow.

The most common type of flow in sewer systems is a hybrid of these two: *part-full pipe flow*, in which a liquid flows in a pipe by gravity, with a free surface. The liquid only fills the pipe area when the flow rate equals or exceeds the designed capacity, and the bed of the pipe slopes down in the direction of flow. Traditionally the theories used are most closely related to those for full pipes, though part-full pipe flow is actually a special case of open-channel flow.

In this chapter we deal with the types of flow in the following order: pipe flow (Section 7.3), part-full pipe flow (7.4), and open-channel flow (7.5). Aspects of hydraulics are developed where appropriate in other chapters, for example, those relating to special features in Chapter 8; to the design of sewers in Chapters 9, 10, and 12; storage in Chapter 13; pumped systems in Chapter 14; and flow models in Chapter 19.

This chapter presents the main principles of hydraulics relevant to urban drainage (Section 7.2). It is intended as an introduction to the subject for those who have not studied it before, or as a refresher course for those who have. For more information, a number of sources are referred to in the text. For general reference, Chadwick et al. (2013) give a practical engineering treatment of all aspects. Hamill (2011) provides helpful explanations, and Kay (2008) uses a more descriptive approach.

7.2 BASIC PRINCIPLES

7.2.1 Pressure

Pressure is defined as force per unit area. The common units for pressure are kN/m^2 or bars (1 bar = 100 kN/m^2).

Absolute pressure is pressure relative to a vacuum, and *gauge pressure* is pressure relative to atmospheric pressure. Gauge pressure is used in most hydraulic calculations. Atmospheric pressure varies but is approximately 1 bar.

In a still liquid, pressure increases with vertical depth:

$$\Delta p = \rho g \Delta y \qquad (7.1)$$

where Δp is the increase in pressure (N/m²), ρ is the density of liquid (for water, 1000 kg/m³), g is the gravitational acceleration (9.81 m/s²), and Δy is the increase in depth (m).

Pressure at a point is equal in all directions.

7.2.2 Continuity of flow

In a section of pipe with constant diameter and no side connections (Figure 7.1), in any period of time the mass of liquid entering (at 1) must equal the mass leaving (at 2). Assuming that the liquid has a constant density (mass per unit volume), the volume entering (at 1) must equal the volume leaving (at 2).

In terms of flow rate (volume per unit time, Q),

$$Q_1 = Q_2 \qquad (7.2)$$

The common units for flow rate are m³/s or L/s (1 L [Litre] = 10^{-3} m³).

The velocity of the liquid varies across the flow cross section, with the maximum for a full pipe in the centre. "Mean velocity" (v) is defined as flow rate per unit area (A) through which the flow passes:

$$v = \frac{Q}{A} \qquad (7.3)$$

The common units for velocity are m/s (see Example 7.1).

Equation 7.2 can be rewritten as

$$v_1 A_1 = v_2 A_2$$

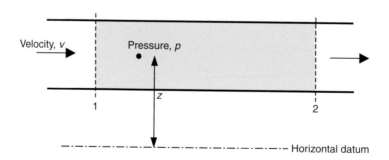

Figure 7.1 Continuity and definition of symbols for pipe flow.

EXAMPLE 7.1

The diameter of a pipe flowing full increases from 150 to 200 mm. The flow rate is 20 L/s (0.020 m³/s). Determine the mean velocity upstream and downstream of the expansion.

Solution

$$\text{Upstream}(1): A_1 = \frac{\pi 0.15^2}{4} \quad \text{so} \quad v_1 = \frac{Q}{A_1} = \frac{0.02 \times 4}{\pi 0.15^2} = 1.13 \, \text{m/s}$$

$$\text{Downstream}(2): A_2 = \frac{\pi 0.2^2}{4} \quad \text{so} \quad v_2 = \frac{Q}{A_2} = \frac{0.02 \times 4}{\pi 0.2^2} = 0.64 \, \text{m/s}$$

7.2.3 Flow classification

In hydraulics there are two terms for "constant": *uniform* and *steady*. Uniform means constant with distance, and steady means constant with time. The words have negative forms: *nonuniform* means not constant with distance; *unsteady* means not constant with time.

Hydraulic conditions in urban drainage systems can be

- Uniform steady: The flow cross-sectional area is constant with distance, and flow rate is constant with time.
- Nonuniform steady: The flow area varies with distance, but flow rate is constant with time.
- Uniform unsteady: The flow area is constant with distance, but flow rate varies with time.
- Nonuniform unsteady: The flow area varies with distance, and flow rate varies with time.

Flow in sewers is generally unsteady to some extent: wastewater varies with the time of day, and storm flow varies during a storm. However, in many hydraulic calculations, it is not necessary to take this into account, and conditions are treated as steady for the sake of simplicity. Most of the rest of this chapter considers only steady flow. In some cases, unsteady effects are significant and must be considered: for example, storage effects (considered in Chapter 13), sudden changes in pumping systems (referred to in Section 14.4.3), and storm waves in sewers (see Sections 10.6 and 19.4).

7.2.4 Laminar and turbulent flow

The property of a fluid that opposes its motion is called *viscosity*. Viscosity is caused by the interaction of the fluid molecules creating friction forces between the layers of fluid travelling at different velocities. Where velocities are low, fluid particles move in straight, parallel trajectories – *laminar* flow. Where velocities are high, fluid particles follow more chaotic paths, and the flow is *turbulent*.

Flow can be identified as laminar or turbulent using a dimensionless number, the Reynolds number (R_e), defined for a pipe flowing full as

$$R_e = \frac{vD}{\nu} \tag{7.4}$$

where v is the mean velocity (m/s), D is the pipe diameter (m), and ν is the kinematic viscosity of liquid (for water, typically 1.1×10^{-6} m²/s, dependent on temperature).

EXAMPLE 7.2

For the pipe in Example 7.1, determine the Reynolds number for both sides of the change in diameter. Is the flow laminar or turbulent? Assume that the pipe carries water, kinematic viscosity 1.1×10^{-6} m²/s.

Solution

$$\text{Reynolds number upstream (1)} = \frac{v_1 D_1}{\nu} = \frac{1.13 \times 0.15}{1.1 \times 10^{-6}} = 154100$$

$$\text{Reynolds number downstream (2)} = \frac{v_2 D_2}{\nu} = \frac{0.64 \times 0.2}{1.1 \times 10^{-6}} = 116400$$

Both are well into the turbulent zone.

When $R_e < 2000$, the flow in the pipe is laminar; when $R_e > 4000$, flow is turbulent. In most urban drainage applications, flow is firmly in the turbulent region (see Example 7.2).

7.2.5 Energy and head

A flowing liquid has three main types of energy: pressure, velocity, and potential. In hydraulics, the most common way of expressing energy is in terms of *head*, energy per unit weight (common units, m).

The three types of energy expressed as head are

$$\text{pressure head} \left(\frac{p}{\rho g} \right) \quad \text{velocity head} \left(\frac{v^2}{2g} \right) \quad \text{potential head } z$$

The symbols p, v, and z are illustrated in Figure 7.1.

Total head (H) is the sum of the three types of head, given by the *Bernoulli equation*:

$$H = \frac{p_1}{\rho g} + \frac{v^2}{2g} + z \tag{7.5}$$

When a liquid flows in a pipe or a channel, some head (h_L) is "lost" from the liquid. So, for water flowing between sections 1 and 2 in the full pipe on Figure 7.1:

$$H_1 - h_L = H_2 \tag{7.6}$$

or

$$\frac{p_1}{\rho g} + \frac{v_1^2}{2g} + z_1 - h_L = \frac{p_2}{\rho g} + \frac{v_2^2}{2g} + z_2 \tag{7.7}$$

If flow in Figure 7.1 is uniform, and the pipe is horizontal, then

$$v_1 = v_2 \quad \text{and} \quad z_1 = z_2$$

Therefore, Equation 7.7 becomes

$$h_L = \frac{p_1}{\rho g} - \frac{p_2}{\rho g}$$

The head loss is equal to the difference in pressure head.

7.3 PIPE FLOW

7.3.1 Head (energy) losses

The head or energy losses in flow in a pipe are made up of *friction losses* and *local losses*. Friction losses are caused by forces between the liquid and the solid boundary (distributed along the length of the pipe), and local losses are caused by disruptions to the flow at local features like bends and changes in cross section. Total head loss h_L is the sum of the two components.

The distribution of losses, and the other components in Equation 7.7, can be shown by two imaginary lines.

The *energy grade line* (EGL) is drawn a vertical distance from the datum equal to the total head.

The *hydraulic grade line* (HGL) is drawn a vertical distance below the energy grade line equal to the velocity head.

The two lines are drawn for a pipe flowing full on Figure 7.2. The lines allow all the terms in Equation 7.7 to be identified.

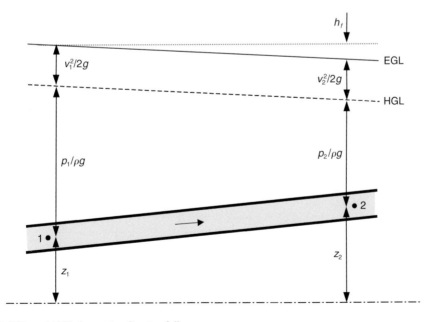

Figure 7.2 EGL and HGL for a pipe flowing full.

7.3.2 Friction losses

A fundamental requirement in the hydraulic design and analysis of urban drainage systems is the estimation of friction loss. The basic representation of friction losses, valid for both laminar and turbulent flow, is the Darcy-Weisbach equation:

$$h_f = \frac{\lambda L}{D} \cdot \frac{v^2}{2g} \tag{7.8}$$

where h_f is the head loss due to friction (m) (Figure 7.2), λ is the friction factor (no units), L is the pipe length (m), and D is the pipe diameter (m).

The term h_f/L is the gradient of the energy grade line, and (for uniform flow) of the hydraulic grade line, and is often referred to as the *hydraulic gradient* or *friction slope*.

7.3.3 Friction factor

The friction factor λ is one of the most interesting aspects of pipe hydraulics. Is it constant for a particular pipe or does it vary? Does it depend on pipe roughness or not?

As mentioned, velocity varies over the flow area. In turbulent flow (the type of flow of most significance in urban drainage), velocity levels are similar across most of the cross section but fall rapidly near the pipe wall (Figure 7.3). Very near to the wall, a boundary layer exists, where the velocity is low and laminar conditions occur: a *laminar sub-layer*.

Frictional losses are affected by the thickness of the laminar sub-layer relative to the "size" of the roughness of the pipe wall. In commercial pipes, the wall roughness is measured in terms of an equivalent sand roughness size (k_s) and can be thought of as the mean projection height of the roughness from the pipe wall.

Figure 7.4 presents the Moody diagram. This is a plot of the friction factor λ, against Reynolds number R_e for a range of values of relative roughness k_s/D. The Moody diagram demonstrates the relative effects of the thickness of the laminar sub-layer and the size of the roughness.

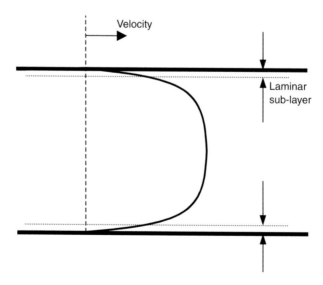

Figure 7.3 Velocity profile in turbulent flow, with laminar sub-layer.

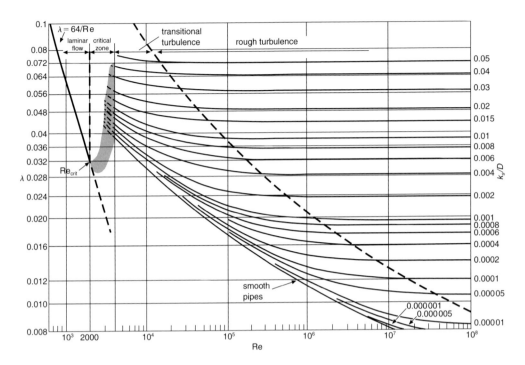

Figure 7.4 Moody diagram. (Reproduced from Chadwick A. et al. 2013. *Hydraulics in Civil and Environmental Engineering*, 5th edn, CRC Press. With permission of E & FN Spon.)

When the roughness projections are small compared with the thickness of the laminar sub-layer, the friction losses are independent of pipe roughness and dependent on Reynolds number. Flow is said to be "smooth turbulent." And λ is a function of R_e, but not of k_s/D.

When the roughness projections are much greater in height, losses are linked to pipe roughness only (not Reynolds number), and conditions are known as "rough turbulent." Then λ is a function of k_s/D, but not of R_e.

Between these two conditions lies a transitional turbulent region where conditions are dependent on roughness height and Reynolds number. Most urban drainage flows are rough or transitionally turbulent flows.

Whichever is the greater, the thickness of the laminar sub-layer or the size of the roughness, the friction factor can be determined mathematically from the Colebrook-White equation:

$$\frac{1}{\sqrt{\lambda}} = -2\log_{10}\left(\frac{k_s}{3.7D} + \frac{2.51}{R_e\sqrt{\lambda}}\right) \tag{7.9}$$

The Moody diagram illustrates the relationship between the basic hydraulic variables and their relative importance, but it does not directly represent the variables routinely used in engineering practice: flow rate, velocity, pipe diameter, roughness, and gradient.

The Colebrook-White equation can provide an explicit expression for velocity by substitution of λ using Equation 7.8 and R_e using Equation 7.4, giving

$$v = -2\sqrt{2gS_fD}\,\log_{10}\left(\frac{k_s}{3.7D} + \frac{2.51v}{D\sqrt{2gS_fD}}\right) \tag{7.10}$$

where k_s is the pipe roughness (m), S_f is the hydraulic gradient or friction slope, h_f/L (–), and ν is the kinematic viscosity (m²/s).

As illustrated by Example 7.3, this equation can be easily solved for v. However, it is rather intractable if the variable of concern is the hydraulic gradient (S_f) or pipe diameter (D). Three possible options are available: to formulate an iterative solution using a computer or programmable calculator, to use design charts or tables, or to rely on approximate equations.

7.3.4 Wallingford charts and tables

7.3.4.1 Charts

A popular graphical method consists of a series of charts produced by Hydraulics Research, Wallingford (Hydraulics Research, 1990), an example of which is given as Figure 7.5.

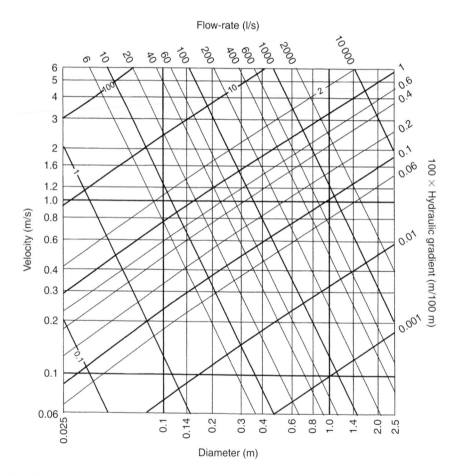

Figure 7.5 Wallingford chart for $k_s = 0.6$ mm. (Based on Hydraulics Research 1990. *Charts for the Hydraulic Design of Channels and Pipes*, 6th edn, Hydraulics Research, Wallingford. With permission of HR Wallingford Ltd.)

In these charts, the three dependent variables (D, Q, and S_f) are graphically related to velocity v, with each chart being valid for a particular roughness height k_s. Using the chart for a particular pipe roughness, it is possible to determine any two of the variables if the other two are known (Examples 7.3 and 7.4). The charts are drawn for water at 15°C, for which kinematic viscosity, $v = 1.141 \times 10^{-6} \text{m}^2/\text{s}$.

EXAMPLE 7.3

1. A 600 mm diameter circular pipe is used to pump stormwater. Calculate the mean velocity and flow rate in the pipe if the hydraulic gradient is 0.01 by solving the Colebrook-White equation. Assume the roughness height k_s is 0.6 mm and kinematic viscosity is $1.14 \times 10^{-6} \text{m}^2/\text{s}$.
2. What value of hydraulic gradient is suggested by the Moody diagram in conjunction with the Darcy-Weisbach equation? (Use the value of v already determined, together with the remaining data given.) Are conditions smooth, transitional, or rough turbulent?
3. Check v and Q using the appropriate Wallingford chart.

Solution

1. Velocity can be calculated by direct substitution into the Colebrook-White Equation 7.10 given

$$D = 0.6\,\text{m}, \ S_f = 0.01, \ k_s = 0.6\,\text{mm} \quad \text{and} \quad v = 1.14 \times 10^{-6}\,\text{m}^2/\text{s}$$

$$v = -2\sqrt{2g0.01 \times 0.6}\,\log_{10}\left(\frac{0.06 \times 10^{-3}}{3.7 \times 0.6} + \frac{2.51 \times 1.14 \times 10^{-6}}{0.6\sqrt{2g0.01 \times 0.6}}\right)$$

$$= 2.43\,\text{m/s}$$

and flow rate by continuity (7.3):

$$Q = vA = 2.43 \times \frac{\pi 0.6^2}{4} = 688\,\text{l/s}$$

2. $\dfrac{k_s}{D} = \dfrac{0.6}{600} = 0.001 \quad R_e = \dfrac{2.43 \times 0.6}{1.14 \times 10^{-6}} = 1279000$

from Moody diagram (Figure 7.4), $\lambda = 0.019$ (rough), so from Equation 7.8:

$$\frac{h_f}{L} = \frac{\lambda}{D}\frac{v^2}{2g} = \frac{0.019}{0.6}\frac{2.43^2}{2g} = 0.01$$

3. The Wallingford chart (Figure 7.5) can be used by finding the point of intersection of the hydraulic gradient line (read downwards sloping right to left) with the diameter line (read vertically upwards).

Hydraulic gradient $= 0.01$, or 1 in 100, $D = 0.6$ m
This gives

$$v \simeq 2.4\,\text{m/s} \ Q \simeq 700\,\text{L/s}$$

EXAMPLE 7.4

A circular storm sewer is to be designed to run just full when conveying a flow rate of 0.07 m³/s. If the mean velocity is specified at 1 m/s, calculate the required pipe diameter and gradient using an appropriate method. Assume the roughness height is 0.6 mm and kinematic viscosity 1.141×10^{-6} m²/s.

Solution

In principle, this problem could be solved using the Colebrook-White equation. However, this would require an iterative solution, best suited to a computational method. A direct solution can be achieved by reading from the Wallingford chart for $k_s = 0.6$ mm (Figure 7.5):

$$Q = 70 \text{ L/s}, v = 1.0 \text{ m/s}$$

This gives

$$D = 0.3 \text{ m, and gradient} = 0.42 \text{ in 100, or 1 in 240}$$

7.3.4.2 Tables

Output from the Colebrook-White equation can also be presented in tables. Table 7.1 is an example, giving Q and v for a variety of values of D and S_f, for k_s of 1.5 mm and kinematic viscosity of 1.14×10^{-6} m²/s. Comprehensive sets of tables that fulfil the same function as the Wallingford charts are also commercially available (Wallingford and Barr, 2006).

7.3.5 Approximate equations

A number of approximate solutions to the Colebrook-White equation have been developed over the last four decades (Romeo et al., 2002). Barr (1975), for example, has developed a series of explicit approximate equations that enable determination of D and S_f directly with only minor loss of accuracy. For determination of S_f the following equation may be used (in terms of v, D, and k_s):

$$S_f \approx \frac{v^2}{8gD\left[\log_{10}\left\{\dfrac{k_s}{3.7D} + \left(\dfrac{6.28v}{vD}\right)^{0.89}\right\}\right]^2} \tag{7.11}$$

and for determination of D:

$$D \approx \frac{v^2}{8gS_f\left[\log_{10}\left\{\dfrac{1.558}{(v^2/2gS_fk_s)^{0.8}} + \dfrac{15.045}{(v^3/2gS_fv)^{0.73}}\right\}\right]^2} \tag{7.12}$$

Table 7.1 Table of output from the Colebrook-White equation, for $k_s = 1.5$ mm Q (L/s) in **bold**; v (m/s) in *italic*

Diameter (m)	$S_f = 0.001$		$S_f = 0.002$		$S_f = 0.005$		$S_f = 0.01$		$S_f = 0.02$		$S_f = 0.05$	
0.2	**10**	*0.33*	**15**	*0.47*	**23**	*0.75*	**33**	*1.06*	**47**	*1.50*	**75**	*2.38*
0.3	**31**	*0.43*	**43**	*0.62*	**69**	*0.98*	**98**	*1.38*	**139**	*1.96*	**220**	*3.11*
0.375	**55**	*0.50*	**79**	*0.71*	**125**	*1.13*	**177**	*1.60*	**250**	*2.27*	**396**	*3.59*
0.75	**347**	*0.79*	**492**	*1.11*	**780**	*1.76*	**1104**	*2.50*	**1562**	*3.54*	**2472**	*5.60*

These have the advantage of being tractable, but the disadvantage of being rather lengthy to solve.

7.3.6 Roughness

Typical values of roughness k_s for use in the Colebrook-White equation, related to sewer type and age, are given in Table 7.2. The values described as "new" are appropriate for new, clean, and well-aligned pipes. They are appropriate in the design of stormwater pipes (where excessive sediment content is not expected) or in establishing the initial flow conditions in newly laid foul or combined sewers. "Old" values are generally preferred in the design or analysis of foul and combined sewers, where roughness is related more to the effect of biological slime than the pipe material.

For preliminary design purposes, or where existing pipe conditions are unknown, a value of $k_s = 0.6$ mm is suggested for storm sewers and $k_s = 1.5$ mm for foul sewers (irrespective of pipe material). Sewers subject to sediment deposition can have k_s values in the order of 30–60 mm (see Chapter 16 for further discussion). As an example of the effect of roughness height, for a 150 mm diameter pipe conveying 10 L/s, a change in the value of k_s from 0.6 to 1.5 mm would result in an increase in flow depth and a decrease in flow velocity of about 10%.

For rising mains, roughness can be empirically related to velocity (Forty et al., 2004; Lauchlan et al., 2005):

$$k_s \approx av^{-2.34} \tag{7.13}$$

where $a = 0.45$ for $v = 1.0$ m/s, $a = 0.17$ for $v = 1.5$ m/s, and $a = 0.09$ for $v = 2.0$ m/s.

The Wallingford tables (Wallingford and Barr, 2006) recommend that for rising mains in "normal" condition, k_s should be taken as 0.3 mm for mean velocities up to 1.1 m/s, and 0.15 mm for velocities between 1.1 and 1.8 m/s.

7.3.7 Local losses

Local losses occur at points where the flow is disrupted, such as bends, valves, and changes of area. In certain circumstances (e.g., where there are many fittings in a short length of pipe), these can be equal to or greater than the friction losses. Local losses are usually expressed in terms of velocity head as follows:

$$h_{local} = k_L \frac{v^2}{2g} \tag{7.14}$$

where h_{local} is the local head loss (m) and k_L is a constant for the particular type of fitting (–).

Table 7.2 Typical values of roughness (k_s)

Pipe material	k_s range (mm)	
	New	Old
Clay	0.03–0.15	0.3–3.0
PVC–U (and other polymers)	0.03–0.06	0.15–1.50
Concrete	0.06–1.50	1.5–6.0
Fibre cement	0.015–0.030	0.6–6.0
Brickwork – good condition	0.6–6.0	3.0–15
Brickwork – poor condition	—	15–30
Rising mains	0.03–0.60	

Table 7.3 Local head loss constants

Fitting	k_L
Pipe entry (sharp edged)	0.50
Pipe entry (slightly rounded)	0.25
Pipe entry (bell mouthed)	0.05
Pipe exit (sudden)	1.0
90° pipe bend ("elbow" – sharp bend)	1.0
90° pipe bend (long)	0.2
Straight manhole on gravity sewer (part-full)	<0.1
Straight manhole on gravity sewer (surcharged)	0.15
Manhole with 30° bend (surcharged)	0.5
Manhole with 60° bend (surcharged)	1.0

In gravity sewers, local losses occur at manholes, but these are only usually significant when the system is surcharged.

Examples of the value of k_L used in design practice are given in Table 7.3.

For design cases when velocity is unknown and calculations are based on the Wallingford charts, a useful alternative to Equation 7.14 is to express local losses as an *equivalent length* of pipe. The sum of this equivalent length and the actual pipe length is then multiplied by the hydraulic gradient (taken from the chart) to give the total (friction + local) losses.

It is helpful to relate this to the friction factor λ. Combining Equations 7.8 and 7.14 gives

$$\text{Total losses} = \frac{\lambda L}{D}\frac{v^2}{2g} + k_L \frac{v^2}{2g}$$

EXAMPLE 7.5

A surcharged manhole with a bend has local loss constant $k_L = 1$. Determine L_E/D (assuming that it is independent of velocity) if this feature occurs in a pipe with D a diameter of (1) 300 mm or (2) 600 mm (k_s 1.5 mm for both). For both cases determine the conditions for which the equivalent length is independent of velocity. (Assume kinematic viscosity = $1.14 \times 10^{-6}\,\text{m}^2/\text{s}$.)

Solution

1. $\dfrac{k_s}{D} = \dfrac{1.5}{300} = 0.005$

   Equivalent length is independent of velocity in rough turbulence. From Moody diagram (Figure 7.4). $\lambda = 0.03$, so (Equation 7.15)

   $$\frac{L_E}{D} = \frac{k_L}{\lambda} = \frac{1.0}{0.03} = 33$$

   This is valid if R_e is greater than 200,000—that is, velocity greater than 0.76 m/s.

2. $\dfrac{k_s}{D} = \dfrac{1.5}{600} = 0.0025$

   Equivalent length is independent of velocity in rough turbulence. From Moody diagram (Figure 7.4),

$$\lambda = 0.025, \text{ so } \frac{L_E}{D} = \frac{k_L}{\lambda} = \frac{1.0}{0.025} = 40$$

This is valid if R_e is greater than 500,000—that is, velocity greater than 0.95 m/s.

So, if an equivalent length L_E is to be added to L in order to replace the separate term for local losses,

$$\frac{L_E}{D} = \frac{k_L}{\lambda} \qquad (7.15)$$

In rough turbulence (Figure 7.4) L_E is independent of v, but in transitional or smooth turbulence it is not (because λ is affected by R_e). As well as being available from the Moody diagram, the value of λ can be inferred from the Wallingford chart, since

$$\lambda = \frac{S_f D 2g}{v^2}$$

we have

$$L_E = \frac{k_L}{S_f} \frac{v^2}{2g}$$

Determination of equivalent length is shown in Example 7.5. Its use is demonstrated in Example 8.2.

7.4 PART-FULL PIPE FLOW

The common flow condition in urban drainage pipes is part-full pipe flow. The presence of the free surface must be taken into account in hydraulic computations.

7.4.1 Normal depth

In uniform steady gravity flow, an equilibrium exists along a part-full pipe or channel. The energy consumed by friction between the liquid and the pipe wall is in balance with the fall along the pipe length. If pipe slope could be increased for the same flow rate, additional energy would be available to the flow, resulting in higher velocity and lower depth. The equilibrium depth is referred to as the *normal depth*.

Since depth of flow and velocity are constant when conditions are uniform, and pressure at the surface is atmospheric, the EGL and HGL are parallel to the bed, and the HGL coincides with the water surface (Figure 7.6).

7.4.2 Geometric and hydraulic elements

When water flows along a part-full pipe, a number of properties, shown in Figure 7.7, can be defined as in Table 7.4.

Hydraulic radius can be related to geometrical properties, and for a circular pipe running full (for example),

$$R = \frac{A}{P} = \frac{\pi D^2/4}{\pi D} = \frac{D}{4} \qquad (7.16)$$

where D is the pipe diameter (m) (see Example 7.6).

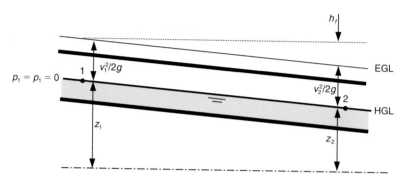

Figure 7.6 EGL and HGL for a pipe flowing part-full.

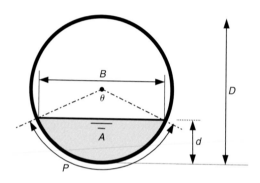

Figure 7.7 Definition of geometric elements for a circular pipe.

A circular cross-sectional shape is most common for sewers and drains. The need to understand the hydraulic conditions at a range of depths results from the wide variations in flow experienced by sewers during their working life.

Figure 7.7 shows the cross section of a pipe of diameter D with flow of depth d. The angle subtended at the pipe centre by the free surface is θ. From geometrical considerations, θ is related to proportional depth of flow d/D as follows:

$$\theta = 2\cos^{-1}\left[1 - \frac{2d}{D}\right] \tag{7.17}$$

Table 7.4 Geometric elements

Property	Symbol	Definition	Common units
Depth	d	Height of water above the channel invert	m
Area	A	Cross-sectional area of flow	m^2
Wetted perimeter	P	Portion of the flow area's perimeter that is in contact with the channel	m
Hydraulic radius	R	A per unit P	m
Top width	B	Flow width at the water surface	m
Hydraulic mean depth	d_m	A per unit B	m

EXAMPLE 7.6

A circular sewer of diameter 300 mm is flowing half full. Determine d, A, P, R, B, and d_m.

Solution

Calculations are simply based on the property definitions (Table 7.4) and the geometry of a circle.

$$D = 0.3 \text{ m}$$

So,

$$d = 0.15 \text{ m}$$

$$A = \frac{1}{2} \cdot \frac{\pi D^2}{4} = \frac{1}{2} \cdot \frac{\pi 0.3^2}{4} = 0.0353 \text{ m}^2$$

$$P = \frac{\pi D}{2} = \frac{\pi 0.3}{2} = 0.471 \text{ m}$$

$$R = \frac{A}{P} = \frac{0.0353}{0.471} = 0.075 \text{ m}$$

$$B = D = 0.3 \text{ m}$$

$$d_m = \frac{A}{B} = \frac{0.0353}{0.3} = 0.118 \text{ m}$$

Expressions for area (A), wetted perimeter (P), hydraulic radius (R), top width (B), and hydraulic mean depth (d_m), based on D and θ, are given in Table 7.5.

Using these relationships, Figure 7.8 gives dimensionless relationships for part-full flow depth, cross-sectional area, wetted perimeter, and hydraulic radius as a proportion of the full-depth value (that is, d/D, A/A_f, P/P_f, and R/R_f). Also shown are lines indicating the velocity ratio v/v_f and flow ratio Q/Q_f (where v and Q are part-full velocity and flow rate, and v_f and Q_f are the full-depth values). These latter lines are slightly dependent on the friction loss equation; the Colebrook-White equation is used here.

Table 7.5 Expressions for geometric elements in a part-full circular pipe

Parameter	Expression (θ in radians)
A	$\dfrac{D^2}{8}(\theta - \sin\theta)$
P	$D\theta/2$
R	$\dfrac{A}{P} = \dfrac{D}{4}\left[\dfrac{\theta - \sin\theta}{\theta}\right]$
B	$D\sin(\theta/2)$
d_m	$\dfrac{A}{B} = \dfrac{D(\theta - \sin\theta)}{8\sin\theta/2}$

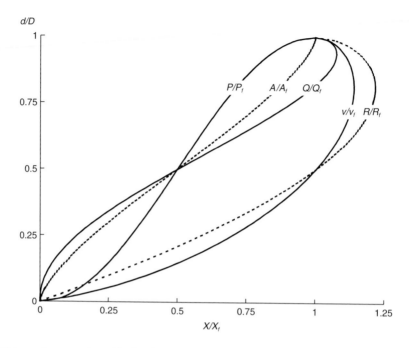

Figure 7.8 Values of geometric and hydraulic elements for part-full pipe flow.

In part-full pipes, maximum flow velocity and flow rate do not occur when the pipe is running full; they occur when it is slightly less than full. This is because the circular shape affects the relative magnitudes of the flow area and the wetted perimeter (which determines the magnitude of the frictional resistance). At low flows, the wetted perimeter is high compared with flow area, resulting in low velocities. Velocity increases with depth (see Figure 7.8) until at the highest depths the increase in the wetted perimeter is again high compared with the flow area, and this results in a fall in velocity. It follows that, eventually, the flow rate will also fall, since flow rate is the product of cross-sectional area and velocity. Table 7.6 summarises these effects.

The effect of a free surface within a pipe is largely understood, and conventionally the hydraulic radius (R) is substituted for pipe diameter (D) in the pipe flow equations to model the varying depth of flow (using Equation 7.16). The substitution appears to predict the effects quite accurately, but there is evidence to suggest that it overestimates flow rate and velocity in rough pipes by a few percent.

Graphs of the velocity ratio v/v_f and flow-rate ratio Q/Q_f in the form of Figure 7.8 can be used in conjunction with the Wallingford pipe flow chart to determine v and Q for any part-full case. This approach is appropriate for problems such as

- Given Q, S_f, and d/D, find D and v.
- Given Q, D, and S_f, determine d/D and v.

Table 7.6 Proportional flow rate and velocity at various depths

	d/D = 0.5	d/D = 1.0	Maximum
Q/Q_f	0.5	1.0	1.08 at d/D = 0.94
v/v_f	1.0	1.0	1.14 at d/D = 0.81

Example 7.7 gives an illustration.

EXAMPLE 7.7

Given $Q = 70$ L/s, $S_f = 1:200$, $d/D = 0.25$, and $k_s = 0.6$ mm, find the necessary diameter D and corresponding velocity.

Solution

From Figure 7.8, $Q/Q_f = 0.14$ (for $d/D = 0.25$), so $Q_f = 500$ L/s. From Figure 7.5, $D = 600$ mm and $v_f = 1.7$ m/s. From Figure 7.8 again, $v/v_f = 0.7$ so $v = 1.2$ m/s.

7.4.3 Butler-Pinkerton charts

An alternative, more direct method for part-full pipe problems is provided by the Butler-Pinkerton charts (Butler and Pinkerton, 1987). These are based on a shape correction factor ψ, which can be used to modify pipe diameter rather than replace it. Usually D is replaced by $4R$ (from Equation 7.16), so incorporating the shape correction factor gives

$$\psi D \equiv 4R$$

$$\psi = \frac{4R}{D}$$

$$\psi = \frac{(\theta - \sin\theta)}{\theta} \tag{7.18}$$

By substituting the shape correction factor ψ into the Colebrook-White equation, an expression valid for any proportional depth of flow is obtained:

$$v = -2\sqrt{2gS_f\psi D}\,\log_{10}\left(\frac{k_s}{3.7\psi D} + \frac{2.51v}{\psi D\sqrt{2gS_f\psi D}}\right) \tag{7.19}$$

An example of a Butler-Pinkerton chart is given as Figure 7.9. A separate chart is required for each roughness value, and for each pipe diameter. Once the correct chart has been

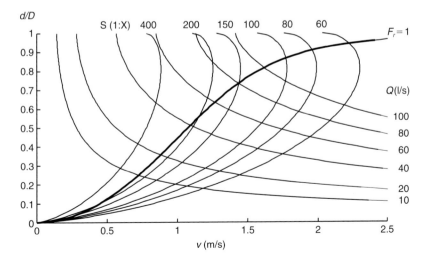

Figure 7.9 Butler-Pinkerton chart (D 300 mm, k_s 0.6 mm).

chosen, definition of any two of the remaining variables allows determination of the other two. This is particularly useful to cope with questions such as

- Given Q, and constraints for minimum v and maximum d/D, find minimum S_f and D (the design case).
- Given D and S_f, find Q and v at various values of d/D (the analysis case).

The charts consist of two families of curves: the S-curves represent the modified Colebrook-White Equation 7.19, and Q-curves represent the continuity equation:

$$v = \frac{8Q}{\psi \theta D^2} \tag{7.20}$$

EXAMPLE 7.8

Given $Q = 60$ L/s, $d/D = 0.75$, minimum $v = 0.75$ m/s and $k_s = 0.6$ mm, find the minimum gradient of a 300 mm diameter pipe.

Solution

The Butler-Pinkerton chart (Figure 7.9) can be used by estimating the point of intersection of the Q-curve (read from the right sloping upwards) with the hydraulic gradient S-curve (read downwards sloping first right then left).
Thus, at $d/D = 0.75$, $S_f = 1{:}280$ ($v = 1.05$ m/s).

The intersection of each S- and Q-curve gives the resulting normal flow depth and steady-state velocity (see Example 7.8). A curve labelled $F_r = 1$ is also shown and is explained in Section 7.5.4.

7.4.4 Non-circular sections

Circular pipes are by far the most common in shape. They have the shortest circumference per unit of cross-sectional area and so require the least wall material to resist internal and external pressures. They are also easy to manufacture. However, other shapes have been adopted in the past for sewers and drains – most commonly egg-shaped pipes, but also rectangular, trapezoidal, U-shaped, oval, horseshoe (arch), and compound types. Figure 7.10 shows a range of shapes, their geometries, and hydraulic elements.

Provided cross-sectional shape does not differ much from the circular, the basic equations for pipe flow can be utilised by substituting $4R$ for D.

7.4.5 Surcharge

Surcharging refers to pipes designed to run full or part full, conveying flow under pressure. This can occur, for example, when flood flows exceed the design capacity, and it is therefore likely that all storm sewers will become surcharged at some time during their operational life.

A sewer pipe can surcharge in one of two ways, normally referred to as "pipe surcharge" and "manhole surcharge."

Figure 7.11a shows a longitudinal vertical section along a length of sewer running part-full (without surcharge). As explained in Section 7.4.1, the hydraulic gradient coincides with the water surface (and is parallel to the pipe bed). If there is an increase in flow entering the sewer, the consequence will be that the depth of flow in the pipe will increase.

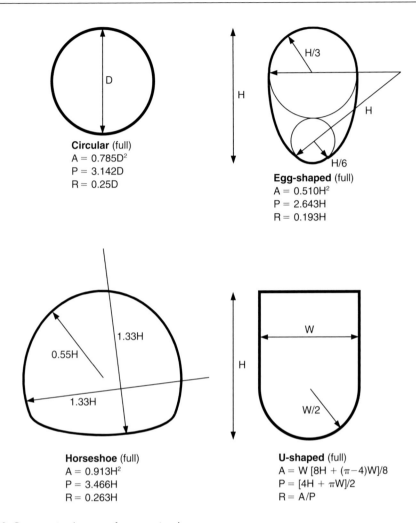

Figure 7.10 Geometric elements for non-circular sewers.

Now imagine the sewer in Figure 7.11b carrying the maximum flow rate (just less than full). If there is an increase in flow entering the sewer, the carrying capacity of the pipe can no longer be increased by a simple increase in depth. The capacity of a pipe is a function of diameter, roughness, and hydraulic gradient. To increase capacity, the only one of these that can change automatically (in response to "natural forces") is hydraulic gradient. It follows that the new hydraulic gradient must be greater than the old (equal to the gradient of the pipe), and the result – pipe surcharge – is shown on Figure 7.11b. Increased local losses at manholes (Table 7.3) will further increase energy losses.

If inflow continues to increase, the hydraulic gradient will increase. The obvious danger is that the hydraulic gradient will rise above ground level. This may cause manhole covers to lift and the flow to flood onto the surface – "manhole surcharge."

The transition from conditions in Figure 7.11a to those in Figure 7.11b is sudden. As shown in Table 7.6, maximum flow is carried when the pipe is less than full. If the pipe is running at this maximum level, a further slight increase in flow rate or small disturbance will result in a sudden increase in pipe flow depth, not only filling the pipe completely, but also establishing a hydraulic gradient in excess of S_o.

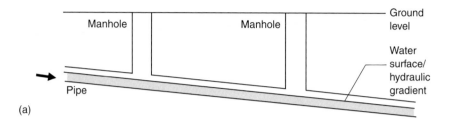

(a)

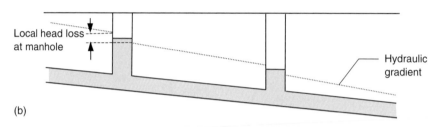

(b)

Figure 7.11 (a) Part-full pipe flow without surcharge. (b) Pipe flow with surcharge.

7.4.6 Velocity profiles

As discussed briefly in Sections 7.2.2 and 7.3.3, velocity varies over the cross section of a pipe. Velocity is at a minimum at the boundary and increases towards the centre. Maximum velocity may be at the surface when the flow depth is low, or a little below it when the flow depth is higher (Figure 7.12). The presence of a sediment bed in the invert of the pipe also affects the profile.

These profiles are significant when considering the transport of types of solids that are found only in specific parts of the cross section (floating solids close to the surface, or heavy solids close to the bed). This is discussed further in Chapters 9 and 16.

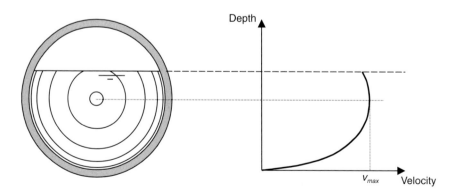

Figure 7.12 Velocity profile in a part-full pipe.

7.4.7 Minimum velocity

It is important for sewers to be able to convey wastewater or stormwater without long-term deposition of solid material. This is normally achieved by specifying a minimum mean velocity, a "self-cleansing" velocity, at a particular flow condition (e.g., pipe-full capacity) or for a particular frequency of occurrence (e.g., daily).

A common design criterion is to specify a minimum velocity when the pipe runs full. A value of 1 m/s is typical. The basis of this is that, although the pipe may never flow precisely full, mean velocities exceed the pipe-full velocity for flow rates greater than $0.5Qf$ (see Figure 7.8). This method has the advantage of simplicity in computation, but lacks precision. The other common approach is to specify self-cleansing velocities at some specified depth of flow (e.g., 0.75 m/s at $d/D = 0.75$).

EXAMPLE 7.9

The metric equivalent of "Maguire's Rule" states that an appropriate self-cleansing pipe slope is given by $S_o = 1/D$ (D in mm). Interpret this in terms of a minimum shear stress standard.

Solution

Assuming $\rho = 1000$ kg/m^3

$$\tau_o = \rho g R S_o = 1000g \cdot \frac{D}{4} \cdot \frac{1}{1000D} \approx 2.5 \, \text{N/m}^2$$

In the past, the velocity criterion has sometimes been relaxed for larger-diameter sewers (say > 750 mm). More recent research has shown this to be a mistake, and there is even evidence to suggest that higher self-cleansing velocities should be specified for larger diameters (see Chapter 16).

7.4.8 Minimum shear stress

Another potentially important parameter related to solid deposition/erosion is boundary shear stress. As water flows over the rigid boundary of the pipe channel, it exerts an average shear stress or drag τ_o (N/m^2) in the direction of flow, given by

$$\tau_o = \rho g R S_o \tag{7.21}$$

Substituting Equation 7.8 and assuming $D = 4R$ gives

$$\tau_o = \frac{\rho \lambda v^2}{8} \tag{7.22}$$

indicating that the applied shear stress varies linearly with friction factor and as the square of velocity of flow. Shear stress is not uniform around the boundary because of the variations in velocity.

7.4.9 Maximum velocity

Historically, many sewerage systems were designed so that velocity would not exceed a specified maximum. This was no doubt a sensible criterion for early brick sewers with relatively weak lime-mortar joints. However, research has shown that abrasion is not normally a problem with modern pipe materials. Perkins (1977) has suggested that no fixed maximum limit is required, but where velocities are high (>3 m/s) careful attention needs to be given to

- Energy losses at bends and junctions
- Formation of hydraulic jumps leading to intermittent pipe choking
- Cavitation (see Section 14.3.5) causing structural damage
- Air entrainment (significant when $v = \sqrt{5gR}$)
- The possible need for energy dissipation or scour prevention
- Safety provisions

7.5 OPEN-CHANNEL FLOW

7.5.1 Uniform flow

Part-full pipe flow (covered in the last section) is the most common condition in sewer systems. Design methods, as we have seen, tend to be related to those for pipes flowing full. However, in hydraulic terms, part-full pipe flow is a special case of open-channel flow, the basic principles of which are considered in this section.

The concept of normal depth, and the nature of the energy grade line and hydraulic grade line, explained in Section 7.4.1 for part-full pipe flow, apply to all cases of open-channel flow.

7.5.1.1 Manning's equation

A number of purely empirical formulas for uniform flow in open channels have been developed over the years, a common example of which is Manning's equation:

$$v = \frac{1}{n} R^{2/3} S_o^{1/2} \tag{7.23}$$

where n = Manning's roughness coefficient; typical values are given in Table 7.7 (units are not usually given, but to balance Equation 7.23 the units of $1/n$ must be $m^{1/3}$ s^{-1}). And S_o = bed slope (–).

If Manning's equation is plotted on the Moody diagram, it gives a horizontal line indicating the equation is only applicable to rough turbulent flow. Koutsoyiannis (2008) has developed a generalised form of the equation valid over a greater range of flow conditions.

Ackers (1958) has shown that if k_s/D is in the typical range of 0.001–0.01, the values of k_s and n are related (to within 5%) by the following relationship:

$$n = 0.012 k_s^{1/6} \tag{7.24}$$

where k_s is in mm, and n is as defined for Equation 7.23.

Table 7.7 Typical values of Manning's n

Channel material	n range
Glass	0.009–0.013
Cement	0.010–0.015
Concrete	0.010–0.020
Brickwork	0.011–0.018

7.5.2 Nonuniform flow

As stated, in uniform free surface flow, when the flow depth is normal, the total energy line, hydraulic grade line, and channel bed (or pipe invert) are all parallel. In many situations, however, such as changes in pipe slope, diameter, or roughness, nonuniform flow conditions prevail, and these lines are not parallel. In sewer systems, it is likely that there will be regions of uniform flow interconnected with zones of nonuniform flow. Methods of predicting conditions in nonuniform flow are presented in the following sections.

7.5.3 Specific energy

If the Bernoulli Equation 7.5 is redefined so that the channel bed is used as the datum (in place of a horizontal plane), we have, with reference to Figure 7.13,

$$\text{total head} = \frac{p}{\rho g} + \frac{v^2}{2g} + z = \frac{\rho g h}{\rho g} + \frac{v^2}{2g} + x = h + x + \frac{v^2}{2g}$$

This gives *specific energy*, E:

$$E = d + \frac{v^2}{2g} \tag{7.25}$$

or

$$E = d + \frac{Q^2}{2gA^2}$$

Thus, for a given flow rate, E is a function of depth only (as A is a function of d). Depth can be plotted against specific energy (Figure 7.14) showing that there are two possible depths at which flow may occur with the same specific energy. The depth that actually occurs depends on the channel slope and friction, and on any special physical conditions in the channel. At the critical depth, d_c, the specific energy is a minimum for a given Q.

7.5.4 Critical, subcritical, and supercritical flow

The non-dimensional Froude number (F_r) is given by

$$F_r = \frac{v}{\sqrt{g d_m}} \tag{7.26}$$

where d_m is the hydraulic mean depth, as defined in Section 7.4.2.

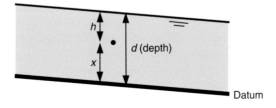

Figure 7.13 Terms for derivation of specific energy.

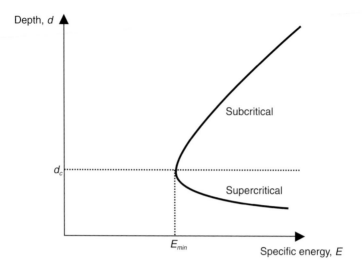

Figure 7.14 Depth against specific energy, for constant flow.

It can be shown that $F_r = 1$ at critical depth. If $F_r < 1$, flow is *subcritical*; the depth is relatively high and the velocity relatively low. This flow is sometimes referred to as "tranquil" flow. If $F_r > 1$, flow is *supercritical*; velocity is relatively high, and depth low. This flow is also called "rapid" or "shooting" flow.

The critical velocity v_c is given by

$$v_c = \sqrt{gd_m} \tag{7.27}$$

For rectangular channels, this leads directly to Equation 7.28 for critical depth:

$$d_c = \sqrt[3]{\frac{q^2}{g}} \tag{7.28}$$

where q is the flow rate per unit width (flow rate divided by channel width).

For circular channels, there is no simple analytical solution. As with the Colebrook-White equation, a solution can be achieved computationally, graphically, or by approximation.

EXAMPLE 7.10

What are the critical depth, velocity, and gradient in a 0.3 m circular sewer if the critical flow rate is 50 L/s? If the pipe is actually discharging 80 L/s, determine the depth of flow (assuming it to be uniform) and comment on the flow conditions ($k_s = 0.6$ mm).

Solution

The Butler-Pinkerton chart (Figure 7.9) can be used by estimating the point of intersection of the Q-curve (read from the right sloping upwards) with the $F_r = 1$ curve.

$$Q = 50\,\text{L/s}, D = 300\,\text{m}$$

This gives

$$d_c/D = 0.57, v_c = 1.2\,\text{m/s}, S_c = 1:200$$

The same charts can be used to find proportional depth of flow that is read at the intersection of the Q-curve and the relevant S-curve (read downwards sloping first right then left):

$$Q = 80\,\text{L/s}, S_f = 1:200$$

This gives

$$d/D = 0.84$$

As the intersection is above the F_r curve, flow must be subcritical.

Critical proportional depth can also be found using Straub's empirical equation:

$$\frac{d_c}{D} = 0.567 \frac{0.05^{0.506}}{0.3^{1.264}} = 0.57$$

Critical conditions ($F_r = 1$) have been plotted on the Butler-Pinkerton chart given (see Figure 7.9), giving critical depth and critical slope for each flow rate. Subcritical conditions exist in the region above this line and supercritical below it.

As a good approximation, the critical depth in a circular pipe (d_c) can be determined from the following empirical equation (Straub, 1978):

$$\frac{d_c}{D} = 0.567 \frac{Q^{0.506}}{D^{1.264}} \tag{7.29}$$

where $0.02 < d_c/D \leq 0.85$ (units for Q, m³/s). See Example 7.10.

Normal depth may be subcritical or supercritical. A *mild* slope is defined as one in which normal depth is greater than critical depth (so uniform flow is subcritical), and a *steep* slope is defined as one in which normal depth is less than critical depth (so uniform flow is supercritical).

Most sewer designs are for subcritical flow. Flow in the supercritical state is acceptable but has the disadvantage that if downstream conditions dictate the formation of subcritical conditions, a hydraulic jump will form. This effect is described later in the chapter. Close to critical depth ($0.7 < F_r < 1.5$), flow tends to be somewhat unstable with surface waves and would not be an appropriate position for a flow monitor (Section 19.5.4), for example (Hager, 2010).

7.5.5 Gradually varied flow

When variations of depth with distance must be taken into account, detailed analysis is required. This is done by splitting the channel length into smaller segments and assuming that the friction losses can still be accurately calculated using one of the standard equations such as Colebrook-White.

The general equation of gradually varied flow can be derived as

$$\frac{d(d)}{dx} = \frac{S_o - S_f}{1 - F_r^2} \tag{7.30}$$

where d is the depth of flow (m), x is the longitudinal distance (m), S_o is the bed slope (–), S_f is the friction slope, h_f/L as defined in Section 7.3.2 (–), and F_r is the Froude number (–).

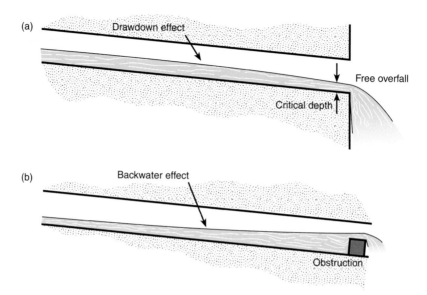

Figure 7.15 Drawdown and backwater effects (in a pipe).

Examples of gradually varied flow in sewer systems are shown in Figure 7.15. Figure 7.15a shows flow ending at a "free overall" – a sudden drop at the end of the pipe or channel such as the inflow to a pumping station. Close to the end of the pipe, conditions are critical, and for a long distance upstream the depth will be subject to a "drawdown" effect (provided flow is subcritical). The effect is most pronounced for flatter pipes.

Figure 7.15b shows flow backing up behind an obstruction. As flow approaches the obstruction, the depth increases: a "backwater" effect.

7.5.6 Rapidly varied flow

When supercritical flow meets subcritical flow, a discontinuity called a *hydraulic jump* is formed (Figure 7.16) at which there may be considerable energy loss.

There is no convenient analytical expression for the relationship between $d1$ and $d2$ on Figure 7.16 in a part-full pipe. Straub (1978), however, has developed an empirical approach using an approximate value for the Froude number:

$$F_{r1} = \left(\frac{d_c}{d_1}\right)^{1.93} \tag{7.31}$$

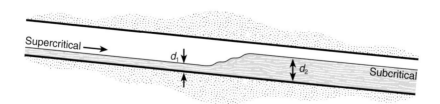

Figure 7.16 Hydraulic jump (in a pipe).

where F_{r1} is the upstream Froude number. For cases where $F_{r1} < 1.7$, the depth d_2 is given by

$$d_2 = \frac{d_c^2}{d_1} \qquad (7.32)$$

and for $F_{r1} > 1.7$,

$$d_2 = \frac{d_c^{1.8}}{d_1^{0.73}} \qquad (7.33)$$

Hydraulic jumps are generally avoided in drainage systems because they have the potential to cause erosion of sewer materials due to turbulence and the release of sewer gases (Section 17.5.1). If they are unavoidable, their position should be determined so that suitable scour protection can be provided.

EXAMPLE 7.11

A 600 mm pipe flowing part-full has a slope of 1.8 in 100 ($k_s = 0.6$ mm). Flow depth (in uniform conditions) is 0.12 m. Confirm that flow is supercritical using Equation 7.29. An obstruction causes the flow downstream to become subcritical; therefore, a hydraulic jump forms. Determine the depth immediately downstream of the jump.

Solution

$$\frac{d}{D} = \frac{0.12}{0.6} = 0.2$$

From Figure 7.8,

$$\frac{Q}{Q_f} = 0.1$$

From Figure 7.5, $Q_f = 900$ L/s, so $Q = 90$ L/s $= 0.09 \text{m}^3/\text{s}$:

$$\frac{d_c}{D} = 0.567 \frac{Q^{0.506}}{D^{1.264}} = 0.567 \frac{0.09^{0.506}}{0.6^{1.264}} = 0.32$$

so $d_c = 0.19$ m and $d < d_c$, so flow is supercritical. From Equation 7.31,

$$F_{r1} = \left(\frac{d_c}{d_1}\right)^{1.93} = \left(\frac{0.19}{0.12}\right)^{1.93} = 2.43$$

so from Equation 7.32,

$$d_2 = \frac{d_c^{1.8}}{d_1^{0.73}} = \frac{0.19^{1.8}}{0.12^{0.73}} = 0.24 \text{m}$$

PROBLEMS

7.1 A pipe flowing full, under pressure, has a diameter of 300 mm and roughness k_s of 0.6 mm. The flow rate is 100 L/s. Use the Moody diagram to determine the friction factor λ and the nature of the turbulence (smooth, transitional, or rough). Determine the friction losses in a 100 m length. Check this by determining the hydraulic gradient using the appropriate Wallingford chart. (Assume kinematic viscosity of $1.14 \times 10^{-6} \, \text{m}^2/\text{s}$.) [0.024, transitional, 0.8 m]

7.2 A pipe is being designed to flow by gravity. When it is full, the flow rate should be at least 200 L/s and the velocity no less than 1 m/s. Use a Wallingford chart to determine the minimum gradient for a 600 mm diameter pipe (k_s 0.6 mm). What will the pipe-full flow rate actually be? At what part-full depth would velocity go below 0.8 m/s? [0.18 in 100, 300 L/s, 180 mm]

7.3 A surcharged manhole with a 30° bend has a local loss constant $k_L = 0.5$. Determine the pipe length, L_E, equivalent to this local loss (assuming that it is independent of velocity) for a pipe with a diameter of 450 mm and k_s of 1.5 mm. If velocity is 1.3 m/s, is the assumption above valid?
(Assume kinematic viscosity = $1.14 \times 10^{-6} \, \text{m}^2/\text{s}$.) [8.7 m, yes]

7.4 A gravity pipe has a diameter of 600 mm, slope of 1 in 200, and when flowing full has a flow rate of 610 L/s and velocity of 2.2 m/s. Flowing part-full at a depth of 150 mm, what are the velocity, flow rate, area of flow, wetted perimeter, hydraulic radius, and applied shear stress?
[1.5 m/s, 80 L/s, 0.055 m², 0.63 m, 0.09 m, 4.4 N/m²]

7.5 A 300 mm diameter pipe is being designed for the following: maximum flow rate 80 L/s, minimum allowable velocity 1 m/s, roughness k_s 0.6 mm. Determine the following, using the Butler-Pinkerton chart:
   a. The gradient required based on the pipe running full
   b. The depth at which it will actually flow at that gradient
   c. The minimum velocity that will be achieved if the working flow rate is 10 L/s
   d. The gradient at which the sewer would need to be constructed to just ensure that the minimum velocity is achieved at that flow rate. [1:190, 250 mm, 0.78 m/s, 1:95]

7.6 A pipe, diameter 450 mm, k_s 0.6 mm, slope 1.5 in 100, is flowing part-full with a water depth of 100 mm. Are conditions subcritical or supercritical? [supercritical]

7.7 If a hydraulic jump takes place in the pipe in Problem 7.6, such that conditions upstream of the jump are as in 7.6, what would be the depth downstream of the jump? [0.18 m]

KEY SOURCE

Chadwick, A., Morfett, J., and Borthwick M. 2013. *Hydraulics in Civil and Environmental Engineering*, 5th edn, CRC Press.

REFERENCES

Ackers, P. 1958. *Resistance of Fluids in Channels and Pipes*, Hydraulics research paper No. 2, HMSO.
Barr, D.I.H. 1975. Two additional methods of direct solution of the Colebrook-White function, TN128. *Proceedings of the Institution of Civil Engineers, Part 2*, 59, December, 827–835.
Butler, D. and Pinkerton, B.R.C. 1987. *Gravity Flow Pipe Design Charts*, Thomas Telford.

Forty, E.J., Lauchlan, C., and May, R.W.P. 2004. *Flow Resistance of Wastewater Pumping Mains*. Report SR641, H.R. Wallingford.

H.R. Wallingford and Barr, D.I.H. 2006. *Tables for the Hydraulic Design of Pipes, Sewers and Channels*, 8th edn, Volume 1, Thomas Telford.

Hager, W.H. 2010. *Wastewater Hydraulics. Theory and Practice*. 2nd edn, Springer.

Hamill, L. 2011. *Understanding Hydraulics*, 3rd edn, Palgrave Macmillan.

Hydraulics Research. 1990. *Charts for the Hydraulic Design of Channels and Pipes*, 6th edn, Hydraulics Research, Wallingford.

Kay, M. 2008. *Practical Hydraulics*, 2nd edn, E & FN Spon.

Koutsoyiannis, D. 2008. A power-law approximation of the turbulent flow friction factor useful for the design and simulation of urban water networks. *Urban Water Journal*, 5(2), 107–115.

Lauchlan, C., Forty, J., and May, R. 2005. Flow resistance of wastewater pumping mains. *Proceedings of the Institution of Civil Engineers, Water Management*, 158(WM2), 81–88.

Perkins, J.A. 1977. *High Velocities in Sewers*, Report No. IT165, Hydraulics Research Station, Wallingford.

Romeo, E., Royo, C., and Monzon, A. 2002. Improved explicit equations for the estimation of friction factor in rough and smooth pipes. *Chemical Engineering Journal*, 86(3), 369–374.

Straub, W.O. 1978. A quick and easy way to calculate critical and conjugate depths in circular open channels. *Civil Engineering (ASCE)*, December, 70–71.

Chapter 8

Hydraulic features

Urban drainage systems contain a wide range of different hydraulic features. In this chapter, we explore these features in detail. Section 8.1 covers flow controls of various types, and 8.2 deals with weirs. Sewer drops and inverted siphons are explained in Sections 8.3 and 8.4, respectively. The last two sections (8.5 and 8.6) discuss gully spacing methods and culvert design.

8.1 FLOW CONTROLS

Flow controls can be used to limit the inflow to, or outflow from, elements in an urban drainage system. Typical uses include restricting the continuation flow at a combined sewer overflow (CSO) to the intended setting (Chapter 12), and controlling water level in tanks to ensure that the storage volume is fully exploited (Chapter 13). Flow controls can also be used to limit the rate at which stormwater actually enters the sewer system in the first place, deliberately backing up water in planned areas like car parks to prevent more damaging floods downstream in a city centre (Chapter 21).

Flow controls can be fixed, always imposing the same relationship between flow rate and water level, or adjustable, where the relationship can be changed by adjustment of the device.

8.1.1 Orifice plate

The simplest way of controlling inflow to a pipe is by an orifice plate. This forces the flow to pass through an area less than that of the pipe (Figure 8.1).

An orifice plate is fixed to the wall of the chamber where the inlet to the pipe is formed, and it usually either creates a smaller circular area (Figure 8.2a) or covers the upper part of the pipe area (Figure 8.2b). The area of the opening can only be changed by physically detaching and replacing or repositioning the plate.

Hydraulic analysis of an orifice is a simple application of the Bernoulli equation (Chapter 7). Comparing the total head at points 1 and 2 on Figure 8.1a, and assuming there is no loss of energy, we can write

$$\frac{p_1}{\rho g} + \frac{v_1^2}{2g} + z_1 = \frac{p_2}{\rho g} + \frac{v_2^2}{2g} + z_2$$

Now $p_1 = p_2 = 0$ (gauge pressure) and $z_1 - z_2 = H$, so, assuming that the velocity at 1 is negligible, we have

$$H = \frac{v_2^2}{2g}$$

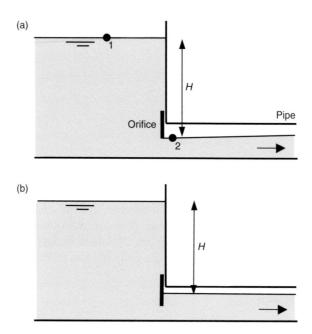

Figure 8.1 Orifice plate (vertical section): (a) free outfall and (b) drowned.

or

$$v_2 = \sqrt{2gH}$$

So, flow rate (Q) is as follows:

$$Q = A_o = \sqrt{2gH}$$

where A_o is the area of the orifice (m²).

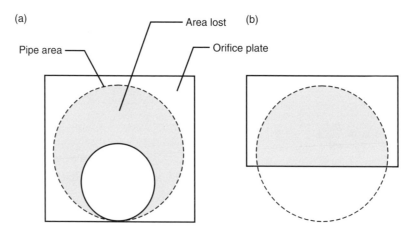

Figure 8.2 Orifice plate arrangements.

The assumptions made above affect the accuracy of the answer, and this is compensated for by an "orifice coefficient," C_d giving

$$Q_{actual} = C_d A_o \sqrt{2gH} \tag{8.1}$$

This is sometimes written as $Q = CA_o \sqrt{gH}$, where C includes the $\sqrt{2}$.

Conditions downstream may cause the orifice to be "drowned" – the downstream water level to be above the top of the orifice opening. H in Equation 8.1 should now be taken as the difference in the water levels, as on Figure 8.1b. The minimum value of H for which the orifice will be not drowned (H_{min}) can be determined from Figure 8.3 (in which D_o is the diameter of the orifice, and D is the diameter of the pipe). Use of Figure 8.3 requires the value of the water level in the pipe downstream (d), which can be calculated using the properties of

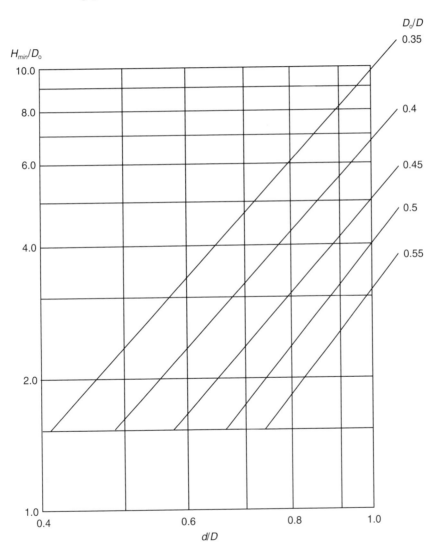

Figure 8.3 Chart to determine H_{min} for non-drowned orifice. (Based on Balmforth, D.J. et al. 1994. *Guide to the Design of Combined Sewer Overflow Structures*, Report FR 0488, Foundation for Water Research. With permission of Foundation for Water Research, Marlow.)

part-full pipe flow described in Section 8.4. Example 8.1 demonstrates the calculation. For $H < H_{min}$ the orifice will be drowned.

For an orifice that is not drowned, C_d in Equation 8.1 generally has a value between 0.57 and 0.6. For a drowned orifice, C_d can be estimated from

$$C_d = \frac{1}{1.7 - (A_o/A)} \tag{8.2}$$

where A is the flow area in the pipe (m^2).

8.1.2 Penstock

A penstock is an adjustable gate that creates a reduction in area at the inlet to a pipe in the manner of Figure 8.2b. The position of the penstock can be raised or lowered either manually or mechanically, by means of a wheel or a motorised actuator turning a spindle.

A penstock is more elaborate than an orifice plate. The advantage is that it can be adjusted to suit conditions – either, in the case of manual adjustment, to an optimum position to suit operational requirements, or, in the case of mechanical adjustment with remote control, to respond to changing requirements, perhaps as part of a real-time control system (described in Chapter 22).

EXAMPLE 8.1

The following arrangement is proposed. A tank will have an outlet pipe with diameter 750 mm, slope 0.002, and k_s 1.5 mm. Flow to the outlet pipe will be controlled by a circular orifice plate, diameter 300 mm. Determine the flow rate when water level in the tank is 2 m above the invert of the outlet.

Solution

First assume that the orifice is not drowned.
So $H = 2.0 - 0.3 = 1.7$ m (see Figure 8.1a).
Equation 8.1 gives $Q_{actual} = C_d A_o \sqrt{2gH}$. Assuming $C_d = 0.6$,

$$Q_{actual} = 0.6\pi \frac{0.3^2}{4}\sqrt{2\,g1.7} = 0.244\,m^3/s \text{ or } 244\,L/s$$

Now check the assumption that the orifice is not drowned, using Figure 8.4.
What is the flow rate in the outlet pipe flowing full?
Use the chart for $k_s = 1.5$ mm, or Table 7.1: $Q_f = 492$ L/s. This gives $Q/Q_f = 0.5$, so (from Figure 7.8) $d/D = 0.5$. For the orifice,

$$\frac{D_o}{D} = \frac{0.3}{0.75} = 0.4$$

Note that the depth of uniform part-full flow in the outlet pipe would be above the top of the orifice. This does not mean that the orifice is necessarily drowned since conditions are nonuniform. For

$$\frac{d}{D} = 0.5 \text{ and } \frac{D_o}{D} = 0.4, \quad \text{Figure 8.3 gives } \frac{H_{min}}{D_o} = 1.7, \text{ so } H_{min} \text{ is } 0.51\,m,$$

which is less than the actual H of 1.7 m, and so the orifice is not drowned. Therefore, the flow rate calculated above (244 L/s) applies.

(a) (b)

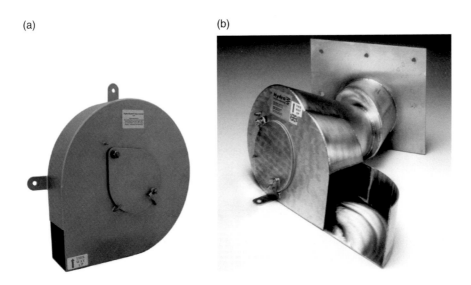

Figure 8.4 Vortex regulator for (a) stormwater and (b) wastewater. (Courtesy of Hydro International.)

Blockage is a potential problem with both an orifice plate and a penstock, and both should be designed to allow a 200 mm diameter sphere to pass.

8.1.3 Vortex regulator

In a similar way to an orifice plate or penstock, a vortex regulator constricts flow, usually with the purpose of exploiting a storage volume; the magnitude of the flow rate passing through the device depends on the upstream water depth. The regulator consists of a unit (see Figure 8.4) into which flow is guided tangentially, creating (at sufficiently high flow rates) a rotation of liquid inside the chamber. This creates a vortex with high peripheral velocities and large centrifugal forces near the outlet. These forces increase with upstream head until an air core occupies most of the outlet orifice creating a back-pressure opposing the flow.

This type of device has a distinctive head-discharge curve as shown in Figure 8.5. The "kickback" occurs during the formation of a stable vortex in rising flow. The shape of the curve depends on the detailed geometry of the regulator and the downstream conditions. During falling flow, there can be a slight kickback, but not as pronounced as for rising flow.

The main advantage of the vortex regulator is that it provides a degree of throttling only possible with an orifice of a much smaller opening. Hence, regulators can avoid problems of blockage or ragging that would occur on small-diameter orifices. In addition, it has been demonstrated that the discharge through a vortex regulator is not directly related either to its inlet or outlet cross-sectional area (Butler and Parsian, 1993). Therefore, the impact of any ragging of the openings is less pronounced than might be expected in comparison with an orifice. The same study also showed that, in all cases, the retention of single solids within the device led to an increase in the discharge until the solid was eventually ejected. An additional advantage is that, since the head-discharge curve is initially flatter than an equivalent orifice, some savings can be made in the volume of storage required for flow balancing.

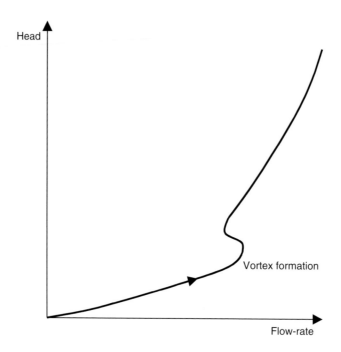

Figure 8.5 Head-discharge relationship for vortex regulator.

8.1.4 Throttle pipe

With a throttle pipe, it is the pipe itself that provides the flow control. Flow rate through the pipe depends on its inlet design, length, diameter, and hydraulic gradient. If the pipe is short, or has a steep slope or large diameter, it may be "inlet controlled"; the flow is controlled by an orifice equivalent to the diameter of the pipe. However, if the pipe is long, the friction loss along its length will be the governing factor. This condition is known as outlet control.

A common throttle pipe application is as the continuation pipe of a stilling pond CSO (to be described in detail in Chapter 12). Figure 8.6 shows that, when the weir is operating, the throttle pipe will be surcharged and thus flow rate will be related to the hydraulic gradient

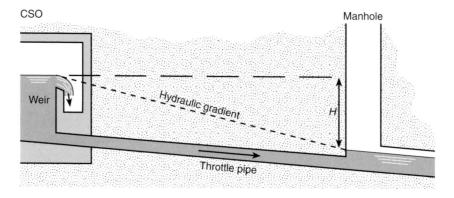

Figure 8.6 Throttle pipe (vertical section).

(not the pipe gradient). There will also be local losses (not shown on Figure 8.6) that may be significant. So, with reference to Figure 8.6,

$$H = S_f L + k_L \frac{v^2}{2g} \tag{8.3}$$

where S_f is the friction slope, given by pipe design chart/table (–), and $k_L(v^2/2g)$ is the local losses (as defined in Section 7.3.7 and Table 7.3)

In throttle pipe calculations, it is sometimes convenient to represent local losses by an equivalent pipe length, as explained in Section 7.3.7 (and demonstrated in Example 8.2).

To prevent blockage, the diameter of the throttle pipe should not be less than 200 mm. Clearly the length of the throttle pipe plays an important part in creating the flow control, and in design cases where it is inappropriate to reduce the diameter, the desired hydraulic control may be achieved by increasing the length (subject to restrictions in site layout). The diameter of an outlet-controlled throttle pipe will certainly be larger than that of the orifice plate giving equivalent flow control.

8.1.5 Flap valve

A flap valve is a hinged plate at a pipe outlet that restricts flow to one direction only. A typical application is at an outfall to receiving water with tidal variation in level. When the level of the receiving water is below the outlet, the outflow discharges by lifting the flap (Figure 8.7a). When the outlet is flooded, the flap valve prevents tidal water entering the sewer (Figure 8.7b). In these circumstances, any flow in the sewer will back up in the pipe, and if the energy grade line rises above the tidal water level, there will be outflow. The flap (which may have considerable weight) will then create a local head loss. Methods of estimating this loss are proposed by Burrows and Emmonds (1988).

8.1.6 Summary of characteristics of flow control devices

Table 8.1 gives a summary of the characteristics of the flow control devices considered above (excluding the flap valve, which has a different function from the other devices).

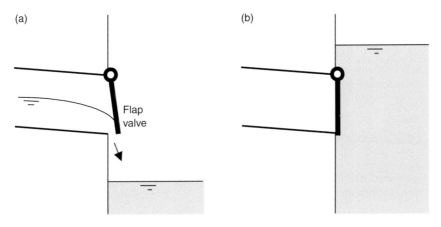

Figure 8.7 Flap valve operation.

Table 8.1 Summary of characteristics of flow control devices

Orifice plate	Simple, cheap.
	Flow control can only be adjusted by physically detaching and replacing or repositioning the plate.
Penstock	Easily adjusted. When automated, can be used for real-time control.
Vortex regulator	Controls flow with larger opening than equivalent orifice.
Throttle pipe	Larger opening than equivalent orifice.
	Significant construction costs.

EXAMPLE 8.2

A throttle pipe carrying the continuation flow from a stilling pond CSO will have a length of 28 m. When the weir comes into operation, the water level in the CSO will be 1.8 m above the water level at the downstream end of the throttle pipe, 1.4 m above the soffit at the pipe inlet. Under these conditions the continuation flow (in the throttle pipe) should be as close as possible to 72 L/s. The roughness, k_s, of the pipe material is assumed to be 1.5 mm, and local losses are taken as $1.4(v^2/2g)$. Determine an appropriate diameter for the throttle pipe. Confirm that this throttle pipe is not "inlet controlled."

If, as an alternative, there were no throttle pipe and flow control was achieved by an orifice, what would be its diameter (assuming that it would not be drowned)?

Solution

Solve by trial and error—a 200 mm pipe gives the following.
   Represent local losses by equivalent length:

$$\frac{k_s}{D} = \frac{1.5}{200} = 0.0075$$

assume rough turbulent, Moody diagram (Figure 7.4) gives $\lambda = 0.034$

so from equation 7.15, $\dfrac{L_E}{D} = \dfrac{k_L}{\lambda} = \dfrac{1.4}{0.034} = 41$

therefore $L_E = 8\,\text{m}$

total length $= 28 + 8 = 36$ m
hydraulic gradient $= 1.8/36 = 0.05$

   Chart for $k_s = 1.5$ mm, or see Table 7.1, gives flow rate (for 200 mm diameter) of 75 L/s. So 200 mm diameter is suitable.

   If the throttle pipe is inlet controlled, control comes solely from the inlet acting as an orifice. Apply orifice formula (assuming $C_d = 0.59$):

$$Q_{actual} = C_d A_o \sqrt{2gH} = 0.59 \times \pi \frac{0.2^2}{4} \sqrt{2g1.4} = 97\,\text{L/s}$$

so control does not solely come from the inlet: the pipe is not inlet controlled. Consider use of an orifice plate

$$Q_{actual} = C_d A_o \sqrt{2gH}$$

What orifice diameter would give the same control as the throttle pipe?

$$0.075 = 0.59 \times \pi \frac{D_o^2}{4} \sqrt{2g1.4} \text{ giving } D_o = 0.175 \text{ m (unacceptably small)}$$

8.2 WEIRS

8.2.1 Transverse weirs

Standard analysis, using the Bernoulli equation, of flow over a rectangular weir gives the theoretical equation for the relationship between flow rate and depth as

$$Q_{theor} = \frac{2}{3} b \sqrt{2g} H^{3/2}$$

where Q_{theor} is the flow-rate (m³/s), b is the width of weir (Figure 8.8) (m), and H is the height of water above weir crest (Figure 8.8) (m).

Several assumptions are made in the analysis, and it is necessary to introduce a discharge coefficient to relate the theoretical result to the actual flow rate:

$$Q = C_d \frac{2}{3} b \sqrt{2g} H^{3/2} \tag{8.4}$$

where C_d = discharge coefficient. With this form of equation, C_d has a value between 0.6 and 0.7; C_d is sometimes written so that it incorporates some of the other constants in the equation.

The value of C_d and the accuracy of the equation depend partly on whether the weir crest fills the whole width of a channel or chamber, or is a rectangular notch that forces the flow to converge horizontally. For the former, an empirical relationship by Rehbock can be used:

$$Q = C_d \frac{2}{3} b \sqrt{2g} [H + 0.0012]^{3/2} \tag{8.5}$$

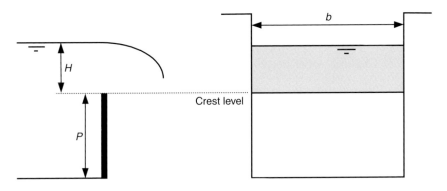

Figure 8.8 Rectangular weir.

in which

$$C_d = 0.602 + 0.0832 \frac{H}{P}$$

and where P is the height of weir crest above the channel bed (Figure 8.8) (m).

8.2.2 Side weirs

The flow arrangements for side weirs are more complex than for transverse weirs because flow rate in the main channel is decreasing with length (as some flow is passing over the weir) and conditions are nonuniform.

The possible flow conditions at a side weir are normally classified into five types as illustrated in Figure 8.9. These conditions can be analysed by assuming that specific energy is constant along the main channel. The standard curve of depth against specific energy (for constant flow rate), introduced as Figure 7.14, is reproduced as Figure 8.10 with the curve for a slightly decreased flow rate added.

The classification of flow types is based partly on the slope of the channel. *Mild* and *steep* slopes are defined in Section 7.5.4.

Type I

Channel slope: mild   Weir crest below critical depth

Depth along the weir is supercritical as a result of the fact that the weir crest is below critical depth. As the flow rate decreases, we move from point 1 to 2 in Figure 8.10, and the depth (d) decreases.

Type II

Channel slope: mild   Weir crest above critical depth

Depth along the weir is subcritical as a result of the fact that the weir crest is above critical depth. As the flow rate decreases, we move from point 3 to 4 in Figure 8.10, and the depth (d) increases.

Type III

Channel slope: mild   Weir crest below critical depth

At the start of the weir, conditions are as Type I. However, conditions downstream are such that a hydraulic jump forms before the end of the weir.

Type IV

Channel slope: steep   Weir crest below critical depth

Conditions are similar to those for Type I, except that supercritical conditions would prevail in the main channel in any case because it is steep.

Type V

Channel slope: steep   Weir crest below critical depth

Conditions are similar to those for Type III, except that supercritical conditions prevail before the start of the weir because the channel is steep.

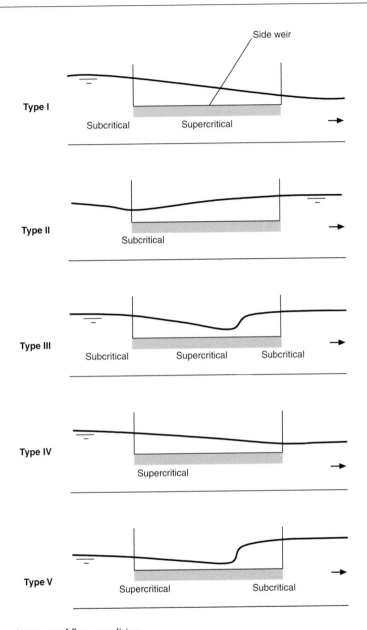

Figure 8.9 Side weir: types of flow condition.

For all types, the variation of water depth with distance, derived from standard expressions for spatially varied flow, is given by

$$\frac{d(d)}{dx} = \frac{Q_c d[-(dQ_c/dx)]}{gB^2 d^3 - Q_c^2} \tag{8.6}$$

where d is the depth of flow (m), x is the longitudinal distance (m), Q_c is the flow rate in the main channel (m³/s), and B is the width of the main channel (m).

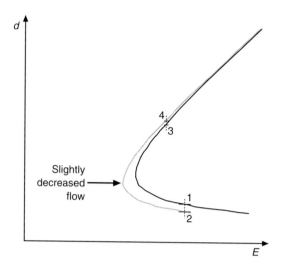

Figure 8.10 Flow parallel to side weir: depth against specific energy.

$[-(dQ_c/dx)]$ is the rate at which flow rate in the main channel is decreasing – that is, the rate at which flow passes over the weir per unit length of weir. Therefore, Equation 8.4 for flow over a weir can be adapted to give

$$-\frac{dQ_c}{dx} = C_d \frac{2}{3} \sqrt{2g} H^{3/2} \qquad (8.7)$$

Note that H is water depth relative to the weir crest, whereas d is water depth relative to the channel bed. In a double-side weir arrangement (one weir on either side of the main channel), the right-hand side of Equation 8.7 is doubled. It has been found that the Rehbock expression, Equation 8.5, gives appropriate values of C_d for side weirs, even though it was originally proposed for transverse weirs.

For methods of solution of these equations, see Chow (1959) and Balmforth and Sarginson (1978). May et al. (2003) present a simple formula for total flow discharged over a side weir, backed up by charts for determining coefficients. They also provide general guidance on design and construction. For high side weir overflows (Type II flow conditions), design calculations can be based on charts presented by Delo and Saul (1989). One of these is given as Figure 8.11 where Q_u is the inflow (m³/s), Q_d is the continuation flow rate (m³/s), B_u is the upstream chamber width (m), B_d is the downstream chamber width (m), P_u is the upstream weir height (m), P_d is the downstream weir height (m), Y_u is the upstream water depth (m), Y_d is the downstream water depth (m), and L is the length of weir (m).

8.3 SEWER DROPS

Simple drop manholes were introduced in Chapter 6. We now consider more significant cases of sewer drops, arising for two main reasons. The first is that the vertical drop may have been specified as an alternative to using much steeper slopes in the approach sewer; this would avoid high sewer velocities, and the risks outlined in Chapter 7. The second is that topography, or the use of a deep tunnel for collection, means that significant vertical drops are unavoidable. A number of alternative drop arrangements can be used.

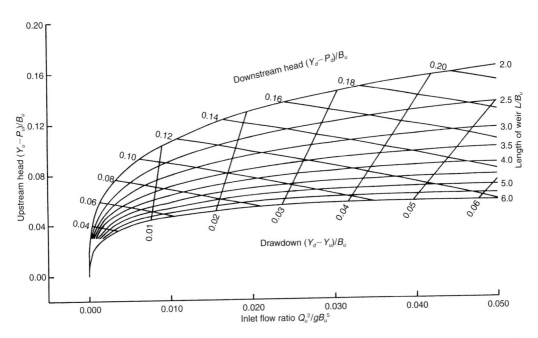

Figure 8.11 Chart for side weir design: double-side weir (horizontal weirs and channel bed), $Q_d/Q_u = 0.1$, $B_d/B_u = 1$, $P_u/B_u = P_d/B_u = 0.6$, $n = 0.010$. (Reproduced from Delo, E.A. and Saul, A.J. 1989. *Proceedings of the Institution of Civil Engineers*, Part 2, 87, June, 175–193. With permission of Thomas Telford Publishing.)

In a *plunge-flow* drop shaft, the flow falls down freely as a jet, plunging into a pool at the base of the shaft, or hitting the opposite wall of the shaft, or landing at the entry to the downstream sewer itself (Chanson, 2004a). Air entrainment can be a significant issue. Air may be entrained by the free-falling jet, or when the jet lands in the pool, or hits the shaft wall, or if a hydraulic jump forms in the downstream sewer. If there is not sufficient ventilation to compensate for the air entrainment, sub-atmospheric conditions may occur in some parts of the system, causing backing up in manholes and sewers. Air entrainment may also increase the risk of downstream choking caused by a sudden transition to pressurised flow, or cause a reduction in pipe discharge capacity due to bulking of the flow by the entrained air (Granata et al., 2011).

The main alternative to the plunge-flow drop shaft, particularly for drops greater than 7–10 m, is the vortex drop shaft.

8.3.1 Vortex drop shafts

In a vortex drop shaft, fluid flows helically downwards in contact with the inside of a circular wall, with a core of air at the centre. The intake must induce vortex flow in the drop shaft. A *scroll intake* is a common arrangement, with design traditionally based on the analysis by Ackers and Crump (1960). The inlet channel is volute shaped in plan (Figure 8.12), inducing a vortex in the vertically downwards flow.

Del Guidice and Gisonni (2010), in the context of the historical need for drop shafts to cope with level differences in the city of Naples, discuss alternative inlet arrangements, and present hydraulic analysis and experimental results relating to scroll intakes. Similarly, Echavez and Ruiz (2008), this time based on experiences in Mexico City, give an overview of drop shaft analysis and also present a hydraulic study of scroll intakes.

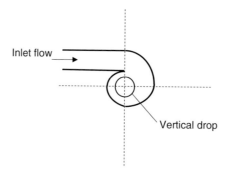

Figure 8.12 Scroll intake to vortex drop.

An alternative arrangement, common in the United States and being used increasingly in the United Kingdom, is the *tangential intake*. This is more compact and simpler to construct than the scroll intake. Riisnaes et al. (2014) compare the dimensions and requirements for a tangential intake with those for a scroll intake, for a vortex drop located at a very restricted site in London. They explain the advantages of a tangential intake for this case.

We now consider technical aspects of the tangential vortex intake. The typical layout is shown in plan on Figure 8.13a and in vertical section on Figure 8.13b. Hydraulic analysis, based on theory and physical modelling, is set out by Yu and Lee (2009), and is summarised here.

As can be seen from Figure 8.13, flow enters via a tapering and steeply sloping channel. At relatively small flow rates, water depth in the approach channel and the sloping intake is determined by the fact that critical depth (introduced in Chapter 7) occurs at section 1 (the start of the steeply sloping section). Conditions within the inlet upstream of this point

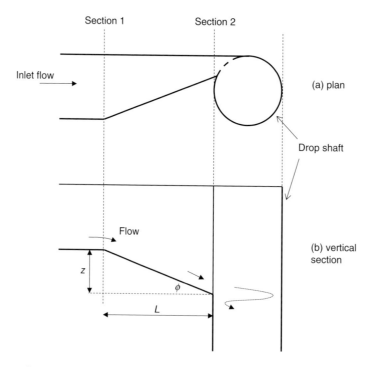

Figure 8.13 Tangential vortex intake.

are subcritical, and conditions downstream are supercritical. This depth can be determined from the standard expression for critical depth in a rectangular channel (Equation 7.28), given here as Equation 8.8:

$$d_{c1} = \sqrt[3]{\frac{q_1^2}{g}} \qquad (8.8)$$

where d_{c1} is the critical depth at section 1 (m), q_1 is the flow rate per unit width at section $1 = Q/B_1$ (m²/s), Q is the flow rate (m³/s), and B_1 is the channel with at section 1 (m).

At much larger flow rates, critical depth occurs at point 2. This drowns the control at point 1, and conditions throughout are subcritical. The expression for this depth is similar (Equation 8.9), but the vertical angle of the inlet channel must be included because depth is measured perpendicular to the bed.

$$d_{c2} = \sqrt[3]{\frac{q_2^2}{g \cos\phi}} \qquad (8.9)$$

where d_{c2} is the critical depth at section 2 (m), q_2 is the flow rate per unit width at section $2 = Q/B_2$ (m²/s), B_2 is the channel width at section 2 (m), and ϕ is the vertical slope angle of intake channel (see Figure 8.13).

Based on specific energy in the approach channel (see Section 7.5.3) and assuming no overall loss of energy, the flow rate at which this shift in control occurs (Yu and Lee, 2009) can be determined from Equation 8.10:

$$Q_c = \frac{\sqrt{g}B_2[2z/3]^{3/2}}{((\cos\phi)^{2/3} - [B_2/B_1]^{2/3})^{3/2}} \qquad (8.10)$$

where Q_c is the flow rate at which the shift in control occurs (m³/s), and for z refer to Figure 8.13.

Conditions at the top of the vortex may cause disruption to the flow in the inlet channel as a result of re-entering the inlet or disturbing the inflow jet. Based on the geometry of the inlet, Yu and Lee (2009) give the maximum flow rate for which "free drainage" (no disruption) can be assumed, as

$$Q_f = \left[\tan\phi \frac{\pi D}{1 - B_2/D}\right]^{3/2} \sqrt{g}B_2(\cos\phi)^2 \qquad (8.11)$$

where Q_f is the maximum flow rate for which free discharge can be assumed (m³/s), and D is the diameter of drop shaft (m).

In design, Q_c should be less than Q_f so that the shift in control can occur when there is free discharge (without disruption to flow in the inlet channel caused by conditions at the top of the drop shaft), and Q_f should exceed the design maximum flow rate (Q_{max}).

For sizing of the drop shaft, Yu and Lee (2009) propose using $k = 1.2$ within Equation 8.12:

$$D = k\left[\frac{Q_{max}^2}{g}\right]^{1/5} \qquad (8.12)$$

A suitable value for B_2 is given as $0.25D$.

EXAMPLE 8.3

a. A vortex drop has a tangential intake with an inlet width (B_1) of 2.2 m. The design maximum flow rate is to be taken as 11 m³/s. The length available for the intake channel is 8 m. Propose acceptable dimensions for the intake.
b. For your selected dimensions, determine a known depth in the intake channel at the design flow rate. For a flow rate 20% greater than the design flow rate, determine a known depth.

Solution

a. Basing the drop shaft diameter on Equation 8.12, with $k = 1.2$, we have

$$D = 1.2 \left[\frac{11^2}{g} \right]^{1/5} = 1.98 \text{ m, so make } D = 2 \text{ m}$$

Base the value of B_2 on $0.25D$, so B_2 should be 0.5 m.

Trying a value for z of 4 m, and using the whole of the 8 m for L, gives $\phi = \tan^{-1}(4/8) = 26.6°$.

Equation 8.10 gives the value of Q_c as 16.5 m³/s.

And Equation 8.11 gives the value of Q_f as 10.8 m³/s.

This is not a good design because Q_f is not only less than Q_c, but it is also less than Q_{max}.

To increase the value of Q_f, we could increase the value of D, but that would be a significant and costly change to the design. Let's see first if increasing ϕ could be effective. We do this by decreasing L; as we do not need an increase in Q_c we also reduce z.

Try $L = 4.25$ m with $z = 3$ m. This gives $\phi = 35.2°$.

Equation 8.10 now gives the value of Q_c as 12.5 m³/s.

And Equation 8.11 gives the value of Q_f as 15 m³/s.

This is a satisfactory design because Q_c is less than Q_f, and Q_f exceeds Q_{max}.

b. The design flow rate of 11 m³/s is less than Q_c, so the depth control is at section 1 in Figure 8.13.

$$q_1 = \frac{Q}{B_1} = \frac{11}{2.2} = 5 \text{ m}^2/\text{s}$$

$$d_{c1} = \sqrt[3]{\frac{5^2}{g}} = 1.37 \text{ m}$$

The known depth for a flow rate of 11 m³/s is at section 1: 1.37 m.

A 20% increase gives $Q = 13.2$ m³/s. This is greater than Q_c, so the depth control moves to section 2:

$$q_2 = \frac{Q}{B_2} = \frac{13.2}{0.5} = 26.4 \text{ m}^2/\text{s}$$

$$d_{c2} = \sqrt[3]{\frac{26.4^2}{g \cos 35.2}} = 4.43 \text{ m}$$

The known depth for a flow rate of 13.2 m³/s is at section 2: 4.43 m (perpendicular to the bed).

Although air entrainment is not considered to be as significant within a vortex drop shaft as it is in a plunge-flow shaft, it can still be significant, and sufficient airflow is clearly an important element in maintaining a stable air core.

The outlet arrangement for a vortex drop shaft must fulfil a number of functions. It must direct the outflow into a horizontal conduit, it must facilitate the dissipation of energy, and it must allow de-aeration to take place as flow enters the outlet pipe (Zhao et al., 2006). Outlet arrangements are heavily dependent on the context of the sewer drop, and, like other aspects of the design of a specific vortex drop shaft of significant size, are often best determined through physical modelling (Riisnaes et al., 2014).

There are other hydraulic considerations in vortex drop shaft design. The risk of blockage must be considered, and this may define a minimum channel width (B_2, the width at section 2 on Figure 8.13a, in the case of a tangential intake). This, in turn, may define the minimum drop shaft diameter. Also, the approach conditions in the channel leading to the intake should be as stable as possible with no bends, steps, or other obstacles for a significant length upstream of the intake.

8.3.2 Other sewer drop arrangements

A variation on the concept of the vortex drop shaft is a hybrid arrangement described by Andoh et al. (2008), which uses an air intake control system to allow both vortex and vertical pipe-full flow, and removes the need for a vortex-inducing inlet.

Granata (2016) discusses the use of a *drop shaft cascade*: a series of plunge-flow drop shafts connected by lengths of sewer. The design is presented as an optimisation problem involving selection of the configuration and number of drop shafts that minimise construction cost. This can be a more efficient solution than a single drop shaft for the same total drop height, provided the sewer drop does not need to be at a single location.

Another arrangement for a sewer drop made up of a number of vertical drops is a stepped cascade. This is a simple approach to energy dissipation but requires significantly more plan area than the use of drop shafts. Flow passes down a series of simple connected steps, each providing a small vertical drop. The nature of the flow conditions is dependent on the value of the flow rate. At relatively low flows, a nappe flow regime exists in which there is a free overfall from each step to the next (Figure 8.14). Assuming subcritical conditions upstream, the depth at the edge of each step is equal to the critical depth (Section 7.5.4). On the next step, the impact of the free-falling jet (or *nappe*) is followed by a hydraulic jump, then subcritical flow, then critical flow again at the edge of that step, and so on. At larger flow rates, a skimming flow regime is established where flow hits the edge of each step but does not form a particular flow pattern on each individual step. Principles for sizing cascade steps are given by Chanson (2004b).

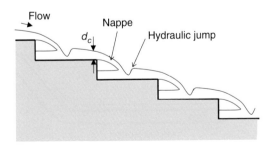

Figure 8.14 Stepped cascade (nappe flow regime).

8.4 INVERTED SIPHONS

Inverted siphons carry flows under rivers, canals, roads, and so on (e.g., Figure 8.15). They are necessary when this crossing cannot be made by means of a pipe-bridge, or by having the whole sewer length at a lower level. Unlike normal siphons, inverted siphons do not require special arrangements for filling; they simply fill by gravity. However, they do present some problems and are avoided where possible.

Inverted siphons are an interesting case from a hydraulic point of view, and are dissimilar from most other flow conditions in sewers. As discussed in Chapter 7, the majority of sewers flow part-full, and when the flow rate is low, the depth is low. When sewer flows are pumped, the pipe flows full and the pumps tend to deliver the flow at a fairly constant rate, but not continuously (as described in Chapter 14). In contrast, inverted siphons flow full, and they flow continuously. At low flow, the velocity can be very low, which, unfortunately, creates the ideal conditions for sediment deposition.

The most important aim in design is to minimise silting. Some silting is virtually inevitable at low flows, but at higher flows the system should be self-cleansing. It is normally assumed that this will be achieved if the velocity is greater than 1 m/s. (This subject is considered further in Chapter 16.) The higher the velocity, the lower is the danger of silting.

EXAMPLE 8.4

An existing single-pipe inverted siphon, carrying wastewater only, is to be replaced with a twin-barrelled siphon because of operational problems caused by sedimentation. The required length is 70 m; available fall (invert to invert) is 0.85 m. Determine the pipe sizes required for an average dry weather flow (DWF) of 90 L/s and a peak flow of 3 DWF. Assume the inlet head loss is 150 mm, the self-cleansing velocity is 1 m/s, and $k_s = 1.5$ mm.

Solution

One approach: use one pipe to carry DWF, second pipe to carry excess.
Available hydraulic gradient = (0.85−0.15)/70 = 0.01.
Use chart for $k_s = 1.5$ mm, or Table 7.1 for pipe calculations.
A 300 mm pipe carries 98 L/s at velocity 1.38 m/s. Velocity is sufficient.
Excess flow = (3 × 90) − 98 = 172 L/s.
A 375 mm pipe carries 177 L/s at velocity 1.6 m/s. Velocity is sufficient.
So, use pipe diameters 300 mm and 375 mm.

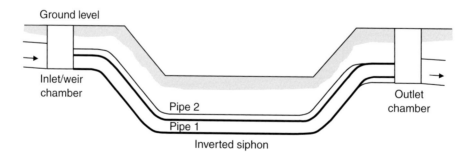

Figure 8.15 Inverted siphon for wastewater, vertical section (schematic).

Many siphons consist of multiple pipes as a means of minimising siltation. The low flows are carried by one pipe, smaller than the sewers on either side of the siphon. At higher flows, this pipe is self-cleansing, and an arrangement of weirs allows overflow into other pipes. In separate systems, two pipes for wastewater are usually enough (Figure 8.15); in combined sewers, a third much larger pipe is usually needed.

Other devices for avoiding siltation are sometimes needed. On small systems, a penstock upstream can be used to back up the flow and create an artificial flushing wave. Silt can be removed directly by providing penstocks or stop boards for isolating sections of pipe, and access for removing silt. An independent washout chamber can be provided and used in conjunction with a system for pumping out silt.

8.5 GULLY SPACING

Several approaches to establishing the required spacing of road gullies have been proposed. The simplest are mentioned in Chapter 6, but in this section more sophisticated methods are outlined.

8.5.1 Road channel flow

The typical geometry of flow in a road channel is as given in Figure 8.16.

For channels of shallow triangular section, Manning's equation (Equation 7.23) can be simplified by assuming the top width of the channel flow (B) equals the wetted perimeter (P), to give

$$Q = 0.31 C y^{8/3} \qquad (8.13)$$

where Q is the channel flow rate with "channel criterion" C (fixed for the road):

$$C = \frac{z S_o^{1/2}}{n} \qquad (8.14)$$

where y is the flow depth (m), z is the side slope (1:z), S_o is the longitudinal slope (–), and n is the Manning's roughness coefficient (m$^{-1/3}$s).

Manning's n for roads ranges from 0.011 for smooth concrete to 0.018 for asphalt with grit. Example 8.5 demonstrates use of these equations.

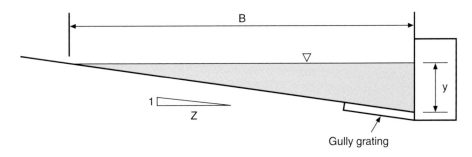

Figure 8.16 Geometry of road channel flow (exaggerated vertical scale).

8.5.2 Gully hydraulic efficiency

The hydraulic efficiency of a gully depends on the depth of water in the channel immediately upstream, the width of flow arriving, and the geometry of the grating. A typical efficiency curve is given in Figure 8.17. This shows that at low flows, gullies are approximately 100% efficient and all flow is captured. Once the approach flow Q exceeds $\overline{Q}_l$, efficiency drops off rapidly. When approach flow is plotted against captured flow (as in Figure 8.18), it is clear that the captured flow $\overline{Q}_l$ corresponding to 100% efficiency is not the maximum flow that the gully can capture. Higher approach flows result in an increase in captured flow due to the greater flow depths over the grating. Thus, the capacity of a gully Q_c can be increased by allowing a small bypass flow. May (1994) suggests that an optimum value is 20%.

EXAMPLE 8.5

Determine the flooded width of a concrete road ($n = 0.012$) when the flow rate is 20 L/s. The road has a longitudinal gradient of 1% and a crossfall of 1:40.

Solution

From Equation 8.14, calculate the channel criterion:

$$C = \frac{40 \times 0.01^{\frac{1}{2}}}{0.012} = 333.3$$

Rearranging Equation 8.13 gives

$$y = \left(\frac{Q}{0.31C}\right)^{3/8} = \left(\frac{0.02}{0.31 \times 333.3}\right)^{3/8} = 0.041\,\text{m}$$

Thus, the depth of flow is 41 mm leading to a width of flow $B = yz = 0.041 \times 40 = 1.62$ m.

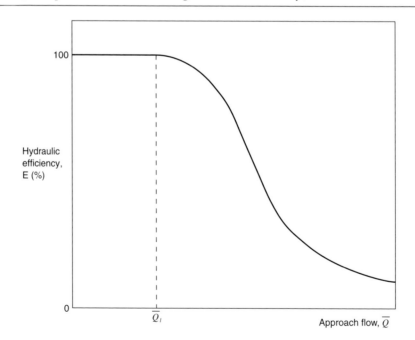

Figure 8.17 Typical gully efficiency curve. (After Davis, A. et al. 1996. *Journal of Chartered Institution of Water and Environmental Management*, 10, April, 118–122.)

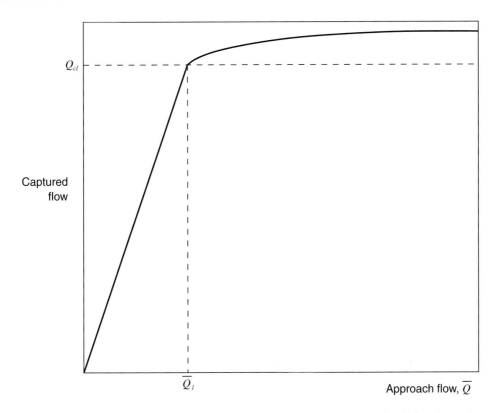

Figure 8.18 Relationship between approaching and captured flow for a typical gully. (After Davis, A. et al. 1996. *Journal of Chartered Institution of Water and Environmental Management*, 10, April, 118–122.)

Thus, the hydraulic capture efficiency E for an individual gully grating is

$$E = \frac{\overline{Q_c}}{\overline{Q}} \tag{8.15}$$

where E is a function of grating type, water flow width, road gradient, and crossfall.

Data on the efficiency of a number of grating types can be found in TRRL Contractor Report CR2 (Hydraulics Research Station, 1984). An example is given in Table 8.2.

Table 8.2 Example gully efficiencies (E) at standard 1:20 crossfall

Flow width	Longitudinal gradient (1:X)				
B (m)	20	30	50	100	300
0.5	100	100	100	100	100
0.75	87	94	97	99	100
1.0	63	75	82	93	96
1.5	33	43	47	60	76

Source: Adapted from Hydraulics Research Station 1984. *The Drainage Capacity of BS Road Gullies and a Procedure for Estimating Their Spacing.* TRRL Contractor Report CR2, Transport and Road Research Laboratory, Crowthorne.

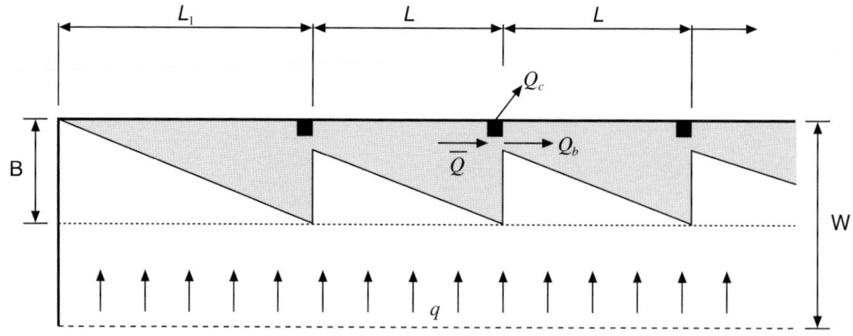

Figure 8.19 Spacing of initial and intermediate gullies.

8.5.3 Spacing

The basic approach to gully hydraulic design is to make sure that they are sufficiently closely spaced to ensure that the flow-spread in the road channel is lower than the allowable width (B). Figure 8.19 shows a schematic of the flow conditions along a road of constant longitudinal gradient and crossfall subject to constant inflow. Gullies are spaced at a distance L apart, except the first gully that is at a distance of L_1. The inflow per unit length q is generated by constant intensity rainfall. The flow bypassing each gully must be included in the flow arriving at the next inlet.

8.5.3.1 Intermediate gullies

The maximum flooded width B and flow rate $\overline{Q}$ occurs just upstream of a road gully and consists of the sum of the runoff $Q_r = qL$ and the bypass flow Q_b from the previous gully:

$$\overline{Q} = Q_b + Q_r$$

And from the Rational Method equation (described more fully in Chapter 10),

$$Q_r = iWL$$

where i is the rainfall intensity for a storm duration equal to the time of entry and assuming complete imperviousness (runoff coefficient, $C = 1$), and W is the road width contributing flow to the gully. The flow arriving at the gully is either captured or bypasses it, so

$$\overline{Q} = Q_b + Q_c$$

$$Q_c = Q_r$$

Thus, the captured flow is equal to the runoff generated between gullies. Hence, substituting Equation 8.10 gives

$$E\overline{Q} = iWL$$

$$L = \frac{E\overline{Q}}{iW} \tag{8.16}$$

8.5.3.2 Initial gullies

The most upstream gully in the system is a special case as it does not have to handle carryover from the previous gully; thus, $Q_b = 0$ and $\bar{Q} = Q_r$, so

$$L_1 = \frac{\bar{Q}}{i\mathrm{W}} \tag{8.17}$$

Example 8.6 shows how gully spacing can be calculated.

A second special case is the terminal gully that can have no carryover. These act as weirs under normal conditions and as orifices under large water depths. Methods to design such gullies are given in Contractor Report CR2 (Hydraulics Research Station, 1984).

8.5.3.3 Potential optimisation

A UK study by Doncaster et al. (2012) into the potential for gully optimisation suggests that it may be appropriate for gullies to be spaced more widely than traditionally specified. In a particular study catchment, they found there was a good case for a 50% reduction in the number of gullies, provided gully gratings could be made to perform more effectively.

8.6 CULVERTS

8.6.1 Culverts in urban drainage

Where urban drainage systems include open channels, culverts may be needed to carry the flow under a road or railway. Culverts are also common on natural watercourses, though the practice of culverting long lengths is now recognised as having a negative impact on amenity and biodiversity. Comprehensive practical advice on culvert design is provided by Balkham et al. (2010).

EXAMPLE 8.6

Determine the spacing of the initial and subsequent gullies on a road in the London area. The road is 5 m wide with a crossfall of 1:20 and a longitudinal gradient of 1%. The road surface texture suggests a Manning's n of 0.010 should be used. A design rainfall intensity of 55 mm/h is to be used at which the flood width should be limited to 0.75 m.

Solution

Allowable flow depth = 0.75/20 = 0.0375 m
  Channel criterion (Equation 8.14),

$$C = \frac{20 \times 0.01^{1/2}}{0.010} = 200$$

Maximum flow rate (Equation 8.13),

$$\bar{Q} = 0.31 \times 200 \times 0.0375^{8/3} = 0.010\,\mathrm{m^3/s}$$

Thus, the spacing of the initial gully should be (Equation 8.17)

$$L_1 = \frac{0.010 \times 3600 \times 10^3}{55 \times 5} = 131 \, \text{m}$$

Read from Table 8.2, $E = 0.99$
$L = EL_1 = 130$ m
Allow a 20% reduction of capacity for potential blockage.
Maximum gully spacing is approximately 100 m.

Hydraulically, if a culvert is not flowing full, it simply behaves as an open channel. The approach is usually to adopt the principles of open-channel flow even when culverts **are** flowing full. This is in contrast with the usual approach to part-full pipes (Section 7.4), which is based on the principles of pipe flow even though there is a free surface.

8.6.2 Flow cases

Different longitudinal water surface profiles occur within a culvert depending on conditions (Figure 8.20). It is assumed that if the depth upstream of the culvert is less than 1.2 times the culvert height, the culvert behaves as an open channel. Under these conditions the shape of the longitudinal water surface profile is influenced by two other factors: whether the slope of the culvert is hydraulically mild or steep (Section 7.5.4), and whether conditions downstream exert an influence on the water depth within the culvert. If they do, this is termed *downstream surcharge*.

If the depth upstream of the culvert is greater than 1.2 times the culvert height, the flow rate in the culvert may be limited either by the properties of the inlet acting as an orifice (Section 8.1.1) or by the friction and local losses in the culvert. These two conditions can be termed *inlet controlled* or *losses controlled*. For the losses-controlled case, there may or may not be downstream surcharge.

These seven possibilities are listed on Table 8.3 together with the basis for the standard calculation procedure. There is more detail in Examples 8.7 and 8.8.

The resulting surface profiles are presented in Figure 8.20.

Table 8.3 Conditions for different water surface profiles

Flow condition	See Figure 8.19	Downstream surcharge?	Basis for calculations
Open channel – mild slope	(a)	No	d_n determined from Manning's equation
	(b)	Yes	S_f determined from Manning's equation
Open channel – steep slope	(c)	No	d_n determined from Manning's equation
	(d)	Yes	d_n determined from Manning's equation with hydraulic jump in culvert
Inlet controlled	(e)		Orifice Equation 8.1
Losses controlled	(f)	No	S_f determined from Manning's equation (full) and local losses
	(g)	Yes	S_f determined from Manning's equation (full) and local losses

Note: d_n normal depth, d_c critical depth, d_u upstream depth.

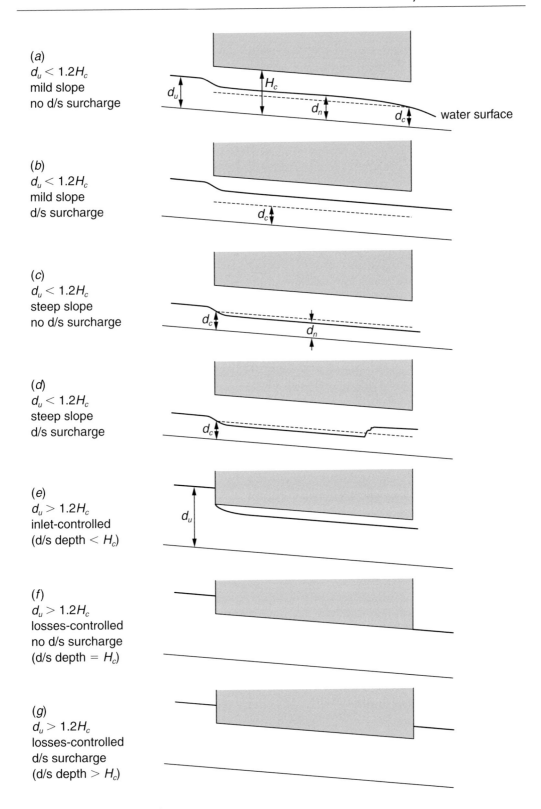

Figure 8.20 Flow cases for culverts.

EXAMPLE 8.7

A rectangular cross-section culvert has a width 1.2 m, height (H_c) 0.6 m, slope 1:100, and Manning's n 0.013. For a significant length within the culvert, flow is at the normal depth, 0.15 m. At the inlet, local losses can be calculated using $k_L = 0.5$. It can be assumed that upstream velocity is negligible. There is no surcharge downstream.

Determine the depth immediately upstream of the culvert (if the bed of the channel and the invert of the culvert are the same at the inlet).

Solution

There is open channel flow in the culvert since 0.15 m < 0.6 m:

$$A = 1.2 \times 0.15 = 0.18 \text{ m}^2$$

$$R = 0.18 \div 1.5 = 0.12 \text{ m}$$

Manning's Equation 7.23: $v = (1/0.013)0.12^{2/3}0.01^{1/2} = 1.87 \text{ m/s}$
$Q = v \times A$ (from Equation 7.3), so $Q = 1.87 \times 0.18 = 0.337 \text{ m}^3/\text{s}$
Is the slope of the culvert hydraulically steep or mild?
To determine d_c critical depth, use Equation 7.27 $v_c = \sqrt{gd_m}$.
From the definition in Table 7.4, it is clear that for a rectangular channel, d_m equals the actual depth. So at critical depth,

$$v_c = \sqrt{gd_c} \text{ or (using Equation 7.3)} \frac{0.337}{1.2 \times d_c} = \sqrt{g \times d_c}$$

this gives $d_c = 0.2$ m, which is greater than normal depth, 0.15 m, so slope is **steep** (Section 7.5.4).

In the culvert, specific energy (Equation 7.25)

$$E = y + \frac{v^2}{2g} = 0.15 + \frac{1.87^2}{2g} = 0.328 \text{ m}$$

Loss at entry (Equation 7.14) $h_L = (0.5 \ v_c^2/2g)$.
Using Equation 7.3, $v_c = 0.337/(1.2 \times 0.2) = 1.4 \text{ m/s}$, so entry loss $= 0.05 \text{ m}$
Upstream velocity is negligible, so
$d_u = $ upstream $E = 0.328 + 0.05 = 0.378$ m

EXAMPLE 8.8

A rectangular cross-section culvert has width 1.2 m, height (H_c) 0.6 m, Manning's n 0.013, and slope 1 in 100. Depth immediately upstream (d_u) is 0.9 m. The length of the culvert is 30 m.

Determine whether or not the culvert is surcharged upstream. Determine whether flow is inlet controlled or losses controlled. Determine the flow rate.

Assume that upstream velocity is negligible, and that there is no surcharge downstream (i.e., depth downstream does not affect depth in the culvert).

For entry loss, use $K = 0.5$. If the entry is acting as an orifice, $C_d = 0.62$.

Solution

$d_u > H_c$, so the culvert is either inlet controlled or losses controlled.

1. Assume inlet controlled:

Equation 8.1, $Q = C_d A \sqrt{2gH} = 0.62 \times 1.2 \times 0.6 \sqrt{2 g 0.6} = 1.53 \text{ m}^3/\text{s}$

2. Assume losses controlled:

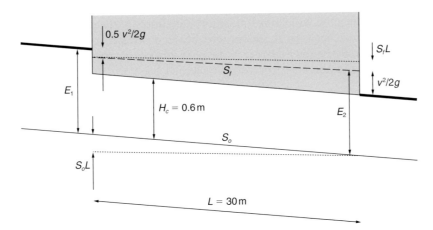

Subscripts 1 and 2 denote conditions at upstream and downstream ends of the culvert, respectively.

E_1 (specific energy) $= d_u = 0.9$ m, since upstream velocity is negligible.

The dashed line represents the hydraulic gradient, S_f.

$v_1 = v_2$; velocity all along the culvert is written as v below.

For this case (with no free surface), the Manning equation should be written

$$v = \frac{1}{n}R^{2/3}S_f^{1/2} \text{ or } S_f = \left[\frac{vn}{R^{2/3}}\right]^2$$

$$R = \frac{1.2 \times 0.6}{1.2 + 0.6 + 1.2 + 0.6} = 0.2 \text{ m}$$

since the wetted perimeter includes the top of the culvert. This gives $S_f = 1.445 \times 10^{-3} \times v^2$.

The height of the upstream water surface above the dotted line can be expressed two ways, so

$$0.9 + S_o L = 0.5v^2/2g + S_f L + v^2/2g + 0.6$$

$$0.9 + 0.01 \times 30 = 1.5v^2/2g + 1.445 \times 10^{-3} \times v^2 \times 30 + 0.6$$

this gives $v = 2.24$m/s so $Q = 1.61$ m³/s.

Losses limit the capacity of the culvert to 1.61 m³/s, but the inlet acting as an orifice limits the flow rate to 1.53 m³/s.

So, the flow rate cannot exceed 1.53 m³/s, and the culvert is inlet controlled.

PROBLEMS

8.1   An orifice plate is being designed for flow control at the outlet of a detention tank. The outlet pipe has a diameter of 450 mm, slope of 0.0018, and roughness k_s of 0.6 mm. The water level in the tank at the design condition varies between 1.5 and 1.7 m above the outlet invert, and the desired outflow is 100 L/s. Select an appropriate orifice diameter. (Assume orifice $C_d = 0.6$; check that the orifice will not be drowned.)   [200 mm, not drowned]

8.2   A throttle pipe to control outflow (to treatment) from a CSO is being designed. The pipe will have a length of 25 m, and diameter 200 mm. A check is being carried out to see how well the pipe will control the flow (roughness of the pipe, $k_s = 0.6$ mm). When the difference between the water level at the upstream and downstream end is 2.5 m, what would be the flow rate in the pipe considering friction losses only? In this condition, is the pipe inlet controlled (assume $C_d = 0.6$)?

[120 L/s, no]

8.3   A rectangular transverse weir in a CSO has a width equal to the width of the chamber itself: 2.2 m. The weir crest is 1.05 m above the floor of the chamber. When the water level is 0.15 m above the crest, determine the flow rate over the weir.

[0.235 m³/s]

8.4   A tangential vortex intake has $Q_{max} = 15$ m³/s, $B_1 = 2.75$ m, $L = 9$ m, $z = 4$ m, $B_2 = 0.56$ m, $D = 2.25$ m (using the symbols defined in Section 8.3). Comment on the appropriateness of this design.
[$Q_c = 18$ m³/s and $Q_f = 16.5$ m³/s, so while Q_c and Q_f exceed Q_{max}, $Q_c > Q_f$, so the design is not OK.]

8.5   Estimate the flow rate in the channel of a road with a longitudinal gradient of 0.5% and a crossfall of 1:40 if the width of flow is 2.5 m. Assume $n = 0.013$.   [42 L/s]

8.6   A rectangular cross-section culvert has a width 1.2 m, height 0.6 m, slope 1:1000, and Manning's n 0.013. For a significant length within the culvert, flow is at the normal depth, 0.2 m. At the inlet, local losses can be calculated using $K = 0.5$. It can be assumed that upstream velocity is negligible.
Determine the depth immediately upstream of the culvert (if the bed of the channel and the invert of the culvert are the same at the inlet). Sketch the shape of the water surface from just upstream of the culvert to just downstream.   [0.236 m]

8.7   A rectangular cross-section culvert has width 1.2 m, height 0.6 m, Manning's n 0.013, and slope 1 in 1000. Depth immediately upstream is 1 m. The length of the culvert is 30 m.
Determine whether or not the culvert is surcharged upstream. Determine whether flow in the culvert is inlet controlled or losses controlled. Determine the flow rate.
Assume that upstream velocity is negligible, and that there is no surcharge downstream (i.e., depth downstream does not affect depth in the culvert).
For entry loss, use $K = 0.5$. If the entry is acting as an orifice, $C_d = 0.62$.

[losses controlled; 1.37 m³/s]

REFERENCES

Ackers, P. and Crump, E.S. 1960. The vortex drop. *Proceedings of the Institution of Civil Engineers*, 16, August, 433–442.

Andoh, R., Osei, K., Fink, J., and Faram, M. 2008. Novel drop shaft system for conveying and controlling flows from high level sewers into deep tunnels. *World Environmental and Water Resources Congress*, ASCE, Honolulu, Hawaii, May, on CD.

Balkham, M., Fosbeary, C., Kitchen, A., and Rickard, C.E. 2010. *Culvert Design and Operation Guide*, CIRIA C689.

Balmforth, D.J. and Sarginson, E.J. 1978. A comparison of methods of analysis of side weir flow. *Chartered Municipal Engineer*, 105, October, 273–279.

Balmforth, D.J., Saul, A.J., and Clifforde, I.T. 1994. *Guide to the Design of Combined Sewer Overflow Structures*, Report FR 0488, Foundation for Water Research.

Burrows, R. and Emmonds, J. 1988. Energy head implications of the installation of circular flap gates on drainage outfalls. *Journal of Hydraulic Research*, 26(2), 131–142.

Butler, D. and Parsian, H. 1993. The performance of a vortex flow regulator under blockage conditions. *Proceedings of the 6th International Conference on Urban Storm Drainage, Niagara Falls*, Canada, 1793–1798.

Chanson, H. 2004a. Hydraulics of rectangular dropshafts. *Journal of Irrigation and Drainage Engineering*, 130(6), 523–529.

Chanson, H. 2004b. *The Hydraulics of Open Channel Flow: An Introduction*, 2nd edn. Chapter 20, Elsevier.

Chow, V.T. 1959. *Open-Channel Hydraulics*, McGraw-Hill.

Davis, A., Jacob, R.P., and Ellett, B. 1996. A review of road-gully spacing methods. *Journal of Chartered Institution of Water and Environmental Management*, 10, April, 118–122.

Del Giudice, G. and Gisonni, C. 2010. Vortex dropshafts: History and current applications to the sewer system of Naples (Italy). *Proceedings of the First IAHR European Congress*, Edinburgh, May.

Delo, E.A. and Saul, A.J. 1989. Charts for the hydraulic design of high side-weirs in storm sewage overflows. *Proceedings of the Institution of Civil Engineers*, Part 2, 87, June, 175–193.

Doncaster, S., Blanksby, J., Shepherd, W., and Sailor, G. 2012. *Gulley Optimisation – An Investigation into the Potential for Gulley Optimisation to Reduce Maintenance Requirement and to Reduce Surface Water Flood Risk*. Research Report, SKINT (North Sea Skills Integration and New Technologies).

Echavez, G. and Ruiz, G. 2008. High head drop shaft structure for small and large discharges. *Proceedings of the 11th International Conference on Urban Drainage*, Edinburgh, September, on CD.

Granata, F. 2016. Dropshaft cascades in urban drainage systems. *Water Science and Technology*, 73.9, 2052–2059.

Granata, F., de Marinis, G., Gargano, R., and Hager, W.H. 2011. Hydraulics of circular drop manholes. *Journal of Irrigation and Drainage Engineering*, 137(2), 102–111.

Hydraulics Research Station. 1984. *The Drainage Capacity of BS Road Gullies and a Procedure for Estimating Their Spacing*, TRRL Contractor Report, CR2, Transport and Road Research Laboratory, Crowthorne.

May, R.W.P. 1994. Alternative hydraulic design methods for surface drainage. *Road Drainage Seminar*, H.R. Wallingford, Wallingford, November.

May, R.W.P., Bromwich, B.C., Gasowski, Y., and Rickard, C.E. 2003. *Hydraulic Design of Side Weirs*, Thomas Telford.

Riisnaes, S., Poole, B., Cooper, M., Thornton, C., Digman, C., and Marples, N. 2014. Building vortex drop shafts in dense urban areas – A different approach. *Chartered Institution of Water and Environmental Management Urban Drainage Group Autumn Conference*.

Yu, D. and Lee, J.H.W. 2009. Hydraulics of tangential vortex intake for urban drainage. *Journal of Hydraulic Engineering*, 135(3), 164–174.

Zhao, C.-H., Zhu, D.Z., Sun, S.-K., and Liu, Z.-P. 2006. Experimental study of flow in a vortex drop shaft. *Journal of Hydraulic Engineering*, 132(1), 61–68.

Chapter 9

Foul sewers

9.1 INTRODUCTION

Separate foul sewers (also known as sanitary sewers or wastewater sewers) form an important component of many urban drainage systems. The emphasis in this chapter is on the design of such systems (Section 9.2). In particular, the distinction is made between large (Section 9.3) and small (Section 9.4) foul sewers and their different design procedures. Solids transport is introduced in Section 9.5. Analysis of existing systems using computer-based methods is covered in Chapters 19 and 20. Design of non-pipe-based (sanitation) systems is discussed in Chapter 23.

9.1.1 Flow regime

All foul sewer networks physically connect wastewater sources with treatment and disposal facilities by a series of continuous, unbroken pipes. Flow into the sewer results from random usage of a range of different appliances, each with its own characteristics. Generally, these are intermittent, of relatively short duration (seconds to minutes), and hydraulically unsteady. At the outfall, however, the observed flow in the sewer will normally be continuous and will vary only slowly (and with a reasonably repeatable pattern) throughout the day. Figure 9.1 gives an idealised illustration of these conditions.

The sewer network will have zones with continuously flowing wastewater, as well as areas that are mostly empty but are subject to flushes of flow from time to time. It is unlikely, even under maximum continuous flow conditions, that the full capacity of the pipe will be utilised. Intermittent pulses feed the continuous flow further downstream, and this implies that somewhere in the system there is an interface between the two types of flow. As the usage of appliances varies throughout the day, the interface will not remain at a single fixed location.

9.2 DESIGN

This chapter shows how foul sewers can be designed to cope with conditions as previously described. A general approach to foul (and storm) sewer design is illustrated in Figure 9.2. This should be read in conjunction with Figure 6.8.

Design is accomplished by first choosing a suitable *design period* and *criterion of satisfactory service*, appropriate to the foul *contributing area* under consideration. The type and number of buildings and their population (the maximum within the design period) are then estimated, together with estimates of the unit water consumption. This information is used to calculate *dry weather flows* in the main part of the system. Flows in building drains and small sewers are assessed in a probability-orientated discharge unit method, based on usage

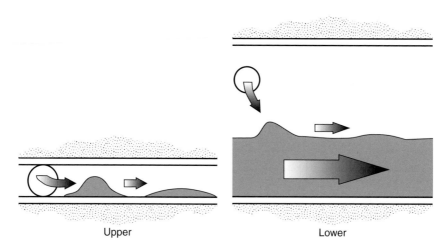

Figure 9.1 Hydraulic conditions in foul sewers in dry weather (schematic).

of domestic appliances. *Hydraulic design* of the pipework is based on safe transportation of the flows generated using the principles presented in Chapter 7. Broader issues of sewer layout including horizontal and vertical alignment are covered in Chapter 6.

9.2.1 Choice of design period

Urban drainage systems have an extended life span and are typically designed for conditions 25–50 years into the future. They may well be in use for very much longer. The choice of design period will be based on factors such as

- Useful life of civil, mechanical, and electrical components
- Feasibility of future extensions of the system
- Anticipated changes in residential, commercial, or industrial development
- Financial considerations

It is necessary to make estimates of conditions throughout the design period that are as accurate as possible.

9.2.2 Criterion of satisfactory service

The degree of protection against wastewater "backing-up" or flooding is determined by consideration of the specified criterion of satisfactory service. This protection should be consistent with the cost of any damage or disruption that might be caused by flooding. In practice, cost-benefit studies are rarely conducted for ordinary urban drainage projects; a decision on a suitable criterion is made simply on the basis of judgment and precedent. Indeed, this decision may not even be made explicitly, but nevertheless it is built into the design method chosen.

The design choice of the peak-to-average flow ratio implicitly fixes the level of satisfactory service in large foul sewers. For small sewers, the criterion can be used explicitly to determine flows, though standard (and therefore fixed) values are routinely used.

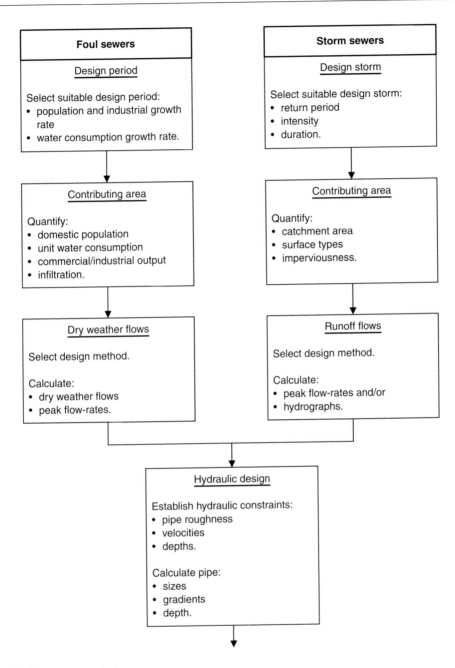

Figure 9.2 Sewer system design.

9.3 LARGE SEWERS

In this text, a distinction is drawn between large and small foul sewers. This is only for convenience, as there is no precise definition to demarcate between the two types. The same pipe may act as both large and small at different times of the day (measured in hours) or at different times in its design period (measured in years).

Flow in large foul sewers is mostly open channel (although in exceptional circumstances this may not be the case), continuous, and quasi-steady. Changes in flow that do occur will be at a relatively slow rate and in a reasonably consistent diurnal pattern. In large sewers, we can say that the inflows from single appliances are not a significant fraction of the capacity of the pipe and that there is substantial baseflow (see Figure 9.1).

9.3.1 Flow patterns

The pattern of flow follows a basic diurnal pattern, although each catchment will have its own detailed characteristics. Generally, low flows occur at night with peak flows during the morning and evening. This is related to the pattern of water use of the community but also has to do with the location at which the observation is made. Figure 9.3 illustrates the impact of three important effects. The inflow hydrograph (a) represents the variation in wastewater generation that will, in effect, be similar all around the catchment (see Chapter 3). If the wastewater was collected at one point and then transported from one end of a long pipe to the other, flow *attenuation* due to in-pipe storage would cause a reduction in peak flow, a lag in time to peak, and a distortion of the basic flow pattern (b). Normal sewer catchments are not like this and consist of many-branched networks with inputs both at the most distant point on the catchment and adjacent to the outfall. Thus, the time for wastewater to travel from the point of input to the point under consideration is variable, and this *diversification* effect causes a further reduction in peak and distortion in flow pattern (c). Additional factors that can influence the flow pattern are the degree of infiltration and the number and operation of pumping stations. These effects can be predicted in existing sewer systems using computational hydraulic models (as described in Chapter 19), but they also need to be predicted in the design of new systems.

The flow is usually defined in terms of an average flow (Q_{av})—or *dry weather flow* (DWF)—and peak flow. The magnitude of the peak flow can then be related to the average flow (see Figure 9.4). A minimum value can also be defined.

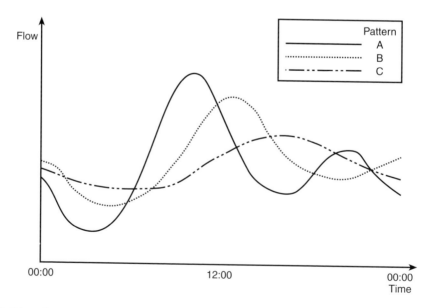

Figure 9.3 Diurnal wastewater flow pattern modified by attenuation and diversification effects.

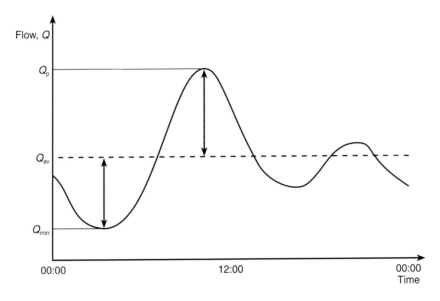

Figure 9.4 Definition of diurnal wastewater flow pattern.

Large sewer design therefore entails estimating the average DWF in the sewer by assuming a daily amount of wastewater generated per person (or per dwelling, or per hectare of development) contributing to the flow, multiplied by the population to be served at the design horizon. Commercial and industrial flows must also be estimated at the design horizon. Allowance should be made for infiltration. The peak flow can be found by using a suitable multiple or peak factor.

9.3.2 Dry weather flow

When the wastewater is mainly domestic in character, DWF is defined as

> The average daily flow ... during seven consecutive days without rain (excluding a period which includes public or local holidays) following seven days during which the rainfall did not exceed 0.25 mm on any one day.

> *(IWEM, 1993)*

If the flow contains significant industrial flows, DWF should be measured during the main production days. Ideally, flows during summer and winter periods should be averaged to obtain a representative DWF.

DWF is therefore the average rate of flow of wastewater not immediately influenced by rainfall; it includes domestic, commercial, and industrial wastes, and infiltration, but excludes direct stormwater inflow. The quantity is relevant both to foul and combined sewers. DWF can be expressed simply in the following manner (Ministry of Housing, 1970):

$$DWF = PG + I + E \qquad (9.1)$$

where *DWF* is expressed in litres per day (L/d), *P* is the population served, *G* is the average *per capita* domestic water consumption (L/hd.d), *I* is infiltration (L/d), and *E* is the average industrial effluent discharged in 24 hours (L/d).

The current definition of dry weather flow does have its weaknesses, particularly the difficulty of finding suitable dry periods, and the lack of direct linkage with Equation 9.1. A method by Bramley and Heywood (2005) involving statistical analysis of daily foul flows regardless of rainfall has the advantages of ease of calculation (without the need for rainfall data) and reduced year-on-year variability.

9.3.3 Domestic flow (PG)

The domestic component of DWF is the product of the population and the average *per capita* water consumption.

9.3.3.1 Population (P)

A useful first step in predicting the contributing population that will occur at the end of the design period is to obtain as much local, current, and historical information as possible. Official census information is often available and can be of much value. Additional data can almost certainly be obtained at the local planning authority, and officers should be able to advise on future population trends, and also on the location and type of new industries. Housing density is a useful indicator of current or proposed population levels.

9.3.3.2 Per capita water consumption (G)

In Chapter 4 we discussed in detail the relationship between water use and wastewater production. We also considered typical *per capita* values and discussed that there will be changes in *per capita* water consumption that are independent of population growth.

Where typical discharge figures for developments similar to those under consideration are available, these should be used. In the absence of such data, the European Standard on *Drain and Sewer Systems Outside Buildings* (BS EN 752: 2008) states that a daily *per capita* figure of between 150 and 300 L should be used. A figure of 220 L (200 L + 10% infiltration) has been widely used in the past in the United Kingdom. However, UK Building Regulations indicate potable water consumption in new buildings should be designed to be no greater than 125 L/hd.d or even 110 L/hd.d where required by planning permission (Approved Document G, 2015).

Specific design allowance can be made for buildings such as schools and hospitals as given in Table 9.1. See also Example 9.1.

Table 9.1 Daily volume and pollutant load of wastewater produced from various sources

Category	Volume (L/d)	BOD$_5$ load (g/d)	Per
Day schools	50–100	20–30	Pupil
Boarding schools	150–200	30–60	Pupil
Hospitals	500–750	110–150	Bed
Nursing homes	300–400	60–80	Bed
Sports centre	10–30	10–20	Visitor

EXAMPLE 9.1

Estimate the average daily wastewater flow (L/s) and BOD$_5$ concentration (mg/L) for an urban area consisting of residential housing (100,000 population), a secondary school (1,000 students), a hospital (1,000 beds) and a central shopping centre (50,000 m²).

Solution

Area	Magnitude	Unit flow (m³/unit.d)	Flow rate (m³/d)	Unit BOD$_5$ load (kg/unit.d)	BOD$_5$ load (kg/d)
Residential	100,000 population	0.20	20 000	0.06	6000
School	1,000 students	0.10	100	0.03	30
Hospital	1,000 beds	0.75	750	0.15	150
Shopping	50,000 m²	0.004	200	0.0015	75
Total			21 050		6255

Average daily wastewater flow = (21,050 × 1000)/(24 × 3600) = 244 L/s
Average BOD$_5$ concentration = (6255 × 1000)/21,050 = 297 mg/L

9.3.4 Infiltration (I)

The importance of groundwater infiltration and the problems it can cause are discussed in Section 4.4. As mentioned above, the conventional approach in design is to specify infiltration as a fraction of DWF—namely 10%. Thus, for a design figure of 200 L/hd.d, 20 L/hd.d would be specified. More recent evidence (Ainger et al., 1997) suggests this may be too low. The suggestion is made that for new systems in high groundwater areas, infiltration figures as high as 120 L/hd.d should be used.

There is a difficulty, however, in making such a large design allowance for infiltration. If an allowance is used, this will increase the design flow rate and may in turn increase the required pipe diameter. A bigger sewer will have a larger circumference and joints, potentially allowing more infiltration to enter the system. Thus, the allowance may well have actually caused more infiltration.

Is there a solution to this dilemma? It is suggested that rather than building-in large design allowances that may cause larger pipes to be chosen, it would be a better investment to ensure high standards of pipe manufacture, installation, and testing.

9.3.4.1 Measurement

The infiltration component of DWF can be estimated in a number of different ways. The simplest way is to assume that night-time flows (Q_{min} in Figure 9.4) represent infiltration. However, with an unknown number of appliances running overnight (e.g., washing machines, dishwashers, dripping taps), this assumption is increasingly unsafe. There is also the difficulty of accounting for attenuation of flows at different points in the network.

Other approaches include using artificial tracers or inferring infiltration based on measuring commonly sampled parameters for wastewater quality, such as temperature, conductivity, or nutrients. De Benedittis and Bertrand-Krajewski (2005) found in a study of French sewers that the computed value of the infiltration fraction varied in a range of up to 20% of DWF depending on which technique was used. However, this variability was considered acceptable and still allowed accurate identification of infiltration at sub-catchment scale.

9.3.4.2 Prevention

The conventional way of addressing infiltration problems in existing systems is by sewer rehabilitation (see Chapter 18). However, this can be an uneconomic approach especially when the source or sources are unknown or occur infrequently (see Chapter 3). Alternatives include short term solutions such as tankering of flows and over-pumping into a nearby watercourse. A longer term approach is to more clearly identify sources and address these on a case-by-case basis.

9.3.5 Non-domestic flows (E)

Background information on non-domestic wastewater flows can be found in Section 4.3. In design, probably the most reliable approach is to make allowance for flows on the basis of experience of similar commerce or industry elsewhere. If these data are not available, or for checking what is known, the following information can be used. Table 9.2 shows examples of daily wastewater volume produced by a variety of commercial sources. Table 9.3 provides areal allowance for broad industrial categories. Henze et al. (2002) present data for a wide range of industries.

Most commercial and industrial premises have a domestic component of their wastewater, and ideally, the estimation of this should be based on a detailed survey of facilities and their use. Mann (1979) suggests that a figure of 40–80 L/hd.(8 hour shift) may be appropriate.

9.3.6 Peak flow

Two approaches to estimating peak flows are used. In the first, typically used in British practice, a fixed DWF multiple is used. In the second, a variable peak factor is specified. Both methods aim to take account of diurnal peaks and the daily and seasonal fluctuations in water consumption together with an allowance for extraneous flows such as infiltration.

Table 9.2 Daily volume and pollutant load of wastewater produced from various commercial sources

Category	Volume (L/d)	BOD$_5$ load (g/d)	Per
Hotels, boarding houses	150–300	50–80	Bed
Restaurants	30–40	20–30	Customer
Pubs, clubs	10–20	10–20	Customer
Cinema, theater	10	10	Seat
Offices	750	250	100 m^2
Shopping centre	400	150	100 m^2
Commercial premises	300	100	100 m^2

Table 9.3 Design allowances for industrial wastewater generation

Category	Volume (L/s.ha)	
	Conventional	Water saving[a]
Light	2	0.5
Medium	4	1.5
Heavy	8	2

[a] Recycling and reusing water where possible.

BS EN 752: 2008 recommends that a multiple up to 6 be used. This figure is most appropriate for use in sub-catchments subject to relatively little attenuation and diversification effects. For larger sewers, a value of 4 is more realistic. A still lower figure (2.5) is relevant for predicting DWFs in combined sewers, because this flow will determine velocity, not capacity.

Sewers for Adoption (Water Research Centre [WRc], 2012) suggests that a design flow of 4000 L/unit dwelling.day (0.046 L/s per dwelling) should be used for foul sewers serving residential developments. This approximates to three persons/property discharging 200 L/hd.d with a peak flow multiple of 6 and 10% infiltration.

Opinions and practice differ on whether the DWF to be multiplied should include or exclude infiltration. If DWF is determined from Equation 9.1, the most satisfactory form of applying a multiple of 4 (for example) is: 4(DWF–I) + I.

Peak flows may also be determined by the application of variable peak factors. Figure 9.3 shows that attenuation and diversification effects tend to reduce peak flows, and so the ratio of peak to average flow generally decreases from the "top" to the "bottom" of the network. Thus, peak factor varies depending on position in the network (see Figure 9.5). Location is usually described in terms of population served or the average flow rate at a particular point.

The relationship between peak factor (P_F) and population can be described algebraically with equations of the following form:

$$P_F = \frac{a}{P^b} \tag{9.2}$$

where P is the population drained (in thousands) and a,b are constants.

However, there are a number of other such equations, and some of the most well known are listed in Table 9.4.

Example 9.2 illustrates that the numerical values produced by different equations can vary significantly. Thus, any of the formulae available should be used with caution.

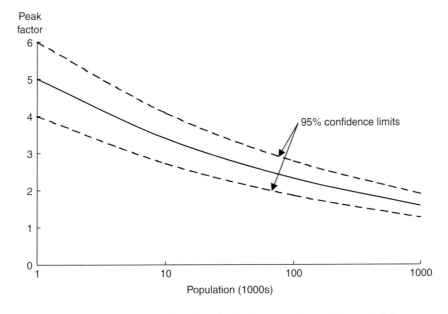

Figure 9.5 Ratio of peak flows to average daily flow (with 95 percentile confidence limits).

Table 9.4 Peak factors

Reference	Method	Notes	Equation
Harman (1918)	$1 + \dfrac{14}{4 + \sqrt{P}}$	1	(9.3)
Gifft (1945)	$\dfrac{5}{P^{1/6}}$	1	(9.4)
Babbitt (1953)	$\dfrac{5}{P^{1/5}}$	1	(9.5)
Fair and Geyer (1954)	$1 + \dfrac{18 + \sqrt{P}}{4 + \sqrt{P}}$	1	(9.6)
—	$4Q^{-0.154}$	2	(9.7)
Gaines (1989a)	$2.18Q^{-0.064}$	3	(9.8)
Gaines (1989b)	$5.16Q^{-0.060}$	3	(9.9)
BS EN 752–4	6	—

[1] Population P in 1000 s.
[2] Flow Q in 1000 m^3/d.
[3] Flow Q in litres per second (L/s).

One of the reasons for the disparity in the peak factor predictions is the general variability in diurnal flow patterns. The degree of uncertainty is also illustrated by the confidence limits (dashed lines) in Figure 9.5. Zhang et al. (2005) have used Buchberger and Wu's (1995) Poisson rectangular pulse model for instantaneous residential water demands to provide a theoretical framework to predict the shape and form of Figure 9.5 and a means to estimate the confidence limits. While strictly speaking this work is applicable to water distribution networks (diversification effects are captured but attenuation effects are not), agreement is still reached with several of the equations listed in Table 9.4.

EXAMPLE 9.2

A separate foul sewer network drains a domestic population of 250,000. Estimate the peak flow rate of wastewater at the outfall (excluding infiltration) using both Babbitt's and Gaines's formulas. The daily *per capita* flow is 145 L.

Solution

Average daily flow = (250,000 × 145)/(3600 × 24) = 420 L/s
Babbitt (Equation 9.5):

$$P_F = \frac{5}{P^{1/5}} = \frac{5}{250^{1/5}} = 1.66$$

Peak flow = 1.66 × 420 = 697 L/s
Gaines (Equation 9.8):

$$P_F = 2.18Q^{-0.064} = 2.18 \times 420^{-0.064} = 1.48$$
Peak flow = 1.48 × 420 = 622 L/s

9.3.7 Design criteria

9.3.7.1 Capacity

Foul sewers should be designed (in terms of size and gradient) to convey the predicted peak flows. It is common practice to restrict depth of flow (typically to $d/D = 0.75$) to ensure proper ventilation.

9.3.7.2 Self-cleansing

Once the pipe size has been chosen based on capacity, the pipe gradient is selected to ensure a minimum "self-cleansing" velocity is achieved. The self-cleansing velocity is that which avoids long-term deposition of solids, and should be reached at least once per day. BS EN 752: 2008 recommends a minimum of 0.7 m/s for sewers up to DN300. Higher velocities may be needed in larger pipes (see Chapter 16). *Sewers for Adoption* (WRc, 2012) requires a velocity of 0.75 m/s to be achieved at the typical diurnal peak of one-third the design flow (i.e., 2 DWF). Some engineers prefer to specify a higher self-cleansing velocity to be achieved at full-bore flow. Figure 7.8 shows how this allows for the reduction in velocity that occurs in pipes that are flowing less than half full. In practice, the pipe size and gradient are manipulated together to obtain the best design.

9.3.7.3 Roughness

For design purposes, it is conservatively assumed that the pipe roughness is independent of pipe material. This is because in foul and combined sewers, all materials will become slimed during use (see Chapter 7). BS EN 752: 2008 recommends a k_s value of 0.6 mm (for use in the Colebrook-White equation) where the peak DWF exceeds 1 m/s, and 1.5 mm where it is between 0.76 and 1 m/s.

9.3.7.4 Minimum pipe sizes

The minimum pipe size is generally set at DN75 or DN100 for house drains and DN100 to DN150 for the upper reaches of public networks, and this choice is based on experience.

9.3.8 Design method

The following procedure should be followed for foul sewer design:

1. Assume pipe roughness (k_s).
2. Prepare a preliminary layout of sewers, including tentative inflow locations.
3. Mark pipe numbers on the plan according to the convention described in Chapter 6.
4. Define contributing area DWF to each pipe.
5. Find cumulative contributing area DWF.
6. Estimate peak flow (Q_p) based on average DWF and peak factor/multiple factor analysis.
7. Make a first attempt at setting gradients and diameters of each pipe.
8. Check $d/D < 0.75$ and $v_{max} > v > v_{min}$.
9. Adjust pipe diameter and gradient as necessary (given hydraulic and physical constraints), and return to step 5.

Example 9.3 illustrates the design of a simple foul sewer network.

EXAMPLE 9.3

A preliminary foul sewer network is shown in Figure 9.6. Design the network using fixed DWF multiples (6 for domestic flows, and 3 for industrial) based on the availability of an average grade of 1:100. The inflow, Q_a is 30 L/s at peak. For the sake of simplicity, infiltration can be neglected.

Data from the network are contained in columns (1), (2) and (6) of the table below. Maximum proportional depth is 0.75, and minimum velocity is 0.75 m/s. Pipes roughness is $k_s = 1.5$ mm.

Solution

Using the raw data on land use, peak inflow rates are calculated. It is assumed that the commercial and industrial rates specified are peak rates.

(1) Pipe number	(2) Number of houses	(3) Peak flow rate $(Q_3)^a$ (L/s)	(4) Commercial area (ha)	(5) Peak flow rate $(Q_5)^b$ (L/s)	(6) Industrial area and type (ha)	(7) Peak flow rate $(Q_7)^c$ (L/s)	(8) Total peak flow rate $(Q_3 + Q_5 + Q_7)$ (L/s)
1.1	200	8.4	—	0	1.65M	19.8	28.2
2.1	250	10.5	—	0	1.70L	10.2	20.7
1.2	140	5.9	1.10	1.1	0.60L	3.6	10.6
1.3	500	21.0	2.80	2.8	—	0	23.8

ᵃ Based on three persons per house, 200 L/hd.d and DWF multiple of 6 ($Q_3 = 0.042$ L/s.house).
ᵇ Based on 300 L/d.100 m² and DWF multiple of 3 ($Q_5 = 1$ L/s.ha).
ᶜ Based on 2 and 4 L/s.ha for **L**ight and **M**edium industry, respectively, and DWF multiples of 3 ($Q_7 = 6$ or 12 L/s.ha).

Pipe velocities and depths are calculated using the Colebrook-White equation or can be read from Butler-Pinkerton charts (e.g., Figure 7.9). The pipe/gradient combination chosen is shown in bold.

(1) Pipe number	(2) Peak flow [L/s]	(3) Cumulative peak flow [L/s]	(6) Assumed pipe size (mm)	(7) Minimum gradient (1:x)	(8) Proportional depth of flow	(9) Velocity (m/s)	Comments
1.1	28.2	58.2	250	90	0.75	1.45	Depth limited
			300	**240**	**0.75**	**1.04**	Depth limited
			375	600	0.67	0.75	Velocity limited
2.1	20.7	20.7	150	47	0.75	1.45	
			225	**270**	**0.64**	**0.75**	
1.2	10.6	89.5	300	95	0.75	1.60	
			375	**320**	**0.75**	**1.02**	
1.3	23.8	113.3	**375**	**200**	**0.75**	**1.27**	
			450	500	0.75	0.90	

9.4 SMALL SEWERS

As described earlier, small sewers are subject to random inflow from appliances as intermittent pulses of flow, such that peak flow in the pipe is a significant fraction of the pipe capacity, and there is little or no baseflow.

As an appliance empties to waste, a relatively short, highly turbulent pulse of wastewater is discharged into the small sewer. As the pulse travels down the pipe, it is subject to

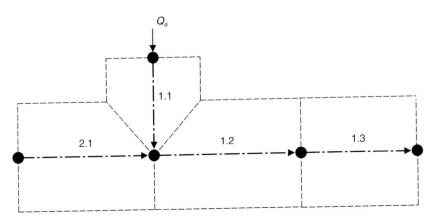

Figure 9.6 System layout (Example 9.3).

attenuation resulting in a reduction in its flow rate and depth, and an increase in duration and length (see Figure 9.1).

9.4.1 Discharge unit method

Building drainage and small sewerage schemes are often designed using the Discharge Unit Method as an alternative to the methods previously described. Using the principles of probability theory, discharge units are assigned to individual appliances to reflect their relative load-producing effect. Peak flow rates from groups of mixed appliances are estimated by addition of the relevant discharge units. The small sewer can then be designed to convey the peak flow. This approach is now explained in more detail.

9.4.1.1 Probabilistic framework

Consider a single type of appliance discharging identical outputs that have an initial duration of t' and a mean interval between use of T'. Hence, the probability p that the appliance will be discharging at any instant is given by

$$p = \frac{duration\ of\ discharge}{mean\ time\ between\ discharges} = \frac{t'}{T'} \tag{9.10}$$

EXAMPLE 9.4

Calculate the probability of discharge of a single WC that discharges for 10 seconds every 20 minutes at peak times. What percentage of time will the WC be loading the system?

Solution

From Equation 9.10,

$$p_{WC} = 10/1200 = 0.0083$$

The WC will be loading the system 0.8% of the time (at peak) and hence will *not* be discharging for 99.2% of the time.

In most systems, however, there will be more than one appliance. How can we answer a question such as, "what is the probability that r from a total of N appliances will discharge *simultaneously*?" Application of the binomial distribution states that if p is the probability that an event will happen in any single trial (i.e., the probability of success) and $(1-p)$ is the probability that it will fail to happen (i.e., the probability of failure), then the probability that the event will occur exactly r times in N trials $(P[r,N])$ is

$$P(r,N) = {}^N C_r p^r (1 - p)^{N-r} \tag{9.11}$$

or

$$P(r,N) = \frac{N!}{r!(N-r)!} p^r (1-p)^{N-r} \tag{9.12}$$

Thus, to use the binomial probability distribution in this application, we must assume that

- Each trial has only two possible outcomes: success or failure – that is, an appliance is either discharging or it is not.
- The probability of success (p) must be the same on each trial (i.e., independent events), implying that t' and T' are always the same.

Neither of these assumptions is fully correct for discharging appliances, but they are close enough for design purposes. Example 9.5 illustrates the basic use of Equation 9.11.

EXAMPLE 9.5

What is the probability that 20 from a total of 100 WCs ($p = 0.01$) will discharge simultaneously?

Solution

N = number of trials = total number of connected appliances = 100
p = probability of success = probability of discharge = 0.01
Using the binomial expression with the above data gives (Equation 9.12)

$$P(20,100) = \frac{100!}{20!80!} 0.01^{20} 0.99^{80} = 2.4 \times 10^{-20}$$

In other words, this eventuality is extremely unlikely.

9.4.1.2 Design criterion

While this type of basic information is of interest, it is not of direct use. In design, we are concerned with establishing the probable number of appliances discharging simultaneously against some agreed standard. Practical design is carried out using a confidence level approach or "criterion of satisfactory service" (J) as introduced in Section 9.2.2. For small sewers, this is defined as the percentage of time that up to c appliances out of N will be discharging. So,

$$\sum_{r=0}^{c} P(r,N) \geq J \tag{9.13}$$

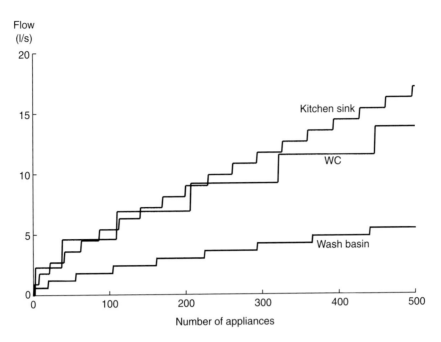

Figure 9.7 Simultaneous discharge of WC, sink, and basin at 99% criterion of satisfactory service.

Table 9.5 Typical UK appliance flow and domestic usage data

Appliance	Flow rate q (L/s)	Duration t' (s)	Recurrence use interval T' (s)	Probability of discharge p
Toilet (9 L)	2.3	5	1,200	0.004
Wash basin	0.6	10	1,200	0.008
Kitchen sink	0.9	25	1,200	0.021
Bath	1.1	75	4,500	0.017
Washing machine	0.7	300	15,000	0.020

Source: Wise, A.F.E. and Swaffield, J.A. 2002. *Water, Sanitary and Waste Services for Buildings*, 5th edn, Taylor and Francis.

In design terms, we are trying to establish the value of c for a given J. A typical value for J would be 99%, implying actual loadings will only exceed the design load for less than 1% of the time (see Example 9.6).

At a given criterion of satisfactory service, each individual appliance will therefore have a unique relationship between

- The number of connected appliances and the number discharging simultaneously
- The number of connected appliances and flow rate (because the discharge capacity of each appliance is known, and assumed constant)

Figure 9.7 illustrates the relationship between number of connected appliances and simultaneous discharge for three common devices, prepared using the binomial distribution and data from Table 9.5. The stepped appearance of the plots does not reflect the resolution of the calculations used to produce them but is inherent in the calculations.

EXAMPLE 9.6

For a criterion of satisfactory service of 99%, determine the number of water widgets discharging simultaneously from a group of five, if their probability of discharge is 20% (unusually high, but used for illustrative purposes). If each widget discharges $q = 0.5$ L/s, find the design flow.

Solution

Now, $N = 5$, $p = 0.2$, and $J = 0.99$. Using Equation 9.12 for increasing values of r, we get

r	$P(r,N)$	$\Sigma\, P(r,N)$	$\Sigma\, q$ (L/s)
0	0.327	0.327	0
1	0.410	0.737	0.5
2	0.204	0.941	1.0
3	0.051	0.992	1.5

So, since at $r = 3$, $\Sigma\, P(r,N) > 0.99$, up to three water widgets will be found discharging 99% of the time, and more than three will discharge just 1% of the time (i.e., during one peak period every hundred days). Design for $c = 3$ simultaneous discharges, $q = 1.5$ L/s.

9.4.1.3 Mixed appliances

In a practical design situation, there will be a mix of appliance types rather than the single types previously discussed. The basic binomial distribution does not take into account the interactions in a mixed system between appliances of different frequency of use, discharge duration, and flow rate.

To overcome this problem, the *discharge unit (DU) method* has been developed based on the premise that the same flow rate may be generated by a different number of appliances depending on their type. DUs are, therefore, attributed uniquely to each appliance type, and the value will depend on

- The rate and duration of discharge
- The criterion of satisfactory service

Recommended values are given in Table 9.6. Note, in particular, that the discharge volume of the WC is 6 L (as compared with 9 L in Table 9.5) to reflect the maximum allowable

Table 9.6 Discharge unit ratings for domestic appliances

Appliance	Discharge units, DU
WC (6 L)	1.2–1.8
Wash basin	0.3–0.5
Sink	0.5–1.3
Bath	0.5–1.3
Washing machine (up to 6 kg)	0.5–0.8

Source:　BS EN 12056–2: 2000. *Gravity Drainage Systems Inside Buildings. Part 2: Sanitary Pipework, Layout and Calculation.*

Table 9.7 Frequency of use factors

Frequency of use	k_{DU}
Intermittent: dwellings, guest houses, offices	0.5
Frequent: hospitals, schools, restaurants	0.7
Congested: public facilities	1.0

Source: BS EN 12056–2: 2000. *Gravity Drainage Systems Inside Buildings. Part 2: Sanitary Pipework, Layout and Calculation.*

in the UK under the current Water Supply (Water Fittings) Regulations (Defra, 1999). It is possible to express all appliances in terms of DUs using a family of design curves, based only on intensity of use. BS EN 12056–2: 2000 recommends a power law be used to approximate the relationship between design flow rate Q and the cumulative number of discharge units DU, so

$$Q = k_{DU}\sqrt{\Sigma n_{DU}} \tag{9.14}$$

where Q is peak flow (L/s), k_{DU} is the dimensionless frequency factor, and n_{DU} is the number of discharge units. The value of k_{DU} depends on the intensity of usage of the appliance(s) and is given in Table 9.7. Design curves are given in Figure 9.8 and are used in Example 9.7.

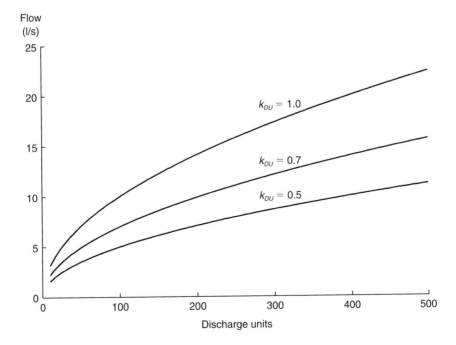

Figure 9.8 Relationship between appliance discharge units and design flow rate.

EXAMPLE 9.7

A residential block is made up of 20 flats, each fitted with a WC, wash basin, sink, bath, and washing machine. It is estimated that in any one flat, between 08:00 and 09:00, all of the appliances are likely to be in use on a Monday. Calculate the design flow rate using the DU method.

Solution

Taking the most conservative values, the discharge units for all appliances $= 1.8 + 0.5 + 1.3 + 1.3 + 0.8 = 5.7$. Hence, for 20 flats the total of discharge units is 114. Assuming $k_{DU} = 0.5$, from Equation 9.14 or Figure 9.8,

$$Q = 0.5\sqrt{114} = 5.3 \text{ L/s}$$

9.4.2 Design criteria

In small sewers and drains, design criteria relate principally to the capacity of the pipe and the requirements of self-cleansing. Sewers are normally designed (BS EN 752: 2008) so that the design flow (at the relevant confidence level) can be conveyed with a proportional depth $d/D < 0.7$. This is done assuming steady, uniform flow conditions as described in Chapter 7.

In small sewers, where solids are transported by being pushed along the pipe invert, self-cleansing is difficult to assess on a theoretical basis (as considered further in Section 9.5). Even if flow is assumed to be steady and uniform (which is not), such low flows may require quite steep gradients to achieve self-cleansing velocities. At the heads of runs, the pipe gradient is usually based on "accepted practice" and can be "relaxed" somewhat (as shown in Table 9.8) to a minimum gradient and number of connected WC, depending on the required pipe size. This is in recognition of the flush wave produced by the WC in transporting solids.

The implication of Table 9.8 is that for a public sewer with diameter 150 mm or greater, the maximum gradient that needs to be used is 1:150, provided there are at least five connected dwellings. *Sewers for Adoption*, however, recommends a minimum of 10 connected dwellings together with a minimum diameter of 100 mm for pipes serving those 10 dwellings (WRc, 2012).

The major factors influencing minimum pipe diameter are its ability to carry gross solids and its ease of maintenance. Large solids frequently find their way into sewers, either accidentally or deliberately, particularly via the WC and property access points. The minimum pipe size is as set out in Section 9.3.7.4.

An application of the small sewer design method is given in Example 9.8.

Table 9.8 BS EN 752: 2008 deemed to satisfy self-cleansing rules for small sewers

Design flow (L/s)	DN (mm)	Gradient	Connected WCs
<1	≤100	≥1:40	–
>1	100	≥1:80	1
	150	≥1:150	5

9.4.3 Choice of methods

As mentioned earlier in the chapter, the two different design methods (for large and small sewers) represent the different flow regimes in foul sewers. If a large network is to be designed in detail, there comes a point where a change must be made from one method to another. The point at which the change takes place depends on local circumstances, but its location is important as it has an impact on pipe sizes and gradients, and hence cost.

A suggested approach is to interpret the probability of appliance discharge as a measure of flow intermittency. So if $Np > 1$, flow is continuous, and the large sewer approach can reasonably be used, otherwise the small sewer method is appropriate (see Example 9.9). Alternatively, BS EN 752: 2008 advises the population-based method should be used if the probability-based method gives a pipe size larger than DN 150.

9.5 SOLIDS TRANSPORT

It is surprising that the transport of gross solids is not routinely and explicitly considered in the design of large or small sewers. However, research has begun to fill the gaps in our understanding of the movement of solids in the different hydraulic regimes encountered, and is giving some important feedback to practical design and operation.

The main characteristics of gross solids transport in sewers are as follows:

- There is a wide variety of solids, and the physical condition of some types varies widely, influencing the way the solids are transported.
- Some solids change their condition as they move through the system, as a result of physical degradation and contact with other substances in the sewer.
- In some hydraulic conditions, solids are carried with the flow, yet at lower flow rates they may be deposited.
- During movement, solids do not necessarily move at the mean water velocity.
- Some solids affect the flow conditions within the sewer.

EXAMPLE 9.8

Design the foul sewer diameter and gradients for the small housing estate shown in Figure 9.9. Data on the network are shown in columns (1) and (2) of the table below. Use the following design data:

Minimum diameter (mm):	150
Minimum velocity (m/s):	0.75
Minimum gradient:	1:150 (provided number of WCs $\geq$5)
Maximum proportional depth of flow:	0.75
Pipe roughness (mm):	0.6

Solution

For each sewer length, use the minimum pipe diameter and calculate the minimum gradient required to achieve the necessary capacity + self-cleansing. Use Tables 9.6–9.8, Equation 9.14, and the Butler-Pinkerton charts.

Assume each dwelling has (WC + basin + sink):

$$DUs = 1.8 + 0.4 + 1.3 = 3.5$$

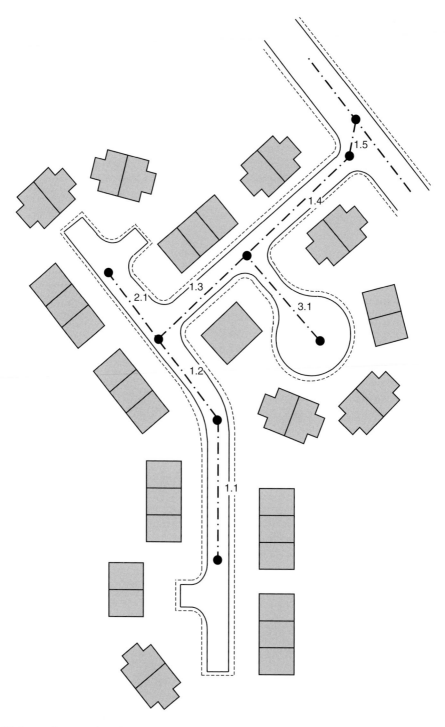

Figure 9.9 System layout and catchment plan (Example 9.8).

For individual pipe lengths draining at least five dwellings, reduce the gradient to 1:150. Take $k_{DU} = 0.5$.

(1) Pipe number	(2) Number of houses	(3) Number of discharge units	(4) Cumulative number of discharge units	(5) Design flow rate (L/s)	(6) Assumed pipe size (mm)	(7) Minimum gradient (1:x)	(8) Proportional depth of flow	(9) Velocity (m/s)	Comments
1.1	4	14	14	1.9	150	55	0.21	0.75	
1.2	9	31.5	45.5	3.4	150	85	0.32	0.75	*
						150	0.37	0.62	
2.1	10	35	35	3.0	150	75	0.28	0.75	*
						150	0.34	0.59	
1.3	1	3.5	84	4.6	150	100	0.37	0.75	*
						150	0.43	0.65	
3.1	6	21	21	2.3	150	70	0.23	0.75	*
						150	0.29	0.54	
1.4	5	17.5	122.5	5.5	150	120	0.43	0.75	*
						150	0.46	0.69	
1.5	2	7	129.5	5.7	150	125	0.45	0.75	*
						150	0.47	0.70	

* Gradient relaxed to 1:150 as $Q > 1$ L/s, number of WCs ≥ 5.

EXAMPLE 9.9

Using the data in Table 9.5, calculate the discharge probability of an equivalent single appliance for one household. Use this to estimate the minimum number of connected households that will generate continuous wastewater flow at peak times. What will be the expected flow rate?

Solution

Appliance	Flow rate q (L/s)	Probability of discharge p	pq
WC (9 L)	2.3	0.004	0.0092
Wash basin	0.6	0.008	0.0048
Kitchen sink	0.9	0.021	0.1890
Bath	1.1	0.017	0.0187
Washing machine	0.7	0.020	0.0140
Total	5.6		0.0656

Assuming a household has one of each appliance, a single equivalent appliance will have a probability of discharge, $p = 0.0656/5.6 = 0.0117$.
If continuous flow occurs when $Np = 1$, $N = 1/0.0117 = 85$ households.
Expected peak flow rate $= Npq = 5.6$ L/s.

9.5.1 Large sewers

When solids are advected (moved while suspended in the flow) in large sewers, forces acting on the solids position them at different flow depths depending on their specific gravity and on the hydraulic conditions. Figure 9.10 indicates how some solids can be carried along at

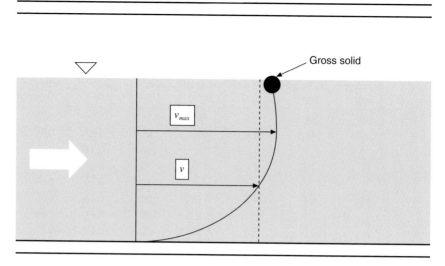

Figure 9.10 Movement of gross solids in large sewers.

levels where the local velocity is greater than the mean velocity (v). This means that solids may "overtake" the flow and arrive at combined sewer overflows and wastewater treatment plants before the peak water flow.

Figure 9.11 shows laboratory results for a solid plastic cylinder (artificial faecal solid) plotted as longitudinal solid velocity against mean water velocity, for two contrasting gradients. A linear relationship fits all the data well ($R^2 = 0.98$), and this was found to be the case for all the artificial solids studied and for various "real" gross solids (Butler et al., 2003). This linear relationship can be expressed as

$$v_{GS} = av + \beta \qquad (9.15)$$

where v_{GS} is the velocity of a particular gross solid (m/s); v is the mean water velocity (m/s); and α, β are coefficients. Laboratory results indicate β to typically be small enough to neglect, but α varies from 0.98 to 1.27 depending on solid type, with lower specific-gravity solids generally having the higher values. It has also been recommended (Davies et al., 1996) that for the modelling of solids movement in unsteady flow, the relationship between the mean water velocity and the average velocity of any solid type can be assumed to be the same in unsteady (gradually varied) flow as it is in steady (uniform) flow.

Generally, solid size has not been found to be an important variable, except at low flow depths. In this case, larger solids tend to be retarded more than smaller ones by contact with the pipe wall.

Under certain hydraulic conditions (typically low flows, such as overnight), solids may be deposited. Davies et al. (1996) found that a solid's propensity to deposit is based on critical hydraulic parameters of flow depth and mean velocity. They argued that (at least for modelling purposes) deposition of solids takes place when the value of *either* depth or mean velocity goes below the critical value, and re-suspension takes place when that level is subsequently exceeded. Figure 9.12 shows a graph of mean velocity plotted against depth, with points representing the conditions for deposition of a sanitary towel observed in a laboratory study. The dotted lines indicate suitable values for the critical depth (vertical) and velocity (horizontal). Above and to the right of the dotted lines are conditions in which these

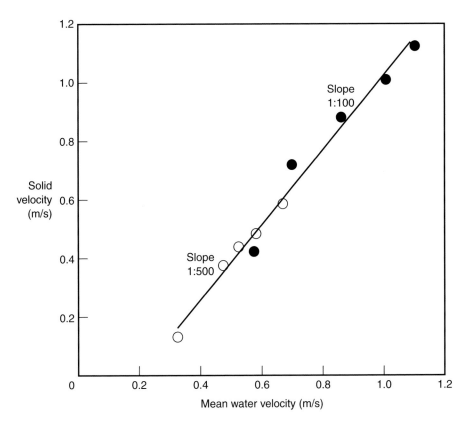

Figure 9.11 Artificial faecal solid velocity versus mean velocity, with linear fit. (After Butler, D., et al. 2003. *Proceedings of Institution of Civil Engineers, Water, Maritime and Energy,* 156[WM2], 165–174.)

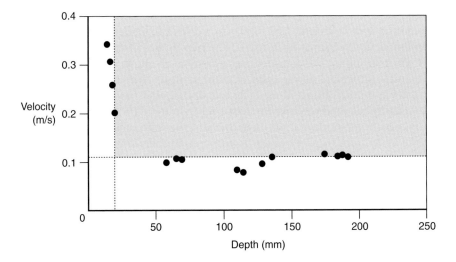

Figure 9.12 Hydraulic conditions for deposition of solids (sanitary towels). (After Butler, D., et al. 2003. *Proceedings of Institution of Civil Engineers, Water, Maritime and Energy,* 156[WM2], 165–174.)

Table 9.9 Critical depth/velocity for various solid types

Solid type	Critical depth (mm)	Critical velocity (m/s)
Solid plastic cylinders:		
Length 80 mm, diameter 37 mm	30	0.20
Length 44 mm, diameter 20 mm	22	0.13
Length 22 mm, diameter 10 mm	20	0.10
Cotton wool wipe	10	0.08
Sanitary towel	20	0.11

types of solid are carried by the flow (both depth and velocity exceeding the critical value). Below or to the left of the dotted lines are conditions in which they would be deposited. Table 9.9 gives depth/velocity values results for various gross solid types.

With the increasing use of water-saving devices and the consequent reduction in water volumes entering the sewer, what are the implications for solid transport in large foul sewers? Blanksby (2006) indicates that the main impacts are the possibility of increased gross solids deposition and the prospect of increased sedimentation in flatter sewers. Some evidence of this has been found for combined sewers (see Box 24.3). However, model studies have shown that although in some scenarios sewer flows, velocities, and proportional depths may be reduced, sewer *blockage* rates are not expected to increase significantly (Penn et al., 2013).

9.5.2 Small sewers

The movement of solids in small sewers is somewhat different than that in large sewers. Laboratory experiments demonstrate that there are two main modes of solid movement: *floating* and *sliding dam*. The floating mechanism occurs when the solid is small relative to the pipe diameter and flush wave input. The solid moves with a proportion of the wave velocity and has little effect on the wave itself. Solids that are large compared with the flush wave and pipe diameter move with a sliding dam mechanism (Littlewood and Butler, 2003). In this case, the flush wave builds up behind the solid, which acts as a dam in the base of the pipe. When the flow's hydrostatic head and momentum overcome the friction between solid and pipe wall, the solid begins to move along the pipe invert. The amount of movement that occurs depends on how "efficient" the solid is as a dam: the higher the efficiency, the further the solid will move for the same flush wave. The two modes of movement are illustrated in Figure 9.13. Photograph (a) shows toilet tissue alone in the flow, and photograph (b) shows toilet tissue and an artificial faecal solid in combination. Note the pool of water forming behind the solid and propelling it along. The role of the toilet tissue in forming the "dam" is also noteworthy. Solids tend to move farthest in the sliding dam mode.

Eventually, whichever mode of movement prevails, the solid will deposit on the pipe invert, some distance away from its entry point. It will remain there until another wave enters the pipe, travels along to meet the stranded solid, and resuspends it. The solid will move further downstream, but for a distance less than the initial movement. The distance moved under the influence of each subsequent flush decreases, until the solid is no longer moved at all by the attenuated flush wave (Swaffield and Galowin, 1992). Thus, each solid, flush wave, pipe diameter, and gradient combination has a *limiting solid transport distance* (LSTD). Figure 9.14 shows that for a 6 L flush volume WC, the solid is not moved much more than 16 m even after 20 flush waves have been passed down the pipe. In fact, very little further movement is noted beyond 10 flushes. On the basis of extensive laboratory tests, Walski et al. (2011) concluded that gross solids in small sewers are more likely to be transported

(a) (b)

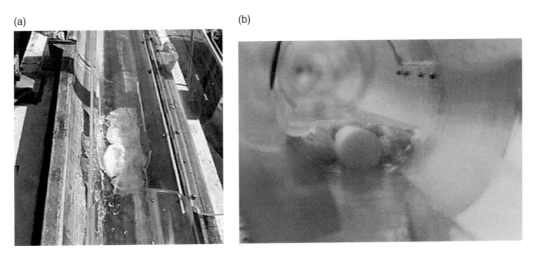

Figure 9.13 Floating (a) and sliding dam (b) mechanisms of solid movement. (Courtesy of Dr. Richard Barnes.)

when there is more flow and flow of longer duration, when the carrying pipe slope is greater, and when the solids have a lower specific gravity.

A similar question to that asked for large sewers can also be posed for small ones: what are the implications for the more widespread adoption of water-saving devices, especially WCs? Tests have shown (Memon et al., 2007) that, for example, when a 6 L WC discharges into a 1:100 gradient, 100 mm diameter pipe, the LSTD is 16.5 m, but for a 3 L WC discharging to a similar drain configuration, LSTD is reduced to 8.5 m (Figure 9.14). The interpretation of this is not clear-cut but suggests that the *propensity* for blockage formation is increased at lower flush volumes. Drinkwater et al. (2008) agree and argue that "available data suggests that a reduction from six- to three-litre flushes in a conventional WC could pose a significant problem for current drainage systems." Lauchlan et al. (2004) suggest that

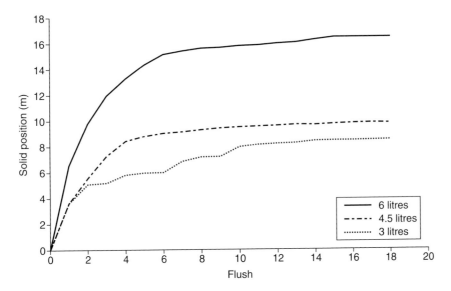

Figure 9.14 Limiting solid transport distance for a gross solid in a 100 mm diameter pipe at a gradient of 1:100 for various toilet flush volumes. (After Memon, F., et al. 2007. *Water Science and Technology*, 55[4], 85–91.)

in 150 mm diameter pipes any such problems would only manifest themselves at gradients of 1:150 or lower. Gormley et al. (2013) go further and assert that on the basis of laboratory studies, under low water use conditions, shallow gradients actually promote solid transport and hence minimise blockage propensity.

9.5.3 Sewer blockage

Most sewer blockages occur in small sewers operating under intermittent flow as previously described. In general, it should therefore be expected that there will be stranded solids at intervals along these pipes, and indeed this should be regarded as normal operation. Individually these solids will not normally be large enough to cause a blockage. The question therefore is how these stranded solids come together to form a blockage. Based on the literature, laboratory results, and field studies, key risk factors include hydraulic conditions within the pipe, the standard of workmanship, and the type of material entering the system. Further detail on these factors and ways to deal with them are discussed in Chapter 17.

PROBLEMS

9.1   Explain how you would go about the preliminary investigation and design of a foul sewer network for a large housing development.

9.2   What are the main differences in the hydraulic regime between large and small foul sewers? What implications do these have on the design procedures adopted?

9.3   Explain the main factors affecting the shape of the DWF diurnal profile.

9.4   Explain what is meant by *dry weather flow*. Define how it is measured, and discuss the limitations of the current approach.

9.5   An urban catchment is drained by a separate foul sewer network and has an area of 500 ha and a population density of 75/hd.d. At the outfall of this catchment, calculate the following:

   a. The average DWF (in L/s) assuming water consumption is 160 L/hd.d, trade effluent is 10 m³/ha.d over 10% of the catchment, and infiltration is 20 L/hd.d
        [84 L/s]

   b. The peak DWF using Babbitt's formula.   [201 L/s]

9.6   If the outfall sewer in Problem 9.5 is 500 mm in diameter with a gradient of 1:200, calculate the following:

   a. The depth of peak flow, assuming $k_s = 1.5$ mm   [325 mm]

   b. The additional population that could be served, assuming that proportional depth does not exceed 0.75.   [2922]

9.7   Redesign the foul sewer network specified in Example 9.3 on a steep site with an inflow of $Q_a = 45$ L/s.

9.8   Explain how the binomial probability distribution forms the basis of the DU small sewer design method.

9.9   It has been estimated that in an office block, each WC is used at peak times every 5 minutes and discharges for 10 seconds. In a group of five WCs, calculate the maximum number discharging simultaneously at the 99.9% confidence level.   [2]

9.10 Redesign the foul sewer network specified in Example 9.3 to serve the residential housing only, using the DU method.

9.11 Calculate the total number of dwellings that can be drained by a 150 mm diameter pipe ($k_s = 1.5$ mm) running with a proportional depth of 0.75 at a gradient of 1:300, using both large and small sewer design methods. Assume 3.5 DUs or 0.046 L/s per dwelling.   [174, 73]

9.12 How many connected households would there need to be on a network to produce continuous flow in the early hours of the morning if at that time the household single equivalent appliance $pq = 0.01$?   [560]

9.13 Explain the main differences in the way gross solids are transported in large and small sewers. How would you expect more widespread use of low flush toilets to affect solid transport?

9.14 What causes sewers to block?

KEY SOURCES

Bizier, P. Ed. 2007. *Gravity Sanitary Sewer Design and Construction*, ASCE Manual and Report No. 60, WEF Manual No. FD-5.

Butler, D. and Graham, N.J.D. 1995. Modeling dry weather wastewater flow in sewer networks. *American Society of Civil Engineers, Journal of Environmental Engineering Division*, 121(2), 161–173.

Swaffield, J.A. and Galowin, L.S. 1992. *The Engineered Design of Building Drainage Systems*, Ashgate.

REFERENCES

Ainger, C.M., Armstrong, R.A., and Butler, D. 1997. *Dry Weather Flow in Sewers*, CIRIA R177.

Approved Document G. 2015. Sanitation, Hot Water Safety and Water Efficiency, The Building Regulations 2010, HM Government.

Babbitt, H.E. 1953. *Sewerage and Sewage Treatment*, 7th edn, John Wiley and Sons.

Blanksby, J. 2006. Water conservation and sewerage systems, Chapter 5, in *Water Demand Management* (eds. D. Butler and F.A. Memon), IWA publishing, Blackpool, UK.

Bramley, E. and Heywood, G. 2005. Towards a new definition of DWF. *WaPUG Autumn Meeting*, Blackpool, UK.

BS EN 752: 2008. *Drain and Sewer Systems Outside Buildings*.

BS EN 12056-2: 2000. *Gravity Drainage Systems Inside Buildings. Part 2: Sanitary Pipework, Layout and Calculation*.

Buchberger, S.G. and Wu, L. 1995. Model for instantaneous residential water demands. *American Society of Civil Engineers, Journal of Hydraulic Engineering*, 121(3), 232–246.

Butler, D., Davies, J.W., Jefferies, C., and Schütze, M. 2003. Gross solids transport in sewers. *Proceedings of Institution of Civil Engineers, Water, Maritime and Energy*, 156(WM2), 165–174.

Davies, J.W., Butler, D., and Xu, Y.L. 1996. Gross solids movement in sewers: Laboratory studies as a basis for a model. *Journal of the Institution of Water and Environmental Management*, 10(1), 52–58.

De Benedittis, J. and Bertrand-Krajewski, J.-L. 2005. Infiltration in sewer systems: Comparison of measurement methods. *Water Science and Technology*, 52, 219–227.

Defra. 1999. *The Water Supply (Water Fittings) Regulations. Statutory Instrument 1999 No. 1148*. HMSO, London.

Drinkwater, A., Chambers, B., and Carmen, W. 2008. *Less Water to Waste. Impact of Reductions in Water Demand on Wastewater Collection and Treatment Systems*. Environment Agency Science Report SC060066.

Fair, J.C. and Geyer, J.C. 1954. *Water Supply and Waste-Water Disposal*, John Wiley and Sons.

Gaines, J.B. 1989. Peak sewage flow rate: Prediction and probability. *Journal of Pollution Control Federation*, 61, 1241.

Gifft, H.M. 1945. Estimating variations in domestic sewage flows. *Waterworks and Sewerage*, 92, 175.

Gormley, M., Mara, D.D., Jean, N., and McDougall, I. 2013. Pro-poor sewerage: Solids modelling for design optimization. *Municipal Engineer*, 166(ME1), 24–34.

Harman, W.G. 1918. Forecasting sewage in Toledo under dry-weather conditions. *Engineering News-Record*, 80, 1233.

Henze, M., Harremoes, P., Arvin, E., and la Cour Jansen, J. 2002. *Wastewater Treatment – Biological and Chemical Processes*, 3rd edn., Springer-Verlag.

Institution of Water and Environmental Management (IWEM). 1993. Glossary. Handbooks of UK Wastewater Practice.

Lauchlan, C., Griggs, J., and Escarameia, M. 2004. *Drainage Design for Buildings with Reduced Water Use*, BRE Information paper, IP 1/04.

Littlewood, K. and Butler, D. 2003. Movement mechanisms of gross solids in intermittent flow. *Water Science and Technology*, 47(4), 45–50.

Mann, H.T. 1979. *Septic Tanks and Small Sewage Treatment Works*, WRc Report No. TR107.

Memon, F., Fidar, A., Littlewood, K., Butler, D., Makropoulos, C., and Liu, S. 2007. A performance investigation of small-bore sewers. *Water Science and Technology*, 55(4), 85–91.

Ministry of Housing and Local Government. 1970. *Technical Committee on Storm Overflows and the Disposal of Storm Sewage*, Final Report, HMSO.

Penn, R., Schütze, M., and Friedler, E. 2013. Modelling the effects of on-site greywater reuse and low flush toilets on municipal sewer systems. *Journal of Environmental Management*, 114, January, 72–83.

Walski, T., Falco, J., McAloon, M., and Whitman, B. 2011. Transport of large solids in unsteady flow in sewers, *Urban Water Journal*, 8(3), 179–187.

Wise, A.F.E. and Swaffield, J.A. 2002. *Water, Sanitary and Waste Services for Buildings*, 5th edn, Taylor and Francis.

Water Research Centre (WRc). 2012. *Sewers for Adoption – A Design and Construction Guide for Developers*, 7th edn., Water UK.

Zhang, X., Buchberger, S.G., and van Zyl, J.E. 2005. A theoretical explanation for peaking factors. *Proceedings of American Society of Civil Engineers, World Water and Environmental Resources Congress: Impacts of Global Climate*, Anchorage, May, 51.

Chapter 10

Storm sewers

10.1 INTRODUCTION

Separate storm sewers (or surface water sewers) form a key component of many urban drainage systems. The emphasis in this chapter is on the design of such pipe-based systems (Section 10.2). Section 10.3 covers the concept of contributing area, and this leads to a detailed discussion of the key stormwater runoff design methods: rational (Section 10.4), time-area (Section 10.5), and hydrograph (Section 10.6). The chapter concludes with a section (Section 10.7) introducing the estimation of runoff from undeveloped sites. Computer-based analysis of existing systems is covered in Chapters 19 and 20, and flooding in Chapter 11. Design of non-pipe-based systems is covered in Chapters 21 and 23.

10.1.1 Flow regime

All storm sewer networks (also called surface water sewer networks) physically connect stormwater inlet points (such as road gullies and roof downpipes) to a discharge point, or outfall, by a series of continuous and unbroken pipes. Flow into the sewer results from the random input over time and space of rainfall-runoff. Generally, these flows are intermittent, of relatively long duration (minutes to hours), and hydraulically unsteady.

Separate storm sewers (more than foul sewers) will stand empty for long periods of time. The extent to which the capacity is taken up during rainfall depends on the magnitude of the event and conditions in the catchment. During low rainfall, flows will be well below the available capacity, but during very high rainfall the flow may exceed the pipe capacity inducing pressure flow and even surface flooding. Unlike in foul sewer design (see Chapter 9), no distinction is made between large and small sewers in the design of storm systems.

10.2 DESIGN

The magnitude and frequency of rainfall are unpredictable and cannot be known in advance, so how are drainage systems designed? The general method has been illustrated in Figure 9.2 (Chapter 9) as a flowchart, and should be read in conjunction with Figure 6.8.

Design is accomplished by first choosing a suitable *design storm*. The physical properties of the storm *contributing area* must then be quantified. A number of methods of varying degrees of sophistication have been developed to estimate the *runoff flows* resulting from rainfall. *Hydraulic design* of the pipework, using the principles presented in Chapter 7, ensures sufficient, sustained capacity. Broader issues of sewer layout including horizontal and vertical alignment are covered in Chapter 6.

10.2.1 Design storm

The concepts of statistically analysed rainfall and the design storm were introduced in Chapter 4. These give statistically representative rainfall that can be applied to the contributing area and converted into runoff flows. Once flows are known, suitable pipes can be designed.

The choice of design storm return period therefore determines the degree of protection from stormwater flooding provided by the system. This protection should be related to the cost of any damage or disruption that might be caused by flooding. In practice, cost-benefit studies are rarely conducted for ordinary urban drainage projects, a decision on design storm return period is made simply on the basis of judgement and precedent.

Standard practice in the United Kingdom (Water Research Centre [WRc], 2012) is to use storm return periods of 1 year or 2 years for most schemes (for steeper and flatter sites, respectively), with 5 years being adopted where property in vulnerable areas would be subject to significant flood damage. Current practice is to design systems such that surface flooding is prevented for storms with return periods up to and including 30 years (WRc, 2012). Flooding from combined sewers into housing areas is likely to be more hazardous than storm runoff flooding of open land, so the type of flooding likely to occur will influence selection of a suitable return period.

Although we can assess and specify design of the rainfall return period, our greatest interest is really in the return period of flooding. It is normally assumed that the frequency of *rainfall* is equivalent to the frequency of *runoff*. However, this is not completely accurate. For example, antecedent soil moisture conditions, areal distribution of the rainfall over the catchment, and movement of rain all influence the generation of stormwater runoff (see Chapters 4 and 5). These conditions are not the same for all rainfall events, so rainfall frequency cannot be identical to runoff frequency. However, comprehensive storm runoff data are less common than rainfall records, and so the assumption is usually the best reasonable approach available.

It is certainly *not* the case, however, that frequency of *rainfall* is equivalent to the frequency of *flooding*. Sewers are almost invariably laid at least 1 m below the ground surface and can, therefore, accommodate a considerable surcharge before surface flooding occurs (see Chapter 7). Hence, the capacity of the system under these conditions is increased above the design capacity, perhaps even doubled. Inspection of Figure 4.2 in Chapter 4 illustrates that a 10-year storm will give a rainfall intensity approximately twice that of a 1-year storm for most durations. It follows, therefore, that where sewers have been designed to a 1-year standard, a surcharge may increase that capacity up to an equivalent of a 10-year storm without surface flooding.

Table 10.1 shows the recommendations made by the relevant European Standard (BS EN 752: 2008) for design storm frequency or return period related to the location of the area

Table 10.1 Recommended design frequencies

Location	Design storm return period (years)	Design flooding return period (years)
Rural areas	1	10
Residential areas	2	20
City centres/industrial/commercial areas:		
• With flooding check	2	30
• Without flooding check	5	—
Underground railways/underpasses	10	50

Source: Adapted from BS EN 752: 2008. Drain and Sewer Systems Outside Buildings.

to be drained. It suggests that a design check should be carried out to ensure that adequate protection against flooding is provided at specific sensitive locations. Design flooding frequencies are also given in the table. More detailed information and advice on designing for surface flooding (exceedance flows) is given in Chapter 11.

10.2.2 Optimal design

Most design is carried out by trial and error, as described in the rest of this chapter. System properties (e.g., pipe diameter and gradient) are proposed and then tested for compliance with design constraints or rules (e.g., self-cleansing velocity). If the rules are violated, a new design is proposed and retested iteratively until satisfactory performance is demonstrated. No real attempt is made to achieve an optimum design, rather one that is fit for purpose. There is a body of academic literature on optimal design of storm sewer systems over many years (e.g., Argaman et al., 1973; Diogo and Graveto, 2006), but the concepts and techniques developed have not, as yet, found their way into routine practice. For a thorough review of previous work and an analysis of why formal optimisation is rarely used, see Guo et al. (2008).

10.2.3 Return period and design life probability of exceedance

As mentioned in Chapter 4, the T-year return period of an annual maximum rainfall event is defined as the long-term average of the intervals between its occurrence or exceedance. Of course, the actual interval between specific occurrences will vary considerably around the average value T, with some intervals being much less than T and others greater.

The probability that an annual event will be exceeded during the design life of the drainage system is derived as follows. The probability that, in any one year, the annual maximum storm event of magnitude X is greater than or equal to the T-year design storm of magnitude x is

$$P(X \geq x) = \frac{1}{T} \tag{10.1}$$

So, the probability that the event will not occur in any one year is

$$P(X < x) = 1 - P(X \geq x) = 1 - \frac{1}{T}$$

and the probability it will not exceed the design storm in N years must be

$$P^N(X < x) = \left(1 - \frac{1}{T}\right)^N$$

The probability r that the event will equal or exceed the design storm at least once in N years is, therefore,

$$r = 1 - \left(1 - \frac{1}{T}\right)^N \tag{10.2}$$

If the design life of a system is N years, there is a probability r that the design storm event will be exceeded at some time in this period. The magnitude of the probability is given by Equation 10.2. Example 10.1 explains how they may be used.

10.3 CONTRIBUTING AREA

The following characteristics of a contributing area are significant for storm sewers: physical area, shape, slope, soil type and cover, land use, roughness, wetness, and storage. Of these, the catchment area and land use are the most important for good prediction of stormwater runoff.

EXAMPLE 10.1

What is the probability that at least one 10-year storm will occur during the first 10-year operating period of a drainage system? What is the probability over the 40-year lifetime of the system?

Solution

First 10 years: $T = 10$, $N = 10$.
 The answer is not

$$r = 1/T = 0.1$$

or

$$r = 10 \times 1/T = 1.0$$

It is (Equation 10.2)

$$r = 1 - (1 - 0.1)^{10} = 0.651$$

Thus, there is a 65% probability that at least one 10-year design storm will occur within 10 years. In fact, it can be shown that for large T, there is s 63% risk that a T-year event will occur within a T-year period.
 Lifetime: $T = 10$, $n = 40$.

$$r = 1 - \left(1 - \frac{1}{10}\right)^{40} = 0.985$$

In general, a very high return period is required if probability of exceedance is to be minimised over the lifetime of the system.

10.3.1 Catchment area measurement

The boundaries of the complete catchment to be drained can be defined with reasonable precision either by field survey or the use of contour maps. They should be positioned such that any rain that falls within them will be directed (normally under gravity) to a point of discharge or outfall.

After the preliminary sewer layout has been produced, the catchment can be divided into sub-catchment areas draining towards each pipe or group of pipes in the system. The sub-areas can then be measured using a geographic information system (GIS)–based package, normally integrated within sewer simulation software. Aerial photographs may also be used. For simplicity, it is assumed that all flow to a sewer length is introduced at its head (that is, at the upstream manhole).

10.3.2 Land use

Once the total catchment area has been defined, estimates must be made of the extent and type of surfaces that will drain into the system. The *percentage imperviousness* (PIMP) of each area is measured by defining impervious surfaces as roads, roofs, and other paved surfaces (Equation 5.5). Measurement can be carried out using maps, aerial photographs, or satellite images (Castelluccio et al., 2015; Finch et al., 1989; Scott, 1994). Table 10.2 and Figure 10.1 illustrate a land-use classification in London.

Alternatively, the percentage imperviousness (PIMP) can be related approximately to the density of housing development using the following relationship:

$$PIMP = 6.4\sqrt{J} \quad 10 < J < 40 \tag{10.3}$$

where J is the housing density (dwellings/ha).

10.3.3 Urban creep

Urban creep is the term used to denote the gradual increase in imperviousness (*PIMP*) of *existing* urban areas. This is caused by the paving over of residential front gardens (typically for parking), and the construction of patios, extensions, conservatories, and so on. Perry and Nawaz (2008) found that between 1971 and 2004 the imperviousness of an existing suburban area of Leeds, UK, increased by 13%, with 75% of this due to the paving of front gardens. The average rate of urban creep in five UK areas between 1999 and 2006 was estimated as 0.38–1.09 m²/dwellings/year (Allitt et al., 2010). It has also been shown to be related to the existing housing density (see Figure 10.2).

The impact of this increased imperviousness will clearly be felt on the magnitude of runoff from any given storm or series of storms (Kelly, 2016). In the Leeds study, the creep was estimated to result in an average annual runoff increase of 12% (Perry and Nawaz 2008). Wright et al. (2011) point out that urban creep can also be expected to have an impact on water quality, due to higher levels of deposited pollutants being washed off during rainfall events.

Table 10.2 Approximate percentage imperviousness of land-use types in London

Land-use category	PIMP
Dense commercial	100
Open commercial	65
Dense housing	55
Flats	50
Medium housing	45
Open housing	35
Grassland	<10
Woodland	<10

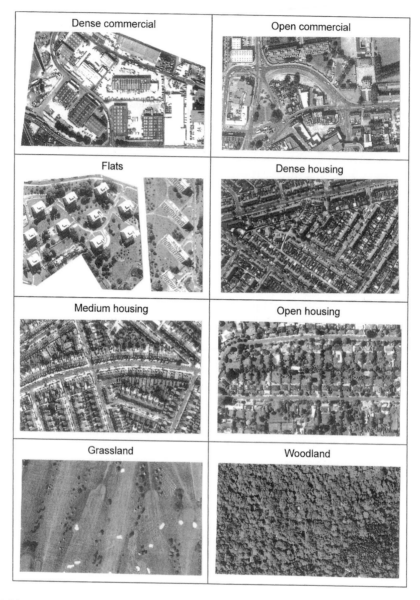

Figure 10.1 Various land use categories in London. Developed by Thames Water Utilities Ltd on their Beckton and Crossness sewerage modelling projects in association with consulting engineers BGP Reid Crowther Ltd and Montgomery Watson Ltd and reproduced with permission.

In the absence of any local evidence or regulatory requirement, a design allowance of a 10% increase in the catchment impervious area (A_i) can be assumed to account for future urban creep.

10.3.4 Runoff coefficient

The dimensionless *runoff coefficient* C is defined in Chapter 5 as the proportion of rainfall that contributes to runoff from the surface. Early workers such as Lloyd-Davies (1906)

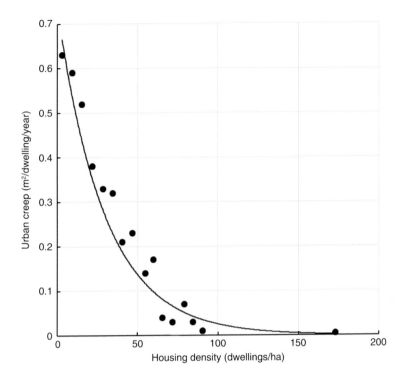

Figure 10.2 The relationship between urban creep and housing density. Based on (HRWallingford. 2012. *Development of Spatial Indicators to Monitor Changes in Exposure and Vulnerability to Flooding and the Uptake of Adaptation Actions to Manage Flood Risk in England*, Report TN-MCS0743-01 R2, www.theccc.org.uk.)

assumed that 100% runoff came from impervious surfaces and 0% from pervious surfaces, so $C = PIMP/100$, and this assumption is still commonly adopted.

However, the coefficient actually accounts for the initial runoff losses (e.g., depression storage) and continuing losses (e.g., surface infiltration), and implicitly accounts for the hydrodynamic effects encountered as the water flows over the catchment surface. Therefore, C must be related to $PIMP$, but does not necessarily have to equal it—some runoff will come from pervious surfaces, for example. Equation 5.4 in Chapter 5 shows clearly that $C = PR/100$ is related to $PIMP$ plus soil type and antecedent conditions. So, considerable knowledge of the catchment is required for accurate determination.

For design purposes, standard values of C such as those in Table 10.3 are often used. Weighted average coefficients are needed for areas of mixed land use.

Table 10.3 Typical values of runoff coefficient in urban areas

Area description	Runoff coefficient	Surface type	Runoff coefficient
City centre	0.70–0.95	Asphalt and concrete paving	0.70–0.95
Suburban business	0.50–0.70	Roofs	0.75–0.95
Industrial	0.50–0.90	Lawns	0.05–0.35
Residential	0.30–0.70		
Parks and gardens	0.05–0.30		

Source: Adapted from Urban Water Resources Council. 1992. *Design and Construction of Urban Stormwater Management Systems*, ASCE Manual No. 77, WEF Manual FD-20.

10.3.5 Time of concentration

An important term used in storm sewer design is *time of concentration* (t_c). It is defined as the time required for surface runoff to flow from the most remote part of the catchment area to the point under consideration. Each point in the catchment has its own time of concentration. It has two components: the overland flow time, known as the *time of entry* (t_e), and the channel or sewer flow time, the *time of flow* t_f. Thus,

$$t_c = t_e + t_f \qquad (10.4)$$

10.3.5.1 Time of entry

The time of entry varies with catchment characteristics such as surface roughness, slope and length of flow path, together with rainfall characteristics. Table 10.4 shows ranges of values dependent on storm return period; rarer, heavier storms produce more water on the catchment surface and, hence, faster overland flow.

10.3.5.2 Time of flow

Velocity of flow in the sewers can be calculated from the hydraulic properties of the pipe, using one of the methods described in Chapter 7. Pipe-full velocity is normally used as a good approximation over a range of proportional depths. If sewer length is known or assumed, time of flow can be calculated.

10.4 RATIONAL METHOD

The Rational Method has a long history dating back to the middle of the nineteenth century. The Irish engineer Mulvaney (1850) was probably the first to publish the principles on which the method is based, although Americans tend to credit Kuichling (1889) and the British credit Lloyd-Davies (1906) for the method itself. The method and its further development are described below.

10.4.1 Steady-state runoff

Consider a simple, flat, fully impervious rectangular catchment with area A. A depth of rain, I, falls in a time, t. If there were also an impervious wall along the edges of the catchment, and no sewers, this rain would simply build up over the area to a depth, I. The volume of water would be $I \times A$.

Now imagine that the runoff is flowing into a sewer inlet at point X with steady-state conditions: water is landing on the area, and flowing away, at the same rate. The sewer will

Table 10.4 Time of entry

Return period (year)	Time of entry (minute)
1	4–8
2	4–7
5	3–6

Source: After Department of Energy/National Water Council (DoE/NWC).

carry the volume of rain ($I \times A$) at a steady, constant rate over the time (t) of the rainfall. So for flow rate (Q),

$$Q = \frac{IA}{t}$$

and since the intensity of rain, $i = I/t$,

$$Q = iA$$

Now, since catchments are not totally impervious, and there will be initial and continuing losses, the runoff coefficient C can be introduced, to give

$$Q = CiA \tag{10.5}$$

Adjusting for commonly used units gives

$$Q = 2.78CiA \tag{10.6}$$

where Q is the maximum flow rate (L/s), i is the rainfall intensity (mm/h), and A is the catchment area (ha).

10.4.2 Critical rainfall intensity

For this method to be used for design purposes, the rainfall intensity that causes the catchment to operate at steady state needs to be known. This should give the maximum flow from the catchment. The Rational Method states that a catchment just reaches steady state when the duration of the storm (and hence intensity i) is equal to the time of concentration of the area.

But why is this? Figure 10.3a shows a hydrograph resulting from uniform rainfall with duration less than the time of concentration. Figure 10.3b gives the hydrograph for the same catchment, resulting from the same uniform rainfall intensity, but this time with infinite duration. The peak flow in Figure 10.3a is the lower because the entire catchment is not contributing together (at steady state): contributions from remote parts of the catchment are

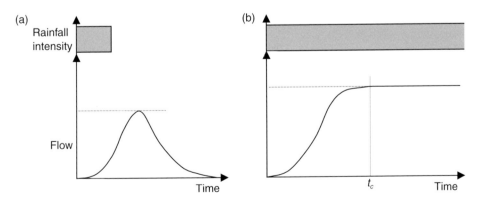

Figure 10.3 Hydrograph response to different duration rainfall of same intensity.

still arriving after contributions from near parts have ceased. The maximum flow is reached when all the catchment contributes together—that is, when time is equal to or greater than the time of concentration t_c, as in Figure 10.3b.

The basis of the Rational Method is, therefore, an engineering "worst case." The duration of the storm must be at least the time of concentration; otherwise, the maximum flow would not be reached. However, it should not be longer, because storms with longer durations have statistically lower intensities (see Figure 4.2). Therefore, the worst case is when the duration is equal to the time of concentration (see Example 10.2).

10.4.2.1 Small areas

BS EN 752: 2008 recommends use of a fixed rainfall intensity of 0.014 L/s/m² (equivalent to 50 mm/h) for small areas (main sewer length <200 m). This avoids using inappropriately high intensities calculated using very low concentration times. It also simplifies design using the Rational Method, since it is not necessary to know the time of concentration for each sub-catchment in order to identify the critical rainfall intensity (as described above).

EXAMPLE 10.2

A new housing estate is to be drained by a separate storm sewer network. The estate is rectangular in plan, 1200 × 900 m, and will consist of approximately 30% paved and roofed surfaces. Determine the maximum capacity required of the sewer carrying storm-water from the whole area. The longest branch leading to this point is 1350 m.

Assume that the average velocity is 1.5 m/s and the time of entry is 4 minutes. Use Equation 4.2 for rainfall calculation where a = 750, b = 10, α = 1, and β = 1.

Solution

$$t_c = 4 + \frac{1350}{1.5 \times 60} = 19 \text{ min} \qquad i = \frac{750}{19 + 10} = 26 \text{ mm/h}$$

$PIMP = 30\%$.

Assume 100% runoff from impervious areas and 0% from pervious areas; hence,

C = PIMP/100 = 0.3
Q = CiA

$$Q = 0.3 \frac{26 \times 10^{-3}}{60 \times 60} 1200 \times 900 = 2.34 \text{ m}^3/\text{s}$$

10.4.3 Modified Rational Method

Increased understanding of the rainfall-runoff process has led to further development of the Rational Method. The Modified Rational Method is recommended in the *Wallingford Procedure* (Department of Energy/National Water Council [DoE/NWC], 1981) and is shown to be accurate for UK catchment sizes up to 150 ha.

In this approach, the runoff of rainfall is disaggregated from other routing effects. Thus, the runoff coefficient C is considered to consist of two components:

$$C = C_v C_R \tag{10.7}$$

where C_v is the volumetric runoff coefficient (–), and C_R is the dimensionless routing coefficient (–).

10.4.3.1 Volumetric runoff coefficient (C_v)

This is the proportion of rainfall falling on the catchment that appears as surface runoff in the drainage system. The value of C_v depends on whether the whole catchment is being considered, or just the impervious areas alone. Assuming the latter,

$$C_v = \frac{PR}{PIMP} \tag{10.8}$$

where PR is given by Equation 5.4 in Chapter 5, and $PIMP$ is given by Equation 5.5. Under summer rainfall conditions, C_v ranges from 0.6 to 0.9, with the lower values pertaining to rapidly draining soils and higher values to heavy clay soils. Note that in this method C_v is calculated, not assumed.

10.4.3.2 Dimensionless routing coefficient (C_R)

The dimensionless routing coefficient C_R varies between 1 and 2, and accounts for the effect of rainfall characteristics (e.g., peakedness) and catchment shape on the magnitude of peak runoff. A fixed value of 1.30 is recommended for design. So, for peak flow Q_p,

$$Q_p = 2.78 \times 1.30 C_v i A_i$$

where A_i is the impervious area (ha):

$$Q_p = 3.61 C_v i A_i \tag{10.9}$$

10.4.4 Design criteria

It is good practice to follow a number of basic criteria during the design process.

10.4.4.1 Capacity

Storm sewers should be designed (size and gradient) to convey the predicted peak flows. It is conventional to design the pipes to just run full (e.g., to $d/D = 1.0$) but not surcharged. The small extra capacity that can be achieved at just below full flow (see Chapter 7) is neglected.

10.4.4.2 Self-cleansing

In addition to capacity, the pipe should also be designed to achieve self-cleansing. This is achieved by ensuring a specific velocity is reached at the design flow (1 m/s is typically used). In practice, the pipe size and gradient are manipulated together to obtain the best design.

For sewers designed for capacity with long return period storms (e.g., 10+ years), a more conservative approach is to design for self-cleansing with more frequent events (e.g., 1 year).

10.4.4.3 Roughness

As with foul sewers, it is conservatively assumed for design purposes that the pipe roughness will be independent of pipe material, although sliming is not a major issue. BS EN 752-4 recommends a k_s value of 0.6 mm for storm sewers.

10.4.4.4 Minimum pipe sizes

The minimum pipe size is generally set at similar levels to those for foul sewers (see Chapter 9).

10.4.5 Design method

The following procedure should be followed for the Modified Rational Method:

1. Assume design rainfall return period (T), pipe roughness (k_s), time of entry (t_e), and volumetric runoff coefficient (C_v).
2. Prepare a preliminary layout of sewers, including tentative inlet locations.
3. Mark pipe numbers on the plan according to the convention described in Chapter 6.
4. Estimate impervious areas for each pipe.
5. Make a first attempt at setting gradients and diameters of each pipe.
6. Calculate pipe-full velocity (v_f) and flow rate ($Q_f = \pi r D^2 v_f / 4$).
7. Calculate time of concentration from Equation 10.4. For downstream pipes, compare alternative feeder branches and select the branch resulting in the maximum t_c.
8. Read rainfall intensity from intensity-duration-frequency (IDF) curves (see Chapter 4) for $t = t_c$ (for design T).
9. Estimate the cumulative contributing impervious area.
10. Calculate Q_p from Equation 10.9.
11. Check $Q_p < Q_f$ and $v_{max} > v_f > v_{min}$.
12. Adjust pipe diameter and gradient as necessary (given hydraulic and physical constraints) and return to step 5.

This is essentially a manual calculation procedure, although software packages are available to automate the repetitive calculations. Example 10.3 illustrates how the method may be used to design a simple storm sewer network.

10.4.6 Limitations

The Rational Method is based on the following assumptions.

1. The rate of rainfall is constant throughout the storm and uniform over the whole catchment.
2. Catchment imperviousness is constant throughout the storm.
3. Contributing impervious area is uniform over the whole catchment.
4. Sewers flow at constant (pipe-full) velocity throughout the time of concentration.

Assumption 1 can underestimate, as can assumption 3 (this is explored further in the next section). But, assumption 2 tends to overestimate, as does assumption 4—sewers do not always run full, and storage effects reduce peak flow. Fortunately, in many cases, these inaccuracies cancel each other out, producing a reasonably accurate result. Thus, the Rational Method and its modified version are simple, widely used approaches suitable for first approximations in most situations and are appropriate for full design in small catchments (<150 ha).

EXAMPLE 10.3

A simple storm sewer network is shown in Figure 10.4. Appropriate rainfall data are given in Table 10.5 and network data in columns (1)–(4) of Table 10.6. Assume pipe gradients are fixed.

Design the network using the Modified Rational Method for a 2-year return period storm using a volumetric runoff coefficient of 0.9 and a time of entry of 5 minutes. Pipe roughness is 0.6 mm.

Solution

Pipe-full velocities and capacities are calculated using the Colebrook-White equation. The design is completed in Table 10.6.

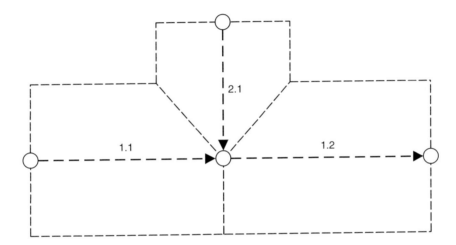

Figure 10.4 System layout (Example 10.3).

Table 10.5 Rainfall intensities (mm/h) at network site (Example 10.3)

Duration (minutes)	I	Return period (years) 2	5
6.0	50.0	61.7	81.6
6.2	49.3	60.7	80.4
6.4	48.5	59.8	79.2
6.6	47.8	58.9	78.1
6.8	47.1	58.0	77.0
7.0	46.4	57.2	75.9
7.2	45.8	56.4	74.9
7.4	45.2	55.6	73.9
7.6	44.5	54.8	72.9
7.8	44.0	54.1	71.9
8.0	43.4	53.4	71.0
8.2	42.8	52.7	70.1
8.4	42.3	52.0	69.3
8.6	41.8	51.4	68.4
8.8	41.2	50.7	67.6
9.0	40.8	50.1	66.8

Table 10.6 Design table for Modified Rational Method (Example 10.3)

Return period	2 years
pipe roughness	0.6 mm
time of entry	5 minutes
coefficient, C_v	0.9

(1) Pipe number	(2) Pipe length (m)	(3) Pipe gradient (1:x)	(4) Impervious area (ha)	(5) Sum impervious area (ha)	(6) Assumed diameter (mm)	(7) Pipe-full velocity (m/s)	(8) Pipe capacity (l/s)	(9) Time of flow (minutes)	(10) Time of concentration (minutes)	(11) Rainfall intensity (mm/h)	(12) Calculated discharge (l/s)	(13) Comments
1.1	120	200	0.4	0.4	225	0.9	35	2.2	7.2	56.4	73.3	$Q_p > Q_f$
					300	1.1	80	1.8	6.8	58.0	75.4	OK
2.1	100	200	0.6	0.6	300	1.1	80	1.5	6.5	59.4	115.8	$Q_p > Q_f$
					375	1.3	140	1.3	6.3	60.3	117.5	OK[a]
1.2	150	400	0.8	1.8	600	1.2	350	2.1	8.9	50.4	295	OK[a]

[a] More cost-effective solutions could be arrived at by varying pipe gradient.

Two points about the methodology are stressed. First, it is important that calculations are carried out for each pipe in turn, and that area and time of concentration refer to the whole upstream contributing area not just to the local subcatchment area. The second point is that each pipe will be designed for a *different* (critical) design storm, with shorter-duration, higher-intensity storms used for upstream pipes (because they have a shorter time of concentration) and longer-duration, lower-intensity storms used for downstream sections.

10.5 TIME-AREA METHOD

10.5.1 The need

Area is treated as a constant in the Rational Method. In reality, the contributing area is not constant. For example, during the beginning of rainfall, the area builds up with time, closest surfaces contributing first, more distant ones later. In many cases, the Rational Method is appropriate even though it does not take this type of effect into account; in others it is not. This can be demonstrated by considering the three simple cases that follow.

Case (1)
The main storm sewers for a proposed industrial estate are shown in plan on Figure 10.5a. The capacity required at X is calculated using the Rational Method. Assume a catchment of area $A = 100{,}000 \text{ m}^2$, $C = 0.6$, $t_e = 4$ minutes, $v_f = 1.5$ m/s, and length of the longest sewer is 450 m. Now, by using Equation 4.2 (a = 750, b = 10, $\alpha = 1$, and $\beta = 1$), for rainfall intensity (i) related to duration (D),

$$i = \frac{750}{D + 10}$$

We have

$$t_c = 4 + [450/(1.5 \times 60)] = 9 \text{ minutes}$$

$$i = \frac{750}{9 + 10} = 39 \text{ mm/h}$$

$$Q = \frac{39}{1000 \cdot 60 \cdot 60} \times 100{,}000 \times 0.6 = 0.65 \text{ m}^3/\text{s}$$

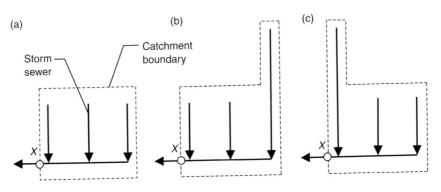

Figure 10.5 Layout of storm sewers.

Case (2)

An alternative layout (Figure 10.5b) is now considered in which the area is increased and one of the sewers is extended. How much additional flow will there be at X?

The same rainfall formula is used, C is still 0.6, the assumed velocity and time of entry are unchanged. However, now A has been increased to 112,000 m² and the length of longest sewer increased to 720 m. This gives

$$t_c = 4 + 720/(1.5 \times 60) = 12 \text{ minutes}$$

$$i = \frac{750}{12 + 10} = 34 \text{ mm/h}$$

$$Q = \frac{34}{1000 \cdot 60 \cdot 60} \times 112{,}000 \times 0.6 = 0.63 \text{ m}^3/\text{s}$$

It appears that the flow at X is now less, even though the area has increased. So, it is the increase in the time of concentration, and consequent reduction in *i*, rather than the increase in area, which has had the greatest effect on Q.

In this case, the Rational Method is inappropriate because it does not consider the "worst case." If rainfall lasting 9 minutes fell on this catchment, the whole of the original 100,000 m² would contribute together, and a flow of 0.65 m³/s, as determined for case (1) would be produced. The value for flow of 0.63 m³/s would underestimate the capacity required.

Case (3)

Now consider another alternative. In Figure 10.5c, A has been increased to 112,000 m² (again), but because of the shape of the extended catchment, the length of longest sewer is no greater than it was for case (1). Therefore, *i* remains as 39 mm/h, so

$$Q = \frac{39}{1000 \cdot 60 \cdot 60} \times 112{,}000 \times 0.6 = 0.73 \text{ m}^3/\text{s}$$

This is an increase on 0.65 m³/s, which makes more sense. The difference is that in case (3), the extra area contributes rapidly (because it is close to X) and does not increase the time of concentration. So the key to whether the Rational Method gives appropriate answers or not is *the way the contributing area builds up with time.*

10.5.2 Basic diagram

A time-area diagram attempts to overcome one of the main limitations of the basic Rational Method by representing the rate of contribution of area (CA). Now for case (1), if the contributing area discharging towards point X builds up at a constant rate, the time-area diagram will be as shown in Figure 10.6a.

It is possible to determine whether or not the Rational Method is appropriate for any particular shape of the time-area diagram by rearranging the basic rational expression:

$$Q = CAi = CA\left[\frac{750}{D + 10}\right] = 750\left[\frac{CA}{D + 10}\right]$$

Thus, the higher the value of $[(C \cdot A)/(D + 10)]$, the greater is the value of Q.

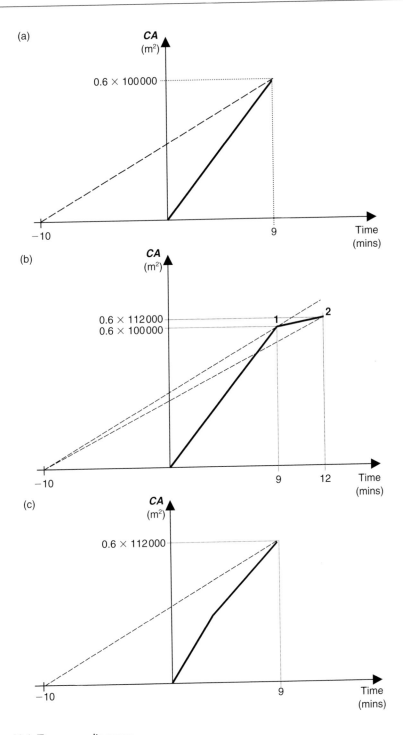

Figure 10.6 Time-area diagrams.

How can we visualise $[(C \cdot A)/(D+10)]$? The gradient of the straight line representing the time-area relationship on Figure 10.5a is $[(C \cdot A)/D]$, but if the point $(D = -10, A = 0)$ is connected to any point on the time-area relationship, a line is produced with a gradient of $[(C \cdot A)/(D+10)]$. The highest value of this gradient, and therefore the highest value of Q, is given by the dashed line on Figure 10.6a.

The time-area diagram for case (2) is given in Figure 10.6b. If the dashed line is drawn through point 1 rather than point 2, it will have the steepest gradient; therefore, the value of Q will be greatest.

The line through 1 looks like a tangent and is the basis of a method called the *Tangent Method* (Reid, 1927), which has now been largely superseded. However, it still provides a means for determining whether the Rational Method gives the worst case for a particular time-area diagram. When a tangent of this type can be drawn, it means that the Rational Method has *not* given the worst case.

The time-area diagram for case (3) is given in Figure 10.6c. No tangent can be drawn in this case; therefore, the Rational Method *does* give the worst case. (Note that the tangents in Figure 10.6a–c were drawn from time $= -10$ minutes only because of the particular formula used for rainfall.)

10.5.3 Diagram construction

The diagram is used for storm sewer design by assuming that the time-area plot for each individual pipe sub-catchment is linear. However, the design of each pipe is not only concerned just with the local sub-catchment (in a similar way to the Rational Method), but also with the "concentrating" flows from upstream pipes. The combined time-area diagram for each pipe can be produced using the principle of linear superposition. This is illustrated by the case of a simple two-pipe system (and assuming $C = 1$) in Figure 10.7a.

In this network, the time of concentration of pipe 1 is $t_{c(1)}$, and this is plotted directly onto the time-area plot (Figure 10.7b). Pipe 2 has a sub-catchment time of concentration of $t_{c(2)}$, relative to its own outfall. However, this diagram is not directly overlaid on the previous one but is lagged by the time of flow in pipe 1, $t_{f(1)}$. The ordinates of the two separate diagrams are added to produce the complete diagram.

The resulting time-area diagram is made up of linear segments, but as more individual pipes are added, the shape tends to become non-linear. Example 10.4 illustrates a more complex network.

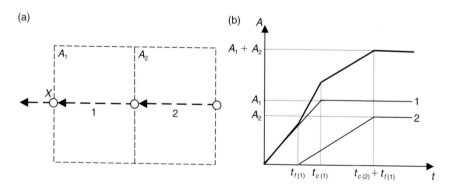

Figure 10.7 Example construction of time-area diagram: (a) simple storm sewer network; (b) time-area diagram for point X.

10.6 HYDROGRAPH METHODS

One of the major weaknesses of the Rational Method is that it only produces worst-case design flow and not a hydrograph of flow against time. Hydrograph methods have been developed to overcome this limitation.

10.6.1 Time-Area Method

The Time-Area Method uses the time-area diagram to produce not only a peak design flow, but also a flow hydrograph. The method also allows straightforward use of time-varying rainfall—the design storm (see Chapter 4).

Equation 5.8 from Chapter 5 is repeated below, and this gives the basic equation for finding flow $Q(t)$ when a continuous time-area diagram is combined with rainfall depth increments, $I_1, I_2, \ldots I_N$.

$$Q(t) = \sum_{\omega=1}^{N} \frac{dA(j)}{dt} I_\omega \qquad (10.10)$$

where $Q(t)$ is the runoff hydrograph ordinate at time t (m³/s), $dA(j)/dt$ is the slope of the time-area diagram at time j (m²/s), I_ω is the rainfall depth in the ωth of N blocks of duration Δt (m), and j is $t - (\omega - 1) \Delta t(s)$.

EXAMPLE 10.4

A storm sewer network has been designed initially with the layout shown in Figure 10.8, and data given (in columns (1)–(5)) in Table 10.7. Construct the time-area diagram at the network outfall assuming a time of entry of 2 minutes and a Manning's n of 0.010.

Solution

For simplicity, assume the pipes run full to calculate flow velocities and, hence, time of flow.

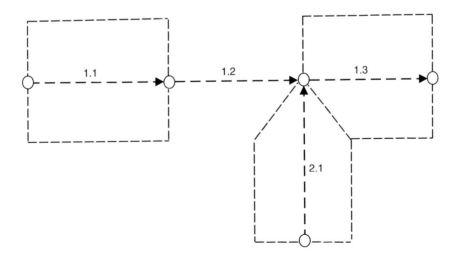

Figure 10.8 System layout (Example 10.4 and 10.5).

Table 10.7 Data for Example 10.4

(1) Pipe number	(2) Diameter (mm)	(3) Pipe length (m)	(4) Pipe gradient (1:x)	(5) Impervious area (ha)	(6) Pipe-full velocity (m/s)	(7) Time of flow (minutes)
1.1	375	1000	300	2.5	1.2	13.9
1.2	450	500	400	0	1.2	7.1
2.1	300	620	250	2.5	1.1	9.2
1.3	600	300	500	1.0	1.3	4.0

Time of flow data and time of entry information are used to derive the time-area diagram (Figure 10.8):

1.3 $t_{c(1.2)} = 6$ minutes, flow starts to contribute at $t = 0$ minutes

2.1 $t_{c(2.0)} = 11.2$ minutes, $t_{f(1.2)} = 4.0$ minutes

1.1 $t_{c(1.0)} = 15.9$ minutes, $t_{f(1.1)} + t_{f(1.2)} = 11.1$ minutes

Assuming linear incremental change in the time-area diagram, $\Delta A_1, \Delta A_2 \ldots \Delta A_j \ldots$ over rainfall time blocks, $\Delta t_1, \Delta t_2 \ldots \Delta t_j \ldots$, runoff is given by

$$Q(t) = \sum_{\omega=1}^{N} \Delta A_j i_{e\omega} \quad (10.11)$$

where i_e is $I/\Delta t$.

This can be expanded to give

$$Q(1) = A_1 i_1$$

$$Q(2) = A_2 i_1 + A_1 i_2$$

$$Q(3) = A_3 i_1 + A_2 i_2 + A_1 i_3$$

...

The method is summarised as follows:

- Select a suitable integer time interval Δt, typically $t_c/10$.
- Prepare a suitable rainfall hyetograph using Δt as the time interval.
- Produce a time-area diagram for each design point under consideration.

Calculate outflow by reading off the relevant rainfall intensity from the hyetograph (i_1) and contributing area (A_j) from the time-area diagram for each time increment (Δ_t). These need to be accumulated for each time step according to Equation 10.11.

Example 10.5 illustrates how the method may be used to design a simple storm sewer network.

10.6.1.1 Limitations

This method has the advantage over the Rational Method in that it takes some account of the shape of the catchment, allowing an output hydrograph to be produced and to include the effects caused by time-varying rainfall. However, the method still only allows linear translation of the flood wave through the catchment and makes no allowance for storage effects.

EXAMPLE 10.5

Using the time-area diagram developed in Figure 10.9, evaluate the outfall flow hydrograph for the sewer network in Figure 10.7 under the following design storm.

Time (minutes)	Effective rainfall depth (mm)
0–5	3
5–10	6
10–15	3

Solution

Using a 5-minute time increment ($\Delta t = 5$ minutes), read off the cumulative contributing area at each time step and then the incremental area. Convert the rainfall depths to intensities.

Δt	Time (minutes)	ΣA (ha)	ΔA (ha)	i_e (mm/h)
1	5	1.4	1.4	36
2	10	2.6	1.2	72
3	15	4.2	1.6	36
4	20	4.9	0.7	
5	25	5.7	0.8	
6	30	6.0	0.3	
			$\Sigma 6.0$	

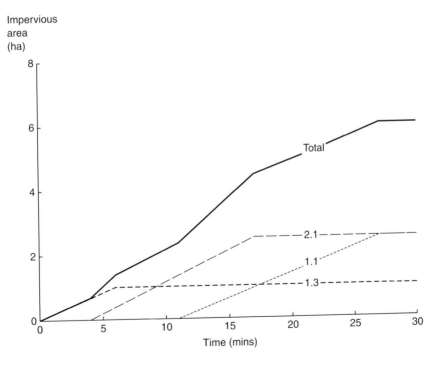

Figure 10.9 Time-area diagram (Example 10.4).

From Equation 10.11,

$Q(1) = 1.4 \times 36 = 50.4$

$Q(2) = 1.2 \times 36 + 1.4 \times 72 = 144.0$

$Q(3) = 1.6 \times 36 + 1.2 \times 72 + 1.4 \times 36 = 194.4$

$Q(4) = 0.7 \times 36 + 1.6 \times 72 + 1.2 \times 36 = 183.6$

$Q(5) = 0.8 \times 36 + 0.7 \times 72 + 1.6 \times 36 = 136.8$

$Q(6) = 0.3 \times 36 + 0.8 \times 72 + 0.7 \times 36 = 93.6$

$Q(6) = 0.3 \times 72 + 0.8 \times 36 = 50.4$

$Q(6) = 0.3 \times 36 = 10.8$

The ordinates should be multiplied by 2.78 to obtain the hydrograph flow in litres per second (L/s), from which the peak flow is

$Q_p \approx 550 \, l/s \, @ \, 17 \, minutes$

Note: this exceeds the capacity of the pipe.

10.6.2 Level pool routing method

In the Rational and basic Time-Area methods, pipe-full velocity is used as the flow routing velocity. However, in reality this will not be the case, and many parts of the network will operate part-full. In this method, storage in the network is taken into account in a relatively simple way using a *level pool* or *reservoir* routing technique (explored further in Section 13.4). The method assumes that the retained water acts as a level reservoir of liquid with outflow being uniquely related to water level or storage. This is accomplished in a pipe network by assuming that proportional depth of flow is identical at all points in the system. This assumption is valid if

- The system is designed with a reasonable degree of taper.
- All of the pipes in the system are geometrically similar.

Both of these requirements should be satisfied in new systems.

Consider an individual circular pipe of length L carrying flow with proportional depth d/D. Thus, $A = f_1(d/D)$ and $V = f_2(d/D)$, where A is the cross-sectional area of flow in the pipe, and V is the volume of water stored. Now, if d/D is constant everywhere, $S = f_3(d/D)$, where S is the whole system retention (storage volume). For a given slope and pipe roughness, outflow at the design point, $O = f_4(d/D)$. Therefore,

$$O = f_5(S) \text{ or } S = f_6(O) \tag{10.12}$$

Thus, there is a unique relationship between total volume of water stored (system retention) and the outflow rate at the design point.

So, knowing the inflow I, flow routing can be performed using the basic storage equation to estimate O. Example 10.6 illustrates application of the method.

EXAMPLE 10.6

Calculate the outflow hydrograph and peak for the sewer network in Figure 10.8 accounting for in-pipe storage.

Solution

Initially, derive a relationship between S and O for circular pipes. Table 7.5 gives

$$A = \frac{D^2}{8}(\theta - \sin\theta) = \frac{\pi D^2}{4}\left(\frac{\theta - \sin\theta}{2\pi}\right)$$

where θ is defined in Equation 7.17.

Now $\pi D^2/4$ is the cross-sectional area of each pipe. So if L is the length of each pipe of diameter D,

$$S = \left(\Sigma L \frac{\pi D^2}{4}\right)\left(\frac{\theta - \sin\theta}{2\pi}\right)$$

Hence, storage can be derived from the system data:

Pipe number	Diameter (mm)	Pipe length (m)	LD^2
1.1	375	1000	140.6
1.2	450	500	101.3
2.1	300	620	55.8
1.3	600	300	108.0
			$\Sigma 405.7$

$$S = 50.7(\theta - \sin\theta)$$

Manning's Equation 7.23 gives

$$O = \frac{1}{n} A R^{2/3} S_o^{1/2}$$

Therefore,

$$O = \frac{1}{n} \frac{D^2}{8} (\theta - \sin\theta) \left(\frac{D}{4}\left[\frac{\theta - \sin\theta}{\theta}\right]\right)^{2/3} S_o^{1/2}$$

$$O = \frac{1}{n} \frac{D^{8/3}}{20.16} \frac{(\theta - \sin\theta)^{5/3}}{\theta^{2/3}} S_o^{1/2} \tag{10.13}$$

For the outfall pipe (1.3), $D = 0.6$ m, $S_o = 0.002$, and $n = 0.010$ can be substituted into Equation 10.13. So, by varying θ from $0 \to 2\pi$ (i.e., d/D: $0 \to 1$), S and O can be plotted together as in Figure 10.10a.

Using a time step of 5 minutes, construct the relationship between $(S/\Delta t) + (O/2)$ and O, also shown (Figure 10.10b).

The calculation now follows the procedure in Example 13.2, with the result shown in Figure 10.11.

The routed peak flow is seen to be

$$O_p \approx 370 l/s @ 25 \text{ minutes}$$

which is a considerable reduction from the previous estimate, and now within the capacity of the pipe.

10.6.3 Limitations

The level pool routing method is an advance over the basic time-area method in that it takes pipe storage effects directly into account. However, the routing assumption is relatively crude, and indeed no account is taken of surface storage effects, and surcharge effects cannot be handled.

Coverage of these aspects effectively requires flow simulation models that have greater power than any of the design-orientated tools described in this chapter. These are described in Chapter 19.

(a)

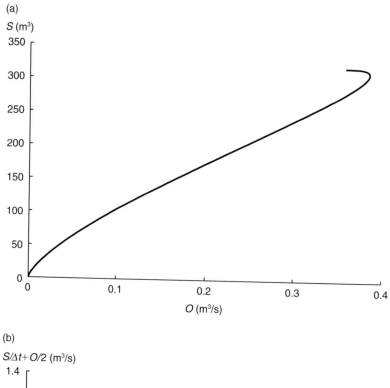

(b)

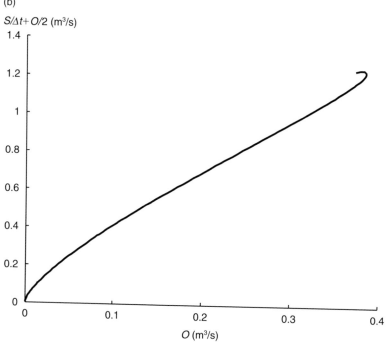

Figure 10.10 (a) Graph of *S* against *O* (Example 10.6). (b) Graph of *S*/Δ*t* 1 *O*/2 against *O* (Example 10.6).

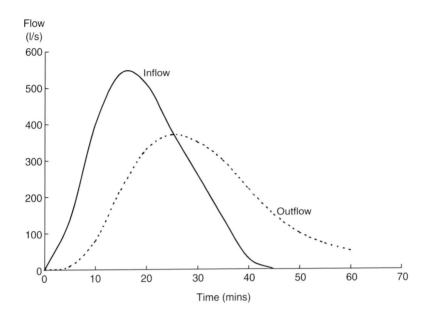

Figure 10.11 Routed hydrograph (Example 10.6).

10.7 UNDEVELOPED SITE RUNOFF

A common requirement when developing a particular site is an estimate of its undeveloped (greenfield) runoff response. This is challenging because different approaches and methods give widely differing results. However, Woods Ballard et al. (2016) provide a comprehensive review of these methods in *The SuDS Manual* and recommend use of the *FEH*-based *Revitalised Flood Hydrograph Model* (ReFH 2.2). Further details of the method, its development, and its application have been provided by Wallingford HydroSolutions (WHS, 2016), and a software package can be downloaded from www.hydrosolutions.co.uk/software/refh-2/. The model draws on both the *FEH* 1999 and 2013 rainfall model data available from the *FEH* Web Service (Chapter 4) and has been integrated with commonly used urban drainage software packages.

PROBLEMS

10.1   Describe and justify the main stages in storm sewer design.

10.2   "The frequency of rainfall is neither equivalent to the frequency of runoff nor of flooding." Discuss this statement with reference to recommended design storm return periods.

10.3   If a storm sewer surcharges during heavy rainfall, was it under-designed? Explain your answer.

10.4   A storm sewer network has been designed based on a 2-year return period storm for a 50-year design life. What is the probability that the network will
    a.   Surcharge at least once in 2 years (the probability of failure)?
    b.   Surcharge at least once during its design life?
    c.   Flood in any one year, assuming flooding is caused by the 10-year return period storm?
    d.   Flood at least once in 10 years?   [0.75, 1.00, 0.10, 0.65]

10.5 What is urban creep, and why is it an important issue for urban drainage?

10.6 What is percentage imperviousness, and how is it related to the runoff coefficient?

10.7 Explain what you understand by time of entry, time of flow, and time of concentration. Why is the duration of the design storm in the Rational Method taken as the time of concentration?

10.8 Demonstrate that a rainfall intensity of 0.014 L/s/m² is equivalent to 50 mm/h.

10.9 Explain the concept of the Rational Method. What are its main limitations?

10.10 A small separate storm sewer network has the following characteristics (Figure 10.12):

Sewer	Length (m)	Contributing impervious area (m²)
1.1	180	2000
2.1	90	6000
3.1	90	9000
1.2	90	4000

Use the Rational Method to determine the capacity required for each pipe in the network. Assume a time of entry of 4 minutes, that the pipe-full velocity in each pipe is 1.5 m/s, and that design rainfall intensities can be determined from Equation 4.2, where a = 750, b = 10, α = 1, and β = 1. Further, assume 100% runoff from impervious areas.

[26, 83, 125, 257 L/s]

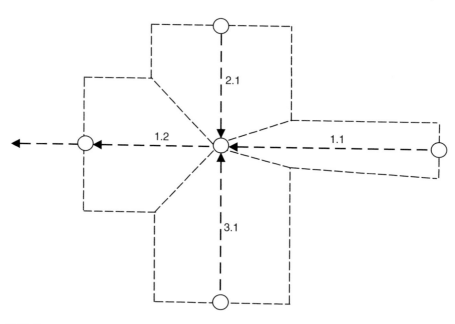

Figure 10.12 System layout (Problem 10.10).

10.11 What is a time-area diagram? Explain how it is used in the Time-Area Method.

10.12 Construct the time-area diagram (at point X) for each of the equally sized and graded catchments (1–3) shown in Figure 10.13.

10.13 For the network in Problem 10.10, draw a time-area diagram and use it to check the capacity of pipe 1.1 using the Tangent Method. [267 L/s]

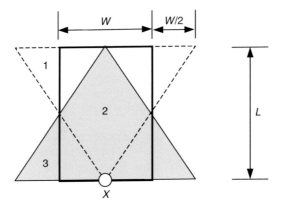

W | W/2

L

X

Catchment 1: dashed triangle
Catchment 2: rectangle
Catchment 3: shaded triangle

Figure 10.13 Three catchments (Problem 10.12).

10.14 Estimate the hydrograph at the outlet to pipe 1.1 (Problem 10.10) for a short storm with the following profile:

Time (min)	0–1	1–2	2–3
Intensity (mm/h)	20	28	64

Use the Time-Area Hydrograph Method. [Peak 127 L/s]

10.15 Describe the basis of the level pool routing method, and discuss its strengths and weaknesses.

10.16 Discuss the challenges in estimating runoff from undeveloped sites.

KEY SOURCES

Department of Energy/National Water Council (DoE/NWC). 1981. *Design and Analysis of Urban Storm Drainage. The Wallingford Procedure*, 4 vol., Standing Technical Committee Report No. 28.

Shaw, E.M., Beven, K.J., Chappell, N.A., and Lamb, R. 2010. *Hydrology in Practice*, 4th edn, CRC Press.

Woods Ballard, B., Wilson, S., Udale-Clark, H., Illman, S., Scott, T., Ashley, A., and Kellagher, R. 2016. *The SuDS Manual*, 5th edn, CIRIA C753.

REFERENCES

Allitt, R., Allitt, M., and Allitt, K. 2010. *Impact of Urban Creep on Sewerage Systems.* Report 10/WM/07/14, UK Water Industry Research, London.

Argaman, Y., Shamir, U., and Spivak, E. 1973. Design of optimal sewerage systems. *American Society of Civil Engineers Journal of Environmental Engineering*, 99(5), 703–716.

BS EN 752: 2008 *Drain and Sewer Systems Outside Buildings.*

Castelluccio, M., Poggi, G., Sansone, C., and Verdoliva, L. 2015. Land Use Classification in Remote Sensing Images by Convolutional Neural Networks. *arXiv.org*, arXiv: 1508.00092 [cs.CV].

Diogo, A.F. and Graveto, V.M. 2006. Optimal layout of sewer systems: A deterministic versus a stochastic model. *American Society of Civil Engineers Journal of Hydraulic Engineering*, 132(9), 927–943.

Finch, J., Reid, A., and Roberts, G. 1989. The application of remote sensing to estimate land cover for urban drainage catchment modelling. *Journal of the Institution of Water and Environmental Management*, 3, 558–563.

Guo, Y., Walters, G., and Savic, D. 2008. Optimal design of storm sewer networks: Past, present and future. *Proceedings of the 11th International Conference on Urban Drainage*, Edinburgh, September, on CD-ROM.

HRWallingford. 2012. *Development of Spatial Indicators to Monitor Changes in Exposure and Vulnerability to Flooding and the Uptake of Adaptation Actions to Manage Flood Risk in England*, Report TN-MCS0743-01 R2, www.theccc.org.uk.

Kelly, D.A. 2016. Impact of paved front gardens on current and future urban flooding, *Journal of Flood Risk Management*, doi: 10.1111/jfr3.12231

Kuichling, E. 1889. The relation between the rainfall and the discharge of sewers in populous district. *Transactions of American Society of Civil Engineers*, 20, 1–56.

Lloyd-Davies, D.E. 1906. The elimination of storm water from sewerage systems. *Proceedings of Institution of Civil Engineers*, 164(2), 41–67.

Mulvaney, T.J. 1850. On the use of self-registering rain and flood gauges in making observations on the relation of rainfall and flood discharges in a given catchment. *Transactions of Institution of Civil Engineers Ireland*, 4(2), 18.

Perry, T. and Nawaz, R. 2008. An investigation into the extent and impacts of hard surfacing of domestic gardens in an area of Leeds, United Kingdom. *Landscape and Urban Planning*, 86(1), 1–13.

Reid, J. 1927. The estimation of storm-water discharge. *Journal of Institution of Municipal Engineers*, 53, 997–1021.

Scott, A. 1994. Low-cost remote-sensing techniques applied to drainage area studies. *Journal of Institution of Water and Environmental Management*, 8, 498–501.

Urban Water Resources Council. 1992. *Design and Construction of Urban Stormwater Management Systems*, ASCE Manual No. 77, WEF Manual FD-20.

Wallingford HydroSolutions (WHS). 2016. *The Revitalised Flood Hydrograph Model. ReFH 2.2: Technical Guidance*, WHS/CEH.

Water Research Centre (WRc). 2012. *Sewers for Adoption—A Design and Construction Guide for Developers*, 7th edn, Water UK, Swindon.

Wright, G.B., Arthur, S., Bowles, G., Bastien, N., and Unwin, D. 2011. Urban creep in Scotland: Stakeholder perceptions, quantification and cost implications of permeable solutions. *Water and Environment Journal*, 25, 513–521.

Chapter 11

Flooding

11.1 INTRODUCTION

In the previous chapter we looked in detail at methods to design piped storm sewers to operate effectively under defined hydrological conditions. The concept of the design storm was introduced, the choice of which determines the degree of protection afforded against sewer surcharging and flooding. What should be the engineer's approach to the situation when design flows are exceeded to the extent that flow either emerges on to the urban surface from the underground system or can no longer enter the already full pipes? Until relatively recently, in UK and European practice, little account has been taken of this phenomenon because the system was considered to "have failed." It is instructive to revisit the limited advice on this topic given in the first edition of this book (2000): "Generally, when piped systems are being designed, care should be taken to 'define' overland flow flood routes to minimise damage to properties."

In recent years much more attention has been given to this topic as a consequence of the following:

- Significant and recent urban flood events (e.g., winter 2015/2016 storms Desmond and Eva flooding 16,000 UK properties)
- An increased understanding of the contribution of stormwater flooding (e.g., over 750,000 properties in England and Wales are estimated to have a 1% annual likelihood of flooding of this type [Environment Agency, 2014])
- A greater understanding of the implications of climate change (as discussed in Chapter 4)
- Increasing urbanisation resulting in more runoff and properties at risk
- Increasing urban creep (as discussed in Chapter 10)
- Stormwater flood damage costs estimated at £270 million *per annum* in England and Wales alone (POST, 2007)

Stormwater flooding is defined as inundation caused by rainfall such that the runoff caused exceeds the conveyance capacity of the urban drainage system, and *exceedance flow* is generated on the urban surface. The terms *surface water flooding* or *pluvial flooding* are sometimes used synonymously with stormwater flooding, or sometimes used to denote urban flooding where there is no sewer network or the network is already at full capacity. It should also be noted that stormwater flooding may occur from combined sewers and so actually be a mixture of stormwater and wastewater.

This chapter covers the concept of exceedance flow and what it means (Section 11.2), it reviews appropriate standards (Section 11.3), and defines the notion of flood risk in urban areas (Section 11.4). Appropriate management strategies are also discussed together with the latest strategic approaches (Section 11.5), with a concluding section (Section 11.6) on flood resilience.

11.2 EXCEEDANCE

Before exceedance flow can be formally defined, it is important to establish that the drainage system is more than just the pipes in the ground. This is emphasised by formally considering the system as having minor and major components. The *minor system* consists of the drainage hardware such as gully inlets, manholes, pipes, and stormwater management systems (e.g., SuDS) used to control more frequent storm flows and generally under the control of the system designer and subsequent owner. The *major system* consists of *surface flood pathways* such as roads and paths, and includes temporary storage areas such as car parks and playing fields. These may not have been specifically designed as such and are known as *default pathways. Design pathways* are constructed or designated at the planning stage and will additionally consist of floodways, retention basins, flood-relief channels, and open and culverted watercourses.

For any particular rainfall event, if the conveyance capacity of the piped or open channel minor system is exceeded (either due to lack of capacity or blockage), the excess flow that appears on the surface in the major system is termed *exceedance flow*. Figure 11.1 shows exceedance in progress as the surcharged minor combined sewer system spills its flow onto the default major system road surface.

11.2.1 Systems interface

A key aspect of exceedance is the interface between the minor and major systems. The points of contact between the two systems and their mutual interactions are as follows:

- Gully inlets: normally ingress points to the minor system for stormwater. If the conveyance capacity is reached, flow will not be able to enter at these points and will be retained on the surface. If conveyance capacity is exceeded, flow may exit at these

Figure 11.1 System surcharge and exceedance flow generation. (Courtesy of Professor Dušan Prodanović.)

points into the major system (see Figure 11.1). Similar effects are caused if the inlet or inlet connection becomes blocked.

- Manholes: normally access points into the minor system. If the conveyance capacity is exceeded, flow may reverse out of the system at these points.
- Open stormwater management systems: include swales and detention basins, which may overtop, spilling flow onto surrounding ground (see Chapter 21 for further details).
- WCs and other household appliances: in extreme cases or if WCs are sited in building basements, surcharged flow may reverse up the (combined) building drains and be forced backward through the toilet and into properties.
- River outfalls: normally used as egress points for the minor system. However, if water levels rise in the receiving watercourse, a backwater can be formed reducing the effective capacity of the minor system and potentially inducing exceedance. In some cases, where no flap valve exists, water can also enter the minor system.

The cause or type of flooding is difficult to diagnose after the event. In many cases, flooding is due to multiple effects at multiple points of contact. However, in recent years, many sewerage undertakers have made a substantial effort to investigate flooding incidents and maintain detailed records. Since 2010, *Lead Local Flood Authorities* (designated in England and Wales under the Flood and Water Management Act) also have a duty to investigate the cause of flooding.

11.2.2 Exceedance flow

Calculating the degree and extent of the interaction between the minor and major system is complex and resource consuming. Recent developments in modelling software have made this easier to represent, but significant effort is still required to validate such models. Defra (2010) recommends three levels of assessment and detail (strategic, intermediate, and detailed) depending on the needs of the particular study or engineering project. While the focus is on existing problems, level 3 could also be applied to new developments. The three levels are as follows:

1. Strategic: indicates where exceedance may accumulate over a large geographical scale (unlikely to include the minor system)
2. Intermediate: identifies not only where exceedance occurs but also more accurately establishes where it comes from (will include the minor system)
3. Detailed: represents the topography and increasing levels of detail applied typically to a township or localized problem

A later section in this chapter discusses risk assessment. Above-ground modelling approaches are covered in more detail in Chapter 19.

11.3 STANDARDS

It is convenient to define three threshold standards (Figure 11.2) when designing or analysing drainage systems, set to avoid the following:

1. Surcharging (minor system)
2. Surface flooding (minor system)
3. Property flooding (major system)

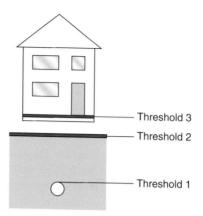

Threshold 3

Threshold 2

Threshold 1

Figure 11.2 Urban drainage flooding threshold level standards.

Threshold 1 standards are those traditionally used in design practice, for the minor system, such as given in BS EN 752: 2008, and listed in Table 10.1 as "design storm return period." This denotes the pipe-full flow hydraulic capacity of the system based on a particular rainfall event return period and assumes that the rainfall return period is equivalent to runoff return period. (See Chapter 10 for a discussion of this.) Once this threshold is crossed, the system will begin to surcharge. *Threshold 2* standards refer to the maximum capacity of the system to convey the stormwater without exceedance flow being generated. Standards such as these are also given in BS EN 752: 2008 and shown in column 2 of Table 10.1, referred to as "design flooding return period." The *threshold 3* standard is much less well developed and agreed. It is not explicitly covered in BS EN 752: 2008 or in *Sewers for Adoption* (Water Research Centre [WRc], 2012). However, in practice, sewerage undertakers use a range of threshold 3 standards depending on the cost of a solution (with the argument being that it is better to at least reduce the probability of flooding rather than provide no solution at all if the threshold is set too high and is unaffordable). National Planning Policy Guidance for England and Wales requires threshold 3 protection of a 100-year return period (Department for Communities and Local Government [DCLG], 2012). This standard effectively denotes the maximum capacity of the major system.

The issue is further complicated by the way that threshold standards are specified. The BS EN explains that level of protection against flooding can be specified using design storm return periods *or* flooding frequencies. However, it is often not clear which is being referred to. *Sewers for Adoption*, for example, specifies a threshold 2 standard requiring no part of a site to flood from the drainage system due to a 30-year return period design storm. In one way, this simplifies the issue in that an appropriate storm event is selected and the network designed to convey it without flooding. But, this approach gives no indication of the actual flood frequency protection provided and does not necessarily or explicitly comply with the BS EN flooding frequency standards.

The key point here is that when designing for exceedance, care should be taken to understand exactly what standard of protection is expected (and how flexible or otherwise it is) and exactly how that standard is formulated. The wider point is that further clarity and agreement in the area are vital.

11.4 FLOOD RISK

Chapter 10 considers the concept of the design storm and how it can be expressed either in terms of return period or exceedance probability. (The two are inversely related as shown in Equation 10.1.) Design is then based on probabilistic thresholds as given in the previous section. However, the management of coastal, river, and (increasingly) stormwater flooding follows a risk-based approach, where both the likelihood *and* outcome of flooding are taken into account. Put simply, the risk of flooding *R* is given by

$$R = P \cdot C \tag{11.1}$$

where *R* is the total risk under management over *n* years (£), *P* is the probability of a particular occurrence; and *C* is an undesirable consequence (or damage) of the occurrence (£).

The basic idea is to move away from considering the probability of a storm event in isolation, and instead to relate it formally and explicitly to the detailed impact or consequence it will have. For example, flooding may occur regularly in one location (hence with a high probability) but have very little impact, so the risk is low. The same probability of flooding may occur at another location, this time causing high impact due to the particular local circumstances, giving high risk. A probability-based approach would not directly distinguish between these two situations, whereas a risk-based approach would. Thus, investment can be targeted towards maintenance and improvement of those assets that contribute most towards risk reduction.

High probability events with high consequences will be classified as high risk; similarly low-probability, low-consequence events will be low risk. As Figure 11.3 shows, however, low probability–high consequence situations are classified as medium risk, as are those with high probability and low consequence. These two cases are likely to be viewed very differently by the public (Merz et al., 2009) and may well need to be managed differently even though they pose similar (medium) risk.

However, to use such a risk assessment approach a more comprehensive analysis is required, which combines hydrological knowledge about the frequency of different storm events with hydraulic modelling information regarding inundation frequency of flood water, and flood damage evaluation.

UK Water Industry Research (UKWIR, 2011) has developed a risk-based framework based on defining the probability and consequence of stormwater flooding for actual and predicted flooding. This includes a method to assess actual incidents, with the probability defined by recorded flood frequency rather than using a hydraulic model. The method can be used to help prioritise investments.

11.4.1 Flood damage

Flood-induced damage can be classified as tangible or intangible, distinguished by the ease or otherwise of assigning a monetary value. Tangible damage can also be subdivided into

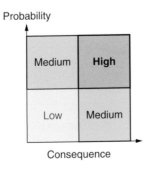

Figure 11.3 The relationship between probability, consequence, and risk.

direct or indirect (Hammond et al., 2015). Damage to property (e.g., floors, walls, ceilings, and all internal items) is direct, whereas interruption and disruption of economic and social activities is indirect, including loss of production and profit, disruption to road traffic, and additional costs for emergency and cleaning work. Intangible damage also includes health and psychological impacts. Whittle et al. (2012) highlight a range of experiences suffered by families including their postflooding vulnerability and emotional recovery. In practice, only tangible effects are considered in any formal analysis, although with consideration of multiple benefits and impact assessment becoming more commonplace, this is likely to change in the future.

11.4.1.1 Flood depth

Although not the only factor related to property damage, flood depth or stage is the key indicator. If the floodwater depth is below ground-floor level, damage is likely to be limited for most properties, although damage may be significant if the water enters basements, cellars, or voids under floors. Once floodwater level rises to 0.5 m above ground-floor level, properties can be significantly affected, including damage to internal surfaces, electrical sockets and equipment, kitchen cupboards, carpets, furniture, and personal belongings. Much greater damage still can occur when flood depth is greater than 0.5 m above ground-floor level, and in extreme cases, the structural integrity of the building can be compromised (Office of the Deputy Prime Minister [ODPM], 2003).

The *Multi-Coloured Handbook* (Penning-Rowsell et al., 2016) contains relationships between flood damage and flood depth, duration, property types, and social categories for both urban and rural areas in the United Kingdom, based on an extensive dataset. Examples of several depth–damage curves are given in Figure 11.4.

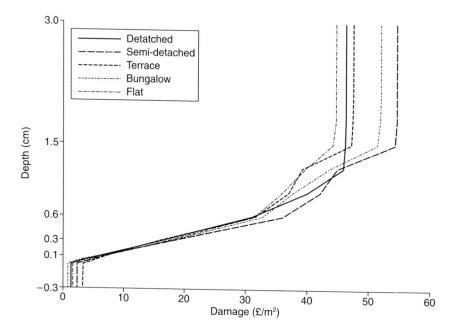

Figure 11.4 Relationship between flood depth and direct damage to residential properties. (After Penning-Rowsell, E. et al. 2003. *The Benefits of Flood and Coastal Defence: Techniques and Data for 2003.* Flood Hazard Research Centre, Middlesex University.)

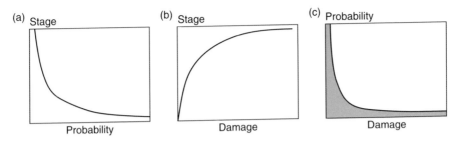

Figure 11.5 Interrelation of flood probability, stage, and damage.

There is a relationship at every location between flood stage and flood probability, and between flood stage and flood damage. These can be combined as shown in Figure 11.5a through c to give a direct flood probability–flood damage relationship. The *annual average risk* of flooding is the integral of the product of probability and damage (Equation 11.1), which is the (shaded) area under the curve in Figure 11.5c.

A further important consideration, over and above direct flood damage, is the consequence of damage or disruption to important facilities. For example, buildings providing essential services (e.g., hospitals, emergency services) or critical infrastructure (e.g., water, electricity, transport) should have a higher consequence attached to being flooded. This could be implemented by setting different standards of protection such as a 1 in 200 annual probability event for national infrastructure with 1 in 1,000 protection for the most critical facilities (Cabinet Office, 2010).

11.4.2 Risk assessment

Defra (2010) proposes a three-level risk assessment approach taking into account contemporary software capabilities. Key steps to applying a risk assessment include the following:

- Identify the level of assessment and understand the appropriateness of the available models and software to determine the risk.
- Select the *storm* return periods to be used as the threshold 2 standard, and determine the critical duration for the network under consideration. Note that for larger systems, this may require more than one duration.
- Begin the assessment with the 30-year return period (for example), critical duration storm, and determine locations that surcharge or flood. Focus modelling efforts in known flooding locations. If no flooding occurs, increase the storm return period.
- Estimate the consequences of the flooding at each location, with the degree of detail linked to the level of study. This may range from enumerating the number of properties impacted, to establishing total damage costs using the depth-damage relationships. A number of different tools are now available to help support this process (e.g., Chen et al., 2016).
- Present the results across an area appropriate for the type of assessment either qualitatively using a risk matrix (Figure 11.3) or quantitatively by monetising the expected annual damage.
- Test solutions or designs repeating the process to check if risk is increased or reduced.

Ryu and Butler (2008) argue that continuous simulation is more appropriate for urban drainage flood risk assessment than considering single storm events. This allows a more complete and accurate evaluation of flood risk (i.e., the area under the curve in Figure 11.5c). However, this thoroughness comes at the expense of computational and physical time. Dawson et al. (2008) have developed an approach based on urban flood risk assessment but with the additional goal of attributing the risk between key stakeholders.

11.5 MANAGEMENT

11.5.1 Options

A key aspect of stormwater flooding is the need to devise effective and sustainable approaches to managing the risk it creates. In terms of new systems, the basic approach is careful design of both the minor and major systems against appropriate threshold standards and equally careful planning of the interaction between them. For existing systems, a range of possible solutions is available, and these are summarised in the following list:

- Reduce or limit inflow (*minor system*): Chapter 21 explains the options available for this, key among them is use of stormwater management measures to infiltrate flow locally or to store and reuse it. An alternative approach is to divert or disconnect flows to alternative systems or outfalls.
- Increase system capacity (*minor system*): Improved cleaning can increase pipe capacity if it is limited by blockage or sediment (Chapter 17). Alternatively, rehabilitation may be considered (Chapter 18), possibly including pipe upsizing.
- Store more flow (*minor or major system*): The key aim is to attenuate flow and therefore reduce peak flow rates. Storage can be provided at source (Chapter 21), within the minor system (Chapter 13) or the major system (see Section 11.5.2).
- Better deploy surface flow features (*major system*): This key area is discussed in Section 11.5.2.
- Improve building flood resilience: This non-structural response is covered in Section 11.5.3.

11.5.2 Surface flow features

The surface conveyance system for exceedance flow can be made up of many elements, including gully inlets, pavements, roads, drives, and other less formal flow pathways. Balmforth et al. (2006) and Digman et al. (2014) provide a variety of examples and case studies.

11.5.2.1 Gully inlets

Road gullies were considered in Chapter 6 and further in Chapter 8 as important parts of the minor urban drainage system. As discussed, they are also a key interface between the major and minor systems. The design of road gully sizing/spacing recognises and indeed utilises the fact that a proportion of the flow passes by, and over, the gully grating. Under extreme rainfall events, higher exceedance flows generate higher depths of flow and greater flow widths, and could also have a negative impact on the hydraulic efficiency of the gully itself. So, in design, attention needs to be given to selecting appropriate capacity and spacing of inlets. In analysing the effectiveness of existing gullies, effort should focus on establishing whether sufficient inlet capacity is available given limiting kerb heights and/or flow widths.

Only in exceptional circumstances will the capacity of the gully grating restrict inflow into the gully pot (unless it is deliberately designed to do so or becomes blocked with debris).

However, the gully outlet, or lateral pipework may be the limiting component. Escarameia and May (1996) have determined this capacity experimentally at 11 L/s and 19 L/s for 100 mm and 150 mm gully outlets, respectively, with results corroborated by Shepherd et al. (2012). Example 11.1 demonstrates the point numerically.

EXAMPLE 11.1

Based on the conditions described in Example 8.5, calculate the captured flow by each gully and compare this with the capacity of a freely discharging 100 mm gully lateral. Comment on the performance of the gully under exceedance flows.

Solution

Captured flow is equal to the runoff generated between gullies. So from Equation 8.10,

$$Q_c = E\overline{Q} = 0.99 \times 0.010 = 9.9\,\text{L/s}$$

This can be compared with the lateral capacity of 11 L/s, indicating a well-matched gully and lateral. However, very little (lateral) capacity is available to cope with extreme flows.

11.5.2.2 Surface pathways

On undeveloped sites the underlying topography defines natural flow pathways. When designing new sites, the most effective approach is to follow natural pathways as far as possible, providing linked, free passage for flowing excess water. This includes identification of points of entry/exit to the drainage system plus the location of effective flow barriers such as kerbs, walls, buildings, and other relevant urban features. These pathways will only rarely convey significant flows and will be used for other purposes day to day. Legal safeguards (e.g., recording structures or features on registers held by the Lead Local Flood Authority) and appropriate maintenance are needed to ensure their continued availability.

The channel forming a surface pathway can be designed by standard application of open channel hydraulics as set out in Chapter 7, typically using Manning's equation. An appropriate case is given later in the book, in Example 23.1. Complications arise when conditions deviate significantly from uniform, steady flow. Details on the design of flow transitions and junctions are given in Balmforth et al. (2006). The concept of using the road as a drain is further explored in Digman et al. (2014) and in Chapter 23.

11.5.2.3 Surface storage

Surface storage capacity can be developed to replace or reduce the need for extensive surface pathways. The key points about major system storage areas are that they are above-ground, multi-functional facilities. Their goal is to retain the flood volume temporarily and release it slowly at a lower flow rate. A wide range of alternatives is available as listed in Table 11.1. The capacity design of such facilities is comparable to minor system storage design, discussed in detail in Chapter 13.

An example of this type of "sacrificial" storage area is a car park, where water can be allowed to accumulate up to kerb height over a defined area. Although vehicle owners may be temporarily inconvenienced, there is unlikely to be any significant structural damage to the site after flooding, safety risks are minimal due to the low depths, and as long as the area can be adequately drained after the event normal use can be quickly resumed. Figure 11.6 illustrates a good example of this from the Netherlands.

Table 11.1 Types of major system storage

Storage type	Description	Maximum water depth (m)
Car parks	Used to temporarily store exceedance flows. Depth restricted due to potential hazard to vehicles, pedestrians, and adjacent property.	0.2
Recreational areas	Hard surfaces used as basketball, five-a-side football and hockey pitches or tennis courts.	Typically 0.5 m, up to 2 m in extreme cases
Minor roads	Roads with speed limits up to 30 mph where depth of water can be controlled by design.	0.1
Playing fields	Used for sport such as football and rugby. Set below the ground level of the surrounding area and may cover a wide area, offering significant flood volume.	Typically 0.5 m, up to 2 m in extreme cases
Parkland	Has a wide amenity use. Often may contain a watercourse. Care needed to keep floodwater separate and released in a controlled fashion to prevent downstream flooding.	Typically 0.5 m, up to 2 m in extreme cases
School playgrounds	Playgrounds can provide significant flood storage. Extra care should be taken to ensure safety of the children.	0.3
Industrial areas	Low-value storage areas. Care should be taken in the selection as some areas could create significant surface water pollution.	0.6

Source: Adapted from Balmforth, D. et al. 2006. *Designing for Exceedance in Urban Drainage—Good Practice*, CIRIA C635.

11.5.2.4 Safety

Allowing stormwater onto the surface does pose the risk of contact with the public and hence safety issues should be addressed. For flow across surfaces that could be used by pedestrians, hazard regimes as a function of depth and velocity should be carefully considered in accordance with the recommendations in Table 11.2. In addition, the depth of flow in any channel should be limited to 0.3 m or 0.2 m if trafficked. See also Example 11.2.

Figure 11.6 Multi-functional flood storage scheme in Benthemplein Water Square, Rotterdam. (Courtesy of C40 Cities.)

Table 11.2 Flood safety criteria

Hazard criteria	Max depth (m)	Depth-velocity (m²/s)
Low hazard for children	0.5	<0.4
Moderate hazard for adults	1.2	<0.8
Significant hazard for adults	1.2	<1.2

Source: Adapted from Russo, B. et al. 2013. *Natural Hazards*, 69(1), 251–265.

In flood storage areas, depth should generally be limited to a maximum of 0.6 m. It may be appropriate to increase this depth up to 1.5 m or even 2 m in extreme circumstances (Woods-Ballard et al., 2015). In all cases, an appropriate design risk assessment should be completed, considering key aspects such as location, signposting, safe access, and egress.

11.5.3 Flood protection of buildings

Buildings are not normally within the province of urban drainage engineers. However, two areas related to urban drainage flooding are particularly pertinent and should be considered: layout and fabric.

EXAMPLE 11.2

A footpath will be used as a flow conveyance route during flooding. The path is of rectangular cross section, 100 mm deep, 2.5 m wide, and with a gradient of 1%. Check the path will be safe for pedestrians. $n = 0.018$.

Solution

Using Manning's Equation 7.23:

$$v = \frac{1}{n} R^{2/3} S_o^{1/2} = \frac{1}{0.018} \left(\frac{2.5 \times 0.1}{2.5 + 2 \times 0.1} \right)^{2/3} 0.01^{1/2} = 1.14 \text{ m/s}$$

We check with Table 11.1:

$$d \cdot v = 0.11 < 0.4 \, \text{m}^2/\text{s}$$

This result indicates the path would still be safe for all pedestrians to use while flooded.

11.5.3.1 Layout

In many cases, the effects of flooding can be minimised by the careful positioning of buildings in relation to the topography and defined flood pathways, and by the sympathetic design of landscaping features. Figure 11.7 shows a small housing development where the above-ground flow pathways have been defined taking into account site topography. A temporary, landscaped flood storage area is identified at the low point of the site.

A second key aspect is correctly setting the level of housing relative to the major system, including threshold levels. Valuable flood protection can be achieved by careful detailing of the property and surrounds including entrance details, driveway slopes, and drop kerbs.

11.5.3.2 Fabric

An important aspect of mitigating flood risk is improving the flood protection of buildings at the design stage. Detailed advice on this has been published (Bowker et al., 2007;

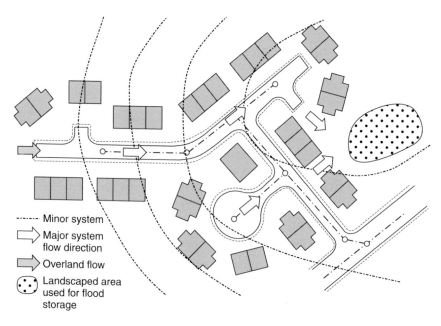

------ Minor system
⊳ Major system
   flow direction
⊳ Overland flow
(⦿) Landscaped area
   used for flood
   storage

Figure 11.7 Identification of major system flood flow pathways.

ODPM, 2003). The basic strategy proposed is that the best approach at low flood depths (<0.3 m outside the building) is to avoid flooding by excluding the building from the flood risk zone (such as by raising or moving the property). Failing this, building flood protection needs to be improved. At low depths (<0.6 m) an attempt should be made to exclude the water, but at high depths water is best allowed to enter and then exit relatively unimpeded. Table 11.3 summarises this.

Table 11.3 Building fabric design strategy

Strategy	Design water depth (m)	Approach	Mitigation measures
Avoidance		• Remove building/development from flood hazard	• Land raising, landscaping, raised thresholds
Resistance and resilience	<0.3	• Attempt to keep water out—"water exclusion strategy"	• Materials and constructions with low permeability
	0.3–0.6	• Attempt to keep water out, in full or in part, depending on structural assessment • If structural concerns exist follow approach below	• Materials with low permeability to at least 0.3 m • Flood-resilient materials and designs • Access to all spaces to permit drying and cleaning
	>0.6	• Allow water through property to avoid risk of structural damage • Attempt to keep water out for low depths of flooding—"water entry strategy"	• Materials with low permeability up to 0.3 m • Accept water passage through building at higher water depths • Design to drain water away after flooding • Access to all spaces to permit drying and cleaning

Source: Bowker, P. et al. 2007. *Improving the Flood Performance of New Buildings. Flood Resilient Construction.* Communities and Local Government, RIBA Publishing.

For existing buildings that have been subjected to flooding, appropriate repair techniques for damaged floors, walls, services, fittings, and building services can be found in Garvin et al. (2005).

11.6 FLOOD RESILIENCE

Resilience is a new concept that has emerged that concerns how systems of all types prepare for, respond to, and recover from shocks (Zhou et al., 2010). Gersonius (2008) argues that flood resilience incorporates four capacities: to avoid damage through the implementation of structural measures, to reduce damage in the case of a flood that exceeds a desired threshold, to recover quickly to the same or an equivalent state, and to adapt to an uncertain future. Djordjević et al. (2011) state resilience is equal to resisting, recovering, reflecting, and responding. Zevenbergen et al. (2008) make the point that the development of flood resilience requires an understanding of how cities respond to flooding across varying spatial and temporal scales from stormwater flooding to river and coastal flooding.

Butler et al. (2016) argue that resilience can be thought of in two separate but inter-related ways: properties and performance. In the first case, the emphasis is on providing attributes of the system that are assumed to build resilience, and in the latter the focus is on what standard or level of service should be provided to achieve acceptable resilience.

11.6.1 Properties

Properties that build or promote resilience may include increasing flexibility, diversity, connectivity, and redundancy (Cabinet Office, 2011; Gersonius et al., 2013). Based on these general properties, it is possible to provide recommendations on how to build flood resilience based on a raft of measures as shown in Table 11.4. Batica and Gourbesville (2014) have developed a Flood Resilience Index that can be applied on a case-by-case basis to indicate the extent and rate change of resilience measures. While appealing and

Table 11.4 Flood resilience measures

Category	Measure	
Capacity building	• Flood maps (inundation and risk) • Informational material (brochures, public presentations, Internet portals) • Communication	• Face-to-face learning • Web-based learning • Training • Collaborative platforms
Land use control	• Spatial planning • Flood risk adapted land use • Building regulations	• Building codes • Zoning ordinances
Flood preparedness	• Flood-resistant buildings • Wet-proofing • Dry-proofing	• Flood contingency plans (local scale) • Infrastructure maintenance
Financial preparedness	• Insurance of residual risk	• Reserve funds
Emergency response	• Evacuation and rescue plans • Forecasting and warning services	• Emergency operations control • Emergency response staff
Emergency infrastructure	• Allocation of temporary containment structures • Telecommunications network	• Transportation and evacuation facilities
Recovery	• Disaster recovery plans	• Governmental pecuniary provisions

Source: Adapted from Batica, J. and Gourbesville, P. 2014. Flood resilience index—Methodology and implementation. *Proceedings of the 11th International Conference on Hydroinformatics,* August, New York.

relatively straightforward to carry out, the effect of specifying particular properties or measures on the performance of a system is uncertain. Increased connectivity, for example, may provide resilience for some threats but actually decreases resilience to others, such as targeted attacks (Albert et al., 2000). Lamond et al. (2015) make the case for how provision of stormwater management measures can develop greater flood resilience (see Chapter 21).

11.6.2 Performance

The second approach requires that resilience performance is agreed and quantified. Butler et al. (2016) suggest this can be defined as "the degree to which the system minimises level of service failure *magnitude* and *duration* over its design life when subject to exceptional conditions." Mugume et al. (2015) and Mugume and Butler (2016) have used this definition to specify an urban drainage resilience index (*Res*) that links failure in terms of above-ground flood magnitude and duration with system residual functionality using a procedure called Global Resilience Analysis (GRA):

$$Res = 1 - (V_F/V_I \cdot t_F/t)$$

(11.2)

where V_F is the total flood volume (m³), V_I is the total inflow volume into the system (m³), t_F is the mean duration of flooding (hours), and t is the elapsed time (hours).

The causes of failure (flooding) in this context can be either functional (e.g., rainfall-induced) or structural (e.g., pipe blockage).

Mugume et al. (2015) have applied the index to a large open channel system in Kampala, Uganda, by modelling the effect of systematically failing (in this context, blocking) multiple channels (links) and computing the flood volume and duration at all nodes. Figure 11.8a illustrates the initial resilience index of the system and how this deteriorates with the degree of failure experienced. Figures 11.8b and 11.8c indicate how two different strategies or measures perform relative to an assumed minimum acceptable resilience level of service threshold of 0.7 (the dashed horizontal line).

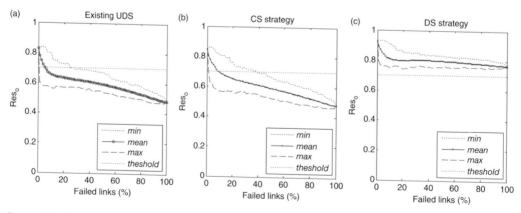

Figure 11.8 Mean, minimum, and maximum values of Res_0 for various degrees of link failure for (a) the existing urban drainage system (b) a centralised storage strategy, and (c) a distributed storage strategy. The horizontal dashed line is an assumed minimum acceptable resilience level of service threshold of 0.7 (reproduced from Mugume et al. (2015) *Water Research*, 81, 15–26, under the terms of the Creative Commons Attribution License CC BY).

The main advantage of the GRA method is that the focus is transferred from the accurate quantification of the probability of occurrence of sewer failures (e.g., Egger et al., 2013) to the evaluation of the effect of different sewer failures, on the ability of an urban drainage system to minimise the resulting flooding impacts (e.g., Kellagher et al., 2009). The main disadvantage of this general performance-based approach is the computational burden it demands.

The whole issue of resilience and its relationship to sustainability is returned to in more detail in Chapter 24.

PROBLEMS

11.1 Why has the issue of urban drainage flooding been neglected for so many years, and why is it now of such interest?

11.2 Explain the concept of flow exceedance. Is it a good thing, or should it be eliminated?

11.3 Three sewer flooding threshold standards are now being used in practice. Explain what they are and comment on the magnitude you think they should be set at.

11.4 A 100-year return period flood has caused extensive property damage of £25 million in area A. Area B is subject to regular flooding on average once a year. Each minor inundation causes an average of £250,000 damage. Which area would you prioritise for investment and why?   [£250,000]

11.5 Explain the differences between tangible and intangible damage, and direct and indirect damage. How would you attempt to quantify each category?

11.6 An extreme storm event causes a residential area to be flooded to an average depth of 0.5 m. For the same storm, by what depth must flooding be reduced to cut the flood risk by a half?   [0.25 m]

11.7 What options are available to reduce urban drainage flood risk? Discuss when and where each solution would be most applicable.

11.8 Calculate the width of flooding just upstream of the initial gully in the road specified in problem 8.5 if the design rainfall intensity is doubled to 100 mm/h. [1.90 m]

11.9 An area has been flooded. Rather than reconstruct the sewer system it is proposed that building modifications are carried out. What types of modification would you specify? How could this reduce flood risk?

11.10 What is resilient urban drainage?

11.11 What measures would you put in place to improve flood resilience and why?

11.12 What is the difference between resilient properties and resilient performance?

KEY SOURCES

Balmforth, D., Digman, C., Kellagher, R., and Butler, D. 2006. *Designing for Exceedance in Urban Drainage—Good Practice*, CIRIA C635.

Digman, C., Balmforth, D., Hargreaves, P., and Gill, E. 2014. *Managing Urban Flooding from Heavy Rainfall: Encouraging the Uptake of Designing for Exceedance*, CIRIA C738.

Evans, E., Ashley, R., Hall, J., Penning-Rowsell, E., Sayers, P., Thorne, C., and Watkinson, A. 2004. *Foresight. Future Flooding. Scientific Summary: Volume II Managing Future Risks*. Office of Science and Technology, London.

Mcbain, W., Wilkes, D., and Retter, M. 2010. *Flood Resilience and Resistance for Critical Infrastructure*, CIRIA C688.

Pitt, M. 2007. *Learning Lessons from the 2007 Floods. An Independent Review*. The Cabinet Office.

Zevenbergen, C., Cashman, A., Evelpidou, N., Pasche, E., Garvin, S., and Ashley, R. 2010. *Urban Flood Management*. CRC Press.

REFERENCES

Albert, R., Jeong, H., and Barabasi, A.L. 2000. Error and attack tolerance of complex networks. *Nature*, 406(6794), 378–382.

Batica, J. and Gourbesville, P. 2012. A resilience measures towards assessed urban flood management—CORFU project. *Proceedings of 9th International Conference on Urban Drainage Modelling*, September, Belgrade.

Batica, J. and Gourbesville, P. 2014. Flood resilience index—Methodology and implementation. *Proceedings of the 11th International Conference on Hydroinformatics*, August, New York.

Bowker, P., Escarameia, M., and Tagg, A. 2007. *Improving the Flood Performance of New Buildings. Flood Resilient Construction.* Communities and Local Government, RIBA Publishing.

BS EN 752: 2008. *Drain and Sewer Systems Outside Buildings.*

Butler, D., Ward, S., Sweetapple, C., Astaraie-Imani, M., Diao, K., Farmani, R. and Fu, G. 2016. Reliable, resilient and sustainable water management: The Safe and SuRe approach. *Global Challenges*, doi:10.1002/gch2.1010.

Cabinet Office. 2010. *Strategic Framework and Policy Statement on Improving the Resilience of Critical Infrastructure to Disruption from Natural Hazards.* London.

Cabinet Office. 2011. *Keeping the County Running: Natural Hazards and Infrastructure. A Guide to Improving the Resilience of Critical Infrastructure and Essential Services.* London.

Chen, A.M., Hammond, M.J., Djordjevic, S., Butler, D., Khan, D.M., and Veerbeek, W. 2016. From hazard to impact: Flood damage assessment tools for mega cities. *Natural Hazards*, 82(2), 857–890.

Dawson, R.J., Speight, L., Hall, J.W., Djordjevic, S., Savic, D., and Leandro, J. 2008. Attribution of flood risk in urban areas. *Journal of Hydroinformatics*, 10(4), 275–288.

DCLG. 2012. *Technical Guidance to the National Planning Policy Framework*, Department for Communities and Local Government.

Defra. 2010. *Surface Water Management Plan Technical Guidance*, Department for Environment, Food and Rural Affairs.

Djordjević, S., Butler, D., Gourbesville, P., Mark, O., and Pasche, E. 2011. New policies to deal with climate change and other drivers impacting on resilience to flooding in urban areas: the CORFU approach. *Environmental Science and Policy*, 14(7), 864–873.

Egger, C., Scheidegger, A., Reichert, P., and Maurer, M. 2013. Sewer deterioration modeling with condition data lacking historical records. *Water Research*, 47, 6762–6779. http://dx.doi.org/10.1016/j.watres.2013.09.010.

Environment Agency. 2014. *Flood and Coastal Erosion Risk Management Long Term Investment Scenarios (LTIS) 2014.*

Escarameia, M. and May, R.W.P. 1996. *Surface Water Channels and Outfalls: Recommendations on Design*, Report SR406, HR Wallingford.

Garvin, S., Reid, J., and Scott, M. 2005. *Standards for the Repair of Buildings Following Flooding*, CIRIA Report C623.

Gersonius, B. 2008. Can resilience support integrated approaches to urban drainage management? *Proceedings of 11th International Conference on Urban Drainage.* Edinburgh.

Gersonius, B., Ashley, R., Pathirana, A., and Zevenbergen, C. 2013. Climate change uncertainty: Building flexibility into water and flood risk infrastructure. *Climatic Change*, 116, 411.

Hammond, M.J., Chen, A.S., Djordjević, S., Butler, D., and Mark, O. 2015. Urban flood impact assessment: A state-of-the-art review. *Urban Water Journal*, 12(1), 14–29.

Kellagher, R.B.B., Cesses, Y., Di Mauro, M., and Gouldby, B. 2009. An urban drainage flood risk procedure—A comprehensive approach. *WaPUG Annual Conference*, Blackpool.

Lamond, J.E., Rose, C.B., and Booth, C.A. 2015. Evidence for improved urban flood resilience by sustainable drainage refit. *Proceedings of the Institution of Civil Engineers, Urban Design and Planning*, 168(DP2), 101–111.

Merz, B., Elmer, F. and Thieken, A.H. 2009. Significance of "high probability/low damage" versus "low probability/high damage" flood events. *Natural Hazards and Earth Systems Science*, 9, 1033–1046.

Mugume, S.N. and Butler, D. 2016. Evaluation of functional resilience in urban drainage and flood management systems using a global analysis approach. *Urban Water Journal*, doi:10.1080/157 3062X.2016.1253754.

Mugume, S.N., Gomez, D.E., Fu, G., Farmani, R., and Butler, D. 2015. A global analysis approach for investigating structural resilience in urban drainage systems, *Water Research*, 81, 15–26.

ODPM. 2003. *Preparing for Floods. Interim Guidance for Improving the Flood Resistance of Domestic and Small Business Properties*. Office of the Deputy Prime Minister.

Penning-Rowsell, E., Johnson, C., Tunstall, S., Tapsell, S., Morris, J., Chatterton, J. et al. 2003. The Benefits of Flood and Coastal Defence: Techniques and Data for 2003. *Flood Hazard Research Centre*, Middlesex University.

Penning-Rowsell, E., Priest, S., Parker, D., Morris, J., Tunstall, S., Viavattene, C. et al. 2016. *Flood and Coastal Erosion Risk Management: Handbook and Data for Economic Appraisal*. Available online only: http://www.mcm-online.co.uk/handbook.

POST. 2007. *Urban Flooding*. Postnote 289, Parliamentary Office of Science and Technology.

Russo, B., Gómez, M., and Macchione, F. 2013. Pedestrian hazard criteria for flooded urban areas. *Natural Hazards*, 69(1), 251–265.

Ryu, J. and Butler, D. 2008. Managing sewer flood risk. *Proceedings of the 11th International Conference on Urban Drainage*, Edinburgh, September, on CD-ROM.

Shepherd, W., Blanksby, J., Doncaster, S., and Poole, T. 2012. Assessment of road gullies. *Proceedings of the 10th International Conference of Hydroinformatics, HIC 2012*, Hamburg.

UK Water Industry Research (UKWIR). 2011. *A Risk Based Approach to Flooding*, Report 11/WM/17/2.

Whittle, R., Walker, M., Medd, W., and Mort, M. 2012. Flood of emotions: Emotional work and long-term disaster recovery. *Emotion, Space and Society*, 5, 60–69.

Woods Ballard, B., Wilson, S., Udale-Clark, H., Illman, S., Scott, T., Ashley, A., and Kellagher, R. 2015. *The SuDS Manual*, CIRIA C753.

Water Research Centre (WRc). 2012. *Sewers for Adoption: A Design and Construction Guide for Developers*, 7th edn, Water UK.

Zevenbergen, C., Veerbeek, W., Gersonius, B., and Van Herk, S. 2008. Challenges in urban flood management: Travelling across spatial and temporal scales. *Journal of Flood Risk Management*, 1(2), 81–88.

Zhou, H., Wang, J., Wan, J., and Jia, H. 2010. Resilience to natural hazards: A geographic perspective. *Natural Hazards*, 53, 21–41.

Chapter 12

Combined sewers and combined sewer overflows

12.1 BACKGROUND

Combined sewer systems are discussed in Chapter 1. A significant percentage of sewer systems in many countries are combined, which makes them an important topic in urban drainage. The essential features of combined sewers are that they carry both wastewater and stormwater in the same pipe. As it is not possible to carry the full combined flow to treatment and there is a need to limit the frequency of flooding, it is necessary for combined sewer overflows to discharge a proportion of the flow to a watercourse. This chapter deals with the special characteristics of combined sewers, and in particular with combined sewer overflows. In Section 12.2 we discuss how flows are accommodated in combined sewers, in 12.3 we go on to describe the specific role of combined sewer overflows (CSOs) leading to Section 12.4, which discusses how and to what extent pollution is controlled using this system. Various CSO designs are explained in Section 12.5, their effectiveness evaluated in Section 12.6, and design details described in Section 12.7.

12.2 SYSTEM FLOWS

The inflow to a combined sewer system consists of both wastewater (see Chapter 3) and stormwater (see Chapter 5). At the point of inflow, the flow rates can be calculated from the methods given in Chapters 9 and 10.

A typical layout of a small combined sewer system including a CSO is given in Figure 12.1. All connections of stormwater and wastewater are made to the single combined sewer. Upstream of the CSO, the pipe carries the full combination of stormwater and wastewater from the upstream catchment. If the combined flow does not exceed the setting of the CSO, all continues to the wastewater treatment plant. If the combined flow exceeds the CSO setting, there will be overflow to the stream, and the flow retained in the system downstream will be determined by the CSO setting. Therefore, at different points throughout a combined sewer system in storm conditions, there can be significant differences in both the rate and composition of flow. In addition, the downstream system can substantially influence the ability of the CSO to pass forward flow.

12.2.1 Low flow rates

A combined sewer pipe has a significantly larger diameter than the foul sewer in a separate system draining a catchment of the same size (since the combined sewer must also have capacity to carry stormwater). This means that in dry weather, when the wastewater flow rate is relatively low, the combined sewer (compared with the foul sewer) will have lower

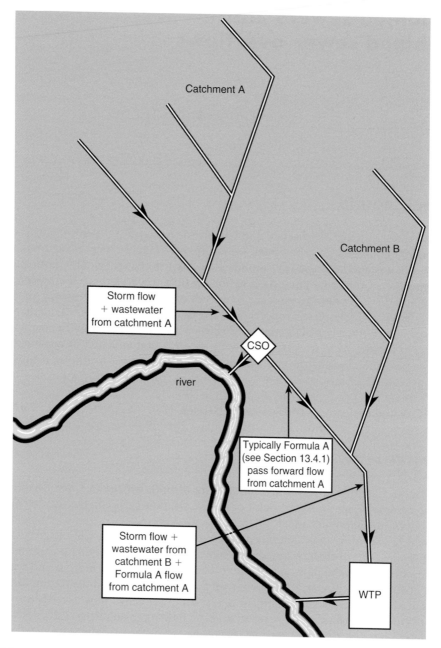

Figure 12.1 Typical layout of combined sewer system (schematic plan).

flow depths and greater contact between the liquid and the pipe wall—both leading to a greater risk of sediment deposition.

It is partly for this reason that egg-shaped sewers were popular in the past—with a smaller diameter in the lower part of the cross section for the low flows, and greater area above for storm flows. Egg-shaped sewers are quite common in older systems.

The effect is illustrated and investigated in Example 12.1.

EXAMPLE 12.1

Determine depth and velocity for a part-full flow rate of 15 L/s in (1) a 225 mm diameter foul sewer, and (2) a 600 mm diameter combined sewer. For both, take the gradient as 1 in 100, and roughness k_s as 1.5 mm.

Solution

1. Chart/table gives flow rate full $Q_f = 45$ L/s
   $Q/Q_f = 0.33$, and from Figure 7.11 $d/D = 0.375$, so depth = 85 mm
   Chart/table gives velocity (full) $V_f = 1.1$ m/s
   from Figure 7.11 for $d/D = 0.375$, $v/v_f = 0.9$, so velocity = 0.99 m/s
2. Chart/table gives flow rate full $Q_f = 610$ L/s
   $Q/Q_f = 0.025$, and from Figure 7.11 $d/D = 0.125$, so depth = 75 mm
   Chart/table gives velocity (full) $V_f = 2.15$ m/s
   from Figure 7.11 for $d/D = 0.125$, $v/v_f = 0.45$, so velocity = 0.97 m/s.

Note that the larger pipe size leads to a slightly lower depth (and greater area of contact with pipe wall), but the effect on velocity is negligible. This is a general result, and demonstrates that egg-shaped sewers do not have such a significant effect on low-depth velocities as might be expected.

12.3 THE ROLE OF CSOs

12.3.1 Flow and pollutants

The main function of a CSO is hydraulic: to take an inflow and divide it into two outflows, one to the wastewater treatment plant (the continuation flow, or flow retained) and one to the watercourse (the spill flow). The normal means of achieving this is a weir. If the surface of the flow passing through the CSO is below the crest of the weir, flow continues to the wastewater treatment plant (WTP) only. As the flow rate increases, so does the level of the water surface, to provide an increase in the hydraulic gradient along the continuation pipe. When the water surface is above the weir crest, some flow passes over the weir, while the rest continues to the WTP. The flow rate over a weir is related to the depth of water above the crest, so if the water surface continues to rise, so does the spill flow. The continuation flow is also likely to rise slightly as a result of the increase in head.

A hydraulic design of a CSO requires care (as considered in more detail later). There could be a number of effects of poor hydraulic design. If a spill took place prematurely, the capacity of the continuation pipe would be under-used, and an unnecessarily large volume of polluted flow would be discharged to the watercourse. But if the weir was set too high, excessive surcharge of the upstream system might be caused, potentially leading to flooding, and too much flow might be forced down the continuation pipe causing flooding elsewhere in the sewer system. In a good hydraulic design, spill will take place at the optimum level, and the continuation flow will not increase greatly while the spill flow is increasing with rising water level.

The other main function of a CSO is related to pollution. The ideal would be that all pollutants continued to the WTP (i.e., were retained within the sewer system), but this is not achieved. CSOs with screens demonstrate success in retaining larger solids, but fine suspended and dissolved material tends to be split between continuation flow and spill flow in the same proportion as the split in the flows.

The impact of CSO discharges on receiving waters is considered in Chapter 2. These impacts are likely to be most serious when CSOs are poorly designed, operating ineffectively, or cannot pass an appropriate amount of flow downstream. Also, sewers that back up as a result of sediment deposition problems may cause CSOs to operate prematurely (before the inflow has reached the CSO setting), or, in extreme cases, to spill even in dry weather conditions. This may cause serious pollution of receiving waters.

12.3.2 First foul flush

In some systems, a significant feature is the first flush in early storm flows, which may contain particularly high pollutant loads. These are likely to have been derived from the following:

1. *Catchment surface washoff and gully pots.* A first flush from this source would be expected as a result of the early rainfall washing off pollutants accumulated on the catchment surface and in gully pots since the last rainfall.
2. *Wastewater flow (including gross solids).* Since the storm wave moves faster than the wastewater *baseflow*, the front of the wave can consist of an ever-increasing volume of overtaken undiluted baseflow. However, this is normally diluted to some extent by the inflow from intermediate branches. Solids entering may float, be neutrally buoyant, or move along the bed (contributing to point 3).
3. *Near-bed solids.* In many sewer systems, high concentrations of organic solids have been observed in a layer moving just above the bed. The added turbulence as the storm flow increases causes these solids to become mixed with the stormwater.
4. *Pipe sediments.* Increasing storm flows provide suitable conditions for re-erosion of the deposited material.

A first foul flush can be identified on hydrographs and pollutographs recorded in the system. An obvious sign would be a sharp increase in pollutant concentration near the start of a storm. In fact, even if concentration remained constant as flow rate increased, this would signify an increase in pollutant load rate. A first flush can also be identified by plotting cumulative load against cumulative flow volume (Figure 12.2). The 45° line indicates that pollutants are uniformly distributed throughout the storm. If the line for a particular storm is above the 45° line, a first flush is suggested. Flushes from different conditions or catchments can be compared in this way. In some catchments, the effect is pronounced.

12.4 CONTROL OF POLLUTION FROM COMBINED SEWER SYSTEMS

12.4.1 CSO settings and permits

12.4.1.1 Technical Committee Formula A

Until 1970 the traditional CSO setting had been 6 × dry weather flow (6DWF) in the United Kingdom. An extensive study in the 1950s and 1960s of the effects of CSOs was carried out by the government-appointed "Technical Committee on Storm Overflows and the Disposal of Storm Sewage," whose final report was published in 1970 (Ministry of Housing and Local Government, 1970). Among the conclusions was that it was illogical to base a CSO setting merely on a multiple of DWF, and that, because of the harmful effects of CSO pollution, the new standard setting should give a "modest improvement" (that is, divert less pollution to watercourses).

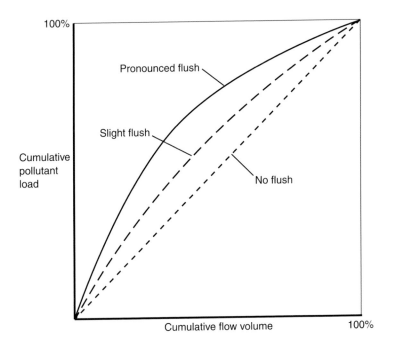

Figure 12.2 Representation of first foul flush.

The setting of 6DWF had allowed for diurnal variations in wastewater flow, plus some stormwater (so that for low-intensity rainfall there would be no overflow). If people in one area happened to use more water than they did in another, that was no reason for more stormwater to be retained in the sewer system. It was considered appropriate for a CSO setting to be based on DWF plus some storm allowance (related to population, but not to water consumption). The committee also felt there were ambiguities about inclusion of infiltration and industrial flows in traditional practice. The proposed new standard CSO setting was given by their Formula A:

$$\text{Setting} = DWF + 1360P + 2E \text{(liters/day)} \quad \text{(Formula A)} \tag{12.1}$$

for which, from Chapter 9

$$DWF = PG + I + E \tag{9.1}$$

where P is the population, G is water consumption per person (L/d), I is the pipe infiltration rate (L/d), and E is average industrial effluent (L/d).

In a catchment where G is 200 L/head.day and there is no infiltration or industrial flow, Formula A gives a setting of 7.8 DWF (see Example 12.2).

The report recommended that the coefficients could be treated with flexibility. In situations where discharge was to a large river, 1360P could be decreased, or it could be increased if to a small stream; industrial effluent of high strength might require an increase in the term 2E.

The report also contained many other detailed recommendations about CSO design.

EXAMPLE 12.2

A combined sewer catchment serves a population of 50,000 and has an impervious area (A_i) of 18 ha. Determine the overflow setting required upstream of the main outfall sewer using the following approaches:

1. 6DWF
2. Formula A
3. River water quality adjacent to the overflow limited to a BOD_5 of 10 mg/L for the 1-year return period, 20-minute duration event.

Additional information:
Wastewater flow = 250 L/hd.d
Rainfall intensity for the 1-year, 20-minute event, $i = 20$ mm/h
River flow upstream of CSO = 1 m³/s, with BOD_5 of 2 mg/L
Overflow $BOD_5 = 500$ mg/L
Pipe infiltration and industrial flows are negligible.

Solution

1. $DWF = 50,000 \times 250 = 12.5 \times 10^6$ L/d or 145 L/s

    $Setting = 6 \times 145 = 870$ L/s

    $Setting = DWF + 1360P + 2E$ (12.1)

    $\qquad = 12.5 \times 10^6 + 1360 \times 50,000$

    $\qquad = 932$ L/s

2. Runoff flow rate $= iA_i = 18 \times 10^4 \times 20/3600 = 1000$ L/s

    BOD_5 load rate: river upstream + overflow = river downstream

    $1000 \times 2 + Q_{overflow} \times 500 = (1000 + Q_{overflow}) \times 10$

    $\qquad\qquad Q_{overflow} = 16$ L/s

    $DWF + runoff = Setting + Q_{overflow}$

    $\qquad 145 + 1000 = Setting + 16$

    $\qquad\qquad Setting = 1129$ L/s

12.4.1.2 Scottish Development Department (SDD)

The report of the Working Party on Storm Sewage (Scotland), *Storm Sewage: Separation and Disposal* (SDD, 1977), gave details of another significant study of CSOs. Guidelines for CSO setting and storage volume were related to the amount of dilution when combined sewer flow was overflowed to a watercourse (a factor not covered explicitly in Formula A). Dilution was defined as the *minimum flow* in the stream (the flow rate exceeded 95% of the time) compared with the sewer dry weather flow. For dilution of more than seven to one, Formula A was considered satisfactory. For dilution of six to one, it was recommended that either the setting should be enhanced to Formula A + 455P, or that Formula A should be used in its original form and storage should be provided. For any lower dilution, increasing amounts of storage were recommended, in conjunction with a Formula A setting.

12.4.1.3 Urban pollution management

The *Urban Pollution Management Manual* (Foundation for Water Research, 2012), sets out procedures for management of wet weather discharges from urban drainage systems

Table 12.1 CSO emission standards

Water use	Standard
Aquatic life	Intermittent, 99 percentile and relevant WFD standards
Shellfish	10 spills per annum Spill for no more than 3% of the time
Bathing	Three spills per bathing season (sufficient/good status) Two spills per bathing season (excellent status)
Amenity	6 mm solids separation[a]

Source: Based on Environment Agency. 2011. *Water Discharge and Groundwater (from point source) Activity Permits (EPR 7.01)*, Environment Agency, Bristol, UK.

[a] Separation of a "significant quantity" of solids greater than 6 mm in any two dimensions.

to receiving waters against the four standards discussed in Section 2.6.3. CSO emission standards based on various water end uses are given in Table 12.1. Three of the standards, aquatic life, bathing, and shellfish, typically relate to the finer dissolved pollutants and require storage or treatment to reach acceptable discharge levels.

12.4.2 Monitoring

There is a continuing trend to monitor CSOs, in particular the major ones, to ensure and demonstrate compliance with the Urban Waste Water Treatment Directive. Permit compliance is usually assessed by site inspection to make sure the required provision is in place, and monitoring to record spill events. As flow rate or total overflow volume is relatively difficult and expensive to monitor, spill event time and duration are more commonly monitored (e.g., by use of level sensors) (Environment Agency, 2011). Measurements are typically made at 2- or 15-minute intervals and the number of spills logged. Flow measurement may be specified to demonstrate over a period of time whether overflows are meeting their permits or where an investigation, and ultimately further enhancements, may be required. The Chartered Institution of Water and Environmental Management's (CIWEM, 2016) good practice guide provides further detail on the practical application.

12.5 APPROACHES TO CSO DESIGN

12.5.1 High side weir

12.5.1.1 Principles

Most modern CSOs are high side weir overflows with screens. We consider first the basic design of a high side weir overflow without screens, used predominantly in the 1980s and 1990s before the development of robust fine screens. An overflow with high weirs, scumboards, and a stilling zone upstream can provide reasonable retention of both floating and sinking solids. Double-side weirs can provide good hydraulic control. High side weirs are often associated with storage. This can be a storage zone downstream of the weir for retention of floating solids, or a large storage volume for retention of the first flush.

12.5.1.2 Dimensions and layout

Recommendations are given by Balmforth et al. (1994) (Figure 12.3) and Wastewater Planning Users Group (WaPUG, 2006). Flow in the chamber must be subcritical. Many

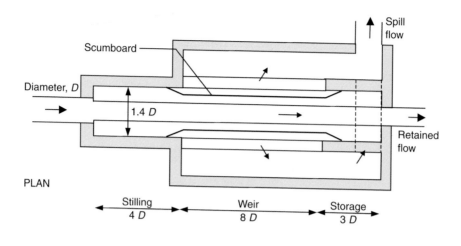

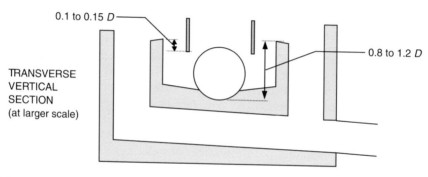

Figure 12.3 High side weir CSO: general arrangement and dimensions.

poorly performing CSOs (with screens) result from unsatisfactory hydraulic conditions due to their basic design leading to an hydraulic jump in the chamber, potentially resulting in severe pollution and expensive rehabilitation. Designing high sided weir chambers and their dimensions are discussed in Section 12.5.2 and in greater detail in Section 12.7 and Example 12.5.

12.5.2 Screens

12.5.2.1 Principles

Screens are incorporated as a matter of routine within CSO structures. They provide a physical barrier to solids with the aim of preventing them from entering the overflow pipe. The focus is on gross solids, defined in this context as being larger than 6 mm in two dimensions and of sewage origin. Common screens are meshes or perforated plates (Figure 12.4), which prevent these solids from being discharged by retaining them in the sewer system. "Static" screens rely on flow patterns within the CSO and the design of the screen itself to prevent blocking or blinding, but typically have to be cleaned manually as part of planned and reactive maintenance. Powered screens are mechanically cleaned during storm flows to prevent blinding, usually by brushing. However, many different approaches to CSO screen design have been tried, as described in the next section.

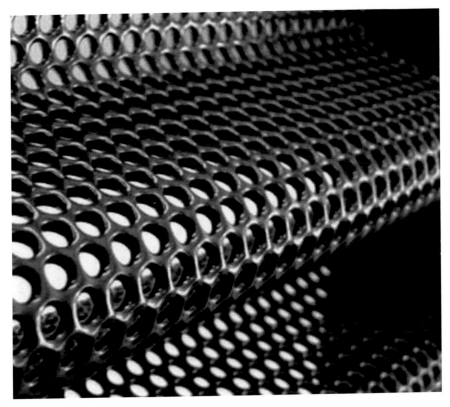

Figure 12.4 Example of a static screen, Hydro-Static Screen. (Courtesy of Hydro International.)

12.5.2.2 Development

A field study of screens at CSOs (Meeds and Balmforth, 1995) concluded that mechanically raked bar screens are unlikely to achieve retention efficiencies of greater than 50% (of all gross solids). Screens typically now consist of a mesh rather than parallel bars, partly as a result of solids separation requirements. A laboratory study (Saul et al., 1993) demonstrated that a mesh screen to be more effective at retaining solids than a bar screen with the same spacing, though a field study using actual wastewater concluded that 6 mm mesh screens are unlikely to achieve retention efficiencies of greater than 60% (Balmforth et al., 1996).

The Urban Pollution Management (UPM) manual (FWR, 2012, Appendix C) contains the results of a testing programme using a dedicated CSO test facility at Wigan Wastewater Treatment Plant. The tests used CSOs fitted with

- 6 mm mesh screens
- 10 mm bar screens
- No screens, but designed according to the principles of the *Guide to the Design of Combined Sewer Overflow Structures* (Balmforth et al., 1994)

The resulting plots of solids separation efficiency are presented as standards against which any alternative or novel approach can be compared using a defined test procedure. The plot for "good engineering design" is given as Figure 12.5, and the plot for 6 mm solids separation is given as Figure 12.6. *Total efficiency* is the percentage of the total mass of solids

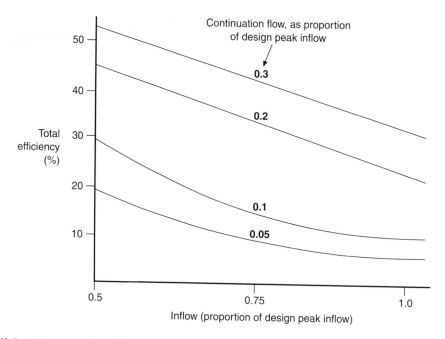

Figure 12.5 Solids separation efficiency for "good engineering design." (Based on FWR. 2012. *Urban Pollution Management Manual,* 3rd edn, Foundation for Water Research, http://www.fwr.org/UPM3. With permission of Foundation for Water Research, Marlow.)

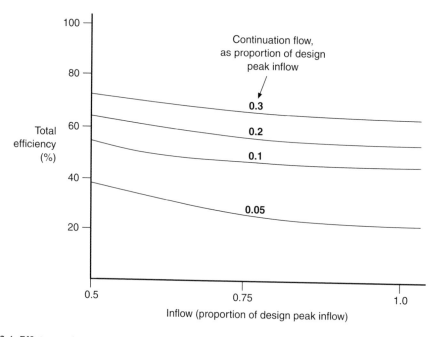

Figure 12.6 Efficiency for 6 mm solids separation. (Based on FWR. 2012. *Urban Pollution Management Manual,* 3rd edn, Foundation for Water Research, http://www.fwr.org/UPM3. With permission of Foundation for Water Research, Marlow.)

entering the CSO that is retained within the sewer system (in the continuation flow or in the chamber). The 6 mm separation does not come near to retaining *all* solids. A wide range of proprietary screen arrangements have since been tested at the same facility and compared with these standards in order to help engineers choose suitable devices (Saul, 2000; UKWIR, 2006).

12.5.2.3 Dimensions and layout

It is possible to fit screens to any of the CSO types described in this chapter. However, the standard UK recommendation (WaPUG, 2006) is for CSOs incorporating screens to be of the high-sided weir type.

The recommended dimensions and layout in the WaPUG guide to *The Design of CSO Chambers to Incorporate Screens* (WaPUG, 2006) are based on physical tests and on computational fluid dynamics (CFD) modelling. The inlet length of the chamber allows flows to turn onto the screen, with a minimum value of $0.4D_{in}$ but must be at least 0.5 m. Outlet lengths enable solids from the screens to be returned to the outgoing flow and continue through the network. Outlet lengths are based on $1.5D_{in}$. The recommended chamber width is $1.4D$ (where D is a diameter of incoming pipe), which is the same as the recommendation by Balmforth et al. (1994) for a high-sided weir CSO without a screen (Figure 12.3). All these dimensions must be sufficient to enable maintenance of screens; therefore, in smaller CSO chambers, it is common for larger than the minimum size chambers being built.

The weir length should be no more than $6D$ (compared with $8D$ by Balmforth et al. [1994] on Figure 12.3), and the stilling and storage lengths are not needed, though shorter inlet and outlet lengths are still recommended. Either double- or single-side weirs are appropriate. The basic layout of a single high side weir overflow suitable for incorporating screens, using the principles of the WaPUG (2006), is given in Figure 12.7.

The guide refers to three possible positions for the screen in a high-sided weir CSO given in Figure 12.8. A fourth alternative position is on the outfall, similar to a screen used at the inlet works to WTP. The three common positions for screens are as follows:

- Mounted vertically on the weir, so that flow over the weir passes horizontally through it
- Mounted horizontally above the main channel, so that flow over the weir must first pass upward through the screen
- Mounted on the downstream face of the weir, so flow that has passed over the weir then passes down through the screen

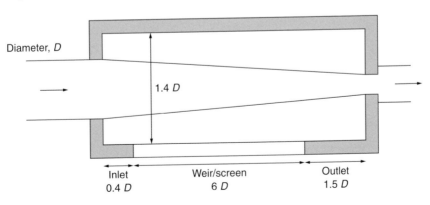

Figure 12.7 Side weir CSO for screens: general arrangement and dimensions with a typical weir/screen length.

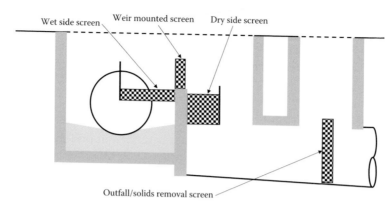

Figure 12.8 Possible positions of CSO screens.

12.5.3 Stilling pond

12.5.3.1 Principles

The main principles of a stilling pond CSO are illustrated in Figure 12.9. In dry weather and low-intensity rain, the flow enters via the inlet pipe, passes along a channel through the overflow, and leaves via the throttle pipe. In heavier rainfall, as the inflow increases, the

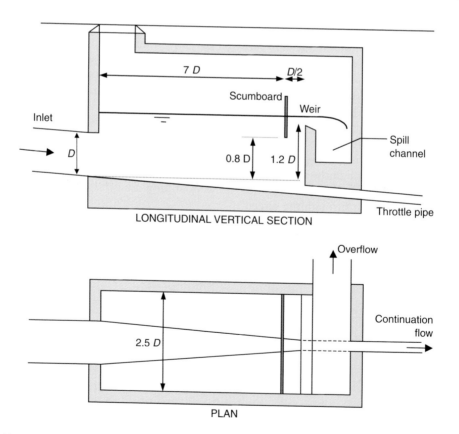

Figure 12.9 Stilling pond CSO: general arrangement and dimensions.

capacity of the throttle pipe is exceeded and flow backs up inside the chamber. The level usually has to rise above the top of the inlet pipe before it reaches the crest of the weir. This causes the inflow to become stilled, which helps to ensure that sinking solids are not carried over the weir. When the water level is above the weir crest, water spills over the weir and out via the spill channel and pipe. The scumboard is positioned to limit the passage of floating solids over the weir. In fact it does more: it sets up a pattern of circulation in the chamber, which brings many floating solids back to the upstream end of the chamber—making it even less likely that they will flow over the weir.

12.5.3.2 Development

The first major investigation was by Sharpe and Kirkbride (1959). The results are much quoted and had a genuine impact on engineering practice. They concluded that the best conditions were achieved when the inlet velocity was low and the upstream sewer was well flooded in order to create stilling conditions in the chamber. The scumboard created a reverse surface flow that took the floating solids away from the weir. Their recommendation was that the distance from the inlet to the scumboard be at least 4.2 times the diameter of the inlet pipe (D). Other recommended dimensions included a chamber width of 2.5D, a distance from the scumboard to the weir of 0.5D, and a weir level similar to the soffit of the incoming pipe.

Frederick and Markland (1967) carried out laboratory studies of model stilling pond arrangements. Many of their conclusions confirmed those of Sharpe and Kirkbride (1959), "in particular, the incoming sewer needs to be surcharged in order to produce a favourable reverse current near the surface." The main difference in their conclusions was in terms of length, which they recommended should be as great as possible, with overall length no less than seven times inlet diameter.

Balmforth (1982) studied the separation of a wide variety of solids in a model stilling pond. A particular aim was to resolve differences in the recommendations for chamber dimensions between Sharpe and Kirkbride and Frederick and Markland. Balmforth confirmed that there were significant advantages in the longer length recommended by Frederick and Markland.

12.5.3.3 Dimensions and layout

Recommendations for chamber dimensions, based on the development described above and best knowledge of CSO operation, are given in the *Guide to the Design of Combined Sewer Overflow Structures* by Balmforth et al. (1994). Their recommended dimensions for a stilling pond are given in Figure 12.9. They are based on the diameter of the incoming pipe. A method of determining this diameter (common to a number of different CSO types) is given in Section 12.7.

A dry weather flow channel runs along the centre of the chamber, contracting in area from the inlet to the throttle pipe. It should have sufficient size and longitudinal slope to carry flow rate equal to the capacity of the throttle pipe and avoid sediment deposition. The base on either side of this channel slopes towards the centre to drain liquid to the DWF channel. The capacity of the throttle pipe is crucial in determining the setting of the CSO. The upstream head is a function of the crest level and characteristic of the weir.

12.5.4 Hydrodynamic vortex separator

12.5.4.1 Principles

Several types of overflow arrangement exploit the separation of solids that occurs in the circular motion of a liquid.

When such a flow is considered in two dimensions, by studying a horizontal section, theory suggests that heavier solids will follow a path towards the outside of the circle. This leads to a design of *vortex overflow* in which the overflow weir is placed on the inside of the chamber and the continuation pipe on the outside (where heavier solids tend to congregate). Floating solids are prevented from flowing over the weir by a baffle.

Studies in three dimensions, with particular chamber shapes, have suggested that heavy solids collect at the bottom of the chamber, in the centre. These have led to designs with the opposite arrangement: the weir on the outside, and the continuation pipe at the centre. Other, more elaborate chamber arrangements have led to flow patterns with even more complex properties.

12.5.4.2 Development

Smisson (1967) carried out extensive work on models and full-scale vortex overflows, giving detailed descriptions of flow patterns, and design recommendations. The weir was positioned on the inside of the vortex and the continuation pipe on the outside.

A different type of vortex chamber was proposed by Balmforth et al. (1984), called a "vortex overflow with peripheral spill." This design "makes use of the known ability of vortex motion to separate settleable solids, but differs from earlier designs in that the foul outlet pipe is set in the centre of the chamber floor, and the overflow occurs over a weir formed in the peripheral (outer) wall."

Modern descendants of the vortex overflow are called *hydrodynamic separators*. A patented design, the Storm King Overflow, in which separation of solids takes place within a complex flow pattern of upwards and downwards helical flow before passing through a static screen, has been applied in the UK and elsewhere. The arrangement is shown in Figure 12.10.

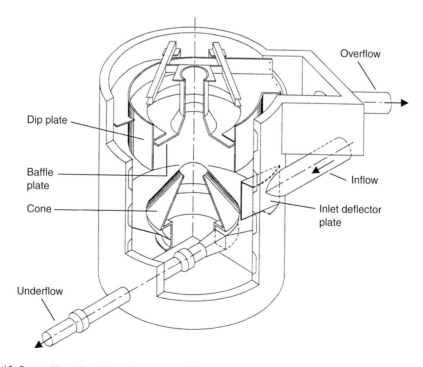

Figure 12.10 Storm King Overflow. (Courtesy of Hydro International.)

Similar use of these principles has been made in other countries, the U.S. Environmental Protection Agency's (EPA) "swirl regulator" (Field, 1974), and the German "Fluidsep" vortex separator (Brombach, 1987, 1992), for example.

Much of the development is related to specific patented devices, but research has continued into the more general principles of devices of this type (e.g., Huebner and Geiger, 1996). Fenner and Tyack (1997, 1998) have proposed scaling protocols for physical models. Saul and Harwood (1998) have studied retention efficiency for full-scale sanitary gross solids. A review of the various types of hydrodynamic separators in use has been given by Andoh (1998). Andoh et al. (2008) describe high-rate treatment and disinfection using advanced hydrodynamic separators.

12.5.4.3 Dimensions and layout

Hydrodynamic separators are designed and fabricated by their manufacturers based on performance specifications.

12.5.5 Storage

12.5.5.1 Principles

The aim of providing storage at a CSO is to retain pollutants in the sewer system rather than allowing them to be overflowed to a watercourse, even after a weir on the main sewer has come into operation during a storm. When flows in the system have subsided after the storm, the polluted flow retained in the storage can be passed onwards to treatment. Clearly the larger (and more expensive) the storage, the lower is the amount of pollution reaching the watercourse. Optimum sizing of storage needs to take into account the fact that polluting loads during storm flow vary with time. It is common (but not universal) that early flows are particularly polluted as a result of a first foul flush, as considered in Section 12.3.2.

Storage has tended to be provided in conjunction with a high side weir arrangement, but storage can be used to supplement any CSO configuration.

Storage can be provided online or offline. In an *online* arrangement, the flow passes through the tank even in dry weather when the capacity of the tank is not being utilised. When flow rate increases during a storm, a downstream control will cause the level to rise, to fill up the storage volume, and eventually overflow at the weir (Figure 12.11a). After the storm, the tank empties by gravity into the continuation pipe. In an *offline* arrangement, flow is diverted to the tank via a weir as the level begins to rise (Figure 12.11b). When the tank is full, a higher weir comes into operation and diverts further flows to the watercourse. Screens may be fitted to either weir but are most commonly found on the diversion weir to reduce solids entering the storage in the first place. After the storm, the tank is emptied into the continuation pipe by gravity or by pumping. The rate at which the storage tank can be emptied is governed by the amount of spare capacity in the pipe and/or treatment plant.

Storage at a CSO can be provided in a number of forms: rectangular chamber, circular vertical shaft, or oversized pipe or tunnel.

12.5.5.2 Development

In the UPM procedures (FWR, 2012), the size of storage is optimised by application of sewer quality modelling (either simple or complex). Because of the UPM emphasis on considering the system as a whole, design rules for sizing individual tanks are not proposed. Example 12.3 is an illustration of the way in which model simulations can be used to investigate

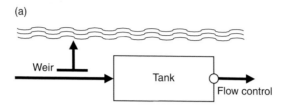

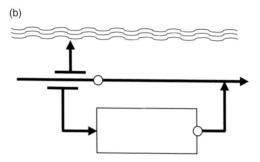

Figure 12.11 Storage tank: (a) online and (b) offline.

possible storage proposals. A decision is not made until the full range of design rainfall patterns has been considered, the costs of alternative storage strategies have been determined, and the effects on the rest of the system have been assessed.

EXAMPLE 12.3

Figure 12.12 gives a simulated flow hydrograph and chemical oxygen demand (COD) pollutograph (concentration and load rate) for a catchment, in response to a particular rainfall pattern. Rainfall in this case started at a low intensity, causing a slight increase in flow rate and dilution of COD concentration (from the dry weather level of 470 mg/L). During the early period, load rate is constant. After 45 minutes, the rain became more intense, causing a significant first flush, apparent from both the concentration and load-rate graphs.

Determine the approximate size of storage that would be needed (1) to retain pollutants until COD concentrations no longer exceed dry weather level, and (2) to retain pollutants until COD load rate no longer exceeds dry weather level. Assume that the detention tank will have outflow via a control that limits flow rate to 100 L/s, and via an overflow that operates when the storage is full.

Solution
1. The volume retained would equal the area under the hydrograph (above the 100 L/s line) up to *a*.
   From the graph, volume ≈ 440 m³.
2. The volume retained would equal the area under the hydrograph (above the 100 L/s line) up to *b*.
   From the graph, volume ≈ 580 m³.

As well as optimising the size of storage, designers need to pay attention to the layout of the chambers to avoid excessive sedimentation. When a tank is full of virtually stationary liquid, conditions are ideal for deposition of suspended solids (and concentrations are likely to be high in the first flush that the tank will have been designed to retain). In large storage

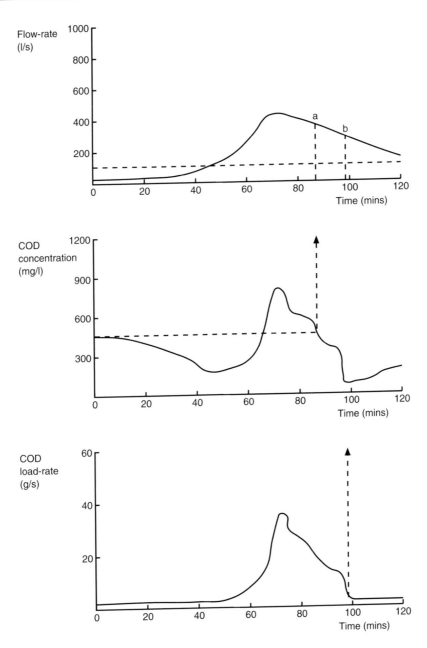

Figure 12.12 Storage: interpreting simulated hydrograph and pollutograph. (Example 12.4)

tanks, it is not uncommon to use CFD to analyse and design the storage tanks and mainte-nance requirements (Stovin and Saul, 2000).

12.5.5.3 Dimensions and layout

Saul and Ellis (1992) found that a DWF channel with a steep longitudinal gradient was help-ful by creating suitably high velocities. Long narrow tanks with a single DWF channel had better self-cleansing properties than wider tanks with multiple channels. Length-to-width ratio should be as high as possible, and width should not exceed 4 m. Guidance is also given by the Water Research Centre (1997). CFD modelling has confirmed the importance of the length-to-width ratio (Stovin and Saul, 2000).

12.6 EFFECTIVENESS OF CSOs

12.6.1 Performance measures

Hydraulic performance can be expressed in terms of liquid volumes, using the term *flow split*:

$$\text{Flow split} = \frac{\text{storm volume retained in the sewer system}}{\text{total storm inflow volume}}$$

The split can also be expressed in terms of pollutant loads (cumulative mass of pollutant), using the term *total efficiency*:

$$\text{Total efficiency} = \frac{\text{storm load retained in the system}}{\text{total storm inflow load}}$$

Both terms are needed to judge the success of a CSO design, combined in the term *treatment factor*:

$$\text{Treatment factor} = \frac{\text{total efficiency}}{\text{flow split}}$$

The success of a CSO design in the separation of pollutants is indicated by the amount by which the treatment factor exceeds 1. A treatment factor less than 1 indicates that a design is unsuccessful in this respect. This is illustrated in Example 12.4.

EXAMPLE 12.4

For a particular storm, a CSO gives a flow split of 20% and a total efficiency of 33%. Comment on its effectiveness. How effective would it have been if the total efficiency had been 15%?

Solution

A flow split of 20% for a particular storm indicates that one-fifth of the total inflow volume was retained in the system, and four-fifths was overflowed. If the total efficiency was 33% (one-third of pollutants retained in the system, two-thirds overflowed), we can deduce that in these conditions the design has some qualities in retaining pollutants, over and above the straightforward split in flow. The treatment factor is 1.65 (33% divided by 20%).

If the total efficiency was only 15%, this would suggest that instead of the desired effect of retaining pollutants, the CSO was giving the opposite effect. The resulting treatment factor would be 0.75.

12.6.2 ROLE OF CFD

CFD modelling is being increasingly used as a tool for assessing and comparing CSO configurations. Advances in hardware and software mean that this software can be used in place of, or when appropriate in conjunction with, physical modelling (Harwood and Saul, 2001). Jarman et al. (2008), presenting an overview of CFD modelling in urban drainage, state that "the most extensive application of CFD to urban drainage system analysis has been in relation to CSO chambers," and predict that opportunities for using CFD in this way will increase in the future. CFD modelling is being used more widely to understand complex flow situations in urban drainage (e.g., Jarman et al., 2015).

12.6.3 Gross solids

A laboratory comparison (Saul et al., 1993) of the ability of large-scale model CSOs (stilling pond, vortex with peripheral spill, hydrodynamic separator, high side weir) to retain sanitary gross solids in steady and unsteady flow came to the disappointing conclusion that the performance of all types of chamber was relatively poor at design flow rate, with treatment factor rarely much above unity. This was later further confirmed by studies at full scale (Saul, 1999). This is a result of the fact that the solids—those most likely to cause problems at actual CSOs—have a density close to that of water and, therefore, have low terminal velocities. Solids retention efficiencies plotted against terminal velocity for many studies have consistently demonstrated a characteristic cusp or "gull's wing" shape (e.g., Figure 12.13). Often the range of terminal velocities of particles studied has been wide, giving quite good efficiencies for the clear "floaters" and "sinkers." But the reality is that the most common, and most aesthetically sensitive solids, with their close-to-neutral buoyancy, show the CSO designs at their worst. The standard engineering solution has become that of incorporating screens (Section 12.5.2).

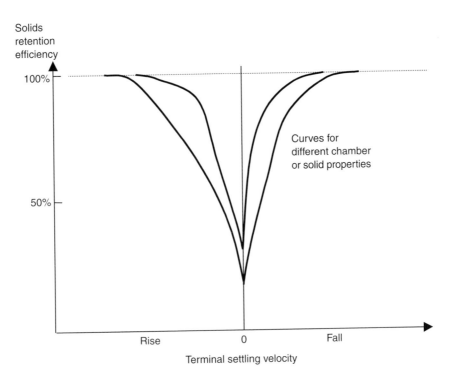

Figure 12.13 Plots of CSO solids separating efficiency (conceptual).

12.6.4 Choice of CSO design

Myerscough and Digman (2008) observe that "the standard WaPUG chamber"—that is, a high side weir overflow fitted with screens using the principles of WaPUG (2006)—is well established as "best practice" but that there is often a need for a "non-standard" design. They point out that physical modelling and CFD play an important role in understanding how CSOs perform, and that such studies can allow these structures to be used with confidence, enable improvements to be made in design, and avoid costly mistakes.

It should be added that we have only been considering the conventional types of CSO in this chapter, and have tended to concentrate on the control of larger solids. Some studies have concentrated on fine suspended particles and dissolved pollutants, and on advanced devices that include high-rate treatment processes (e.g., Andoh et al., 2008; Vetter et al., 2001).

12.7 CSO DESIGN DETAILS

12.7.1 Diameter of inflow pipe

Dimensions of CSO chambers are based on the diameter of the inflow pipe. The minimum diameter of this pipe is determined from

$$D_{min} = KQ^{0.4}$$

(12.2)

where Q is the peak inflow for the design return period (m³/s) and K is the constant taken from Figure 12.14 for a high-sided weir.

A relationship (represented on Figure 12.14) between incoming pipe diameter and flow has been developed by physical testing for a range of high-sided weir heights, c_w, and weir lengths, L_w (where the weir height is greater than 50% of the incoming diameter). This

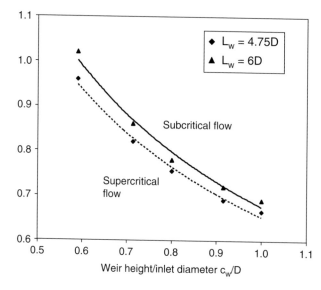

Figure 12.14 Determining values of K to calculate the incoming pipe size from the WaPUG CSO Design Guide. (Courtesy of CIWEM Urban Drainage Group.)

reduces the likelihood of supercritical flow forming in the chamber (subject to understanding the forces in the incoming pipe/inlet to the chamber).

The Design of CSO Chambers to Incorporate Screens (WaPUG, 2006) recommends a design return period of 5 years, which matches the requirements outlined by the Environment Agency (2011). Nearly all CSO chambers are designed with a screen unless they only spill in events with a return period far greater than 5 years.

The general form of Equation 12.1 has its origin in the Darcy-Weisbach Equation 7.8, which can be rearranged to

$$h_f = \frac{\lambda L Q^2}{12.1 D^5} \quad \text{or} \quad D = \left(\frac{\lambda L}{12.1 h_f} \right)^{0.2} Q^{0.4}$$

where D is the diameter (m), λ is the friction factor (–), L is the length (m), h_f is the friction head loss (m), and Q is the flow rate (m³/s). A design example of the inlet pipe and chamber dimensions for a high sided weir CSO with a screen is given in Example 12.5.

12.7.2 Creating good inlet flow conditions

A critical element in the design is to ensure that turbulence, particularly because of a hydraulic jump in the chamber, is not created, and this should always be investigated. Calculating the Froude Number (Equation 7.26) indicates if the flow in the incoming pipe is subcritical or supercritical. Where the flow is subcritical throughout the pipe and chambers, a hydraulic jump will not occur.

If supercritical flow meets subcritical flow, then the position of the hydraulic jump should be investigated using the specific force Equation 12.3 for the water in the pipe entering, Fs_1, and the water in the chamber, Fs_2. The design should check that force in the chamber is greater than the force in the pipe $Fs_2 > Fs_1$, so that the jump occurs upstream of the chamber:

$$\text{Fs} = \rho Q v + \rho g A \bar{z} \tag{12.3}$$

where $\bar{z}$ is the depth to the centroid of flow cross section (m).

Flows, velocities, depths, and areas are calculated for the 5-year incoming flow to the pipe or chamber.

12.7.3 Weirs

The hydraulic characteristics of weirs are considered in Section 8.2. The weir length and height are critical in sizing the incoming pipe. There is a balance in selecting an appropriate weir height: if it is too high, it may increase the risk of flooding, but if it is too low it may create unfavourable conditions in the chamber for the screen to perform.

12.7.4 Selecting, sizing, and accommodating the screen

The chamber must be of sufficient size to accommodate the screen to ensure that

- Unfavourable hydraulic conditions are not created that reduce the effectiveness of the screen, or in serious cases damage the screen and lead to unscreened sewage being discharged.
- Maintenance can be completed safely including the ability to replace parts of the screen when all work is undertaken within the closed chamber.

The selection of the screen and its position depend on a number of factors, including

- Hydraulic headroom—what is happening in the drainage network upstream and downstream
- Design spill flow that will pass through the screen
- Frequency of spill—higher-spilling CSOs require a self-cleaning screen
- Availability of a power supply or ease of providing power supply
- Planning constraints if a kiosk is required for a powered screen
- Space requirements
- Accessibility for maintenance
- Cost to procure, build, and maintain

EXAMPLE 12.5

Determine the basic dimensions of a high-sided CSO with screen. Peak flow with a return period of 5 years is estimated as 900 L/s.

The length of the screen required to accommodate the spill flow indicates that the length of the weir (L_w) should be 4.4 m.

Solution

Try inlet pipe diameter (D) of 0.75 m, and weir height (c_w) of 0.6 m:

$$\frac{L_w}{D} = \frac{4.4}{0.75} = 5.9$$

$$\frac{c_w}{D} = \frac{0.6}{0.75} = 0.8$$

From Figure 12.14 (reading from the line for $L_w = 6D$), K should be no less than 0.8. So,

$$D_{min} = KQ^{0.4} = 0.8 \times 0.9^{0.4} = 0.77 \text{ m}$$

We may need to upsize the inlet pipe. An alternative might be to increase the height of the weir.

If we upsize the inlet pipe to 825 mm, while retaining the weir height of 0.6 m,

$$\frac{L}{D} = \frac{4.4}{0.825} = 5.3$$

$$\frac{c_w}{D} = \frac{0.6}{0.825} = 0.73$$

From Figure 12.14 (now reading between the two lines), K should be no less than 0.83. So, from Equation 12.2.

$$D_{min} = KQ^{0.4} = 0.83 \times 0.9^{0.4} = 0.8 \text{ m}$$

This confirms that an 825 mm diameter inlet pipe, with a weir height of 0.6 m, would be appropriate.

Or, if we retain the 750 mm diameter inlet pipe but increase the weir height to, say, 0.64 m,

$$\frac{L_w}{D} = \frac{4.4}{0.75} = 5.9$$

$$\frac{c_w}{D} = \frac{0.64}{0.75} = 0.85$$

From Figure 12.14 (reading from the line for $L_w = 6D$), K should be no less than 0.76. So,

$$D_{\min} = KQ^{0.4} = 0.76 \times 0.9^{0.4} = 0.73 \text{ m}$$

This confirms that the alternative of a 750 mm diameter inlet pipe, with a weir height of 0.64 m, would also be appropriate.

Since the dimensions of the chamber are generally based on the diameter D, let us select this second option. So, with $D = 0.75$ m, the dimensions should be as follows:

Dimension	Calculation	Value	Notes
Chamber width	1.4 D	1.05 m	May need to be increased to accommodate the screen
Inlet length	0.4 D = 0.3 m	0.5 m	Minimum value of 0.5 m
Outlet length	1.5 D	1.125 m	
Weir length	Based on screen	4.4 m	
Overall length	0.5 + 4.4 + 1.125	6.025 m	

12.7.5 Control of outflow

Control of the continuation flow is an important part of the hydraulic design of a CSO. The setting of the overflow is normally defined as the continuation flow when spill starts—that is, when liquid level reaches the weir crest. As flow rate over the weir increases, so will depth. It is best if the retained outflow does not vary greatly as a consequence. The common methods of control are

- Fixed orifice
- Adjustable (manual or automated) penstock
- Vortex flow regulator
- Throttle pipe

These are described in Section 8.1. The control is typically designed to "freely" discharge; however, at times, it may become drowned, either because the downstream pipe capacity or the downstream network controls the flows.

12.7.6 Chamber invert

Chambers should be as self-cleansing as possible. Deposition of solids can be limited by suitable longitudinal and lateral slopes; lateral benching should slope at between 1:4 and 1:8.

12.7.7 Design return period

As stated above, design of inlet pipe and determination of the main chamber dimensions are based on the peak flow rate with a 5-year return period being used to satisfy regulatory requirements (Environment Agency, 2011). A check should also be made to see how a proposed chamber would respond to more extreme events—including a 30-year return period storm. This is particularly true for the spill channel and outlet pipe, which is the route taken by most of the flow in extreme events.

12.7.8 Top water level

The top water level (TWL) in the chamber can be determined from the design maximum inflow and the hydraulic properties of both outflow pipes. The impact of the CSO on the sewer system upstream and downstream may only be fully understood by using a suitable sewer system flow model. The outfall pipe needs to be checked carefully. If it could be drowned at the downstream end or if it discharges to tidal water, this will also need to be considered. Often more detailed hydraulic checks are appropriate beyond a sewer network model for the chamber performance. The TWL is one consideration in deciding the level of the roof of the structure.

12.7.9 Access

Human access is normally via manhole covers at ground level. Where screens are included, there needs to be appropriately sized and positioned access for vertical installation and removal of any machinery. There should be access to clear potential blockages, especially in throttle pipes. Thorough safety precautions are required during maintenance (considered in Section 17.5).

PROBLEMS

12.1 A combined sewer with diameter 750 mm, slope 0.002, and k_s 1.5 mm, drains a catchment with a DWF of 15 L/s. For a particular rainfall, the flow of stormwater is 750 L/s. Does the sewer have sufficient capacity to carry stormwater and DWF? Would the daily maximum flow in dry weather provide self-cleansing conditions? What are the likely consequences of this? How could the design of this pipe have been improved?   [no, no, $v = 0.44$ m/s]

12.2 What is meant by the "first foul flush"? What may cause it, and what are its implications?

12.3 What are the main functions of a combined sewer overflow? Explain the common alternative overflow configurations.

12.4 Define the term *CSO setting*. Describe the importance of a CSO setting, and ways in which it can be set.

12.5 The population of a catchment is 5,000, average wastewater flow is 180 L/hd.d. Infiltration is 10% of the domestic wastewater flow rate, and average industrial flow is 2 L/s. Determine the DWF and the CSO setting according to Formula A. Express the CSO setting as a multiple of DWF.   [13.5 L/s, 96.2 L/s, 7.1]

12.6 If the receiving water for the CSO in Problem 12.5 offers dilution of 2:1, how much storage should be provided in conjunction with the setting determined in Problem 12.5 (on the basis of the recommendations of the Scottish Development Department, 1977)? If the overflow is operating at this setting and all overflow is diverted to storage, how long would the storage take to fill if the inflow was constant at 500 L/s?   [400 m³, 16.5 minutes]

12.7 Explain approaches to CSO design by which the discharge of gross solids to the environment can be reduced.

12.8   The basic layout of a single high side weir overflow suitable for incorporating screens is being determined using the principles of WaPUG (2006). The diameter of the existing inlet pipe is 525 mm, with an inflow of 480 L/s and weir height of 0.4 m. Confirm the size of the inlet pipe (adjusting the weir height if appropriate), determine the size of the chamber, and propose a suitable weir length.   [diameter 675 mm, weir height 425 mm, width 925 mm, inlet length 500 mm, weir length 3.15 m, and outlet length 1.01 m].

KEY SOURCES

Balmforth, D.J., Saul, A.J., and Clifforde, I.T. 1994. *Guide to the Design of Combined Sewer Overflow Structures*, Report FR 0488, Foundation for Water Research.

Environment Agency. 2011. *Water Discharge and Groundwater (from point source) Activity Permits (EPR 7.01)*, Environment Agency, Bristol, UK.

Foundation for Water Research (FWR). 2012. *Urban Pollution Management Manual*, 3rd edn, http://www.fwr.org/UPM3.

Wastewater Planning Users Group (WaPUG). 2006. *The Design of CSO Chambers to Incorporate Screens*. WaPUG Guide. http://www.ciwem.org/groups/urban-drainage-group.

REFERENCES

Andoh, R.Y.G. 1998. Improving environmental quality using hydrodynamic separators. *Water Quality International*, January/February, 47–51.

Andoh, R.Y.G., Egarr, D.A., and Faram, M.G. 2008. High-rate treatment and disinfection of combined sewer overflows using advanced hydrodynamic vortex separators. *Proceeding of the 11th International Conference on Urban Drainage*, Edinburgh, September, on CD.

Balmforth, D.J. 1982. Improving the performance of stilling pond storm sewage overflows. *Proceedings of the First International Seminar on Urban Drainage Systems*, Southampton, September, 5.33–5.46.

Balmforth, D.J., Lea, S.J., and Sarginson, E.J. 1984. Development of a vortex storm sewage overflow with peripheral spill. *Proceedings of the Third International Conference on Urban Storm Drainage*, Gotenborg, June, 107–116.

Balmforth, D.J., Meeds, E., and Thompson, B. 1996. Performance of screens in controlling aesthetic pollutants. *Proceedings of the Seventh International Conference on Urban Storm Drainage*, 2, Hannover, September, 989–994.

Brombach, H. 1987. Liquid–solid separation at vortex storm overflows. *Proceedings of the Fourth International Conference on Urban Storm Drainage, Topics in Urban Storm Water Quality, Planning and Management*, Lausanne, September, 103–108.

Brombach, H. 1992. Solids removal from CSOs with vortex separators. *Novatech 92, International Conference on Innovative Technologies in the Domain of Urban Water Drainage*, Lyon, November, 447–459.

Chartered Institution of Water and Environmental Management (CIWEM). 2016. *Event Duration Monitoring. Good Practice Guide*, Urban Drainage Group.

Environment Agency. 2011. *Water Discharge and Groundwater (from point source) Activity Permits (EPR 7.01)*, Environment Agency, Bristol, UK

Fenner, R. and Tyack, J.N. 1997. Scaling laws for hydrodynamic separators. *American Society of Civil Engineers. Journal of Environmental Engineering*, 123(10), October, 1019–1026.

Fenner, R. and Tyack, J.N. 1998. Physical modeling of hydrodynamic separators operating with underflow. *American Society of Civil Engineers, Journal of Environmental Engineering*, 124(9), September, 881–886.

Field, R. 1974. Design of a combined sewer overflow regulator/concentrator. *Journal of WPCF*, 46(7), 1722–1741.

Frederick, M.R. and Markland, E. 1967. The performance of stilling ponds in handling solids. Paper No. 5, *Symposium on Storm Sewage Overflows*, Institution of Civil Engineers, May, 51–61.

Harwood, R. and Saul, A.J. 2001. Modelling the performance of combined-sewer overflow chambers. *Journal of the Chartered Institution of Water and Environmental Management*, 15(4), 300–304.

Huebner, M. and Geiger, W. 1996. Influencing factors on hydrodynamic separator performance. *Proceedings of the Seventh International Conference on Urban Storm Drainage*, 2, Hannover, September, 899–904.

Jarman, D.S., Faram, M.G., Butler, D., Tabor, G., Stovin, V.R., Burt, D., and Throp, E. 2008. Computational fluid dynamics as a tool for urban drainage system analysis: A review of applications and best practice. *Proceeding of the 11th International Conference on Urban Drainage*, Edinburgh, September, on CD.

Jarman, D., Butler, D., Tabor, G., and Andoh, R. 2015. CFD modelling of vortex flow controls at high flow rates. *Engineering and Computational Mechanics*, 168(1), 17–34.

Meeds, B. and Balmforth, D.J. 1995. Full-scale testing of mechanically raked bar screens. *Journal of the Chartered Institution of Water and Environmental Management*, 9(6), 614–620.

Ministry of Housing and Local Government. 1970. *Technical Committee on Storm Overflows and the Disposal of Storm Sewage, Final Report*, HMSO, London.

Myerscough, P.E. and Digman, C.J. 2008. Combined sewer overflows—Do they have a future? *Proceedings of the 11th International Conference on Urban Drainage*, Edinburgh, September, on CD.

Saul, A.J. 1999. *CSO Performance Evaluation: Results of a Field Programme to Assess the Solids Retention Performance of Side Weir and Stilling Pond Chambers*. UKWIR Report 97/WW/08/01.

Saul, A.J. 2000. *Screen Efficiency (Proprietary Designs)*. UKWIR Report 99/WW/08/5.

Saul, A.J. and Ellis, D.R. 1992. Sediment deposition in storage tanks. *Water Science and Technology*, 25(8), 189–198.

Saul, A.J. and Harwood, R. 1998. Gross solid retention efficiency of hydrodynamic separator CSOs. *Proceedings of the Institution of Civil Engineers, Water, Maritime and Energy*, 130, June, 70–83.

Saul, A.J., Ruff, S.J., Walsh, A.M., and Green, M.J. 1993. *Laboratory Studies of CSO Performance*, Report UM 1421, Water Research Centre.

Scottish Development Department. 1977. *Storm sewage: Separation and disposal. Report of the Working Party on Storm Sewage (Scotland)*, HMSO, Edinburgh.

Sharpe, D.E. and Kirkbride, T.W. 1959. Storm-water overflows: The operation and design of a stilling pond. *Proceedings of the Institution of Civil Engineers*, 13, August, 445–466.

Smisson, B. 1967. Design, construction and performance of vortex overflows. Paper No. 8, *Symposium on Storm Sewage Overflows*, Institution of Civil Engineers, May, 99–110.

Stovin, V.R. and Saul, A.J. 2000. Computational fluid dynamics and the design of sewage storage chambers. *Journal of the Chartered Institution of Water and Environmental Management*, 14(2), 103–110.

UK Water Industry Research (UKWIR). 2006. *National CSO Test Facility Wigan WWTW: CSO Screen Efficiency 1997–2005*. UKWIR 06/WW/08/14.

Vetter, O., Stotz, G., and Krauth, K. 2001. Advanced stormwater treatment by coagulation process in flow-through tanks. *Novatech 2001, Proceedings of the Fourth International Conference on Innovative Technologies in Urban Drainage*, Lyon, France, 1089–1092.

Water Research Centre (WRc). 1997. *Sewerage Detention Tanks—A Design Guide*, Swindon.

Chapter 13

Storage

13.1 FUNCTION OF STORAGE

This chapter is concerned with the hydraulic design of storage and tanks at any point in the urban drainage system. Section 13.1 looks at the various functions of storage, while Section 13.2 discusses overall design and Section 13.3 sizing. Level pool routing is explained in Section 13.4, and alternative routing procedures are discussed in Section 13.5.

Storage can be provided by construction of below-ground detention tanks and chambers, or may exist within the system without being deliberately provided, especially in pipes with spare capacity. Storage for stormwater can also be created above ground, as discussed in Section 8.1, and is an integral element in stormwater management (Chapter 21). In the context of this chapter, we refer to them all as "tanks."

In an urban drainage system, storage tanks can have the functions of

- Limiting flooding
- Reducing the amount of polluted storm flow discharged to a watercourse
- Controlling flow to treatment
- Providing a resource for reuse
- Forming part of an active system or real-time control

An example is a new development to be drained by a conventional separate sewer system discharging stormwater to a small stream. To reduce the risk of flooding in the stream, the maximum discharge from the new development must be restricted to a low value. If a detention tank is provided to achieve this, the outflow is likely to be via a flow control (as considered in Section 8.1), often in conjunction with a weir (see Section 8.2) to operate at higher flow rates. The typical relationship between inflow and outflow for a case where outflow is controlled, and does not vary significantly with water level, is shown in Figure 13.1a. The volume of water stored for the case illustrated is given by the shaded area. When outflow exceeds inflow, the tank empties.

It is also useful to consider the hydraulic role of storage in more general cases (beyond specific application to detention tanks) where outflow may vary significantly, for example, in reservoirs, or where conceptual "reservoirs" are used to represent more complex systems (as in Sections 5.3.4 and 10.6.2). A general relationship between inflow and outflow is shown in Figure 13.1b. At any value of time, the difference between the inflow and outflow ordinates $(I_t - O_t)$ gives the overall rate at which water in the storage is increasing (if inflow exceeds outflow) or decreasing (if outflow exceeds inflow). The total volume of water entering the storage up to any given time, say t' in Figure 13.1b, is given by the shaded area between the curves.

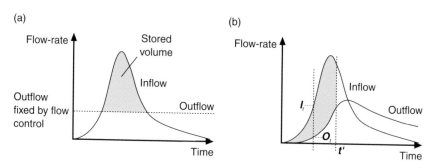

Figure 13.1 Storage: inflow and outflow hydrographs.

13.2 OVERALL DESIGN

Storage devices come in a number of shapes, sizes, and configurations. Small volumes can be provided in manholes or in oversized pipes. Proprietary concrete or glass reinforced plastic (GRP) tanks are also available. An alternative to a conventional tank (Andoh et al., 2001) is a system known as *geocellular storage*, based on a three-dimensional plastic matrix with a high void ratio within which the water is stored, removing the need for the structural function of a tank (Figure 13.2). Various types and forms of geocellular storage with the appropriate structural strength can be used within infiltration trenches (Stephenson, 2008) and pervious pavements (Chapter 21). The Construction Industry Research and Information Association (CIRIA) provides guidance on the testing, structural and geotechnical design, installation, and maintenance of such systems (O'Brien et al., 2016). Larger systems include purpose-built reinforced concrete tanks or multiple-barrelled tank sewers. An important distinction is whether they operate online or offline (see also Chapter 12).

Figure 13.2 Installation of a Stormcell storage device. (Courtesy of Hydro International.)

13.2.1 Online

Online detention tanks are constructed in series with the urban drainage network and are controlled by a flow control at their outlet. Flow passes through the tank unimpeded until the inflow exceeds the capacity of the outlet. The excess flow is then stored in the tank, causing the water level to rise. An emergency overflow is provided to cater for high flows (an online storage tank is depicted in Figure 12.11a in Chapter 12). As the inflow subsides at the end of the storm event, the tank begins to drain down, typically by gravity.

The flow control is normally one of those described in Chapter 8: an orifice, weir, vortex regulator, or throttle pipe. An electrically actuated gate linked to a downstream sensor may also be fitted. This will provide more precise control and also enable tank size to be minimised. Details of sewer system control are given in Chapter 22.

A common arrangement for an online tank is an oversized pipe or rectangular culvert. These *tank sewers* are provided with a dry weather (in combined systems) or low flow channel to minimise sediment deposition (Figure 13.3). Benching with a positive gradient is also provided. Another arrangement uses smaller multiple-barrelled sewers operating in parallel. These provide the necessary storage and have better self-cleansing characteristics.

13.2.2 Offline

Offline tanks are built in parallel with the drainage system as shown in Figure 12.11b. These types of tanks are generally designed to operate at a predetermined flow rate, controlled at the tank inlet. An emergency overflow is provided, as for the online tank. Flow is returned to the system either by gravity or by pumping, depending on the system configuration and levels. A flap valve is normally used for gravity returns.

Offline tanks require less volume than online tanks for equivalent performance and hence less space, but the overflow and throttling devices necessary to divert, regulate, and return flows tend to be more complicated. Maintaining self-cleansing is also more difficult for this type of tank. Regular maintenance is therefore important.

13.2.3 Flow control

The points at which flow control is required for both online and offline tanks are marked in Figure 12.11. Table 13.1 presents a summary of the flow control requirements. The common devices are described in Chapter 8.

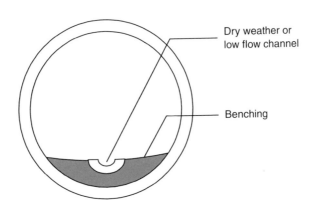

Dry weather or low flow channel

Benching

Figure 13.3 Tank sewer.

Table 13.1 Flow control for tanks

Type of storage	Type of control	Purpose	Common devices
Online	Outlet restriction	To match continuation flow to the capacity of the downstream sewer	Orifice Penstock Vortex regulator Throttle pipe
	Relief	To divert excess flow when capacity of storage (and downstream sewer) has been exceeded	High side weir
Offline	Continuation restriction	To match continuation flow to the capacity of the downstream sewer	Orifice Penstock Vortex regulator Throttle pipe
	Tank inlet	To pass flow into the tank when the downstream capacity has been exceeded	Orifice Penstock High side weir
	Relief	To divert excess flow when capacity of storage (and downstream sewer) has been exceeded	High side weir
	Tank outlet	To return stored flow to the sewer once the storm has passed	Orifice Penstock Vortex regulator Throttle pipe Pump

More information on the use of flow control devices in conjunction with storage tanks is given by the Water Research Centre (WRc, 1997) and related to SuDS in Woods-Ballard et al. (2015). Further details on a variety of flow control devices and their application within larger storage facilities are presented by Hall et al. (1993).

13.3 SIZING

The hydraulic design of a tank or pond serving a new development usually entails limiting the outflow for a specific storm event. So, typical design criteria are

- Rate of outflow—this can be fixed by one of a number of approaches:
  - No greater than estimated values from the undeveloped (or greenfield) site (somewhat problematic as it is difficult to accurately predict runoff flows from small, undeveloped catchments [but see Section 10.7])
  - A value linked to the area of the site and often specified by national or local planning policy (e.g., 4–6 L/s.ha)
  - The capacity of the downstream sewer or watercourse
- Design storm—smaller tanks designed to mitigate flooding are typically designed for 30-year return period storms. For large lakes, much higher return periods may be specified.

When designing storage to reduce water quality impacts, time series rainfall (Chapter 4) is typically used, combined with water quality modelling (Chapter 19).

The questions in design are as follows: what active storage volume is required to achieve the outflow limitation and which is the critical storm that produces the worst case? It is not simply a case of using the Rational Method, as the critical storm is usually of longer duration than the one giving maximum instantaneous flow. The next section describes a simplified approach to tank sizing. This is followed by explanation and examples of flow routing procedures now commonly found in commercial urban drainage models.

13.3.1 Preliminary storage sizing

A preliminary estimate of storage volume requirements for peak flow attenuation (in online tanks) can be obtained by using

$$S = V_I - V_O \tag{13.1}$$

where S is the storage volume (m³), V_I is total inflow volume (m³), and V_O is total outflow volume (m³).

In this case, outflow is via an outlet restriction using one of the devices in Table 13.1.

Figure 13.4 shows a plot of inflow volume, V_I, versus storm duration, D, for a particular return period. Outflow volume, V_O, has also been plotted, assuming a constant discharge. The difference in the ordinates of the two curves gives the storage, S, required for any duration storm. The design storage (S_{max}) is the maximum difference between the curves. Example 13.1 shows how S_{max} can be identified using a tabular approach.

13.3.2 Storage routing

A more accurate assessment of the effect of the storage can be obtained by routing an inflow hydrograph through the tank/pond. This can be done using the level-pool routing technique described in the next section. In practice, most engineers establish storage volume using one of the proprietary models discussed in Chapter 20.

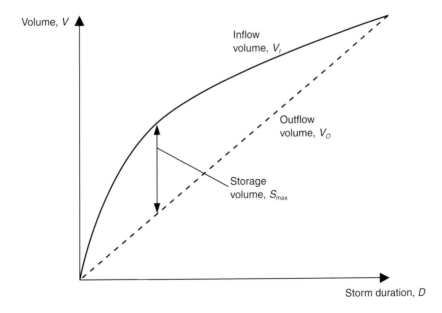

Figure 13.4 Storage volume as a function of storm duration.

13.4 LEVEL POOL (OR RESERVOIR) ROUTING

Calculating the relationship between inflow and outflow as flow passes through storage (e.g., as shown in Figure 13.1b) is called *routing*. A standard calculation method with a wide range of applications is now given.

The difference between inflow and outflow equals the rate at which the volume of water in the storage changes with time, or

$$I - O = \frac{dS}{dt} \tag{13.2}$$

where I is the inflow rate (m³/s), O is the outflow rate (m³/s), S is the stored volume (m³), and t is time (s).

The simplest application is shown in Figure 13.5. Here, there is one outflow controlled by an arrangement such as a weir, giving a simple relationship between O and H (height of water above the weir crest). S in this case is the *temporary storage*, the volume created when there is outflow. The key to the method is that both O and S are functions of H.

EXAMPLE 13.1

A housing development has an impermeable area of 25 ha. Determine approximately the volume of storage required to balance the 10-year return period storm event. Downstream capacity is limited, and a maximum outflow of 100 L/s has been specified. Rainfall statistics are the same as those derived in Example 4.2.

Solution

In the table below, column (3) is the inflow volume, which is derived from the product of (1), (2), and the impermeable area (25 ha). Column (4) is the outflow volume; the product of column (1) and the outflow rate (100 L/s). The storage is the difference between (3) and (4).

(1) Storm duration, D (h)	(2) Intensity, i (mm/h)	(3) $V_I = iA_iD$ (m³)	(4) $V_O = Q_OD$ (m³)	(5) $S = V_I 2V_O$ (m³)
0.083	112.8	2350	30	2320
0.167	80.4	3350	60	3290
0.25	62.0	3875	90	3785
0.5	38.2	4775	180	4595
1	24.8	6200	360	5840
2	14.9	7450	720	6730
4	8.6	8600	1440	7160
6	6.1	9150	2160	6990
10	4.0	10 000	3600	6400
24	2.0	12 000	8640	3360

The maximum storage, S_{max}, is 7160 m³.

We solve Equation 13.2 for fixed time steps and consider "average" conditions over the period of each time step. Therefore, the average inflow during a time step minus the average outflow equals the change in stored volume during the step:

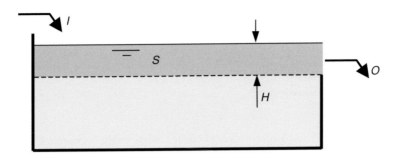

Figure 13.5 Simple application of level pool routing.

$$\frac{I_1 + I_2}{2} - \frac{O_1 + O_2}{2} = \frac{S_2 - S_1}{\Delta t} \tag{13.3}$$

where I_1, O_1, S_1 are inflow, outflow, and stored volume at the start of the time step, respectively; I_2, O_2, S_2 are inflow, outflow, and stored volume at the end of the time step, respectively; and Δt is the time step.

A typical application is to calculate outflow for known values of inflow. In each time step, the unknown will be O_2. Since O and S are related via H, we put S_2 with O_2 on the left-hand side of the equation:

$$\frac{S_2}{\Delta t} + \frac{O_2}{2} = \frac{S_1}{\Delta t} - \frac{O_1}{2} + \frac{I_1 + I_2}{2}$$

It is convenient to have the term $\left[(S/\Delta t) + (O/2)\right]$ on both sides, so we rearrange to

$$\left[\frac{S_2}{\Delta t} + \frac{O_2}{2}\right] = \left[\frac{S_1}{\Delta t} + \frac{O_1}{2}\right] - O_1 + \frac{I_1 + I_2}{2} \tag{13.4}$$

Now we need to incorporate the way both O and S vary with H. The neatest way of doing this is to create a relationship between $\left[(S/\Delta t) + (O/2)\right]$ and O (based on the variations of O and S with H). This is demonstrated by Example 13.2.

13.5 ALTERNATIVE ROUTING PROCEDURE

One disadvantage of the routing method just described is that it is difficult to implement using widely available computational tools such as spreadsheets. However, this can be overcome by transforming the rate of change of storage into the rate of change of head, as follows:

$$\frac{dS}{dt} = A\frac{dH}{dt} \tag{13.5}$$

EXAMPLE 13.2

Outflow from a detention tank is given by $O = 3.5\,H^{1.5}$. The tank has vertical sides and a plan area of 300 m². Inflow and outflow are initially 0.6 m³/s, and then inflow increases to 1.8 m³/s at a uniform rate over 6 minutes. Inflow then decreases at the same rate (over the next 6 minutes) back to a constant value of 0.6 m³/s. Using a time step of 1 minute, determine the outflow hydrograph.

Solution

We first use the way O and S vary with H to create a relationship between $[(S/\Delta t) + (O/2)]$ and O, as in Table 13.2.

We can use the data in Table 13.2 to plot $[(S/\Delta t) + (O/2)]$ against O (Figure 13.6). The calculation now progresses as in Table 13.3. The values of I, and therefore $(I_1 + I_2)/2$, are known. The first value of O is 0.6 m³/s, and from this the first value of $[(S/\Delta t) + (O/2)]$ can be determined via the graph in Figure 13.6 (giving a value of 1.8 m³/s, as indicated). So for the first time step, $[(S_2/\Delta t) + (O_2/2)]$ is calculated from Equation 13.4 giving $1.8 - 0.6 + 0.7 = 1.9$ (circled in Table 13.3). The corresponding value of O is determined from Figure 13.6, working this time from the y axis to the x axis, giving 0.65 m³/s (boxed in Table 13.3).

So now we know O after the first time step. For the next time step, $[(S_1/\Delta t) + (O_1/2)]$ is 1.9, and $[(S_2/\Delta t) + (O_2/2)]$ is calculated again from Equation 13.4: $1.9 - 0.65 + 0.9 = 2.15$.

Table 13.2 Variation with H (Example 13.2)

H	O $3.5\,H^{1.5}$	S $300\,H$	$\dfrac{S}{\Delta t} + \dfrac{O}{2}$
(m)	(m³/s)	(m³)	(m³/s)
0	0	0	0
0.2	0.31	60	1.16
0.4	0.89	120	2.44
0.6	1.63	180	3.81
0.8	2.50	240	5.25

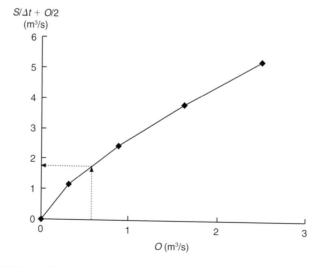

Figure 13.6 Graph of $S/\Delta t + O/2$ against O (Example 13.2).

Table 13.3 Routing calculation (extract) (Example 13.2)

Time (minutes)	I (m³/s)	O (m³/s)	$\dfrac{S}{\Delta t} + \dfrac{O}{2}$ (m³/s)	$\dfrac{I_1 + I_2}{2}$ (m³/s)
0	0.6	0.6	1.8	0.7
1	0.8	0.65	1.9	0.9
2	1.0	0.76	2.15	

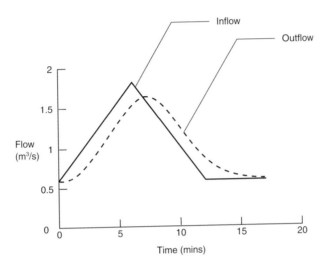

Figure 13.7 Inflow (given) and outflow (calculated) hydrographs (Example 13.2).

O_2 is again determined from Figure 13.6, giving 0.76 m³/s—the outflow at 2 minutes. The calculation proceeds in this way until all the values of O have been determined. The resulting outflow hydrograph is given in Figure 13.7.

Typically, there is a relationship between A (plan area of the storage) and H (head over the downstream control device). For storage ponds with vertical sides, A is constant, but for more complex shapes a function could be used, such as

$$A = \alpha H^{\beta}$$

(13.6)

where α and β are constant.
So, from Equation 13.2,

$$I = O + \frac{dS}{dt}$$

and if the outflow regulator is an orifice outlet (Equation 8.1),

$$O = C_d A_o \sqrt{2gH}$$

then

$$I = C_d A_o \sqrt{2gH} + A \frac{dH}{dt}$$

and, therefore,

$$\frac{dH}{dt} = \frac{I - C_d A_o \sqrt{2gH}}{A} = f(H, t) \tag{13.7}$$

The derivative can be simply represented (using Euler's approximation) as

$$\frac{dH}{dt} \approx \frac{H(t + \Delta t) - H(t)}{\Delta t}$$

So, substituting into Equation 13.7 and solving for $H(t + \Delta t)$ gives

$$H(t + \Delta t) = H(t) + \Delta t . f(H, t) \tag{13.8}$$

which can be solved iteratively as shown in Example 13.3. As Euler's approximation assumes linear change of head over time, it is most accurate when small time increments are used. It is recommended that

$$\Delta t < 0.1 T_p$$

where T_p is the time for the hydrograph to reach peak value.

EXAMPLE 13.3

An online balancing pond is needed to limit the peak storm runoff from the site to 100 L/s. Design a suitable vertically sided storage tank using an orifice plate ($C_d = 0.6$) as the outflow regulator. The maximum head available on the site is 1.5 m. The inflow hydrograph is given below.

Time (h)	0	0.25	0.50	0.75		1.00	1.25	1.50	1.75
Flow (L/s)	0	75	150	225		300	375	450	525
Time (h)	2.00	2.25	2.50	2.75	3.00	3.25	3.50	3.75	4.00
Flow (L/s)	600	525	450	375	300	225	150	75	0

Solution

Determine the required orifice diameter D for the maximum outflow (100 L/s) at the available height (1.5 m):

$$A_o = \frac{O}{C_d \sqrt{2gH}} = \frac{0.100}{0.6\sqrt{2g1.5}} = 0.031 \, \text{m}^2$$

where A_o is the orifice cross-sectional area.

$$D = \sqrt{\frac{4A_o}{\pi}} = 0.197 \text{ m}$$

Thus, use a $D = 200$ mm orifice:

$$O = C_d A_o \sqrt{2gH} = 0.6 \times 0.031 \sqrt{2g} H^{0.5} = 0.082 H^{0.5}$$

What plan area A is needed? Equation 13.7 gives

$$f(H, t) = \frac{I - O}{A}$$

$$\Delta t = 0.25 \text{h} = 900 \text{ s}$$

In the table below, the all of column (2) and the first data point in column (3) refer to data known initially. Column (4) is calculated from the orifice equation. Column (5) is column (2) minus column (4) divided by the plan area of the storage (Equation 13.7). The "new" head in column (6) is the sum of column (3) and column (5) times the time increment (Equation 13.8). Column (3) takes the head from column (6) at the previous time step.

(1)	(2)	(3)	(4)	(5)	(6)
T (h)	I (m^3/s)	$H(t)$ (m)	O (m^3/s)	$f(H,t)$ (m/s)	$H(t + dt)$ (m)
0.00	0.000	0.00	0.000	0.000000	0.00
0.25	0.075	0.00	0.000	0.000030	0.03
0.50	0.150	0.03	0.013	0.000055	0.08
0.75	0.225	0.08	0.023	0.000081	0.15
1.00	0.300	0.15	0.032	0.000107	0.25
1.25	0.375	0.25	0.041	0.000134	0.37
1.50	0.450	0.37	0.050	0.000160	0.51
1.75	0.525	0.51	0.059	0.000187	0.68
2.00	0.600	0.68	0.068	0.000213	0.87
2.25	0.525	0.87	0.076	0.000179	1.03
2.50	0.450	1.03	0.083	0.000147	1.16
2.75	0.375	1.16	0.088	0.000115	1.27
3.00	0.300	1.27	0.092	0.000083	1.34
3.25	0.225	1.34	0.095	0.000052	1.39
3.50	0.150	1.39	0.097	0.000021	1.41
3.75	0.075	1.41	0.097	− 0.000009	1.40
4.00	0.000	1.40	0.097	− 0.000039	1.36
4.25	0.000	1.36	0.096	− 0.000038	1.33
4.50	0.000	1.33	0.095	− 0.000038	1.30

A number of areas were tried iteratively. The above refers to $A = 2500$ m^2.
At this area (volume), $H_{max} = 1.41$ m (<1.5 m) and $O_{max} = 97$ L/s (<100 L/s), which is acceptable.

PROBLEMS

13.1 A development has an impermeable area of 1.8 ha. Basing rainfall estimation on the formula $i = 750/(t + 10)$, where i is rainfall intensity in millimetres per hour (mm/h) and t is duration in minutes, determine the volume of storage needed to limit outfall to 70 L/s. (Try storm durations of 8, 12, and 16 minutes.)   [72.4 m³]

13.2 A detention tank on a sewerage scheme is rectangular in plan: 25 m × 4 m. It is being operated in such a way that the only outflow is over a weir. The flow rate over the weir is given by the standard expression for a rectangular weir, in which $C_D = 0.63$. The width of the weir is 1.5 m. Consider the following case. Initially inflow is zero, and the water surface is at the level of the weir crest. Then inflow increases at a uniform rate over 12 minutes to 0.9 m³/s, and reduces immediately at the same rate back to zero. Determine the peak outflow, using a time step of 2 minutes.   [0.8 m³/s]

13.3 How much did the tank in Problem 13.2 attenuate or delay the hydrograph peak? How would normal operation of the tank differ from that described in Problem 13.2?

REFERENCES

Andoh, R.Y.G., Faram, M.G., Stephenson, A., and Kane, A. 2001. A novel integrated system for stormwater management. *Novatech 2001, Proceedings of the Fourth International Conference on Innovative Technologies in Urban Drainage*, Lyon, France, 433–440.

Hall, M.J., Hockin, D.L., and Ellis, J.B. 1993. *Design of Flood Storage Reservoirs*, CIRIA/Butterworth-Heinemann.

O'Brien, A.S.O., Lile, C.R., Pye, S.W., and Hsu, Y.S. 2016. *Structural and Geotechnical Design of Modular Geocellular Drainage Systems*, CIRIA C737.

Stephenson, A.G. 2008. A holistic hard and soft SUDS system used in the creation of a sustainable urban village community. *Proceeding of the 11th International Conference on Urban Drainage*, Edinburgh, September, on CD-ROM.

Water Research Centre (WRc). 1997. *Sewerage Detention Tanks—A Design Guide*, Swindon.

Woods-Ballard, B., Wilson, S., Udale-Clark, H., Illman, S., Scott, T., Ashley, A., and Kellagher, R. 2015. *The SuDS Manual*, 2nd edn, CIRIA C753.

Chapter 14

Pumped systems

14.1 WHY USE A PUMPING SYSTEM?

As indicated in Chapter 1, the need for urban drainage arises from human interaction with the natural water cycle. Sewers usually drain in the same direction that nature does: by gravity. Gravity systems require relatively little maintenance, certainly when compared with systems involving a significant amount of mechanical equipment or the need to maintain fixed pressures. And while neglect is undesirable, so is unnecessary maintenance, and so gravity sewer systems prevail. This can be seen as the result of an implicit decision to accept high capital costs (for deep, large, and expensive sewers) if they result in low operating costs. More explicit consideration of the balance between capital and operational costs to minimise total expenditure (TOTEX) is now becoming more commonplace. But in some cases, gravity is not enough, usually when it is not cost-effective to provide treatment facilities for each natural sub-catchment. In these circumstances, it is appropriate to pump, and that is the focus of this chapter. The chapter starts in Section 14.2 with discussion on the general arrangement of pumped systems moving to their design in Section 14.3. Rising mains and types of pumps are covered in Sections 14.4 and 14.5, respectively, with pumping station design discussed in Section 14.6. Some engineers favour non-gravity systems for more general application, and these approaches are discussed in Section 14.7 followed by a short introduction to energy consumption in Section 14.8.

14.2 GENERAL ARRANGEMENT OF A PUMPING SYSTEM

Sewer pumping systems have a number of general features:

- In sewer systems based mostly on gravity flow, pumped sections require comparatively high levels of maintenance. Engineers, therefore, prefer to keep the pumped lengths to a minimum to lift the flow as required, and then the system can revert to gravity flow as soon as possible. (Figure 14.1 gives a simple section of a typical arrangement.)
- The liquid being pumped contains solids that vary significantly in size and nature; therefore, pumps must be designed with the risk of clogging, abrasion, and damage in mind. The nature of the liquid also creates risks of septicity, corrosion of equipment, and production of explosive gases (as considered in Chapter 17).
- It is common for centrifugal pumps to deliver flow at a fairly constant rate or, where there are a number of pumps that may work in combination, there may be a number of alternative fixed rates. The exception to this is if variable speed pumps are used. Whatever the flow rate handled by the pumping system, it must generally exceed the rate of flow arriving from the gravity system or be combined with some form of

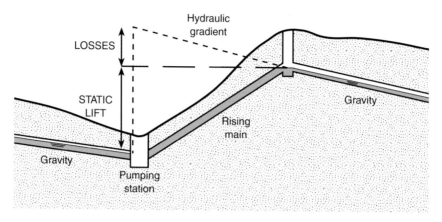

Figure 14.1 Typical sewer pumping arrangement.

storage, otherwise there would be a risk of flooding. So pumping systems tend to work on a stop-start basis, with flow arriving at a reception storage (a *wet well*), as shown in the simplified pumping station arrangement in Figure 14.2. When the pumps are operating, the wet well empties; when the pumps are not operating the wet well fills. The water level in the sump is used to trigger the stop and start of the pumps. Where pumps are used specifically as part of a storage scheme, the downstream receiving conditions may dictate when the pumps start and stop.

14.3 HYDRAULIC DESIGN

14.3.1 Pump characteristics

Hydraulically, the function of a pump is to add energy, usually expressed as head (energy per unit weight) to a liquid. The hydraulic performance of a pump can be summed up by the *pump characteristic curve*, a graph of the head added to the liquid, plotted against flow rate.

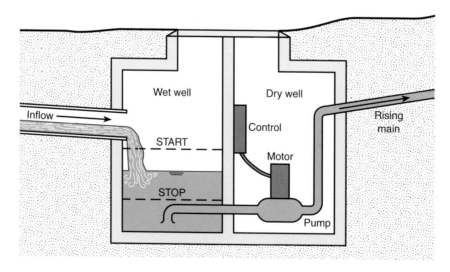

Figure 14.2 Simplified pumping station arrangement.

A typical pump characteristic is given in Figure 14.3a; this shows generally reducing head for increasing flow rate, but it is not a simple relationship—what goes on inside a pump is complex in hydraulic terms. The different types of pumps available are considered in Section 14.5. The characteristics for each type of pump are derived from tests carried out by the manufacturer and are available in the manufacturer's literature.

But at what values of flow rate and head will the pump operate when connected to a particular pipe system? The engineer answers that question at the design stage in the following way.

14.3.2 System characteristics

The pipe system to which the pump will be connected will have a characteristic curve of its own: the *system characteristic*. Water must be given head in order to

- Be lifted physically (the *static lift*—see Figure 14.1).
- Overcome energy losses due to pipe friction and local losses at bends, valves, and so on. As flow rate increases, velocity increases and energy losses increase in proportion to the square of velocity (as set out in Section 7.3).
- Provide velocity head if the water is to be discharged to atmosphere at a significant velocity.

The system characteristic can be determined from

Head = Static lift + Dynamic losses and velocity head

Losses and velocity head are proportional to velocity squared. A typical system characteristic is shown in Figure 14.3b.

So there are two characteristics: the pump characteristic, which gives the head that a pump is capable of producing while delivering a particular flow rate; and the system characteristic, which gives the head that would be required for the system to carry a particular flow rate. If a specific pump is going to be connected to a specific system, there is only one set of conditions where what the pump has to offer can satisfy what the system requires: it is the point where the pump characteristic and the system characteristic cross, as shown on Figure 14.3c. This is called the *operating point* or *duty point*.

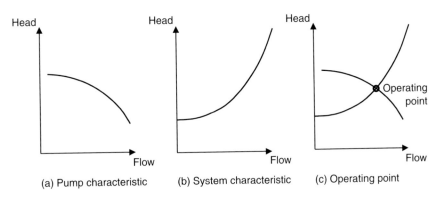

Figure 14.3 Pump and system characteristic curves.

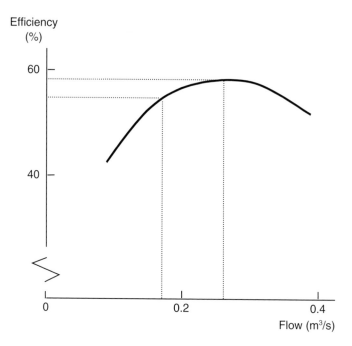

Figure 14.4 Pump efficiency against flow.

14.3.3 Power

The power required at the operating point can be derived from the operating flow rate and head in conjunction with the pump's efficiency.

Power (P), *energy per time*, is the product of *weight per time* $\rho g Q$ and head, *energy per weight*. So,

$$P = \rho g Q H \qquad (14.1)$$

where P is density (taken as 1000 kg/m³ for water); G is gravitational acceleration, 9.81 m/s²; Q is the operating flow rate (m³/s); and H is operating head (m).

A pump gives power to the water, and it receives power ("power supply"), usually in the form of electrical power. The pump and motor are not 100% efficient at converting the power supply into power given to water. Efficiency (power given to the water divided by power supplied to the pump) varies with flow rate and can be taken from the manufacturer's plot (e.g., Figure 14.4). Therefore,

$$\text{Power supplied} = \frac{\rho g Q H}{\eta} \qquad (14.2)$$

where η *is* efficiency (–).

Example 14.1 demonstrates these calculations.

- Note that the water level in the sump is not constant because as the pump drains the sump, the level goes down (and therefore the static head increases).
- This may be significant, and if it is, this must be taken into account in design.

EXAMPLE 14.1

A pump in a sewer system is connected to a rising main with an internal diameter of 0.3 m and a length of 105 m. The rising main discharges to a manhole at a level 20 m above the water level in the sump. The roughness k_s of the rising main is 0.3 mm, and local losses total $0.8 \times v^2/2g$.

The pump has the following characteristics:

Q (m³/s)	0	0.1	0.2	0.3	0.4
H (m)	33	32	29	24	16
efficiency (%)		42	56	57	49

Determine the flow rate, the head, and the power supplied at the operating point.

Solution

The system characteristic is given by

Total head required = Static lift + Friction losses + Local losses + Velocity head

This can be expressed as

$$H = 20 + \frac{\lambda L}{D} \frac{v^2}{2g} + 0.8 \frac{v^2}{2g} + \frac{v^2}{2g}$$

We can find λ from the Moody diagram (Figure 7.4), and its value will be constant, provided flow is rough turbulent. Assuming that it is,

$$\frac{k_s}{D} = \frac{0.3}{300} = 0.001, \text{ giving } \lambda = 0.02$$

So,

$$H = 20 + \frac{v^2}{2g} \left[\frac{0.02 \times 105}{0.3} + 0.8 + 1 \right]$$

$$= 20 + 8.8 \frac{v^2}{2g}$$

Of course, $Q = vA$
So,

$$v = \frac{4Q}{\pi 0.3^2}$$

From this, we can determine the relationship between H and Q for the pipe system (the system characteristic).

Alternatively, we can use Wallingford charts or tables (see Section 7.3.4) to determine the system characteristic. Local losses + velocity head = $(0.8 + 1)\ v^2/2g$, and this can be expressed as an equivalent length using Equation 7.15. So,

$$L_E = D\frac{1.8}{\lambda} = 27 \text{ m}$$

For the system, for any value of Q: $H = 20 + 132 \times S_f$ (from chart or table).

The system characteristic (from either method) is plotted (as in Figure 14.3c) together with the pump characteristic. The operating point is where the lines cross; at this point, flow rate is 0.26 m³/s. This gives a velocity of 3.7 m/s, giving R_e of about 10^6 – in the rough turbulent zone – so the assumption about constant λ is valid.

At the operating point, head is 26 m.

Pump efficiency has been plotted on Figure 14.4. At a flow rate of 0.26 m³/s, efficiency is 57%. So,

$$\text{power supplied} = \frac{\rho g Q H}{\eta} = \frac{\rho g \times 0.26 \times 26}{0.57} = 116 \text{ kW}$$

- In systems with few hydraulic discontinuities (i.e., fittings), the velocity head may be insignificant in relation to the losses and is ignored. In this example, velocity head was significant and so was rightly included.
- In another case, instead of discharging to atmosphere at a manhole, the rising main outlet might be "drowned," for example, submerged in a tank into which the liquid is being pumped. In this case, the static lift must be measured up to the liquid surface in the tank. This surface is unlikely to have a velocity; therefore, velocity head will not be included. However, there will be exit losses at the point where the rising main discharges to the tank.

14.3.4 Pumps in parallel

A common arrangement is for two (or more) pumps to be placed in parallel (Figure 14.5). The additional pump(s) can

- Act as a standby to replace others when there is a fault
- Work in a "duty, standby" arrangement where both pumps are used separately but in turn to deliver the required flow
- Operate in an "assist, standby" arrangement where one pump delivers the normal flow but the other comes into operation to reinforce the first pump when high discharges are needed

When two identical pumps are operating in parallel, each delivers flow rate Q and raises the head by H (Figure 14.5), so overall the flow rate is $2Q$, all experiencing an increase in head of H. The characteristic for two pumps in parallel is given in Figure 14.6. For each value of H, the flow rate is doubled to $2Q$. The new operating point is given by the intersection with the system characteristic (Figure 14.6).

For pumps in parallel, care is needed when determining the efficiency. The operating flow rate in Figure 14.6 is for both pumps together. Half that value gives the flow rate in each pump, and this should be used in determining efficiency from Figure 14.4, as this gives the efficiency for a single pump. Example 14.2 demonstrates this.

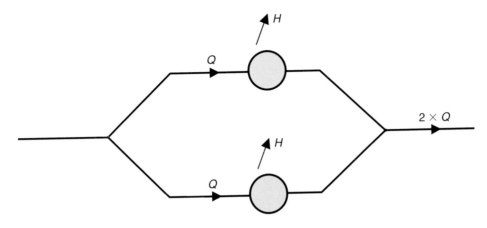

Figure 14.5 Pumps in parallel (schematic).

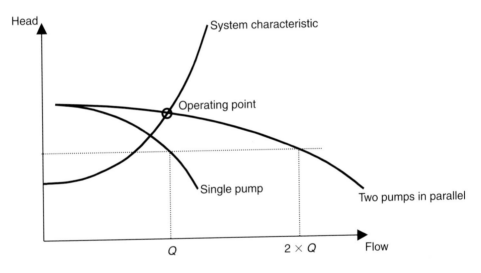

Figure 14.6 Operating point: pumps in parallel.

14.3.5 Suction and delivery pipes

The pipe on the upstream, or inlet side of a pump, is referred to as the *suction pipe*, and the pipe on the downstream, or outlet side, is referred to as the *delivery pipe*. In Examples 14.1 and 14.2, the suction pipe was short and was not considered separately. This arrangement is common in drainage applications. It is also common for the pump volute or body to be below the level of liquid in the sump, as in Figure 14.2, to ensure that the pumps remain "primed" (full of liquid). However, where this is not the case and the suction pipe is long or the pump is at a higher level than the sump level, it is important to ensure that pressure on the suction side of the pump stays well above the vapour pressure of the liquid. This is to avoid cavitation that leads to loss of performance and damage to pumps and ancillaries (Chadwick et al., 2013). Generally, long suction pipes are to be avoided due to the difficulty of maintaining velocity within an appropriate range (see Section 14.4).

EXAMPLE 14.2

For Example 14.1, what would be the flow rate, head, and power supplied at the operating point if an additional pump, identical to the first, was arranged in parallel?

Solution

The pump characteristic for the pumps in parallel is determined by doubling the flow rate for each value of H:

For one pump, Q (m³/s)	0	0.1	0.2	0.3	0.4
Two pumps in parallel, Q (m³/s)	0	0.2	0.4	0.6	0.8
Head, H	33	32	29	24	16

The characteristic for two pumps in parallel can be plotted, together with the system characteristic (and, for the purposes of illustration, the characteristic for one pump) as in Figure 14.6. At the operating point, flow rate is 0.34 m³/s, and head is 30 m. As has already been pointed out, care is needed when handling efficiency for pumps in parallel. The flow rate in each pump is 0.17 m³/s; therefore, the efficiency of each pump (Figure 14.4) is 54%. So,

$$\text{Power supplied} = 2 \times \frac{\rho g \times 0.17 \times 30}{0.54} = 185\,\text{kW}$$

14.4 RISING MAINS

14.4.1 Differences from gravity sewers

It is useful to consider the ways in which rising mains are different from gravity sewers.

14.4.1.1 Hydraulic gradient

Gravity pipes are designed assuming that the hydraulic gradient is numerically equal to the pipe slope. As shown in Section 7.4, this is because, in part-full pipe flow, the hydraulic gradient coincides with the water surface and, therefore, in uniform flow, is parallel to the invert of the pipe. In a rising main, of course, none of this applies. At the pumps, the flow is given a sudden increase in head, and this is "used" to achieve the static lift and overcome losses along the pipe (Figure 14.1). The slope of the hydraulic gradient is the natural one: downwards in the direction of flow, while the rising main does its job: it rises. The pipe can be laid at a constant depth and follow the profile of the ground.

14.4.1.2 Flow is not continuous

At times there may be no flow in the main, and at other times there may be a number of alternative flow rates, depending on how many pumps are operating. When the pumps are not operating, wastewater stands in the rising main. Therefore, it is important that when pumping resumes, the velocities are sufficient to scour deposited solids.

A standard design range for the rising main velocity is between 0.75 m/s and 1.8 m/s (Water UK/WRc, 2012). The minimum suitable internal diameter is usually considered to be 100 mm.

To avoid septicity, wastewater should not be retained in a rising main for more than 12 hours. It is sometimes necessary to arrange for the addition of oxygen or oxidising chemicals to control septicity (considered in Section 17.8).

When the range of flows is high, dual rising mains can be used to maintain velocities high enough to prevent deposition. One can also act as standby; but, in this case, both mains must be used regularly to avoid septicity.

One possible consequence of starting and stopping the flow is extremely high or low pressures resulting from surge waves, considered in Section 14.4.3.

14.4.1.3 Power input

We must provide power to create flow in the system. The power is needed year after year for as long as the system operates. This creates trade-offs in the selection of an economic design, including the operating cost of the pumping station and, if also being constructed, the rising main.

A smaller diameter pipe will be cheaper, and the resulting higher velocities will help to ensure scouring of deposits, but the higher velocities will also create higher head losses (proportion to velocity squared) and, therefore, higher power costs. The economic decision needs to consider design life and time-related comparisons of capital and operating costs.

14.4.1.4 The pipes are under pressure

There is, of course, no open access to rising mains in the way that there is for gravity sewers.

There are some economic advantages of rising mains in comparison with gravity sewers. Because rising mains are under pressure, and always full, the diameter tends to be smaller and the depth of excavation less than a gravity pipe (which is usually not full and must slope downward). The physical gradient of a rising main is not necessarily uniform or even in the same direction (i.e., rising) all the time.

14.4.2 Design features

Common materials for rising mains are ductile iron, steel, and some plastics. Flexible joints are preferable to allow for differential settlement and other causes of underground stress. (There is more detail on pipe material and construction in Chapter 15.)

Valves and other hydraulic features need to be included at key points in a rising main. There should be an isolating valve, normally a "sluice" (or gate) valve, near the start of the rising main, so that the pumping station pipework can be worked on without emptying the main. There must be a non-return (or *reflux*) valve for each pump, which prevents backflow when pumping stops. Summits (local high points) in the rising main should be avoided if possible, but where unavoidable, should be provided with air release valves. Washout facilities (for emptying the main) should be provided at low points in the main.

Thrust blocks or tied pipework may be needed to withstand the forces created when water is forced to change direction. Their design is beyond the scope of this text but is covered by Thorley and Atkinson (1994) and American Water Works Association (2009).

14.4.3 Surge

Pressurised pipelines are typically subject to rapid changes in flow velocity, say due to pump operations, and this gives rise to one particular risk—*surge* (also commonly referred to as water hammer).

A change in flow (velocity) in a liquid is always associated with a change in pressure. If flow changes rapidly, these changes in pressure can be significant. The effect is known as surge, and the consequences of ignoring it at the design stage can be catastrophic, with the creation of pressures high or low enough to cause damage to pipework, fittings, or restraints. Not all pumping systems are likely to suffer from serious surge problems, and many devices for overcoming the problems can be simple, but surge must be considered when a pumping system is being designed, since a rapid change of flow will always occur with pump failure. There are numerous other events that can lead to elevated surge risks, including valve operations, air valve activity, and sectional priming.

A crucial factor is the rate at which the flow (velocity) changes. If the flow changes gradually, the normal assumption that water (or wastewater) is incompressible can be maintained, and the changes in pressure are not significant. If the flow changes rapidly, the compressibility of water must be considered, with significant pressure changes throughout the system propagating at a velocity dependent on the fluid characteristics and distensibility of the pipe material. Methods of predicting the changes in pressure are considered by Chadwick et al. (2013), and in greater detail by Creasey (1977) and Swaffield and Boldy (1993). It is now common practice for detailed computer-based mathematical modelling to be undertaken on schemes where surge is identified as a potential risk.

Most standard (clean water) surge protection devices (e.g., air vessels or surge tanks) are typically (but not always) deemed inappropriate for wastewater application because of the problems of blockage or stagnation of the stored liquid. Commonly employed methods for managing the surge response of a system include controlled pump operations, controlled air flow management via air valves, pump flywheels, as well as a suitable selection of pipe materials and restraint requirements.

14.5 TYPES OF PUMP

As stated, the function of a pump is to add energy to a liquid. There are a number of ways in which this energy can be transferred, but the most common is by a rotating "impeller" driven by a motor (a *rotodynamic pump*).

The most common rotodynamic pump for use with wastewater is a centrifugal pump in which the impeller forces the liquid radially into an outer chamber called a *volute* (Figure 14.7). In effect, the volute converts velocity head into pressure head. The impeller often has a special design to avoid clogging by solids, and this feature means that centrifugal pumps for wastewater tend to have lower specified efficiencies (about 50% to 70%) than centrifugal pumps for clean water (up to 90%). A common requirement is for these pumps to be capable of handling a 100 mm diameter sphere. They are suitable for a wide range of conditions—a single pump with flow rate 7–700 L/s and head 3–45 m will typically operate at speeds of around 1450 rpm within a range of 400–3000 rpm. Centrifugal pumps require priming (filling with water before pumping can begin) and so must normally be installed below the lowest level of wastewater to be pumped.

Axial-flow pumps are simpler than centrifugal pumps and have impellers (acting like a propeller) that force the liquid in the direction of the longitudinal axis (Figure 14.8). Axial flow pumps are suited to relatively high flow rates and low heads with efficiencies of 75%–90%. Unlike centrifugal pumps, axial-flow pumps suffer a rapid decrease in head with increased discharge. In *mixed-flow pumps*, the direction in which the water is forced by the impeller is at an intermediate angle, so flow is partially radial and partially axial. Mixed-flow pumps are recommended for medium heads between 6 and 18 m and are capable of pumping flows up to 10,000 L/s. Axial- and mixed-flow pumps are

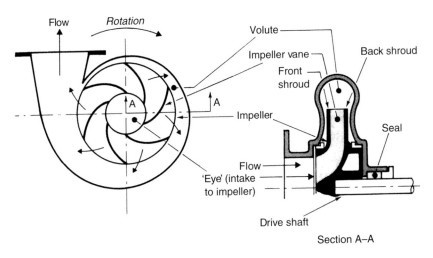

Figure 14.7 Centrifugal pump. (Reproduced from Chadwick, A. et al. 2013. *Hydraulics in Civil and Environmental Engineering*, 5th edn, CRC Press. With permission of E & FN Spon.)

most appropriate for pumping stormwater. *Vortex pumps* are used to handle high solids/grit load where the impeller creates a vortex beyond the volute and into the surrounding water. This helps gather up the solids, including fibrous and stringy material, without coming into contact with the impeller or blocking it. These are relatively low-flow, low-head pumps and are ideal for applications such as providing protection against flooding for individual properties while having the ability to discharge into the downstream system. Figure 14.9 illustrates the shape of pump characteristics for the main types of pumps. In practice, pump selection for a particular application is made by matching requirements to manufacturer's data.

The pump and the motor that drives it are often kept in a "dry well", separate from the wastewater (Figure 14.2). But an alternative is a *submersible pump* in which the motor is encased in waterproof protection and submerged in the wastewater that is to be pumped (Figure 14.10). This greatly simplifies the design of the pumping station and is common for small to medium installations.

For very small installations, a rotodynamic pump may not be suitable, because the risk of clogging places a limit on the smallness of a pump. An alternative system is a *pneumatic*

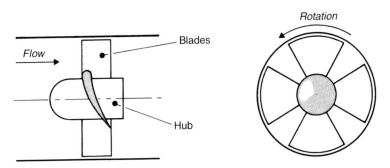

Figure 14.8 Axial flow pump. (Reproduced from Chadwick, A. et al. 2013. *Hydraulics in Civil and Environmental Engineering*, 5th edn, CRC Press. With permission of E & FN Spon.)

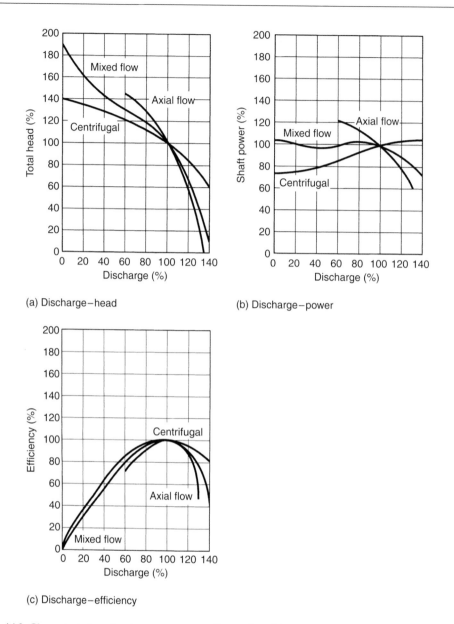

Figure 14.9 Characteristics of various pump types. (Reproduced from Kay, M. 2008. *Practical Hydraulics*, 2nd edn, E & FN Spon. With permission of E & FN Spon and amended.)

ejector, in which the wastewater flows by gravity into a sealed unit and is then pushed out using compressed air. These require little maintenance and are not easily blocked by solids. However, they are of low efficiency and limited capacity (1–10 L/s).

While many pumps operate at a single fixed speed, some types switch between two or more speeds (*multi-speed*), and others can run at continuously variable speeds. The benefit of variable speed pumps is that the pumped outflow from the pumping station can be more closely matched to the inflow (from the system); therefore, less storage volume is required. Also, pumps do not have to be stopped and started so frequently; deposition resulting from

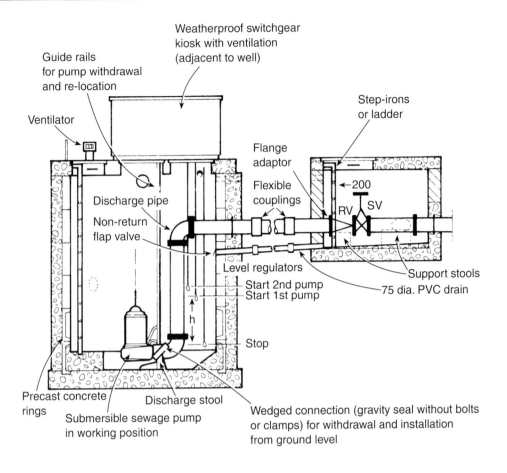

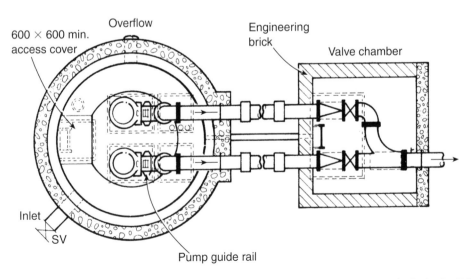

Figure 14.10 Submersible pump. (Reproduced from Woolley, L. 1988. *Drainage Details*, 2nd edn, E & FN Spon. With permission of E & FN Spon.)

liquid lying still in the rising main is reduced; and flow rates, velocities, and therefore losses tend to be lower. However, variable speed pumps are more expensive, require more complex control arrangements, and may be very inefficient at some speeds.

14.6 PUMPING STATION DESIGN

14.6.1 Main elements

The design of pumping stations usually involves the integration of a number of branches of engineering, including civil, structural, mechanical, electrical, instrumentation, control, and automation. Also, a pumping station is one of the few elements of an urban drainage system that may be seen above ground, so there may also be significant architectural aspects (Figure 14.11). They require a planning application, in which noise and odour, as well as appearance, may be issues. The extent of all these aspects will, of course, depend on size—pumping stations in urban drainage schemes may be very small, serving just a few people, or they may be large and complex engineering structures serving large populations.

The basic components of a pumping station were already described in this chapter. Pumps (nearly always more than one) take sewer flow from a reception volume, a sump, and deliver it with increased head into a rising main. The pumps, most commonly centrifugal, are driven by motors, which must be provided with a supply of electrical power. Pumping stations deemed as critical to the sewer network performance may need dual or alternative supplies. There must be arrangements for controlling the pumps, usually related to liquid level in the sump.

Figure 14.11 Pumping station: architectural treatment.

14.6.1.1 Wet well–dry well

When the pumps and motors are kept completely separate from the liquid, the sump is referred to as the *wet well*, and the chamber containing pumps, and so on, is referred to as the *dry well*. The motors may be directly beside the pumps or, to provide further remoteness from moisture and for ease of access, may be at a higher level, connected by a long shaft. Figure 14.12 shows a typical configuration.

14.6.1.2 Wet well only

The wet well–dry well separation is not needed when submersible pumps are used (as illustrated in Figure 14.10). The wet well in which submersible pumps are placed can be of a simple construction, based on precast concrete segmental rings. For inspection or maintenance, the pumps must be lifted out.

14.6.2 Number of pumps

The appropriate number of pumps is a function of

- The need for standby pumps to be available to cover for faults and planned maintenance
- The flow capacity of the pumps, alone and in parallel, determined from the calculations covered in Section 14.3
- The variation in inflow

The simplest pumping station consists of a duty/standby arrangement. It is common, however, in larger installations to have a number of pumps, arranged in parallel (as previously discussed), which are brought into use successively as inflow increases.

14.6.3 Control

In most systems, while the pumps are running, the level in the sump is falling. At a fixed level, the pumps are turned off and the level starts to rise. Subsequently, the level reaches the point at which pumping is resumed.

All pumping stations require some control. The basic requirement is sensing of upper and lower sump levels, and the consequent starting and stopping of the pumps. With more pumps, and more complex starting and stopping procedures, the complexity of the control system increases. Common methods of sensing water level are by ultrasonic detector, float switches, and electrodes. Various level points may be used in order to help prevent "rag rafts" and grit deposits. The safe frequency of operation of the electric motor starter is limited; it is typical to design for between 10 and 15 starts/hour.

14.6.4 Sump volume

To determine the required sump volume (V) between "stop" and "start" levels, the time taken to fill the sump while the pump is idle (t_1) is given by

$$t_1 = \frac{V}{Q_1}$$

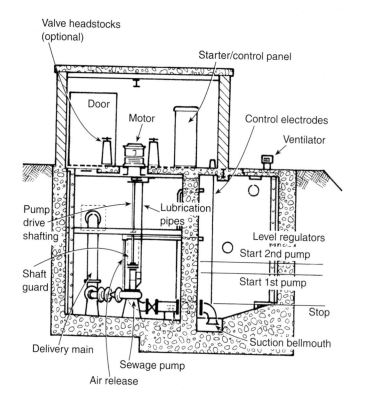

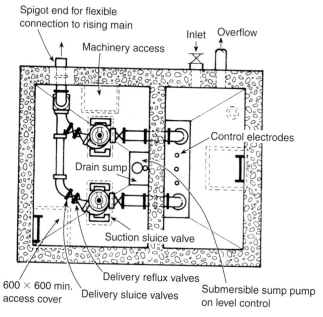

Figure 14.12 Pumping station arrangement. (Reproduced from Woolley, L. 1988. *Drainage Details*, 2nd edn, E & FN Spon. With permission of E & FN Spon.)

where Q_I is the inflow rate. The time taken to pump out the sump (t_2) is

$$t_2 = \frac{V}{(Q_o - Q_I)}$$

where Q_o is the outflow (pump) rate. Thus, the time between successive starts of the pump, the pump cycle (T), is

$$T = \frac{V}{Q_I} + \frac{V}{(Q_o - Q_I)} = \frac{VQ_o}{Q_I(Q_o - Q_I)} \tag{14.3}$$

Now, to find the minimum sump volume required, V is differentiated with respect to Q_I and equated to zero:

$$\frac{dV}{dQ_I} = \frac{T(Q_o - 2Q_I)}{Q_o} = 0 \tag{14.4}$$

$$Q_o = 2Q_I$$

So, the pump sump should be sized for a pumped outflow rate that is double the inflow rate. Substituting Equation 14.4 into Equation 14.3 gives

$$V = \frac{TQ_o}{4}$$

If $n = 3600/T$ is the number of motor starts per hour,

$$V = \frac{900Q_o}{n} \tag{14.5}$$

Thus, the minimum sump working volume is determined from the rate of outflow in conjunction with the allowable frequency of motor starts (see Example 14.3). In practice, there can be other factors that override this, such as the size and arrangement of the pumps. For variable speed drives, this calculation is not strictly required. However, unless the pumping station has an overflow, it is appropriate to specify this volume as a safety measure.

EXAMPLE 14.3

The peak inflow to a sewerage pumping station is 50 L/s. What capacity sump and duty/standby pumps will be required if the number of starts is limited to 10 per hour? How long will the pump operate during each cycle?

Solution

For minimum sump volume, $Q_o = 2Q_I$, so
Capacity of duty pump, $Q_o = 100$ L/s
Capacity of standby pump, $Q_o = 100$ L/s
Pump sump volume (Equation 13.5),

$$V = \frac{900 \times 0.1}{10} = 9\,\mathrm{m}^3$$

Time taken to empty sump,

$$t_2 = \frac{9}{(0.1 - 0.05)} = 180 \text{ s} = 3 \text{ min}$$

14.6.5 Flow arrangements

Within a pumping station, pipework is usually ductile iron with flanged joints. Flexible joints should be placed outside walls to allow for differential settlement. For each pump there should be a sluice valve on the suction and delivery sides for isolating the pump (Figure 14.12).

The base of the sump is usually given quite steep slopes to limit deposition of solids (Figure 14.12). More detail on sump arrangements is given by Prosser (1992) and typically outlined in sewerage undertaker's standards.

In large pumping stations, some form of preliminary treatment to remove large solids, most commonly by means of screens, may be necessary. Alternatively, recirculation pumps may be used to reduce the likelihood of solids settling.

All pumping stations should have an emergency overflow in case of complete failure of the pumps, with storage for wastewater inflow during emergency repairs. In combined systems, it may be necessary to provide an overflow for storm flows. This would be based on the same principles as other combined sewer overflows (CSOs), described in Chapter 13.

14.6.6 Maintenance

A pumping station, with mechanical, electrical, instrumentation, control, and automation equipment, is one part of a sewer network that has obvious maintenance needs. And while it is true of any part of a sewer system, it is particularly important that the maintenance needs of pumping systems are taken into account at the design stage. Care and expense in design may reduce the cost of maintenance, and care and expense in maintenance may limit performance deterioration and reduce the cost of replacement.

Priorities in taking maintenance requirements into account in design are to ensure that

- It is possible to isolate and remove the main elements of pipework and equipment. There must be access to allow the pumps to be lifted out vertically; this is especially true for submersible pumps, which it must be possible to lift out with ease.
- Problems caused by solids can be overcome (by suitable wet well design, appropriate impellers, pump selection, pipe sizes, access to clear blockages)
- Emergencies caused by breakdown, power failure, and so on, can be coped with.

The possibility of power failure needs to be taken into account. Solutions are typically one or more of the following:

- Dual electrical supply
- Standby generator(s)
- Connection points for a temporary generator
- Diesel-powered pumps in larger stations

Maintenance procedures for pumping installations are covered by Wharton et al. (1998) and Sewers and Water Mains Committee (1991).

An important element in maintenance is monitoring performance. Small to medium-sized pumping stations are usually controlled from the wastewater treatment plant that they serve, by telemetry. The types of information likely to be communicated are

- Failure in the electricity supply
- Pump failure
- Unusually high levels in the wet well
- Flooding of the dry well
- Operation of the overflow

The information is needed for effective operation of the system, and in particular to aid decisions about when to attend to operational problems. Pumping stations may also be fitted with flow measurement devices, used to monitor performance, and (potentially) as part of a management system for the catchment.

More detailed guidance on practical aspects of pumping station design is given by Prosser (1992), Wharton et al. (1998), and BS EN 752: 2008, and for smaller installations in *Sewers for Adoption* (WRc, 2012).

14.6.7 Energy demands

One feature of a pumping system is that it has a regular energy requirement, and this contrasts with many other aspects of urban drainage systems. Decisions about the design of sewer systems, and the choices about the use of pumped systems, have always involved a comparison of capital costs against operating costs. When the sustainability of drainage alternatives is also considered, energy consumption must be seen as a significant issue (see Section 14.8).

14.7 NON-GRAVITY SYSTEMS

Where the ground surface is very flat, or where ground conditions make construction of deep pipes difficult, alternative options to gravity drainage are non-gravity systems such as *vacuum sewerage* or *pressure sewerage.*

In vacuum sewerage, wastewater is drained from properties by gravity to collection sumps. When the liquid surface in the sump rises to a particular level, an *interface valve* opens and wastewater is drawn into a pipe in which low pressure (in the order of −0.6 bar) has been created by a pump. After the collection sump has been emptied, the interface valve remains open for a short time to allow a volume of air at atmospheric pressure to enter the pipe. The mixture travels at high velocity (5–6 m/s) towards the vacuum source. The wastewater is then retained in a collection vessel for subsequent pumped removal. Vacuum systems consist of small, shallow pipes with relatively high running costs, compared with the large, deep pipes and low running costs of gravity systems. In the right circumstances, vacuum systems show overall cost advantages (Consterdine, 1995). There are a few vacuum systems in the United Kingdom (see, e.g., Ashlin et al., 1991). Further details are given by Read (2004). Vacuum systems are covered by BS EN 1091: 1997.

In pressure sewerage, wastewater is similarly drained from properties by gravity to local collection sumps. Each sump contains a high head package pump system that discharges into a foul flow only rising main. Systems can be designed with hundreds of properties connected. Pipe sizes are small diameter at shallow depths, and instrumentation control manages when the pumps can discharge. Such systems have become more commonplace, for example, in New Zealand (Carne et al., 2014). Pressure systems are covered by BS EN 1671: 1997.

14.8 ENERGY USE

The water industry is a very large user of energy, approximately 1% of the UK's total (www. ofwat.gov.uk/regulated-companies/improving-regulation/climate-change/). In the UK this amounted to 8290 GWh per annum in 2007/2008 (Water UK, 2008). It is the fourth most energy-intensive sector with costs for bought-in power accounting for over 13% of total production costs (Marshall, 1998). In 2005/2006 the energy required for wastewater services (collection, treatment, discharge, and sludge management) was 634 kWh/Ml (Water UK, 2006). The contribution of urban drainage to this overall figure is not known in detail, but it will be relatively modest compared with treatment use. Reducing energy consumption has always been an industry goal, but added impetus has now emerged because of the need to minimise carbon emissions.

Within urban drainage the key use of energy is at pumping stations. However, many pumping stations do not operate at optimum energy efficiency (as defined for example in Figure 14.4). Yates and Weybourne (2001) argue that by improving operation, energy reductions of 30%–50% can be consistently made. Typical improvements include

- Refurbishing or replacing worn equipment
- Improving the scheduling of multi-pump systems
- Employing high-efficiency motors, efficiency-enhancing coatings, or variable speed drives
- More effectively matching pump performance to demand
- Using condition-based monitoring to understand the pump performance
- Employing continuous monitoring of energy usage

PROBLEMS

14.1 A pumping system has a static lift of 15 m. The pump characteristics are listed below, together with the total losses in the rising main (velocity head can be neglected).

	Q (m³/s)	0	0.05	0.1	0.15	0.2
Pump:	H (m)	25	24	20	14	7
	efficiency (%)		45	55	55	50
Rising main:	losses (m)	0	1	4	9	16

Determine the flow rate, head delivered, and power supplied to the pump at the operating point. If the diameter of the rising main is 250 mm, are conditions suitable for scouring of deposited solids? If the rising main became rougher with age, would the flow rate, and head, increase or decrease?   [0.105 m³/s, 19 m, 36 kW, $v = 2.1$ m/s OK, Q decrease, H increase]

14.2 For the same system as in Problem 14.1, if an additional identical pump is operating in parallel, determine the total flow rate, head delivered, and power supplied to the pumps at the operating point. Which arrangement—one pump or two in parallel— uses power more efficiently?   [0.14 m³/s, 23 m, 64 kW, one pump]

14.3   There are three main categories of rotodynamic pumps. Describe for each category (1) their basic mode of operation, (2) their advantages and disadvantages, and (3) their application in urban drainage.

14.4   A pumping station sump is being designed to suit an inflow of 30 L/s. What rate of pumped outflow would give the minimum sump volume? What sump volume would then be required if the pump was to operate at (1) 6 starts/hour and (2) 12 starts/hour? At 12 starts/hour there is 5 minutes between each start. For how much of that time is the pump operating, and for how long is it idle?   [60 L/s, 9 m³, 4.5 m³, 2.5 minutes]

14.5   Designing a pumping system presents a different set of challenges from designing a gravity system. Explain why.

14.6   Compare and contrast vacuum sewerage and pressure sewerage.

14.7   Considerable quantities of energy are used by the water industry. Identify the main use of energy in urban drainage systems, and discuss ways of reducing it.

KEY SOURCES

Prosser, M.J. 1992. *Design of Low-Lift Pumping Stations*, CIRIA R121.

Tukker, M., Kooij, K., and Posthf, I. 2016. *Hydraulic Design and Management of Wastewater Transport Systems*, IWA Publishing.

American Water Works Association. 2009. *Ductile-Iron Pipe and Fittings*, 3rd edn, Manual 41, AWWA.

REFERENCES

Ashlin, D.E., Bentley, S.E., and Consterdine, J.P. 1991. Vacuum sewerage—The Four Crosses experience. *Journal of the Institution of Water and Environmental Management*, 5(6), 631–640.

BS EN 1091: 1997. *Vacuum Sewerage Systems Outside Buildings*.

BS EN 1671: 1997. *Pressure Sewerage Systems Outside Buildings*.

BS EN 752: 2008. *Drain and Sewer Systems Outside Buildings*.

Carne, S., Salmon, G., and Gamst, C. 2014. Prevention or cure—Designing sewers to achieve low I/I. *Water New Zealand*, 183, March, 59–61.

Chadwick, A., Morfett, J., and Borthwick, M. 2013. *Hydraulics in Civil and Environmental Engineering*, 5th edn, CRC Press.

Consterdine, J.P. 1995. Maintenance and operational costs of vacuum sewerage systems in East Anglia. *Journal of the Chartered Institution of Water and Environmental Management*, 9(6), 591–597.

Creasey, J.D. 1977. *Surge in Water and Sewage Pipelines. Measurement, Analysis and Control of Surge in Pressurized Pipelines*, Technical Report TR51, Water Research Centre.

Kay, M. 2008. *Practical Hydraulics*, 2nd edn, E & FN Spon.

Marshall, L.C. 1998. *Economic Instruments and the Business Use of Energy*, HM Treasury. London.

Prosser, M.J. 1977. *The Hydraulic Design of Pump Sumps and Intakes*, CIRIA with BHRA (Fluid Engineering).

Read, G.F. 2004. Vacuum sewerage, in *Sewers: Replacement and New Construction* (ed. G.F. Read) (Chapter 16, 327–338), Elsevier Butterworth-Heinemann.

Sewers and Water Mains Committee. 1991. *A Guide to Sewerage Operational Practices*, WSA/FWR.

Swaffield, J.A. and Boldy, A.P. 1993. *Pressure Surge in Pipe and Duct Systems*, Avebury Technical.

Thorley, A.R.D. and Atkinson, J.H. 1994. *Guide to the Design of Thrust Blocks for Buried Pressure Pipelines*, CIRIA R128.

Water UK. 2006. *Sustainability Indicators 2005/6*. www.water.org.uk.

Water UK. 2008. *Sustainability Indicators 2007/8*. www.water.org.uk.

Wharton, S.T., Martin, P., and Watson, T.J. 1998. *Pumping Stations, design for improved buildability and maintenance*, CIRIA R182.

Woolley, L. 1988. *Drainage Details*, 2nd edn, E & FN Spon.

WRc 2012. *Sewers for Adoption—A Design and Construction Guide for Developers*, 7th edn, Water UK.

Yates, M.A. and Weybourne, I. 2001. Improving the energy efficiency of pumping systems. *Journal of Water Supply Research and Technology—Aqua*, 50, 101–111.

Chapter 15

Structural design and construction

15.1 TYPES OF CONSTRUCTION

Most sewers are constructed underground. This is achieved by one of three general methods:

- Open-trench construction
- Tunnelling
- Trenchless construction

Open-trench construction consists of excavating vertically along the line of the sewer to form a trench, laying pipes in the trench, and backfilling—see Figure 15.1. It is suitable for a wide range of pipe sizes and depths, and is the common method for small- to medium-scale sewer construction.

Tunnelling typically involves excavating vertically at a particular location for access and then excavating outwards at an appropriate gradient to form the space for the sewer to be constructed. Trenchless construction covers a wide range of technologies that do not include the use of an existing pipeline.

Tunnelling generally involves sizes large enough for human access in which a lining (eventually part of the sewer fabric) is constructed from inside the excavation. Tunnelling tends to be associated with large-scale projects like interceptor sewers.

Underground construction techniques that involve inserting pipes in the ground without a trench are called *trenchless* or "no-dig" methods. They avoid disruption on the surface, and have become increasingly popular as the technology has developed and engineers have become more aware of the costs to business and society of conventional trench construction. As well as its use in construction of new sewers, this type of technology is widely used in sewer rehabilitation, as described in Chapter 18.

This chapter describes these three methods of construction in Sections 15.5–15.7. Before doing so, we consider other physical aspects of sewer pipes and their design. Pipelines must possess a number of physical properties (Section 15.2). They must also give satisfactory performance hydraulically and structurally. Hydraulic performance is considered in Chapter 7. The important issue of structural performance is considered in Section 15.3. Site investigation is covered in Section 15.4, and issues surrounding costs and carbon accounting are introduced in Section 15.8.

15.2 PIPES

15.2.1 General

The nominal size (DN) of a pipe is the diameter of the pipe in millimetres rounded up or down to a convenient figure for reference. In some materials (including clay and concrete)

Figure 15.1 Open-trench construction. (Courtesy of Thames Water.)

the DN refers to the inside diameter, and in some (including plastics), it refers to the outside diameter. The actual diameter may be slightly different from the DN. Of course, the precise diameter must be clear in the manufacturer's product data and must be used in accurate calculations of hydraulic or structural properties.

General requirements for all materials used in gravity sewer systems are given in BS EN 476: 2011—*General Requirements for Components Used in Drains and Sewers.*

15.2.2 Materials

The main characteristics and applications of the common sewer pipe materials are now considered. Relevant British/European standards are listed at the end of this chapter; these provide more detailed guidance on properties, specifications, and structural behavior. A detailed survey of pipe materials is also given in the *Materials Selection Manual for Sewers, Pumping Mains and Manholes* (Sewers and Water Mains Committee, 1993).

In general, the most important physical characteristics of a sewer pipe material are

- Durability
- Abrasion resistance
- Corrosion resistance (with special consideration of the effects of H_2S)
- Imperviousness
- Strength
- Ability to be laid to line and level (particularly important for gravity flow pipes)

15.2.2.1 Clay

Vitrified clay is a commonly used material for small- to medium-sized pipes. Its major advantages are its strength and its resistance to corrosion, making it particularly suitable for foul sewers where the generation of H_2S leads to acidic corrosion of many other pipe materials. However, clay is both heavy and brittle and, therefore, susceptible to damage during handling, storage, and installation.

15.2.2.2 Concrete

Plain and reinforced concrete pipes (and culverts) are generally used for medium- to large-sized pipes. It is particularly suited to use in storm sewers because of its size, abrasion resistance, strength, and cost. There is potential for H_2S-generated corrosion (see Chapter 17), which is combatted in the first instance by the use of sulphate-resisting cement, though generally domestic wastewater is initially not harmful to concrete pipes. Non-circular cross-sectional pipes are available, such as egg-shaped and rectangular culverts. A specific type of prestressed concrete pipe is made for pressure applications such as rising mains.

15.2.2.3 Ductile iron

Ductile (centrifugally spun) iron pipes are used where significant pressures are expected, such as in pumping mains (Chapter 14) and inverted siphons (Chapter 8), or where high strength is required, such as in onerous underground loading cases and above-ground sewers. Ductile iron is susceptible to corrosion and needs both internal and external protection, such as zinc coating, cementitious internal lining, and in some circumstances, extra protection from either epoxy or polyurethane (PU) coatings.

15.2.2.4 Steel

Steel pipes tend to be used in specialist applications where high strength is required. These include sea outfalls, above-ground sewers, and pipe bridges. Steel pipes require protection from corrosion—by internal lining and external coating, often supplemented by cathodic protection.

15.2.2.5 Unplasticised PVC (PVC-U)

PVC-U pipe has found general application in small-size pipes typically <DN300. It is lightweight, making installation straightforward, and is corrosion resistant. However, as the pipe is flexible, strength relies on support from the bedding, and good construction practice is therefore critical. Smaller sizes are routinely used in building drainage applications but are also used to some extent for public sewers. Structured wall PVC-U pipes utilise a variety of shapes of the pipe wall to increase stiffness for the same volume of material. However, these thinner walls impact on the cleaning methods that may be used.

15.2.2.6 Polyethylene (PE)

Polyethylene pipes are used in a wide range of applications for gravity and pressurised systems, being commonly used for stormwater conveyance and storage as well as rising mains. Some PE pipes, especially for gravity systems, have a structured wall. Pipe sizes can typically range between 16 and 3500 mm. Pipes can be welded using different techniques or connected with couplers.

15.2.2.7 Other materials

Other pipe materials in use include structured wall pipe made from polypropylene, glass-reinforced plastic (GRP), pitch fibre, and asbestos cement. Pipes of these materials are known to fail and are increasingly in need of replacement. Significant care is needed with some materials that may contain asbestos fibres, such as pitch fibre and asbestos cement. Many existing sewers are made of brick and remain in good condition, particularly considering their age.

15.2.2.8 Sizes

The range of sizes, and increments in size, depends on the pipe material. For example, clay pipes are available at a number of smaller diameters, with larger diameters at multiples of 100 mm. Traditional sizes for concrete pipes start at 150 mm and increase at 75 mm increments over a wide range. Table 15.1 gives size ranges, together with British/European standards.

15.2.3 Pipe joints

Sewer pipes are usually supplied and laid in the trench in standard straight lengths, and jointed *in situ*. There are several alternative jointing methods, providing either rigid or flexible joints. In most cases, flexible joints are preferred to allow for differential settlement, nonuniform support, drying, or other effects, without introducing unacceptable bending moments and stresses in the pipe.

The standard jointing methods are as follows.

15.2.3.1 Spigot and socket

This joint is illustrated in Figure 15.2a. It is the normal jointing method for concrete pipes, larger clay pipes, and most ductile iron pipes. The spigot is inserted into the socket. A rigid joint can be created using a material such as cement mortar; however, it is much more common for flexibility and watertightness to be provided by an O-ring or shaped seal of rubber, or equivalent material, placed in a groove at the end of the spigot before insertion (Figure 15.2a). Insertion causes suitable compression of the O-ring, and seals have a locking capability.

15.2.3.2 Sleeve

An alternative is to use a separate sleeve, as shown in Figure 15.2b. Clay pipes of smaller diameters are commonly jointed using flexible polypropylene sleeves. Plastic pipes use plastic sleeves, including angled flexible strips or O-rings.

Table 15.1 Pipe materials: size ranges, and British/European standards

Material	Normal size range (mm)	Principal British/European standards (full titles are given in References)
Clay	75–1000	BS EN 295, BS 65
Concrete	150–3000	BS 5911-1, BS 5911-3 to BS 5911-5, BS EN 1916
Ductile iron	80–1600	BS EN 10224, BS EN 545
Steel	60–2235	BS 534
PVC-U	17–630	BS 4660, BS EN 1401, BS EN 13476
PE	16–2100	BS EN 13476, BS EN 12201

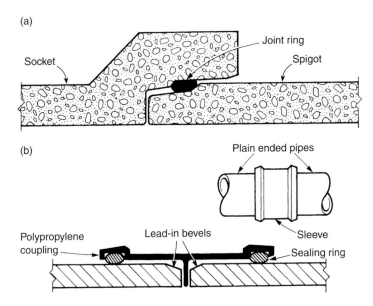

Figure 15.2 (a) Spigot and socket joint (flexible). (b) Sleeve joint. (Reproduced from Woolley [1988] with permission of E & FN Spon.)

15.2.3.3 Bolted flange joints

Simple bolted flanges do not provide flexibility. They are used in rigid installations where exposed pipework (usually ductile iron) needs to be readily dismantled, as in a pumping station (see Section 14.6).

15.2.3.4 Fusion jointing

PE piping can be fused to provide a watertight and continuous pipeline. This is a process where electrically heated plates raise the end of the pipes to a temperature that enables the pipes to be fused together. WIS 4-32-08 (2002) and BS EN 12201-1 2011 provide further information on the specific types of jointing and methods.

15.3 STRUCTURAL DESIGN

15.3.1 Introduction

This section deals with structural design of open-trench sewers. The main components of an open-trench arrangement are shown in Figure 15.3. A pipe laid in a trench has to be strong enough to withstand loads from the soil above it, from traffic and other imposed loads, and from the weight of liquid it carries. With increasing depth of cover, the load from the soil increases, and the load from traffic decreases. The ability to withstand the loads is derived from the strength of the pipe itself, the design of the trench or its absence, and from the nature of the *bedding* on which it is laid. A number of standard "classes" of bedding are defined in Figure 15.4. The material used to support the pipe, together with the depth of support, and the material used to backfill the excavation, are specified. The contribution of the bedding to the overall strength of the system is characterised by a "bedding factor" (given in Figure 15.4, and demonstrated in use in the next section).

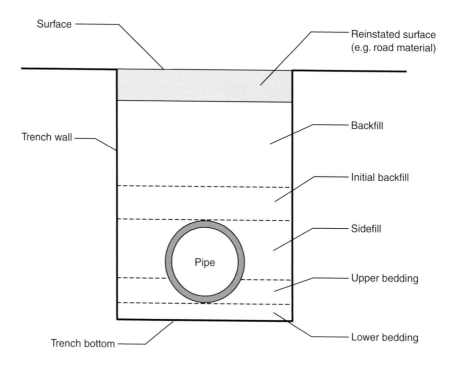

Figure 15.3 Open trench arrangement.

Different calculation procedures are used depending on whether the pipe is considered to be rigid (clay, concrete), semi-rigid (ductile iron), or flexible (plastic), with steel pipe being semi-rigid or flexible depending on its dimensions. There are also differences between trenches considered "narrow" and those considered "wide," and between pipes in trenches and pipes under embankments.

Practising engineers tend to carry out structural design of pipes using standard charts, tables, or software, which relate loads to the properties of the soil and fill material, and the pipe diameter, width of trench, and height of cover. Engineering firms also use their own in-house reference material.

The basis of the procedures is summarised in BS EN 1295-1: 1997 *Structural Design of Buried Pipelines under Various Conditions of Loading.* BS 9295: 2010 *Guide to the Structural Design of Buried Pipelines* provides further information and consolidates previous documents, with a specific focus on UK design methodology. More detailed treatment can be found in Moser and Folkman (2008).

The procedure most commonly appropriate for sewer design is for a *rigid pipe* carrying gravity flow, and that is considered in some detail here. Design of semirigid, flexible, and pressure pipes is not considered in detail, but more information can be found in BS EN 1295–1: 1997, BS 9295: 2010, and Moser and Folkman (2008).

15.3.2 Rigid pipeline design

The total design external load per unit length of pipe (W_e) is given by the sum of the soil load (W_c), the concentrated surcharge load (W_{csu}), and the equivalent external load due to the weight of liquid in the pipe (W_w) (per unit length in each case):

$$W_e = W_c + W_{csu} + W_w \qquad (15.1)$$

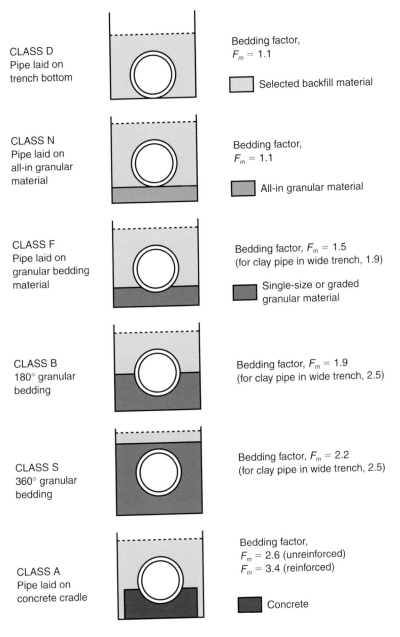

Figure 15.4 Bedding classes and bedding factors.

15.3.2.1 Soil load, W_c

In analysis of a *narrow trench* (Figure 15.5), the soil load is considered to be due to the weight of material in the trench minus the shear between the fill material and the trench sides. W_c—soil load per unit length—is determined from Marston's narrow trench formula, developed using the principles of soil mechanics:

$$W_c = C_d \gamma B_d^2 \qquad (15.2)$$

where

$$C_d = \frac{1 - e^{2K\mu'H/B_d}}{2K\mu'}$$

(15.3)

where K is the Rankine's coefficient: ratio of active lateral pressure to vertical pressure (–), μ' is the coefficient of sliding friction between the fill material and the sides of the trench (–), γ is the unit weight of soil (typically 19.6 kN/m³), B_d is the width of trench at the top of the pipe (m) (Figure 15.5), and H is the depth of cover to crown of pipe (m).

In analysis of a *wide trench* (Figure 15.5), it is assumed that the soil directly above the pipe will settle less than the soil beside it. The soil load is considered to be due to the weight of soil directly above the pipe plus the shear between this and the soil on either side. The effect is considered to reach up only to a certain height, at which there is a "plane of equal settlement."

W_c—soil load per unit length—is determined on the basis of Marston's theory as developed by Spangler, giving

$$W_c = C_c \gamma B_c^2$$

(15.4)

where B_c is the outside diameter of pipe (m).

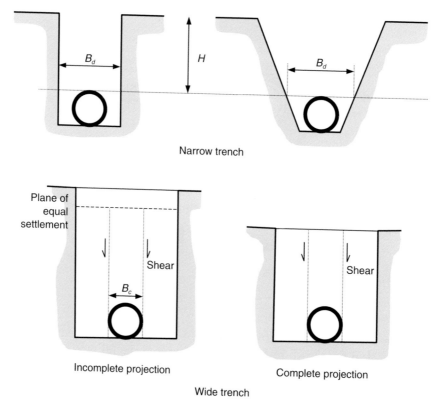

Figure 15.5 Narrow and wide trench.

There are two possible cases, illustrated in Figure 15.5: (1) where the vertical shear planes extend to the top of the cover—*complete projection*, for which

$$C_c = \frac{e^{2K\mu H/B_c} - 1}{2K\mu} \tag{15.5}$$

where μ is the coefficient of friction within the soil mass (–); or (2) where the top of the cover is above the plane of equal settlement—*incomplete projection*—for which C_c is determined from H, B_c, and the product of two terms: r_{sd}, the *settlement deflection ratio*, and p, the *projection ratio*. Expressions for C_c are given in Table 15.2. The term r_{sd} is related to the firmness of the foundation of the trench as given in Table 15.3. The term p is the proportion of the external diameter of the pipe that is above firm bedding.

K, μ', μ, and γ are properties of the fill material and soil. Values of the products $K\mu'$ and $K\mu$ for specific soil types are given in Table 15.4. For a narrow trench, the value of $K\mu'$ used in the calculation should be the lower of the values for the backfill material and for the existing soil in the trench sides. When the soil type is not known, $K\mu'$ is usually taken as 0.13, and $K\mu$ as 0.19. The value of γ is that of the fill material, or a standard value of 19.6 kN/m³.

In design, it is not known in advance whether the case will be one of complete or incomplete projection. Therefore, both cases should be determined and the lower value of C_c used in Equation 15.4.

Similarly, to determine W_c for a trench, the *lower* of the values determined from Equations 15.2 and 15.4 should be used. When the value from Equation 15.2 is used, the trench is defined as narrow, and the specified width must not be exceeded during construction.

Table 15.2 C_c for incomplete projection

$r_{sd}\,p$	Expression for C_c
0.3	$1.39H/B_c$—0.05
0.5	$1.50H/B_c$—0.07
0.7	$1.59H/B_c$—0.09
1	$1.69H/B_c$—0.12

Table 15.3 Values of r_{sd}

Foundation	Rsd
Unyielding (e.g., rock)	1
Normal	0.5 to 0.8
Yielding (e.g., soft ground)	less than 0.5

Table 15.4 Values of $K\mu'$ and $K\mu$

Soil	$K\mu'$ or $K\mu$
Granular, cohesionless materials	0.19
Maximum for sand and gravel	0.165
Maximum for saturated topsoil	0.15
Ordinary maximum for clay	0.13
Maximum for saturated clay	0.11

15.3.2.2 Concentrated surcharge load

The method of determining this term originates from Boussinesq's equation for distribution of stress resulting from a point load at the surface. Some simplification allows W_{csu}—concentrated surcharge load—to be derived from

$$W_{csu} = P_s B_c$$

(15.6)

where P_s is surcharge pressure (N/m^2) and B_c is the outside diameter of the pipe (m).

P_s is a function of the depth of cover and the type of loading (e.g., light road, main road, or different types of railway) as shown in Figure 15.6. Whatever the type of loading when the pipe is in use, it is important to check at the design stage that loadings from construction vehicles will not exceed the concentrated surcharge load predicted.

15.3.2.3 Liquid load

The weight of liquid in the pipe is not strictly an external load, so W_w is the *equivalent* external load due to weight of liquid per unit length of pipe. It is taken as a certain proportion, C_w (water load coefficient), of the weight of liquid held when the pipe is full:

$$W_w = C_w \rho g \pi D^2 / 4$$

(15.7)

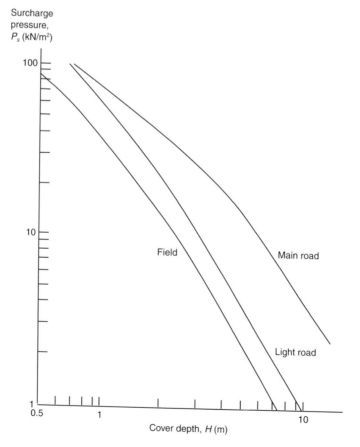

Figure 15.6 Graph of surcharge pressure against cover depth.

where ρ is the density of liquid, in a sewer (1000 kg/m^3); and D is the internal diameter of the pipe (m).

C_w depends on the type of bedding, with a general range of values between 0.5 and 0.8; but for simplicity, the relatively conservative value of 0.75 is often used. W_w does not tend to make a significant contribution to the overall loading for pipes under DN 600.

15.3.2.4 Strength

The strength provided by the chosen combination of pipe material and bedding class is determined by multiplying the strength of the pipe by the factor that indicates the additional strength provided by the bedding, the bedding factor. This overall strength must be sufficient to withstand the total load with an applied factor of safety:

$$W_t F_m \geq W_e F_{se} \tag{15.8}$$

where W_t is crushing strength of the pipe, provided by the manufacturer (N/m); F_m is the bedding factor (–); and F_{se} is the factor of safety, normally 1.25 for clay and concrete pipes.

The normal alternatives for bedding, and the bedding factor for each, have been given in Figure 15.4. Example 15.1 shows a full calculation.

15.4 SITE INVESTIGATION

Site investigation identifies problems with ground conditions and special hazards that may have a significant effect on planning, choice of pipe material, structural design, and construction. The objectives are to provide information useful in the selection of a scheme from a number of alternatives, to inform the detailed design, to estimate costs, and to foresee difficulties. Detailed guidance is given by Whyte (2004).

EXAMPLE 15.1

A sewer pipe has an internal diameter of 300 mm and external diameter 400 mm. It is to be laid under a light road in a trench 0.9 m wide, with a cover depth of 2 m. The original ground is unsaturated clay, and the fill is granular and cohesionless. Assume a value for $r_{sd} p$ of 0.7, use unit weight of soil $\gamma = 19.6$ kN/m^3, and density of liquid $\rho = 1000$ kg/m^3.

The pipe is available with a crushing strength of either 36 or 48 kN/m. Select a suitable class of bedding for each of these pipe strengths.

Solution

First assume a wide trench:

1. Complete projection (Equation 15.5):

$$C_c = \frac{e^{2K\mu H/B_c} - 1}{2K\mu}$$

Fill is granular cohesionless, so from Table 15.4, $K\mu = 0.19$, so

$$C_c = \frac{e^{2 \times 0.19 \times 2/0.4} - 1}{2 \times 0.19} = 15.0$$

2. Incomplete projection:

From Table 15.2, for $r_{sd}p = 0.7$, $C_c = 1.59H/B_c - 0.09 = 7.86$.
Choose the lower value, $C_c = 7.86$, so this is a case of incomplete projection.
Equation 15.4:

$$W_c = C_c \gamma B_c^2 = 7.86 \times 19.6 \times 0.4^2 = 24.6 \, kN/m$$

Now assume a narrow trench:
Equation 15.3:

$$C_d = \frac{1 - e^{-2K\mu'H/B_d}}{2K\mu'}$$

The value of $K\mu'$ used should be the lower of that for the backfill material (0.19) and that for the existing soil in the trench sides (unsaturated clay, from Table 15.4, 0.13).
So,

$$C_d = \frac{1 - e^{-2 \times 0.13 \times 2/0.9}}{2 \times 0.13} = 1.69$$

Equation 15.2:

$$W_c = C_d \gamma B_d^2 = 1.69 \times 19.6 \times 0.9^2 = 26.8 \, kN/m$$

Choose the lower value for a wide and narrow trench, so $W_c = 24.6 \, kN/m$—this is a wide trench case.
Equation 15.6: $W_{csu} = P_s B_c$
From Figure 15.6 (light road) for $H = 2$ m, P_s is 22 kN/m, so

$$W_{csu} = 22 \times 0.4 = 8.8 kN/m$$

Equation 15.7:

$$W_w = C_w \rho g \pi D^2 / 4 = 0.75 \times 1000 \times 9.81 \times \pi \times 0.3^2/4 = 0.5 \, kN/m$$

This uses the usual value of 0.75 for C_w. (W_w is not significant since the diameter is below 600 mm, as suggested.)
Equation 15.1:

$$W_e = W_c + W_{csu} + W_w = 24.6 + 8.8 + 0.5 = 33.9 \, kN/m$$

From Equation 15.8, we require $W_t F_m \geq W_e F_{se}$.
For pipe strength of 36 kN/m, $36 \times F_m \geq 33.9 \times 1.25$.
Bedding factor, F_m, would need to exceed 1.18, so class D or N bedding would not be sufficient, but class F would be.
For 36 kN/m strength pipe, use class F bedding.
For pipe strength of 48 kN/m, $48 \times F_m \geq 33.9 \times 1.25$.
Bedding factor, F_m, would need to exceed 0.88, so class D bedding would be adequate.
For 48 kN/m strength pipe, use class D bedding.

Site investigation is of great importance for all types of sewer construction but may have particular significance for schemes involving construction of sewers in tunnel. In this case, the main areas of interest are geological structure, groundwater, existing services and structures, and special hazards.

Site investigation can typically be divided into three phases:

1. Desk study: analysis of existing data, geological and other maps
2. Site investigation: boreholes, trial excavations, analysis of samples, interpretation
3. During construction: observations and records, probing ahead, further trials and boreholes

15.5 OPEN-TRENCH CONSTRUCTION

15.5.1 Excavation

In urban areas, all excavation must be carried out with great care so as not to damage existing services (e.g., gas, water, and telecommunications), tree roots, and property. In some areas services are densely packed. Their location may be indicated on plans held by the responsible authority and, where necessary, diversions may need to be arranged in advance. Guidance on typical locations is given by the National Joint Utilities Group (www.njug.org. uk). However, there are always problems with the precise location of services in the ground and with services that are unknown, omitted, or wrongly located on the plans. Non-intrusive methods (e.g., electromagnetic location or ground-probing radar [Taylor, 2004]) can be used to locate services from the surface, but these may need to be backed up by trial pits—small excavations dug by hand (since their purpose is to prevent machinery from causing damage).

Pipe trenches are generally excavated mechanically, though hand excavation is needed where access is limited and where existing services are a problem.

The minimum trench width specified in BS EN 1610: 2015 is given in Tables 15.5 and 15.6. The width must not exceed any maximum specified in the structural design, since this might affect the appropriateness of the structural design calculations. The normal maximum depth of a trench is 6 m, but this is less if there are extra surcharge loads on either side of the trench.

Where access is needed to the outside of structures like manholes, a working space of at least 0.5 m needs to be provided. Where more than one pipe is laid in the same trench, working space between the pipes should be 0.35 m for pipes up to 700 mm diameter, and 0.5 m between larger pipes.

The usual system for temporary support of the sides of the trench consists of vertical steel sheets supported by horizontal timber "walings" and adjustable steel struts (Figure 15.1).

Table 15.5 Minimum trench width related to pipe diameter, for supported trench with vertical sides

DN	Minimum trench width (m) (OD = outside diameter)
Below 225	OD + 0.4
225–350	OD + 0.5
350–700	OD + 0.7
700–1200	OD + 0.85
Above 1200	OD + 1

Source: Adapted from BS EN 1610: 2015.

Table 15.6 Minimum trench width related to trench depth

Trench depth (m)	Minimum trench width (m)
Less than 1	No minimum
1–1.75	0.8
1.75–4	0.9
More than 4	1

Source: Adapted from BS EN 1610: 2015.

Whether the sheets form a continuous wall or are placed at a particular separation depends on the condition of the ground and its need for support, as well as on possible inflow of groundwater. Alternatives are ready-made frames, boxes, or trench shields, which are moved progressively as excavation, laying, and backfilling proceed. These supports are generally used to prevent wall collapse for the safe access of operatives, and their effect on trench wall dimensions may need to be considered.

Where there is a significant problem of groundwater entering the trench during construction, dewatering may be necessary—either by pumping from the trench bottom or from remote points—to lower the water table.

15.5.2 Pipe laying

Pipes are delivered to site in bulk, and stored by stacking until needed. They must be stacked carefully to avoid damage, not so high as to cause excessive loads on the pipes at the bottom, and far enough from the trench to avoid any threat to the stability of the excavation. Manufacturers normally specify stacking arrangements.

The nature of the bedding will be specified in the design, and the alternatives have been considered in Section 15.3. Where a pipe is being laid directly on the bed, the trench should be carefully excavated to the correct gradient to ensure that the pipe is supported all along its length. Localised sections of poor ground at the base of the trench must be dug out and replaced by suitable material. Small extra excavations are typically needed to accommodate pipe sockets with some clearance, to ensure that the weight of the pipe is not bearing on the socket. Where granular bedding material has been specified, this must be similarly prepared to the correct gradient to give support all along the pipe.

The safe lifting of pipes varies with both size and material, and the manufacturer's advice should always be followed. It is most common to set out sewer pipes in open-trench using a laser, set up either inside the pipe or above the excavation. The pipe invert (defined in Section 6.3) should be used as the reference point, since this is the most important vertical position from a hydraulic point of view, and most pipes are not sufficiently round for setting out to be carried out by reference to the crown of the pipe.

Pipes are generally laid in the direction of the upwards gradient, so that water in the excavation drains away from the working area. Since the spigot is inserted into the socket, the normal orientation of a pipe is for the socket to be at the upstream end, facing upstream. The specification will indicate tolerances of line and level that must be complied with when the pipe is laid.

Methods of jointing pipes have been described in Section 15.2. Where a pipeline passes through a fixed structure, it is normal to include flexible joints and sometimes short pipe lengths (rocker pipes) close to the wall of the structure. This pipe is generally short and forms the first two joints outside the structure.

Completed sewer sections are tested for leaks by pressure tests using air or water. Criteria for pressure loss with time are given in BS EN 1610: 2015. Sewers that are subsequently to be adopted by the sewerage undertaker may be subject to closed-circuit television (CCTV) inspection (considered further in Chapter 17).

The trench is backfilled in accordance with the design and specification, typically in carefully compacted layers of specified thickness. During backfilling, the trench support is removed progressively. When backfilling is complete, the surface is reinstated.

Good sources of further information on open-trench construction are by Irvine and Smith (1983) and Read (2004).

15.6 TUNNELLING

15.6.1 Lining

Common methods of tunnel lining are capable of withstanding loads over a wide range of conditions. Linings of extra strength can be supplied where necessary. The loads experienced during construction may be more critical than those experienced by the completed sewer.

It is common for sewers in tunnel to have a primary lining, most commonly bolted concrete segments, to support construction and permanent loads, and a secondary lining, often *in situ* concrete, to provide suitably smooth hydraulic conditions. An increasingly common alternative is to use a *smoothbore* lining, which is a primary lining that does not need secondary lining.

15.6.1.1 Primary lining

A precast concrete lining consists of segments that are bolted *in situ* to form a ring, with a narrow key segment at or near the soffit. The excavated area will be slightly larger than the outside of the ring, and the annular space between is filled with grout, injected through holes in the lining.

Standard segments are ribbed and are unsuitable for carrying flow as built. Special concrete blocks can be used to fill the panels between the ribs, but the most common method of achieving a smooth surface is by adding a secondary lining.

15.6.1.2 Secondary lining

In situ concrete can be injected behind circular travelling shutters. Alternatives are ready-made linings of GRPs or fibre-reinforced cement composites, with the annular space between the secondary and primary linings filled with grout.

15.6.2 Ground treatment and control of groundwater

Ground treatment for sewers in tunnels may be needed to control groundwater during construction or to stabilise the ground. The main methods are dewatering, ground freezing, and injection of grouts or chemicals. Groundwater at the tunnel face can be controlled by compressed air.

15.6.2.1 Dewatering

Water is pumped from wells to lower the water table in the area of tunnel construction.

15.6.2.2 Ground freezing

The temperature of the ground is lowered to freeze the groundwater during construction. This is achieved by circulating refrigerated liquids—usually brine or liquid nitrogen—through pipes in the ground.

15.6.2.3 Injection of grouts or chemicals

This may be to reduce permeability, or improve strength of cohesionless soils or broken ground. Injection can be carried out through holes drilled from the tunnel face or from the ground surface. Less drilling may be needed from the tunnel face, but this approach can hold up construction.

15.6.2.4 Compressed air

Groundwater can be held back by balancing the hydrostatic pressure with compressed air inside the tunnel. Pressures are commonly less than 1 atmosphere, but can be higher. The part of the tunnel under pressure is sealed off by air locks, through which personnel and materials pass. Air must be supplied continuously as some escapes through the ground. People working in compressed air must have regular health checks, and work is tightly controlled by health and safety requirements.

15.6.3 Excavation

Tunnels are driven from working shafts, usually supporting drives of roughly equal lengths both upstream and downstream.

Most ground can be excavated by a boring machine or by handheld pneumatic tools. Hard rock may need to be drilled and blasted. If the ground cannot be left unsupported during erection of the primary lining, a tunnel shield, which pushes itself forward from the previously erected primary lining, is used to give continuous support. A tunnel boring machine may combine the functions of shield and mechanical excavator.

Shafts are excavated vertically, mechanically, or by hand, and the ground is usually supported using precast concrete segmental rings similar to those used for tunnels. For sewers in tunnels, working shafts usually become manholes in the completed scheme.

More information on constructing sewers in tunnels is given by Chappell and Parkin (2004a), and detailed information on designing and constructing tunnels is given by Maidl et al. (2014).

15.7 TRENCHLESS METHODS

The choice between open-trench construction and traditional tunnelling can be made on the basis of ease, safety, and cost of construction. Beyond a certain depth and diameter, it is simply cheaper—in purely construction terms—to tunnel, than to excavate and backfill a trench. In contrast, trenchless methods usually become an appropriate alternative to open-trench construction when an additional factor is taken into account: the disruption to business, traffic, and everyday life caused by open-trench construction.

Trenchless methods are also applied commonly to other pipe-laying fields, particularly gas and oil supply, for which some of the techniques were originally developed.

A brief introduction to some of the principal methods is given here; more information can be found in Thomson (1993), Grimes and Martin (1998), Chappell and Parkin (2004b),

Howell (2004a), Howell (2004b), and Syms (2004). The International Society for Trenchless Technology website (www.istt.com) provides a useful source of information of new technologies. Further important applications of trenchless technology—to sewer rehabilitation—are covered in Chapter 18.

15.7.1 Pipejacking

Hydraulic jacks are used to push specially made pipes (without protruding sockets, and strong enough to take the jacking forces) through an excavated space in the ground (Figure 15.7a). Ahead of the pipes is a shield at which excavation takes places either mechanically or by hand tool. The jacks push against a thrust wall located in a specially constructed thrust pit. *An Introduction to Pipejacking and Microtunnelling* (PJA, 2017) summarises the key aspects.

15.7.2 Microtunnelling

This is a form of pipejacking for pipe diameters under 900 mm. Excavation is by unmanned, remote-controlled equipment. PJA (2017) also provides a good summary of the key aspects of this technology.

15.7.3 Auger boring

Soil is removed from the excavated face by an auger (Figure 15.7b), and pipes are jacked into the excavated space. This is generally considered to be a rather inaccurate method and is used for short drives only.

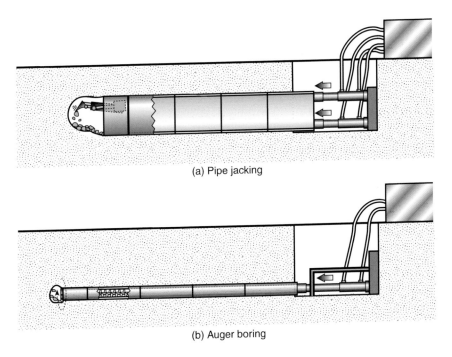

(a) Pipe jacking

(b) Auger boring

Figure 15.7 Pipejacking and auger boring.

15.7.3.1 Directional drilling (DD)

A pilot bore is made using a guided small-diameter drill pipe. The bore is then enlarged to the required size by further mechanical excavation.

15.7.3.2 Impact moling

An earth displacement mole creates a hole in the ground by pushing the earth outwards. A pipe can then be pushed into the space. This method is used for small diameters and short distances only.

15.7.3.3 Pipe ramming

A steel casing is driven through the ground by a pneumatic hammer located in a drive pit. The ground material within the casing is removed either during the driving process or, more commonly, when it is complete. Soil removal may be carried out using compressed air or water, or by mechanical or manual processes.

15.7.3.4 Timber headings

Timber sets support an excavation dug by hand, although steel can also be used. This approach is used where there are short distances to tunnel or where trenchless technologies are not appropriate. It can be used when there is an operational issue, and dig down from the surface is not appropriate. Headings must be appropriately designed, and not less than 1.2 m high by 1 m wide. Further information is available in BS 6164: 2011, and the Health and Safety Executive (HSE, 2017) provides guidance on the acceptable size and drive lengths. Pipes are inserted by hand or on a track, and the surrounding is backfilled, typically by concrete or grout.

15.8 COSTS

15.8.1 Construction costs

Costs for standard construction operations in urban drainage are available from price books and databases, of which the best known is *Spon's Civil Engineering and Highway Works Price Book* (AECOM, 2016). This is updated every year and provides general guidance for estimating construction costs. Costs can be determined either by breaking down the necessary construction operations into the labour, materials, and plant required, or by breaking down the components of the scheme in terms of quantities and unit costs (for volume of a particular grade of concrete, or length of pipe of a particular diameter and material laid in a particular depth of trench, etc.). *Spon's Price Book* provides the up-to-date generic data needed to carry out the calculations. This provides a bottom-up estimate.

In practice, cost estimates for drainage works are also derived from the costs involved in previous schemes, from agreements with partner contractors, and from contact with specialist contractors. This provides a more top-down estimate, although in practice this often does not account for the detailed challenges at a specific location and can be hard to square with bottom-up estimates.

15.8.2 Carbon accounting

Along with understanding the financial cost of capital options and maintenance strategies, it is commonplace to estimate the embodied and operational carbon of a scheme. This is a requirement of the regulator Ofwat for Water and Sewerage Companies in the United Kingdom. Understanding the estimated carbon can influence decision making along with the costs, for example, when selecting enhancements to address environmental drivers or selecting an operational intervention strategy in the drainage network.

Embodied carbon calculations can be made using open access databases for the construction industry, such as those developed by Hammond and Jones (2008). This identified the embodied energy and carbon (dioxide) emissions associated with almost 200 materials, from peer-reviewed literature. Life cycle analysis can generate such values, although recognising that boundary conditions, age of the data sources, and rigour of the analysis can generate differences. Current values can be found in the *CESMM4 Carbon and Price Book* (MottMacDonald, 2013), with estimates built up using the same approach as when financially pricing schemes.

UK Water Industry Research (UKWIR) developed in 2008 and updated in 2012 (Ainger et al., 2012) a framework to undertake the embodied and whole life carbon accounting in a consistent manner for the UK Water Industry. It outlines different stages when accounting may take place and the level of information/detail appropriate.

Operational carbon estimates are often dominated by energy use. While there is often greater focus on carbon associated with treatment, operational interventions are critical in managing urban drainage (outlined in Chapter 17). The UK government (DECC, 2016) provides frequent updates to data to enable conversions of fuels and energy use in carbon ($kgCO_2$). UWKIR (WRc, 2017) developed a carbon accounting workbook for operational emissions, initially in 2005 with regular revisions since. This provides a standardised approach to estimate operational emissions; however, its intention was to report annual emissions in arrears, and therefore care is needed if applied to estimate future emissions. This is particularly important due to the uncertainty linked to future emissions, conversion factors, and the price of carbon. Any future values should be caveated as an estimation, and best practice would be to provide an indication or range, for example, using sensitivity analysis.

Many suppliers are evaluating their own emissions, making environmental product declarations.

PROBLEMS

15.1 Describe the properties of the main sewer pipe materials. What factors affect the selection of a pipe material for a particular case?

15.2 A pipe has an internal diameter of 450 mm and a wall thickness of 50 mm. If it is laid in an open trench of depth 2 m, determine the minimum width of the trench. Similarly, determine the minimum trench width for a pipe of internal diameter 200 mm and wall thickness 22 mm (same depth of trench).   [1.25 m, 0.9 m]

15.3 A pipe with internal diameter 225 mm and external diameter 280 mm is to be laid in a trench of minimum width. The cover depth will be 3 m; the ground is unsaturated clay and the backfill will be granular cohesionless material. Determine the soil load per metre length (W_c) assuming a narrow trench condition. Use unit weight of soil $\gamma = 19.6$ kN/m³.   [35.4 kN/m]

15.4 For the case in Problem 15.3, determine the soil load assuming a wide trench condition. Use the data given in Problem 15.3, and assume a value for $r_{sd}\,p$ of 0.7. Which assumption is appropriate (narrow or wide trench) for design in this case?   [26 kN/m, wide]

15.5 For the case in Problems 15.3 and 15.4, determine the external load per unit length (W_e). The pipe will be under a light road. Use density of liquid $\rho = 1000$ kg/m^3. Pipe is available in this size with a crushing strength of 28, 36, or 48 kN/m. Determine the minimum bedding factor in this case for each of these pipe strengths. Propose two appropriate designs. [29.1 kN/m; 1.3, 1, 0.76; low strength pipe on Class F bedding, medium strength on Class D]

15.6 Describe the main methods of sewer construction. Discuss the factors that influence selection of the most appropriate method in particular cases.

15.7 Describe methods of ground treatment for sewer construction and the circumstances in which they might be used.

15.8 How can carbon be accounted for when costing construction projects?

KEY SOURCES

BS EN 1295: 1997. *Structural Design of Buried Pipelines under Various Conditions of Loading.*
Read, G. 2004. *Sewers: Replacement and New Construction*, Butterworth-Heinemann.

REFERENCES

AECOM (ed.). 2016. *Spon's Civil Engineering and Highway Works Price Book 2017*, CRC Press.
Ainger, C., Grenfell, R., Horton, B., Johnson, A., Jones, C., Lumsden, S., and Smith, C. 2012. *A Framework for Accounting for Embodied Carbon in Water Industry Assets.* UKWIR Report CL01/B207.
Chappell, M., and Parkin, D. 2004a. Tunnel construction, in *Sewers: Replacement and New Construction* (ed. G.F. Read) (Chapter 7, 150–192), Elsevier Butterworth-Heinemann.
Chappell, M., and Parkin, D. 2004b. Pipejacking, in *Sewers: Replacement and New Construction* (ed. G.F. Read) (Chapter 9, 210–223), Elsevier Butterworth-Heinemann.
DECC. Department for Energy and Climate Change. 2016. *Greenhouse Gas Reporting—Conversion Factors 2016*, www.gov.uk/government/publications/greenhouse-gas-reporting-conversion-factors-2016.
Grimes, J.F., and Martin, P. 1998. *Trenchless and Minimum Excavation Techniques, Planning and Selection*, CIRIA SP147.
Hammond, G.P., and Jones, C.I. 2008. Embodied energy and carbon in construction materials. *Proceedings of the Institution of Civil Engineers—Energy*, 161(2), 87–98.
Howell, N. 2004a. Impact moling, in *Sewers: Replacement and New Construction* (ed. G.F. Read) (Chapter 11, 236–253), Elsevier Butterworth-Heinemann.
Howell, N. 2004b. The pipe ramming technique, in *Sewers: Replacement and New Construction* (ed. G.F. Read) (Chapter 12, 254–271), Elsevier Butterworth-Heinemann.
Health and Safety Executive (HSE). 2017. *Tunnelling and Pipejacking: Guidance for Designers*, www.hse.gov.uk/construction/pdf/pjaguidance.pdf.
Irvine, D.J., and Smith, R.J.H. 1983. *Trenching Practice*, CIRIA R97.
Maidl, B., Thewes, M., and Maidl, U. 2014. *Handbook of Tunnel Engineering, Volumes I and II*, Wiley.
Moser, A.P., and Folkman, S. 2008. *Buried Pipe Design*, 3rd edn, McGraw-Hill Professional.
Mott MacDonald. 2013. *CESMM4 Carbon and Price Book 2013*, ICE publishing.
Pipe Jacking Association (PJA). 2017. *An Introduction to Pipe Jacking and Microtunnelling.*
Read, G.F. 2004. Open-cut and heading construction, in *Sewers: Replacement and New Construction* (ed. G.F. Read) (Chapter 6, 132–149), Elsevier Butterworth-Heinemann.
Sewers and Water Mains Committee. 1993. *Materials Selection Manual for Sewers, Pumping Mains, and Manholes*, WSA/FWR.

Syms, B. 2004. Horizontal directional drilling, in *Sewers: Replacement and New Construction* (ed. G.F. Read) (Chapter 14, 297–326), Elsevier Butterworth-Heinemann.

Taylor, N. 2004. Site investigation and mapping of buried assets, in *Sewers: Replacement and New Construction* (ed. G.F. Read) (Chapter 3, 95–99), Elsevier Butterworth-Heinemann.

Thomson, J.C. 1993. *Pipejacking and microtunnelling*, Blackie Academic and Professional.

Whyte, I. 2004. Site investigation, in *Sewers: Replacement and New Construction* (ed. G.F. Read) (Chapter 2, 47–94), Elsevier Butterworth-Heinemann.

Water Research Centre (WRc). 2017. *Workbook for Estimating Operational GHG Emissions*, Version 11. UKWIR Report No. 17/CL/01/25.

BRITISH STANDARDS

BS 65: 1991. *Specification for Vitrified Clay Pipes, Fittings and Ducts, also Flexible Mechanical Joints for Use Solely with Surface Water Pipes and Fittings.*

BS EN 295: 2013. *Vitrified Clay Pipe Systems for Drains and Sewers.*

BS EN 476: 2011. *General Requirements for Components Used in Drains and Sewers.*

BS EN 545: 2010. *Ductile Iron Pipes, Fittings, Accessories and Their Joints for Water Pipelines. Requirements and Test Methods.*

BS EN 1295–1: 1997. *Structural Design of Buried Pipelines Under Various Conditions of Loading.*

BS EN 1401–1: 2009. *Plastics Piping Systems for Non-Pressure Underground Drainage and Sewerage, Unplasticized Polyvinyl Chloride (PVC-U). Specifications for Pipes, Fittings and the System.*

BS EN 1610: 2015. *Construction and Testing of Drains and Sewers.*

BS EN 1916: 2002. *Concrete Pipes and Fittings, Unreinforced, Steel Fibre and Reinforced.*

BS 4660: 2000. *Thermoplastics Ancillary Fittings of Nominal Sizes 119 and 160 for Below Ground Gravity Drainage and Sewerage.*

BS 5911–1: 2002. *Concrete Pipes and Ancillary Concrete Products. Specification for Unreinforced and Reinforced Concrete Pipes (Including Jacking Pipes) and Fittings with Flexible Joints.*

BS 5911–3: 2002. *Concrete Pipes and Ancillary Concrete Products. Specification for Unreinforced and Reinforced Concrete Manholes and Soakaways.*

BS 5911–4: 2002. *Concrete Pipes and Ancillary Concrete Products. Specification for Unreinforced and Reinforced Concrete Inspection Chambers.*

BS 5911–5: 2004. *Concrete Pipes and Ancillary Concrete Products. Specification for Pre-Stressed Non-Pressure Pipes and Fittings with Flexible Joints.*

BS 6164: 2011. *Code of Practice for Health and Safety in Tunnelling in the Construction Industry.*

BS 9295: 2010. *Guide to the Structural Design of Buried Pipelines.*

BS EN 10224: 2002. *Non-Alloy Steel Tubes and Fittings for the Conveyance of Aqueous Liquids Including Water for Human Consumption.*

BS EN 10311: 2005. *Joints for the Connection of Steel Tubes and Fittings for the Conveyance of Water and Other Aqueous Liquids.*

BS EN 12201-1: 2011. *Plastic Piping Systems for Water Supply, and for Drainage and Sewerage Under Pressure. Polyethylene (PE). General.*

BS EN 13476: 2007. *Plastics Piping Systems for Non-Pressure Underground Drainage and Sewerage. Structured-Wall Piping Systems of Unplasticized Polyvinyl Chloride (PVC-U), Polypropylene (PP) and Polyethylene (PE).*

WIS 4-32-08: 2002. *Specification for the Fusion Jointing of Polyethylene Pressure Pipeline Systems Using PE80 and PE100 Materials, Water Industry Specification.*

Chapter 16

Sediments

16.1 INTRODUCTION

Sediment is ubiquitously present in urban drainage systems. It is found deposited on catchment surfaces, in gully pots, and in drains and sewers. Drainage engineers have long recognised its presence in stormwater and the problems it may cause. They have sought to exclude larger, heavier sizes from the piped system by the provision of gully pots and designed sewers to limit in-pipe deposition. The theory is that sediment that does enter the system is carried downstream, where it is eventually trapped and removed at the outlet of the system. This may be the case for newly designed systems, but for older (especially combined) networks, sedimentation in sewers is commonplace. In fact, a review of sediment movement in sewers (Binnie and Partners and Hydraulics Research, 1987) concluded that 80% of UK urban drainage systems had at least some permanent sediment deposits.

The movement of sediment through a drainage catchment is a complex, multi-stage process. Sediment deposited on roads, for example, is initially freed from the surface and then washed transversely by overland flow to the channel, where it is transported parallel to the kerb-line under open channel flow. The sediment is discharged with the flow into the gully inlet under gravity, and is captured in the gully pot (in part) by sedimentation or transferred to the receiving sewer. Once in the sewer, material is transported under open channel flow as suspended or bed-load. Suspended sediment is carried along in the main body of the flow, while bed-load travels more slowly in contact with the invert of the pipe. Some material may be deposited and/or re-eroded as it progresses downstream.

During transport, sediment may be discharged to a watercourse via a combined sewer overflow (CSO) (if in operation) or settled in the wastewater treatment plant (WTP) grit removal device. At points of deposition (surface, gully, sewer), sediment may be extracted from the system by cleaning. A representation of sediment inputs, outputs, and movement through the system is given in Figure 16.1. Further details on the removal of sediment from systems will be given in Chapter 17.

The rate of progress of material through the system depends on factors such as the characteristics of the

- Sediment (physical, chemical)
- Flow (velocity, degree of unsteadiness)
- Drainage network (layout, geometry)

For example, different types of sediment will move through the system in different ways. Particles of very small size or low density may remain in suspension under all normal flow conditions and be transported through the system without being deposited. Sediments with

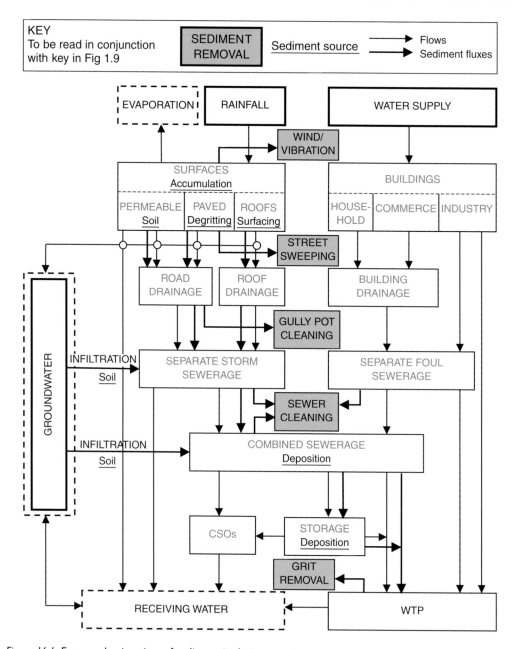

Figure 16.1 Entry and exit points of sediment in drainage systems.

low settling velocities may only form deposits during periods of very low flow, and may readily be re-entrained when higher velocities occur in the pipes as a result of storms or diurnal variations in flow. By contrast, larger and denser particles may only be transported by peak flows that occur relatively infrequently, and in some cases they may form permanent stationary deposits near where they enter the sewer system.

Deposition commonly occurs during dry weather flow periods (particularly during low flow at night) and in decelerating flows during storm recession. Deposits also form at structural and hydraulic discontinuities such as at joints, changes in gradients, and ancillary structures. Only the steepest sewers are immune from deposition.

This chapter reviews the origins (Section 16.2), problems and effects (Section 16.3) of sediment. Sections look at how the sediment is transported through the system (Section 16.4) and at its detailed characteristics (Section 16.5). Sewer self-cleansing design methods are presented in Section 16.6.

16.2 ORIGINS

16.2.1 Definition

Sewer sediments are defined as any settleable particulate material that is found in stormwater or wastewater and is able, under appropriate conditions, to form bed deposits in sewers and associated hydraulic structures. Using the basic solids classification presented in Table 2.1, this would include

- Grit
- Suspended solids
  - Sanitary
  - Stormwater

16.2.2 Sources

Sources of sediment entering sewers are quite diverse. Any particulate-generating material or activity in the urban environment is a potential source. Broadly, three categories can be established: sanitary, surface, and sewer. These are defined in Table 16.1.

Table 16.1 Sources of sewer sediment

Source	Type
Sanitary	• Large faecal and organic matter with specific gravity close to unity
	• Fine faecal and other organic particles
	• Paper and miscellaneous materials flushed into sewers (sanitary refuse)
	• Vegetable matter and soil particles from the domestic processing of food
	• Materials from industrial and commercial sources
Surface	• Atmospheric fall-out (dry and wet)
	• Particles from erosion of roofing material
	• Grit from abrasion of road surfaces or from resurfacing works
	• Grit from de-icing operations on roads
	• Particulates from motor vehicles (e.g., vehicle exhausts, rubber from tyres, wear and tear, etc.)
	• Materials from construction works (e.g., building aggregates, concrete slurries, exposed soil, etc.) and other illegally dumped materials
	• Detritus and litter from roads and paved areas (e.g., paper, plastic, cans, etc.)
	• Silts, sands, and gravels washed or blown from unpaved areas
	• Vegetation (e.g., grass, leaves, wood, etc.)
Sewer	• Soil particles infiltrating due to leaks or pipe/manhole/gully failures
	• Material from infrastructure fabric decay

16.3 EFFECTS

16.3.1 Problems

There are three major effects of sediment deposition leading to a number of serious consequences, and these are summarised in Table 16.2. Table 16.2 also lists parts of the book where further information can be found on the consequences of sediment deposition.

The first effect of deposited sediment is its propensity to initiate blockage. Larger, gross solids and other matter may build up, leading to partial or total blockage of the pipe bore.

The second effect results from the fact that a deposited bed restricts the flow in the sewer, resulting in a loss of hydraulic capacity. The reasons for this effect are discussed in Section 16.3.2, but the result can be pipe or manhole surcharge. Another common effect is the premature operation of CSOs.

The third major effect results from the ability of the deposited sediment to act as a pollutant store or generator. The reasons and possible mechanisms for this are discussed later in the chapter. These pollutants are only stored temporarily, and can be released under flood flow conditions, probably contributing to the commonly observed first foul flush of heavy pollution (see Section 12.3.2). Biochemical changes in the bed of sediment can result in septic conditions, releasing gas that can be highly corrosive to the sewer fabric.

It should be clear from the previous discussion that excessive sediment deposition should be avoided if at all possible at the design stage. Extensive sediment removal is a difficult, recurrent, and expensive process.

16.3.2 Hydraulic

The presence of sediment in sewer flows has three hydraulic effects of varying importance: dissipation of energy in keeping solids in suspension, reduction of flow cross-sectional area, and increased frictional losses due to the texture of the bed.

16.3.2.1 Suspension

In the case of a sewer without deposits, the presence of sediment in the flow or moving along the invert causes a small increase in energy loss, and this is observed as a reduction in discharge capacity of about 1% for rough pipes (Ackers et al., 1996a).

Table 16.2 Effects and consequences of sewer sediment deposition

Effect	Consequences	Further information (Section)
Blockage	• Surcharging	7.4.5
	• Surface flooding	
Loss of hydraulic capacity	• Surcharging	
	• Surface flooding	
	• Premature operation of CSOs	12.3.1
Pollutant storage	• Washout to receiving waters during CSO operation	2.5.1
	• Shock loading on treatment plants	
	• Gas and corrosive acid production	17.8.1

For flows in sewers in which there is a deposited bed of sediment, the energy losses associated with keeping the sediment in motion are relatively insignificant compared with the other effects.

16.3.2.2 Geometry

A deposited bed reduces the cross-sectional area available to convey flow and therefore increases the velocity and head loss for a given discharge and depth of flow. The loss of total area is relatively small (<2%) provided that the depth of sediment is less than 5% of the pipe diameter, but becomes important at sediment depths above about 10%.

16.3.2.3 Bed roughness

Usually of greatest significance is the increase in overall resistance caused by the rough texture of the deposited bed. Above the *threshold of movement*, sediment quickly forms into ripples and dunes, and initially these grow in size as the flow velocity increases. The effective roughness value of dunes kb (k_s in the Colebrook-White equation) can reach 10% or more of the pipe diameter, compared with typical values for the pipe walls that are in the range 0.15–6 mm (depending on the material and the degree of sliming). The value of kb can be estimated (May, 1993) from

$$k_b = 5.62R^{0.61}d_{50}^{0.39} \tag{16.1}$$

where R is hydraulic radius (m) and d_{50} is sediment particle size larger than 50% (by mass) of all particles in the bed (m).

Under these conditions, a 5% depth of deposited sediment with dunes could reduce the pipe-full capacity by 10%–20%. However, at higher velocities the dunes tend to reduce in size until the bed again becomes flat with much lower roughness. The loss of hydraulic capacity due to a deposited bed can, therefore, vary considerably with the flow conditions. The approximate combined effects of shape and roughness are illustrated in Figure 16.2.

16.3.3 Pollutional

Solids of sanitary origin are readily mixed, in combined sewers, with sediments entering from surface sources. The organic material tends to adhere to the heavier inorganic sediment and deposits with them. These deposits form a rough surface that encourages further adherence of organic matter. In such an environment, anaerobic conditions are likely to develop, resulting in partial digestion of the sediment. By-product fatty acids are then liberated into the interstitial liquor (as described further in Chapter 17) with the possibility of substantial biochemical oxygen demand/chemical oxygen demand (BOD/COD) loads being generated. Some evidence suggests that degradation processes in sediment can increase pollutant discharges by up to 400% (Binnie and Partners and Hydraulics Research, 1987). Clearly, the presence of sediment deposits encourages the retention of solids and pollutants during low flows. This increases the potential for degradation before such material is flushed away.

The concept and importance of the first foul flush is discussed in Chapter 12. Sediment deposits are commonly considered to be one of its major causes. Field evidence indicates that up to 90% of the pollution load discharged from CSOs can be derived from erosion of in-sewer deposits (Crabtree, 1989).

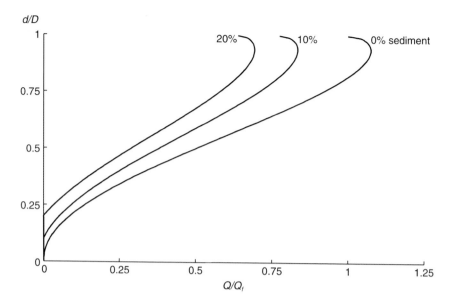

Figure 16.2 Effect of sediment bed on flow capacity of sewers.

16.4 TRANSPORT

The movement of sediment has three broad phases: entrainment, transport, and deposition.

16.4.1 Entrainment

As wastewater flows over a sediment bed in a sewer, hydrodynamic lift and drag forces are exerted on the bed particles. If these two combined forces do not exceed the restoring forces of sediment submerged weight, interlocking and cohesion (if applicable), then the particles remain stationary. If they exceed the restoring force, then entrainment occurs, resulting in movement of the particles at the flow/sediment boundary. Not all the particles of a given size at this boundary are dislodged and moved at the same time, as the flow is turbulent and contains short-term fluctuations in velocity. The limiting condition, below which sediment movement is negligible (the threshold of movement), is usually defined in terms of either a critical boundary shear stress (τ_o) or critical erosion velocity (v). The two are related as shown in Equation 7.22 reproduced below:

$$\tau_o = \frac{\rho \lambda v^2}{8}$$

where ρ is liquid density (kg/m³) and λ is the Darcy-Weisbach friction factor (–).

In storm sewers, sediments are mainly inorganic and non-cohesive, although some deposits may be cementitious and become permanent if undisturbed for long periods. Sediments in foul sewers generally have cohesive-like properties due to the nature of the particles and the presence of greases and biological slimes. In combined sewers, the sediments tend to be a combination of the first two types.

Cohesion will tend to increase the value of shear stress that the flow needs to exert on the deposited bed in order to initiate movement of particles in the surface layer. In laboratory

Table 16.3 Mode of sediment transport

Ws/U*	Mode
>0.6	Suspension
0.6–2	Saltation
2–6	Bed-load

Source: Raudkivi, A.J. 1998. *Loose Boundary Hydraulics*, 4th edn, A.A. Balkema

tests, erosion of synthetic cohesive sediments has been observed to occur at bed shear stresses of 2.5 N/m² (surficial material) and 6–7 N/m² (granular, consolidated deposits). Field studies in large sewers, however, show that lower shear stresses of around 1 N/m² may initiate erosion (Ashley and Verbanck, 1996).

It is possible that cohesion may also alter the way in which the sediment then moves and thus affects the sediment-transporting capacity of the flow. However, experimental research (Nalluri and Alvarez, 1992) using synthetic cohesive sediments suggests that the second factor may not be significant; once the structure of a cohesive bed is disrupted, the particles are stripped away and transported by the flow in a similar way to non-cohesive sediments.

16.4.2 Transport

Once sediment has been entrained into the flow, it travels, as mentioned in the introduction, in suspension or as bed-load. Finer, lighter material tends to travel in suspension and is primarily influenced by turbulent fluctuations in the flow, which in turn are influenced by bed shear. It is advected at mean flow velocity. Heavier material travels by rolling, sliding, or saltating along the pipe invert (or deposited bed) as bed-load. This type of movement is affected by the local velocity distribution, and advection velocities in this mode are considerably lower than the flow mean velocity. In an urban drainage network, with graded materials of differing specific gravity, a combination of these modes exists.

Table 16.3 indicates that the mode of transport depends on the relative magnitude of the lifting effects due to turbulence, as measured by the shear velocity (U^*), and the settling velocity (W_S). Shear velocity is given by

$$U^* = \sqrt{\frac{\tau_o}{\rho}} \tag{16.2}$$

EXAMPLE 16.1

Analysis of the sediment in an urban catchment shows it to consist predominantly of grit with a characteristic settling velocity of 750 m/h. Estimate the mode of transport of this sediment in a 0.15% gradient, 1.5 m diameter sewer flowing half-full.

Solution

$$R = \frac{D}{4} = 0.375 \text{ m}$$

For a pipe flowing half full, the wall shear stress is given by Equation 7.21:

$$\tau_0 = \rho g R S_O = 1000 \times 9.81 \times 0.375 \times 0.0015 = 5.5 \ \text{N/m}^2$$

Shear velocity is given by Equation 16.2:

$$U^* = \sqrt{\frac{5.5}{1000}} = 0.074 \ \text{m/s}$$

So, as

$$\frac{W_s}{U^*} = \frac{750/3600}{0.074} = 2.8 > 2$$

transport will be bed-load.

A large body of knowledge has been built up, including many different predictive equations, for sediment transport, based particularly on loose-boundary channels including rivers (see, e.g., Raudkivi, 1998). Equations are available, normally in terms of the volumetric sediment carrying capacity of the flow, for both suspended and bed-load transport.

Although these equations are useful in highlighting important principles, they should not be used uncritically for sewer design and analysis. Conditions in pipes are different than those in rivers: pipes have rigid boundaries, significantly different and well-defined cross sections, and they transport different material. However, there have been a number of studies particularly focusing on pipes and sewers, both in the laboratory and in the field, and these have been comprehensively reviewed and compared by Ackers et al. (1996a). Recommended transport equations are given in Section 16.6.

16.4.3 Deposition

If the flow velocity or turbulence level decreases, there will be a net reduction in the amount of sediment held in suspension. The material accumulated at the bed may continue to be transported as a stream of particles without deposition. However, below a certain limit, the sediment will form a deposited bed, with transport occurring only in the top layer (the *limit of deposition*). If the flow velocity is further reduced, sediment transport will cease completely (the *threshold of movement*). The flow velocities at which deposition occurs tend to be lower than those required to entrain sediment particles.

16.4.4 Sediment beds and bed-load transport

If an initially clean sewer flowing part-full is subjected to a sediment-laden flow transported under bed-load, but conditions are not sufficient to prevent deposition, a sediment bed will develop. It will increase the bed resistance, causing the depth of flow to increase and the velocity to decrease.

Intuitively, it might be assumed that a reduction in velocity would cause a reduction in the sediment-transporting capacity of the flow, leading to further deposition and possibly blockage. In fact, laboratory evidence has shown (May, 1993) that the presence of the deposited bed actually allows the flow to acquire a *greater* capacity for transporting sediment as bed-load. This is because the mechanism of sediment transport is related to the width of the deposited bed, which can, of course, be much greater than the narrow stream of sediment that is present along the bed of the pipe at the limit of deposition. The effect more than compensates for the reduction in velocity caused by the roughness of the bed. Ultimately, the increased deposited bed depth (and width) and the associated increased sediment transport capacity may balance with the incoming sediment load and prevent further deposition. Thus, in principle, a small amount of deposition may be advantageous in terms of sediment mobility.

16.5 CHARACTERISTICS

16.5.1 Deposited sediment

The characteristics of sewer sediment deposits vary widely according to the sewer type (foul, storm, or combined), the geographical location, the nature of the catchment, local sewer operation practices, history, and customs. Crabtree (1989) proposed that the origin, nature, and location of deposits found within UK sewerage systems could be used to classify sediment under five categories A–E (see Figure 16.3). The characteristics of these deposits are described in the following and are summarised in Table 16.4.

16.5.1.1 Physical characteristics

Type A material is the coarsest material, found typically on sewer inverts. These deposits have a bulk density of up to 1800 kg/m³, organic content of 7%, with some 6% of particles typically smaller than 63 μm. The finer material (type C) is typically 50% organic, with a bulk density of approximately 1200 kg/m³ and some 45% of the particles are smaller than 63 μm. Type E is the finest material, although there is no definite boundary between any of the types A, C, or E. This is perhaps to be expected, as the sediment actually deposited depends on the material available for transport and the flow conditions in specific locations.

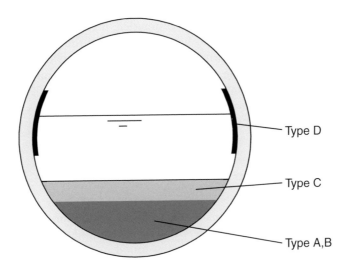

Figure 16.3 Typical sediment deposits in a sewer.

Table 16.4 Physical and chemical characteristics of sewer sediment type classes

Description	Sediment type				
	A	B	C	D	E
Description	Coarse, loose granular material	As A but concreted with fat, tars, etc.	Fine-grained deposits	Organic slimes and biofilms	Fine-grained deposits
Location	Pipe inverts	As A	Quiescent zones, alone or above A material	Pipe wall around mean flow line	In CSO storage tanks
Saturated bulk density (kg/m³)	1720	N/A	1170	1210	1460
Total solids (%)	73.4	N/A	27.0	25.8	48.0
COD (g/kg)*	16.9	N/A	20.5	49.8	23.0
BOD$_5$ (g/kg)*	3.1	N/A	5.4	26.6	6.2
NH$_4^+$–N (g/kg)*	0.1	N/A	0.1	0.1	0.1
Organic content (%)	7.0	N/A	50.0	61.0	22.0
FOG (%)	0.9	N/A	5.0	42.0	1.5

Source: Adapted from Crabtree, R.W. 1989. *Journal of the Institution of Water and Environmental Management*, 3, 569–578.

* Grams pollutant per kilogram of wet bulk sediment.

16.5.1.2 Chemical characteristics

Table 16.4 summarises the mean chemical characteristics of the deposited sediments. A high degree of variability is observed in practice (e.g., coefficient of variation 23%–125%). On a mass-for-mass basis, wall slimes are the most polluting in terms of oxygen demand (49.8 gCOD/kg wet sediment). There is a broad decrease in strength among types, in the order D, E, C, and A, with type A material having mean COD levels of just 16.9 g/kg.

However, this does not show the full significance of the relative polluting potential of each type of deposit. This is illustrated by Example 16.2.

Results from Example 16.2 show that although type D material is of higher unit strength, where small quantities are found in practice, it tends to be relatively insignificant. Type A material clearly shows up as having the bulk of the pollution potential (79% in this case) because of its large volume. The actual value varies depending on location. It should be realised, too, that the total polluting load would only be released under extreme storm flow conditions that erode all the sediment deposits. More routine storm events will probably erode only a fraction of the type A deposits. It is also interesting to note that, in this case, the wastewater itself only represents 10% of the potential pollutant load.

16.5.1.3 Significance of deposits

Type A and B deposits are normally associated with loss of sewer capacity, and type A deposits are the most significant source of pollutants. The nature of the sediment appears to vary between areas, with large organic deposits being found nearer the heads of networks and with more granular material (type A) being found in trunk sewers. Larger interceptor sewers typically have more type C material intermixed with the type A (Ashley and Crabtree, 1992). Pipe wall slimes/biofilms (type D) are important because they are very

Table 16.5 Data for Example 16.2

Type	Depth (mm)	Volume (m³/m)	Bulk density (kg/m³)	BOD₅ (g/kg)	Unit BOD₅ (g/m length)	Percent (%) load
A	0–300	0.252	1720	3.1	1344	79
C	300–320	0.024	1170	5.4	152	9
Wastewater	320–670	0.488	1000	0.35	171	10
D	50 × 10 × 2	0.001	1210	26.6	32	2
Total					1699	100

common, highly concentrated, easily eroded, and typically increase hydraulic roughness (Fouada et al., 2014).

EXAMPLE 16.2

A 1500 mm diameter sewer has a sediment bed (type A) of average thickness 300 mm. Above this is deposited a 20 mm type C layer and above that flows 350 mm of wastewater (BOD₅ = 350 mg/L). Along the walls of the sewer at the waterline are two 50 mm × 10 mm thick biofilm deposits (type D). Calculate the relative pollutant load associated with each element in the pipe.

Solution

For each of the sediment types, the cross-sectional area can be calculated from the pipe geometrical properties (Chapter 7) to give volume per unit length. Combining this information with the bulk density of the sediment and its pollutant strength enables unit pollutional load to be estimated as shown in Table 16.5.

16.5.2 Mobile sediment

16.5.2.1 Suspension

The predominant particles in suspension during both dry and wet weather flows are approximately 40 μm in size, and primarily attributed to sanitary solids. Most of the suspended material in combined sewer flow (≈90%) is organic and is biochemically active with the capacity to absorb pollutants. Settling velocities are usually less than 10 mm/s (Crabtree et al., 1991).

16.5.2.2 Near-bed

Under dry weather flow conditions, sediment particles can form a highly concentrated, mobile layer, "dense undercurrent," or "muddy layer" just above the bed (see Figure 16.4). Solids in this region are relatively large (>0.5 mm), organic (>90%) particles and are believed to be trapped in a matrix of suspended flow (Verbanck, 1995). Concentrations of solids of up to 3500 mg/L have been measured, and the corresponding biochemical pollutants are also particularly concentrated (Ashley and Crabtree, 1992). According to Ashley et al. (1994), typically, 12% of total solids are conveyed in the material moving near the bed. The rapid entrainment of near-bed solids is thought to make a significant contribution to first foul flushes (Chapter 12). This area is still the subject of ongoing research (Larrarte et al., 2017).

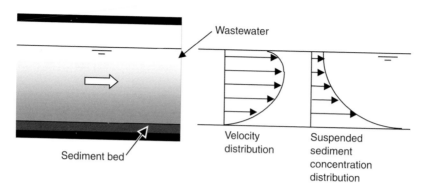

Figure 16.4 Velocity and suspended sediment distributions in dry weather.

16.5.2.3 Granular bed-load

Granular particles (2–10 mm) are transported as "pure" bed-load only in steeper sewers (>2%). In flatter sections (<0.1%), little granular material is observed in motion, presumably because it is deposited.

16.5.2.4 Particle size

In Chapter 5, it was shown how smaller particle sizes tend to be associated with a greater proportion of pollutant than might be expected from their mass. The same is true of mobile sewer sediment as shown in Table 16.6.

16.6 SELF-CLEANSING DESIGN

16.6.1 Velocity based

The need for sewers to be designed to carry sediment has been recognised for many years. Conventionally, this has been done by specifying a minimum "self-cleansing" flow velocity that should be achieved at a particular depth of flow or with a particular frequency of occurrence (see Chapters 9 and 10). Although this approach has apparently been successful in many cases, a single value of minimum velocity, unrelated to the characteristics and concentration of the sediment or to other aspects of the hydraulic behaviour of the sewer, does

Table 16.6 Percentage of total pollution load associated with different particle size fractions

Pollutant	Particle size fraction (μm)		
	<50	50–250	>250
BOD	52	20	28
COD	68	4	28
TKN	16	58	26
Hydrocarbons	69	4	27
Lead	53	34	13

Source: Bertrand-Krajewski, J.-L., Briat, P.M., and Scrivener, O. 1993. *Journal of Hydraulic Research*, 31(4), 435–460.

not properly represent the ability of sewer flows to transport sediment. In particular, it is known that a higher flow velocity is needed to transport a given concentration of sediment in large sewers than in small sewers.

It is also important to appreciate that conditions in gravity sewers are extremely variable. Flow rates and the sediment entering a system can vary considerably with time and location, so a sewer designed to be self-cleansing in normal conditions is still likely to suffer sediment deposition during periods of low flow and/or high sediment load.

16.6.2 Tractive force based

Current U.S. practice (Bizier, 2007) recommends a *tractive force* method for the self-cleansing design of gravity sanitary (foul) sewers. The term is somewhat misleading because what is being referred to is actually a stress rather than a force, with units of Newtons per square metre (N/m^2). To check the impact of specifying a minimum shear stress rather than velocity, the variation of shear stress with depth could be added to Figure 7.8 in Chapter 7, which shows the geometric and hydraulic elements for part-full pipe flow. From Equation 7.21, we can infer that $\tau_o / \tau_{of} \equiv R / R_f$, a plot of which is already shown. As can be seen, the lines for R/R_f and v/v_f are similar, so it does not matter much whether a self-cleansing velocity or a critical shear stress (tractive force) is specified, provided an appropriate value is used. Merritt (2009) suggests using a critical shear stress of 0.87 N/m^2 (associated with a 1 mm diameter particle) for separate foul sewers. Application of this method is illustrated in Example 16.3.

EXAMPLE 16.3

Using the tractive force method, calculate the gradient of a 150 mm foul sewer required to reach a critical shear stress of 1 N/m^2 when the pipe runs half full.

Solution

For a pipe flowing half full, $R = D/4 = 0.0375$ m, and wall shear stress is given by Equation 7.21:

$$1.0 = \rho g R S_O = 1000 \times 9.81 \times 0.0375 \times S_O$$
$$S_O = 0.0269 \approx 1:370$$

16.6.3 The CIRIA method

The CIRIA method (Ackers et al., 1996a; Butler et al., 1996a, 1996b) was developed in an attempt to relate minimum velocity to all the factors that affect it most, namely, pipe size and roughness, proportional flow depth, sediment size and specific gravity, degree of cohesion between the particles, sediment load or concentration, and the presence of a deposited bed.

16.6.3.1 Self-cleansing

The method proposes the following definition of self-cleansing:

An efficient self-cleansing sewer is one having a sediment-transporting capacity that is sufficient to maintain a balance between the amounts of deposition and erosion, with a time-averaged depth of sediment deposit that minimises the combined costs of construction, operation and maintenance.

(Ackers et al., 1996a)

The important aspect of this definition is that it does not necessarily require sewers to be designed to operate completely free of sediment deposits if more economical overall solutions can be achieved by allowing some deposition to occur. This is a viable alternative because, as described in Section 16.4.4, the presence of the deposited bed can significantly increase the sediment transporting capacity of the pipe, despite the adverse effect of the deposits on the geometry and hydraulic roughness.

16.6.3.2 Movement criteria

The results of field and laboratory research indicate that, to achieve self-cleansing performance, sewers should be designed to

1. Transport a minimum concentration of fine-grain particles in suspension
2. Transport coarser granular material as bed-load
3. Erode cohesive particles from a deposited bed

A thorough comparison of the application of mainly laboratory-based equations to typical sewer design situations was carried out, based on these three sediment movement criteria.

16.6.3.2.1 Suspended load transport

Macke's (1982) equation was found to provide a reasonable fit to laboratory data for suspended particles moving at the limit of deposition (no deposition) and was recommended for use as the normal design method:

$$C_v = \frac{\lambda^3 v_L^5}{30.4(S_G - 1)\ W_s^{1.5} A} \tag{16.4}$$

where C_v is the volumetric sediment concentration (discharge rate of sediment/discharge rate of water), v_L is the limiting flow velocity without deposition (m/s), S_G is the sediment specific gravity (−), and A is the flow cross-sectional area (m²).

The equation is valid beyond $\tau = 1.07$ N/m².

Where sediment is to be transported over a sediment bed, the 1973 Ackers-White equation, originally developed for alluvial channels, is advocated. This has been applied to sewer design by Ackers (1991) with a reduced effective sediment bed width as follows:

$$C_v = J\left(W_e \frac{R}{A}\right)^\alpha \left(\frac{d'}{R}\right)^\beta \lambda_c^\gamma \left(\frac{v}{\sqrt{g(S_G - 1)R}} - K\lambda_c^\delta \left(\frac{d'}{R}\right)^\varepsilon\right)^m \tag{16.5}$$

where W_e is the effective width of the sediment bed (m), d' is sediment particle size (m), and λ_c is the friction factor for the pipe and sediment bed (−).

The various empirical coefficients (J, K, M, α, β, γ, δ, and ε) depend on the dimensionless grain size D_{gr} and the mobility parameter at the threshold of movement A_{gr} (defined in more detail by Ackers et al., 1996a).

16.6.3.2.2 Bed-load transport

For bed-load transport at the limit of deposition, no existing equation gave a good fit over a full range of the data, and so a new equation was derived:

$$C_v = 3.03 \times 10^{-2} \left(\frac{D^2}{A}\right) \left(\frac{d'}{D}\right)^{0.6} \left(1 - \frac{v_t}{v_L}\right)^4 \left(\frac{v_L^2}{g(S_G - 1)D}\right)^{1.5} \tag{16.6}$$

$$v_t = \sqrt{0.125 g(S_G - 1)d'} \left(\frac{d}{d'}\right)^{0.47} \tag{16.7}$$

where v_t is the threshold velocity required to initiate movement and d is the depth of flow.

A similar procedure was followed to evaluate equations for bed-load transport in circular pipes with deposited beds. The best fit to the experimental data was provided by a slightly modified version of the method due to May (1993):

$$C_v = \eta \left(\frac{W_b}{D}\right) \left(\frac{D^2}{A}\right) \left(\frac{\theta \lambda_g v^2}{8g(S_G - 1)D}\right) \tag{16.8}$$

where W_b is the sediment bed width (m), λ_g is the friction factor corresponding to the grain shear factor (–), θ is the transition coefficient for particle Reynolds number (–), and η is the sediment transport parameter (–).

Evaluation of this equation is more complicated because of the need to estimate the flow resistance produced by the deposited bed. Full details of the recommended method are given by Ackers et al. (1996a).

Pipes designed in accordance with these equations should allow transport of sediment as bed-load at a rate sufficient to avoid deposition or limit it to a specified depth.

16.6.3.2.3 Bed erosion

The effect of cohesion on the shear stress at the threshold of movement is specifically allowed for in criterion 3. Based on field evidence and experimental investigations into erosion shear stresses, various relationships between pipe diameter and minimum full-bore velocity were identified, depending on the chosen bed shear stress and bed roughness. It was recommended that the design flow conditions needed to erode cohesive particles from a deposited bed should have a minimum value of shear stress of 2 N/m² on a flat bed with a Colebrook-White roughness value of $k_b = 1.2$ mm (based on 1 mm cohesive sediment particles). The required full-bore velocity (v_f) is given by a modification of Equation 7.21:

$$v_f = \sqrt{\frac{8\tau_b}{\rho \lambda_b}} \tag{16.9}$$

$$\lambda_b \approx \frac{1}{4\left[\log_{10}\left(\frac{k_b}{3.7D}\right)\right]^2} \tag{16.10}$$

where τ_b is the critical bed shear stress (N/m²) and λ_b is the friction factor corresponding to the sediment bed (–).

Table 16.7 Typical sewer sediment characteristics and applicability

Sediment class	Normal transport mode	Types of sewer applicable	Parameter	Sediment load		
				Low (L)	Medium (M)	High (H)
Sanitary solids	Suspension	Foul (F)	X (mg/L)	100	350	500
		Combined (C)	d50 (μm)	10	40	60
			S_G	1.01	1.4	1.6
Stormwater solids	Suspension	Foul (F)	X (mg/L)	50	350	1000
		Storm (S)	d50 (μm)	20	60	100
		Combined (C)	S_G	1.1	2.0	2.5
Grit	Bed-load	Storm (S)	X (mg/L)	10	50	200
		Combined (C)	d_{50} (μm)	300	750	750
			S_G	2.3	2.6	2.6

In many cases, the roughness of the deposited bed will be much higher resulting in bed shear stresses that are higher, making this approach conservative.

16.6.3.3 Design procedure

The CIRIA method proposes a detailed procedure in which design tables are worked up from first principles using the equations and information supplied. As an alternative, a simplified procedure is provided in which selected standard values of sediment characteristics and other parameters have been adopted, allowing the presentation of 10 simplified design tables. These cover foul, storm, and combined sewers; medium and high sediment loads (Table 16.7); and criteria based on either no deposition (LOD) or an allowable average deposition of up to 2% of the pipe diameter. Figure 16.5 gives the minimum design velocities for foul and storm sewers based on the design tables (Ackers et al., 1996b). Example 16.4 illustrates the method.

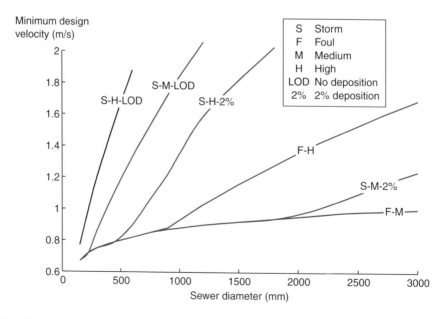

Figure 16.5 Minimum design velocities using simplified procedure.

EXAMPLE 16.4

A 2100 m long storm sewer is to be designed to carry a discharge of 1500 L/s in a location with a maximum available fall of 7 m. There are no specific sediment data available, but the area will be subject to considerable development in the next 10 years. Select an appropriate diameter and gradient circular sewer. Take Manning's n as 0.012.

Solution

Solve Manning's equation with $S0 = 1:300$, and assume the sewer will run full to obtain the required pipe diameter:

$$D = \left(\frac{1.5 \times 0.012 \times 4^{5/3}}{\pi \times (1/300)^{1/2}} \right)^2 = 1.000 \text{ m}$$

Resulting velocity, v, is as follows:

$$v = \frac{1.5 \times 4}{\pi \times 1^2} = 1.9 \text{ m/s}$$

Assume the sediment loading category is high (H) due to anticipated development work, and determine the required velocity to avoid any permanent deposition (LOD).

From curve S-H-LOD: The velocity required for this diameter is not shown, but it is clearly above 2 m/s, which is greater than that available. Within the available gradient of 1/300, the sewer cannot be designed to carry high sediment loading without deposition.

Therefore, allow up to 2% deposition, and determine the required velocity.

From curve S-H-2%: required velocity = 1.35 m/s which is <1.9 m/s and thus achievable.

Note: this velocity is in excess of typical minimum full-bore velocities (e.g., 1 m/s).

16.6.3.4 Limitations

Although the CIRIA method is a significant advance over conventional approaches, it should be noted that it does have limitations (Arthur et al., 1999). The key one is that it is based mainly on data from laboratory pipe-flume experiments using single-sized, granular sediment. This is clearly a simplification of the conditions found in combined sewers (in particular), as described in the rest of this chapter. Some verification with field data has been attempted (May et al., 1996). The benefits of using simplified rather than detailed sediment transport models for sewer sediment management have also been studied with the conclusion that a complementary approach using both methods should provide acceptable levels of uncertainty (Mannina, 2012).

PROBLEMS

16.1 Outline the various sources of sediment. Discuss ways in which sediment generation might be reduced.

16.2 Explain how sediment enters, moves through, and leaves an urban drainage system.

16.3 What are the main problems caused by sediment deposition?

16.4 How does deposited sediment affect the hydraulic characteristics of a sewer system?

16.5 Define and explain the three phases of sediment movement: entrainment, transport, and deposition.

16.6 Sewer sediments have been classified into five types (A–E). Compare and contrast the physical and chemical characteristics of each. What is their significance?

16.7   A 1.5 km long, 1.5 m diameter combined sewer suffers from sediment deposition along its whole length. Measurements reveal an average depth of 300 mm of type A sediment overlain by 20 mm of type C. A storm with a flow of 2.2 m³/s sustained for 30 minutes completely releases the pollutants bound by the sediment bed. Assuming the incoming stormwater has negligible pollution, estimate the average concentration of BOD_5 and COD released.   [567 mg/L, 2993 mg/L]

16.8   Explain the three main modes of sediment transport in sewers. Under what conditions would each mode be most important?

16.9   Calculate the equivalent flow velocity for the sewer designed in Example 16.4 using the tractive force method.   [1.5 m/s]

16.10 Explain the main elements of the CIRIA sewer design procedure (for self-cleansing). What are its advantages over conventional approaches? What are its weaknesses?

16.11 Calculate the minimum velocity needed under CIRIA criterion 3 ($\tau_b = 2$ N/m², $k_b = 1.2$ mm) to cleanse a 1000 mm diameter pipe running half-full. If the pipe has a sediment bed of roughness $k_b = 50$ mm, what will be the actual shear stress generated by this velocity?   [0.88 m/s, 6.9 N/m²]

16.12 Explore the influence and importance of three variables used in the CIRIA equations. What are the implications for sewer design?

KEY SOURCES

Ackers, J.C., Butler, D., and May, R.W.P. 1996a. *Design of Sewers to Control Sediment Problems*, CIRIA R141.

Bizier, P. (ed). 2007. *Gravity Sanitary Sewer Design and Construction*, ASCE Manual and Report No. 60, WEF Manual No. FD-5.

Ashley, R.M., Bertrand-Krajewski, J.-L., Hvitved-Jacobsen, T., and Verbanck, M. (eds). 2003. *Sewer Solids—State of the Art*, Scientific and Technical Report No. 14, IWA Publishers.

Ashley, R., Bertrand-Krajewski, J.-L., and Hvitved-Jacobsen, T. 2005. Sewer solids—20 years of investigation. *Water Science and Technology*, 52(3), 73–84.

Crabtree, R.W., Ashley, R.M., and Saul, A.J. 1991. *Review of Research into Sediments in Sewers and Ancillary Structures*. Report No. FR0205, Foundation for Water Research.

Fan, C.-Y. 2004. *Sewer Sediment and Control. A Management Practices Reference Guide*. U.S. Environmental Protection Agency Report EPA/600/R-04/059.

REFERENCES

Ackers, P. 1991. Sediment aspects of drainage and outfall design. *Proceedings of the International Conference on Environmental Hydraulics*, Hong Kong, 215–230.

Ackers, P. and White, W.R. 1973. Sediment transport: New approach and analysis. *American Society of Civil Engineers Journal of Hydraulic Engineering*, 99(HY11), 2041–2060.

Ackers, J.C., Butler, D., John, S., and May, R.W.P. 1996b. Self-cleansing sewer design: The CIRIA Procedure. *Proceedings of Seventh International Conference on Urban Storm Drainage*, Hannover, September, 875–880.

Arthur, S., Ashley, R., Tait, S., and Nalluri, C. 1999. Sediment transport in sewers—A step towards the design of sewers to control sediment problems. *Proceedings of Institution of Civil Engineers, Water, Maritime and Energy*, 136, March, 9–19.

Ashley, R.M. and Crabtree, R.W. 1992. Sediment origins, deposition and build-up in combined sewer systems. *Water Science and Technology*, 25(8), 1–12.

Ashley, R.M. and Verbanck, M.A. 1996. Mechanics of sewer sediment erosion and transport. *Journal of Hydraulics Research*, 34(6), 753–769.

Ashley, R.M., Arthur, S., Coghlan, B.P., and McGregor, I. 1994. Fluid sediment in combined sewers. *Water Science and Technology*, 29(1–2), 113–123.

Bertrand-Krajewski, J.-L., Briat, P.M., and Scrivener, O. 1993. Sewer sediment production and transport modelling: A literature review. *Journal of Hydraulic Research*, 31(4), 435–460.

Binnie and Partners and Hydraulics Research. 1987. *Sediment Movement in Combined Sewerage and Storm-Water Drainage Systems*, CIRIA PR1.

Butler, D., May, R.W.P., and Ackers, J.C. 1996a. Sediment transport in sewers—Part 1: Background. *Proceedings of Institution of Civil Engineers, Water, Maritime and Energy*, 118, June, 103–112.

Butler, D., May, R.W.P., and Ackers, J.C. 1996b. Sediment transport in sewers—Part 2: Design. *Proceedings of Institution of Civil Engineers, Water, Maritime and Energy*, 118, June, 113–120.

Crabtree, R.W. 1989. Sediments in sewers. *Journal of the Institution of Water and Environmental Management*, 3, 569–578.

Fouada, M., Kishka, A., and Fadela, A. 2014. The effects of scum on sewer flows. *Urban Water Journal*, 11(5), 405–413.

Larrarte, F., Hemmerle, N., Lebouc, L., and Riochet, B. 2017. Additional elements regarding the muddy layer in combined sewers. *Urban Water Journal*, 14(8), 862–867.

Macke, E. 1982. *About Sedimentation at Low Concentrations in Partly Filled Pipes*, Mitteilungen, Leichtweiss—Institut für Wasserbau der Technischen Universität Braunschweig, Heft 76.

Mannina, G., Schellart, A.N.A., Tait, S., and Viviani, G. 2012. Uncertainty in sewer sediment deposit modelling: Detailed vs simplified modelling approaches. *Physics and Chemistry of the Earth*, 42–44, 11–20.

May, R.W.P. 1993. *Sediment Transport in Pipes and Sewers with Deposited Beds*, Report SR 320, HR Wallingford.

May, R.W.P., Ackers, J.C., Butler, D., and John, S. 1996. Development of design methodology for self-cleansing sewers. *Water Science and Technology*, 33(9), 195–205.

Merritt, L. 2009. Tractive force design for sanitary sewer self-cleansing. *Journal of Environmental Engineering*, 135(12), 1338–1347.

Nalluri, C. and Alvarez, E.M. 1992. The influence of cohesion on sediment behaviour. *Water Science and Technology*, 25(8), 151–164.

Raudkivi, A.J. 1998. *Loose Boundary Hydraulics*, 4th edn, A.A. Balkema.

Verbanck, M.A. 1995. Capturing and releasing settleable solids—The significance of dense undercurrents in combined sewer flows. *Water Science and Technology*, 31(7), 85–93.

Chapter 17

Operation and maintenance

17.1 INTRODUCTION

Operation and maintenance (O&M) plays a key role in ensuring the sustainable performance of any urban drainage system. There is a temptation to assume that if there are no immediate problems, there are no O&M needs. However, drainage systems corrode, erode, clog, collapse, and ultimately deteriorate to the point of failure and beyond. So, *planned* O&M is essential for all systems.

This chapter provides a comprehensive overview of the subject beginning in Section 17.2 with an analysis of the drivers affecting the need for O&M and the strategies that can be deployed. Sewer location and inspection methods are described in Section 17.3 and the various cleaning techniques are explained in Section 17.4. An urban drainage system consists of more than just pipes, and maintenance of ancillary structures is covered in Section 17.5 with SuDS maintenance introduced in Section 17.6. The final section (Section 17.7) covers the vital topic of health and safety.

17.2 MAINTENANCE STRATEGIES

There are several reasons why maintenance of urban drainage systems is so important.

17.2.1 Public health

Maintenance of public health is paramount (see Chapter 1), and the continuing good functioning of the system can help to achieve it. In addition, the system should not cause nuisance (e.g., odour) or a health hazard (e.g., rodents) to either its users or its operators.

17.2.2 Asset management

All systems were costly to construct and would be even more costly to replace. High priority must, therefore, be given to maintaining the physical integrity of the assets.

17.2.3 Maintain hydraulic capacity

A primary function of maintenance is to preserve the as-built hydraulic capacity of the system. This will minimise the possibility of wastewater backing up into properties or widespread surface flooding. This can be done by cleaning and ensuring, as far as is practicable, that the system is watertight or has controlled water levels.

17.2.4 Minimise pollution

A key function of urban drainage systems is to minimise environmental pollution. Maintenance has a role in reducing the frequency of discharge of piped systems (especially combined sewers) into the environment, and in avoiding conditions in all drainage systems that cause excessive buildup of pollutants.

17.2.5 Minimise disruption

The water industry is judged by its regulators and customers on the efficiency with which it deals with operation and maintenance. Disruption to the general public should therefore be minimised.

Various degrees of sophistication can be built into maintenance strategies, but there are two main categories: reactive and planned (BS EN 752: 2008).

17.2.6 Reactive maintenance

In reactive maintenance, problems are dealt with on a corrective basis as they arise (i.e., after failure): the *firefighting* approach. This approach will always be required to a certain extent, as problems and emergencies are bound to occur from time to time in every urban drainage system. However, reactive maintenance cannot reduce the number of system failures. To achieve this, a planned approach is needed.

17.2.7 Planned maintenance

In planned maintenance, potential problems are dealt with prior to failure. Unlike reactive maintenance, planned maintenance is proactive and has the objective of reducing the frequency or risk of failure. Central to planned maintenance is a comprehensive monitoring programme and analysis of existing data.

Planned maintenance is not the same as routine maintenance (operations at standard intervals, regardless of need), but involves identifying elements that require maintenance and then determining the optimum frequency of attention.

17.2.8 Operational functions

The major O&M functions are

- Location and inspection
- Cleaning and blockage clearance of pipes
- Chemical dosing
- Rodent baiting
- Ancillary equipment maintenance
- SuDS maintenance
- Fabric rehabilitation

The first six of these functions are described in this chapter. The seventh topic of rehabilitation is considered in Chapter 18.

Maintenance of sewer networks holds some specific challenges, even when compared with other industries. These include

- Geographical size (e.g., dispersion and length of pipework)
- Location (e.g., in private land and under highways)
- Variability and frequency of issues (e.g., can occur with no clear cause and influenced by external factors such as rainfall, gully cleaning, and people's behaviors)
- Physical size of assets (e.g., access, non-man entry)
- Aggressive environment (e.g., hazardous gases)

17.2.9 Role of design

Full consideration should be given to the O&M of systems during the design process. In particular, design should attempt to minimise the degree of maintenance required. That which *is* required should be simplified and allowed for (e.g., safe access). This process can be facilitated by regular communication between design and operations staff, giving benefits to both parties. Whole life costs (TOTEX) can be minimised by determining the optimum balance of O&M costs (OPEX) with construction costs (CAPEX).

17.2.10 Predicting failure

There is a particular operational need to predict the likelihood and consequence of sewer blockage and collapse, as both lead to the same outcome – system failure. For example, in 2016 there were over 350,000 blockages in sewers with about 2% leading to internal flooding (Water UK, 2016). However, the frequency of occurrence and the severity of the consequence can vary significantly depending on the size and location of the asset that fails. Some means of predicting which locations have a higher probability of occurrence than others would be very useful. Typically, due to the paucity and quality of data at smaller diameters, prediction is based on pipe groups or classes rather than individual sewers. Predictions for larger sewers and critical assets are typically based on thorough inspections.

Savic et al. (2006) and Ugarelli et al. (2010) have shown how blockage rate can be related to key asset data. Blockage rate is inversely related to sewer diameter, gradient, and length and varies between internal condition grade (defined in Table 17.1), based on 10,300 blockage events in UK sewers. Collapse rate based on 780 recorded events is shown to be positively correlated with sewer cover depth. Still, care needs to be taken in generalising results due to the significant influence of the local context (Rodríguez et al., 2012).

Statistical models are often used to help estimate the level (and cost) of interventions across an asset base. Erskine et al. (2014) describe such an approach, to estimate the failure frequencies of a water company's assets using a Bayesian tuner. Bailey et al. (2015) present an alternative approach using decision trees to develop blockage prediction models and highlight key variables including property density, sewer length, and flow velocity.

17.3 SEWER LOCATION AND INSPECTION

Sewer location and inspection methods are routine tools of the maintenance engineer. Basic methods for locating sewers and manholes have been available for many years, but these have been improved and newer methods introduced. Inspection, in particular, has been revolutionised by the introduction of remote surveillance equipment. Now, detailed surveys can be carried out in previously inaccessible locations in a cost-effective way such as using laser scanning technology (Roberts and O'Brien, 2016).

Table 17.1 Recommended inspection frequencies for, and extent of, critical sewers

Internal condition grade[+]	Frequency (year)		Extent
	Category A[*]	Category B	%
5	n/a	n/a	2
4	n/a	5	8
3	3	15	13
2	5	20	17
1	10	20	60

Source: Adapted from Sewers and Water Mains Committee. 1991. *A Guide to Sewerage Operational Practices*, Water Services Association/Foundation for Water Research; and National Audit Office. 2004. *Out of Sight – Not Out of Mind. OFWAT and the Public Sewer Network in England and Wales.* Report by the Comptroller and Auditor General. HC 161 Session 2003–2004.

[+] SRM guidance (step 3 Risk Assessment) defines internal condition grades ranging from "collapse" (5) to "acceptable structural condition" (1).
[*] Category A critical sewers are those where collapse repair costs would be highest. Category B sewers, although less critical, would still have substantial collapse costs.

17.3.1 Applications

The main applications of sewer location and inspection are

- Periodic inspection to assess the condition of existing sewers, chambers, and ancillaries including penstocks and screens (planned maintenance)
- Crisis inspection to investigate emergency conditions or the cause of repeated problems along a particular sewer length (reactive maintenance)
- Inspection of workmanship and structural condition of new sewers before "adoption" – see Chapter 6 (quality control)
- Investigation to understand location and position of assets and connections for repair, replacement, or rehabilitation

17.3.2 Frequency

It is impossible to accurately estimate the rate of deterioration of a particular sewer length from a single survey. This is particularly true of insidious problems such as sulphide attack (see Section 17.6). The most reliable way to monitor a sewer's condition is to carry out a series of inspections at given intervals such as those recommended in Table 17.1. The level of inspection chosen reflects an attempt to balance the probability of failure with its consequences.

Repeat inspection data can be used as the basis for deterioration modelling (Micevski et al., 2002; Savic et al., 2006; Wirahadikusumah et al., 2001). Table 17.2 has been prepared based on repeat closed-circuit television (CCTV) data on a UK sewer network to establish the probability of deterioration from one condition grade to another. For example, there is a 77% probability that on subsequent inspection an initially grade 1 sewer will remain in that condition. There is a 23% probability that it will deteriorate to grade 2. The data showed no cases of a greater than one step "fall" in grade. Generally, it is most likely that a sewer at any condition grade will stay at that grade. The data in Table 17.2 assume there have been no interventions between intervening inspections.

Table 17.2 Internal condition grade transition probabilities

Internal condition grade at time t	Internal condition grade at time $t + \Delta t$				
	1	*2*	*3*	*4*	*5*
1	0.77	0.23	0	0	0
2	0	0.83	0.17	0	0
3	0	0	0.90	0.10	0
4	0	0	0	0.95	0.05
5	0	0	0	0	1

Source: Based on Savic, D. et al. 2006. *Proceedings of the Institution of Civil Engineers Water Management*, 159(2), 111–118.

17.3.3 Locational survey

The first steps in a maintenance plan may include checking the accuracy and completeness of existing records of the system, and then initiating a survey on the parts of the system where there is doubt. Before an inspection of any kind can take place, it is necessary to locate the manholes and thereby determine the route of all the sewers in the system under consideration.

Manhole location is usually straightforward, although a metal detector may be required if it is suspected that a cover has been buried. The position and level (cover, soffit, and invert) of each manhole can be determined using standard land surveying techniques. This procedure can now be substantially speeded up using GPS (global positioning satellite) technology, allowing positional data to be logged on-site in seconds.

Techniques available to determine the route of a sewer range from the simple to the sophisticated. Visual inspection of flow directions in manholes is sometimes sufficient, and if not, dye tracing can be carried out. Electronic tracing is also common (Figure 17.1). A probe or "sonde" that emits radio signals is pushed, rodded, jetted, or floated along the sewer and its progress tracked from the surface using a handheld receiver. Using this approach, sewers up to 15 m deep can be traced to a claimed accuracy of ± 10%. Interference from signals generated by other buried metallic assets can, however, cause problems.

Ground-penetrating radar is frequently used to trace the sewer and other buried asset locations. While it is not completely accurate and relies on user interpretation, it provides a non-intrusive alternative to trial holes.

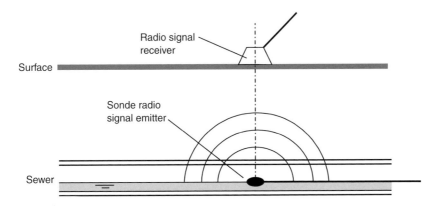

Figure 17.1 Electronic tracing of sewers.

Figure 17.2 CCTV image of a sewer in poor condition. (Courtesy of Telespec Ltd, Guildford.)

17.3.4 Closed-circuit television (CCTV)

In this method, a small TV camera incorporating a light source is propelled through the sewer, and the images are relayed to the surface for viewing and recording (see Figure 17.2).

CCTV is a popular choice for sewer inspection, not least because internal investigation of sewer systems can be carried out quickly, and with minimal disruption, avoiding lengthy shutdowns and unnecessary excavation (Rolfe and Butler, 1990). This method is particularly useful in environments that are too small or hazardous for people to enter. CCTV can also be used to locate and define the cause of a known condition or defect, and to help establish a plan of maintenance. Progress is relatively rapid with typical rates of 400–800 m/d.

The method is commonly used in pipes from 100 to 1500+ mm in diameter, but it is less effective in larger pipes, due to increased lighting requirements and difficulty in achieving adequate resolution of detail with high subject-to-camera distances.

17.3.4.1 Propulsion

Generally, for sewers of <150 mm diameter, the camera is located on the end of a line and can be either pushed through or winched between manholes. For sewers of diameter greater than 150 mm, it is usual for the camera to be mounted on a self-propelled, remote-controlled tractor. The tractor speed varies according to the size of the wheels fitted and will be greater in larger sewers, with typical speeds ranging from 0.1 to 0.2 m/s.

17.3.4.2 Camera operation

The basic camera technology used in sewer inspection came originally from broadcast TV. Development has centred on gathering the maximum amount of information at reasonable cost in very difficult and dirty conditions. Most cameras are now of the charge coupling device (CCD), solid-state type, housed in strong waterproof cases. These have high sensitivity and can be used for surveying very large sewers (over 3 m diameter).

Figure 17.3 A pan-and-tilt in-sewer CCTV camera. (Courtesy of Telespec Ltd, Guildford.)

Lamps attached to the front or sides of the camera head provide lighting. Lenses of several focal lengths are available, including zooms, and focusing can either be pre-set or remotely controlled by the operator. The most useful view for a CCTV camera is usually axially along the sewer, but there are specific occasions (e.g., looking up a house connection or at specific problems) when a lateral view is preferable. Several techniques are available for achieving this, including a wide-angle lens, pan and tilt equipment, and an electronic "fish eye." A typical CCTV camera is illustrated in Figure 17.3.

17.3.5 Manual inspection

Manual inspection is only used in exceptional circumstances and only if inspection cannot be done another way (Rolfe and Butler, 1990). Clearly health and safety issues are paramount with this approach, and these are covered in Section 17.5.

If such inspection *is* necessary, a survey begins at one manhole and works along the length of the sewer in increments (typically of 1 m) to the next. Information can be gathered on mortar loss displaced/missing bricks, cracks, sewer shape, connections, silt/debris, etc. Paper-based methods have traditionally been used to record this information, but portable data loggers or handheld computers with appropriate software are now widely available.

Progress is relatively slow using this procedure (200–400 m/d); it is costly and dangerous. However, high quality information can be obtained.

17.3.6 Other techniques

17.3.6.1 Sonar

Before the advent of in-sewer sonar technology, it was only possible to inspect relatively empty sewers. In many cases, this required the effort and expense of overpumping. Sonar techniques can acquire an image of the profile of a liquid-filled sewer, without the need for a light source (Winney, 1989). The sonar head is controlled from a surface processor and scans through 360° in discrete increments providing information on deposited sediment depths, fat, oil, and grease (FOG) accumulation and pipe deformation (see Figure 17.4).

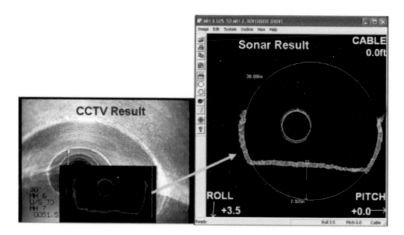

Figure 17.4 Sonar image of a half-submerged sewer with significant sediment bed. (Courtesy of AECOM Water.)

Data derived from the acoustic signal can be displayed on a colour monitor and recorded. Sonar can be combined with conventional methods in TISCIT (Totally Integrated Sonar and CCTV Inspection Technique) systems (Thomson and Grada, 2004).

17.3.6.2 Infrared

Another approach to in-sewer inspection is thermal imagery, in which an infrared camera is used to gather and focus emitted black-body radiation and convert it into a form visible to the human eye. Again, no external light source is necessary. This technique is rather limited in application but can be used to inspect for infiltration, based on the assumption that the wastewater and groundwater are at different temperatures.

17.3.6.3 Sewer profiling

The use of solid-state cameras in conjunction with optoelectronic light-measuring systems enables internal sewer profiles to be accurately mapped. This technique involves a specially configured light head that is attached to the front of the camera, with two light sources that cast a focused circle of light on the internal sewer surface. Any changes in shape can then be detected and accurately quantified using appropriate computer software. This technique is particularly appropriate in monitoring old brick sewers or deformed plastic pipes. Similar profiling techniques using laser beam systems are also available.

17.3.6.4 Alternative approaches

A number of the techniques described above can be combined (e.g., high-definition CCTV, laser profiling, and sonar) to enable inspection and recording without the need to cleanse sewers before inspection. Different-size inspection machines enable both small- and large-diameter trunk sewers to be inspected (e.g., by floating on the water surface). This approach is being used on major inspection programmes to identify issues and to resolve them before they become major incidents, such as seen in the Interceptor Condition Assessment Program in Fort Worth (Underground Construction, 2015).

Some investigation technologies are now autonomous, enabling "robots" to be inserted into the system and controlled remotely from the surface. This minimises the time needed for an inspection chamber to be open during the investigation, saving overall time, inspecting longer distances, and being safer (Longwell, 2010).

Where local and rapid inspections are necessary, such as identifying where a pipe may be partially blocked, acoustic sensors can detect anomalies. Examples include SL-Rat (Utility Engineering, Construction and Maintenance, 2016) and Sewerbatt (WWT, 2016).

Development in this area continues apace with newer devices based on combinations of ground-penetrating radar and electromagnetic techniques able to identify anomalies outside the pipe, such as voids and other services. Other techniques based on broadband electromagnetic systems (originally developed for pressure pipes) are able to detect pipe wall thickness, defects, and other anomalies (Thomson and Grada, 2004).

17.3.7 Data storage and management

CCTV images and video are recorded in digital format and transferred from operator to client by DVD or data transfer over the Internet. This enables multiple sewer lengths to be rapidly processed, facilitating the pinpointing and review of defects.

Sewer surveys generate a great deal of data that require careful and systematic handling. Software packages are now available to aid in data management, allowing defects, coded in a standardised way (WRc, 1993) at relevant locations in the pipe or manhole, to be stored in an easily accessible database. Such information can then be used to assess the structural condition of the system systematically. Most packages produce data files in a format compatible with simulation models. It is common for databases to be part of geographic information systems (GISs), which allow spatial information to be held and graphically displayed. Information on many services can be held on different "layers." Also, sewer record databases can now be linked with computer-aided drawing (CAD) packages to allow speedy production of drawings.

17.4 SEWER CLEANING TECHNIQUES

17.4.1 Objectives

Sewer cleaning is carried out

- Proactively, to remove sediment, FOG, and other blocked material in order to restore hydraulic capacity and limit pollutant accumulation
- Reactively, to deal with blockages or offensive odours
- To permit sewer inspection
- To aid sewer repair/renovation

17.4.2 Problems

Operators face a number of practical problems related to sewer operation that result in pipe blockage, and these are summarised in the following sections. Blockages are defined as full or partial restrictions within the sewer that affect normal operational performance. Some are caused by inappropriate or unintended inputs to the system and some by the design, construction, or condition of the system. The two causes are often interrelated. Most of the problems are more acute in smaller pipes but can escalate and affect larger sewers. The effect ranges from partial loss of capacity to complete stoppage, causing local backing up of wastewater and/or stormwater as a minimum and more widespread sewer flooding and pollution incidents in serious cases.

17.4.2.1 Gross solids (input)

Many cases of blockage are associated with gross solids (see Section 2.3.2), especially those with non-biodegradable content or made of non-woven fabrics. It is often difficult to establish if the solids have caused the blockage or if they are just associated with a problem generated by a system defect.

17.4.2.2 FOG/scale (input)

Solidified fats, oils, and grease (FOG) are often associated with non-domestic properties, particularly restaurants. High-temperature dishwashers often move the FOG from the premises, only for it to cool and solidify further downstream causing loss of hydraulic capacity. However, this reasoning may be oversimplified with the biodegradation of the oil content and water hardness contributing to the buildup of FOG (Williams et al., 2012). In exceptional cases, the buildup can be so severe that "fatbergs" occur causing major blockage (Figure 17.5). Wall scale or encrustation, sometimes found in hard water areas, causes similar problems. FOG can also encourage sediment buildup in all sewers, and cause problems at pumping stations, fine mesh screens, and wastewater treatment plants (Ducoste et al., 2009). Anecdotal evidence suggests some 50% of sewer cleaning activity is associated with FOG problems.

17.4.2.3 Sediment (input)

Sediment is defined as any settleable particulate material that may, under certain conditions, form bed deposits in sewers and associated hydraulic structures. It is normally associated with large, flat sewers. Sedimentation rarely completely chokes the pipe, but it can still have a significant impact on capacity. Chapter 16 discusses the issues associated with sediment deposition in further detail.

Figure 17.5 Brick sewer fatberg. (Courtesy of Andy Brierley.)

17.4.2.4 Defects (system)

System defects are often the catalyst for blockage formation. These can range from poor design, such as severe changes of sewer direction, through to poor construction practice (see the following sections). Also included in this category are defects that have arisen over the course of time such as displaced joints, pipe material (e.g., pitch fibre), and localised adverse gradients.

17.4.2.5 Traps (system)

Traps on smaller upstream sewers can be particularly problematic. These were designed to prevent odour coming back up the system and are typically found on systems serving older properties. However, today they often run dry causing the very odours they were designed to eliminate, and they can also become blocked. It is common for the traps to be removed where they are found to be problematic.

17.4.2.6 Intruding laterals (system)

Intruding laterals or other connections are common defects resulting from poor construction practice (see Section 18.3.4). The intrusion reduces the cross-sectional area of the pipe and can also initiate further blockage.

17.4.2.7 Tree roots (system)

Sewers are susceptible to intrusion of tree roots, which seek out moist conditions. The roots themselves are a nuisance, both in retarding the flow and causing similar problems to intruding laterals.

A number of cleaning techniques and methods are in use, depending particularly on location and severity, including rodding, winching, jetting, flushing, and hand excavation. A combination of more than one method may well be used in any particular locality.

17.4.3 Cleaning approaches

17.4.3.1 Rodding or boring

Rodding or boring is primarily a manual procedure, in which short flexible rods are screwed together and then inserted into the blocked sewer. The principal action is the physical contact of the tools, although the compression of air by plungers and dense brushes can contribute. Other approaches include the use of semi-rigid, coiled glass-reinforced plastic rods supplied in continuous lengths on a reel. The procedure can be mechanised.

This technique is limited to small-diameter pipes (≤ 225 mm) at shallow depths (≤ 2.0 m) and must be close to the access point (≤ 20 m). It is particularly well suited to dislodging blockages and roots. Care must be taken not to damage the existing pipework, especially on more brittle pipe materials.

17.4.3.2 Winching or dragging

Winching or dragging is a technique involving the use of purpose-shaped buckets that are dragged through the sewer collecting sediment, which is emptied out at a manhole. Although the winch can be manually operated, power-driven devices are normally used.

The procedure is capable of removing most materials, even in large pipes, but is most effective in sewers up to 900 mm diameter, up to 50% silted. Care has to be taken that damage is not caused to the sewer fabric.

17.4.3.3 Jetting

Jetting is a widely used technique that relies on the ability of an applied high-pressure (100–350 bar) stream of water to dislodge material from sewer inverts and walls and transport it down the sewer for subsequent removal. Water under these high pressures is fed through a hose to a nozzle containing a rosette of jets sited in such a way that the majority of flow is ejected in the opposite direction to the flow in the hose. The jets propel the hose through the sewer, eroding the settled deposits in the process. A range of nozzles is available to cope with specific situations. Modern combination units incorporate vacuum or air-displacement lifting equipment to remove the material, as well as to dislodge it without the need for man entry.

Jetting is a versatile and efficient procedure for removing a wide range of materials and is widely used in practice. Concern over the possibility of damaging pipes during the jetting process has resulted in the publication of the *Sewer Jetting Code of Practice* (WRc, 2005).

17.4.3.4 Flushing

Flushing is a technique in which short-duration waves of liquid are introduced or created so as to scour the sediment into suspension and, hence, transport it downstream. Waves may be induced by

- Automatic siphons at the heads of sewers
- Dam and release using blow boards/gates
- Hydrant and hose
- Mobile water tanker

Of these, the first is rarely used, and the second is used in bespoke circumstances when flushing is particularly required. All merely move the dislodged material downstream and do not remove it from the system.

17.4.3.5 Hand excavation

Historically, large-diameter sewers were cleared by manually digging out deposited material. Labourers entered the sewer and shovelled sediment into skips that were transported to the surface for emptying. The method is limited in application to larger-size pipes (> 900 mm) and has significant health and safety implications. It is now used only in exceptional circumstances.

17.4.3.6 Invert traps

These are not widely used, but they can significantly assist cleaning in problem locations by creating a preferred location to remove silt and sediment. Invert grit traps are intended to intercept sediments travelling as bed-load in combined sewers. Invert traps in existence are typically large rectangular chambers, which although effective at trapping sediments, also collect other near-bed and suspended solids. Performance of these chambers can be improved by the introduction of a cover with an open slot across the width of the invert designed to collect heavier, inorganic sediment only (Buxton et al., 2001).

Alternatives to traps include proprietary treatment systems, such as hydrodynamic and vortex separators, that help to capture silts, sediments, and other pollutants (Woods-Ballard et al., 2016). These are especially important for SuDS and particularly where geo-cellular structures are used.

17.4.3.7 Gully pots

Gully pots are provided to limit surface sediments entering the urban drainage system. Cleaning is routinely carried out by vacuum lifting equipment on mobile tankers. Cleaning frequency varies from place to place but is normally carried out once or twice a year. The efficiency of cleaning varies but is around 70%. Pots not only trap and retain the sediment for which they are provided, but also other pollutants including oil. Problems arise during dry periods when the retained sediment is degraded anaerobically, allowing NH_4 and COD to build up in the retained liquor. At the next storm event, the liquor and fine and dissolved solids will be mixed and entrained into the flow, adding significantly to the pollutant load (Langeveld et al., 2016).

17.4.4 Comparison of cleaning methods

No one method of cleaning is superior to the others on all occasions; each has its advantages and disadvantages. These are summarised in Table 17.3.

17.4.5 FOG control

A range of approaches is available to control FOG in sewers, including at-source methods, recycling, chemical treatment, and bio-augmentation.

Table 17.3 Relative performance of sewer cleaning techniques

Topic	Rodding	Winching	Jetting	Flushing
Sewer size (mm):				
<375	Good	Fair	Good	Good
450–900	Poor	Good	Fair	Fair
Maximum cleansing distance (m)	25	100	100	50
Number of manholes required	1	2	1	1
Dislodging materials:				
Invert	Fair	Good	Good	Good
Walls	Fair	Fair	Good	Poor
Joints	Fair	Fair	Good	Poor
Materials encountered:				
Silt	Fair	Fair	Good	Good
Sand/gravel	Poor	Good	Good	Good
Rocks	Poor	Good	Fair	Poor
Grease	Fair	Fair	Fair	Poor
Roots	Good	Good	Fair	Poor
Material removed?	No	Yes	Yes	No
Damage potential	Low	Medium	High	Low
Flooding potential?	No	No	No	Yes

Source: Adapted from Lester, J. and Gale, J.C. 1979. *The Public Health Engineer,* 7(3), July, 121–127.

17.4.5.1 At source

The most effective strategy to manage FOG must surely be to employ measures to reduce its inappropriate disposal or capture it before entry into sewer systems. Soft measures include user education (Mattsson et al., 2015), published good practice guides, and provision of authoritative advice. Gelder and Grist (2015) report successes in such approaches with the most substantial reductions in blockages occurring during and after intensive and targeted approaches. Hard measures include the installation and proper maintenance of grease traps and separators. These are typically fitted online, immediately after the sink outlet, and are designed to achieve physical separation of the FOG. Such approaches may be targeted at commercial and domestic consumers.

17.4.5.2 Recycling

A more sustainable option is to collect and recycle the FOG for beneficial incorporation into animal feeds and into soaps and lubricants for industrial use, and for biodiesel production. Recycling options are now being made available to domestic customers as part of kerbside collection arrangements (e.g., http://www.oadby-wigston.gov.uk/pages/household_waste_collection). Collected oils are converted into various fuel sources, such as biodiesel for vehicles.

17.4.5.3 Chemical treatment

A variety of chemical compounds can be used to treat FOG-contaminated wastewater, either before discharge to the sewer or by being introduced directly. Coagulants such as aluminium sulphate, ferric chloride, and lime can be used to dissipate surface charges and break the fat emulsion. Flocculants can be used separately or in combination to enhance settlement.

Many surfactant-based products are available, such as soaps, detergents, and degreasers, that when mixed with FOG-containing water surround the FOG molecules and facilitate emulsification, particularly when water temperature is high.

Caustics and solvents may be used to "cut" FOG. These products essentially raise the water pH and thereby increase the solubility of non-polar substances (grease) in water. However, chemicals such as these are corrosive and can potentially damage the sewer fabric.

Enzymes such as lipase degrade triglycerides, cleaving the fatty acids off the glycerol backbone of fats, thus increasing the solubility of greases. Gary and Sneddon (1999) investigated the effectiveness of enzymes on samples obtained from restaurant grease traps. They found that enzymes without surfactants performed better than those with surfactants. The effectiveness of any particular enzyme is dependent on attributes such as type of grease, its pH, and the grease trap influent flow rate.

A significant drawback of all of these approaches is that they typically only offer a temporary solution in terms of clearing sewers, because the FOG molecules are not completely broken down and can reappear and accumulate downstream where conditions return to "normal."

17.4.5.4 Bio-augmentation

These technologies consist of introducing quantities of naturally occurring or engineered micro-organisms (mainly bacteria), with or without supporting media, to break down the FOG molecules. Bio-augmentation products can be broadly classified into two groups, based on the type of micro-organism used: spore-forming or vegetative bacteria. Drinkwater et al. (2015) outline a protocol for the development of biological dosing into the sewer and pumping stations. Laboratory tests can indicate the effectiveness of compounds in a controlled environment, but the success is very dependent on the on-site conditions. Where manufacturers and water company operatives work together, the results of dosing have been very positive. Bio-augmentation has also been trialled for gully pot sediments but could not be recommended as a replacement of or addition to mechanical cleaning (Scott, 2012).

17.5 ANCILLARY AND NETWORK EQUIPMENT MAINTENANCE

Many sewer networks contain some form of equipment. These may include

- Pumps with valves plus air valves and washouts on rising mains
- Dosing equipment to deal with hydrogen sulphide in the network
- Combined sewer overflows with screens (static and mechanical) and automatic/remote cleaning
- Passive or dynamic flow controls, such as automated penstocks or auto-opening and auto-closing sluices
- Non-return valves may be present to prevent flooding

Maintenance is required to ensure the performance is not compromised, to maximise the asset life, and to avoid sudden and catastrophic failure. Each piece of equipment has a bespoke maintenance regime, including specific tasks and frequencies. These often link to supplier guarantees relating to performance. Sensors and monitors can provide critical information to support the maintenance regime. For example, at pumping stations, condition-based monitoring (e.g., vibration) can help to evaluate the pump performance and whether there is any deterioration.

17.6 SuDS

Table 17.4 summarises the main types of maintenance needed for various SuDS components. These consist in particular of sediment management and vegetation management. More detailed advice on individual SuDS devices is given in Chapter 21, and further general advice can be found in *The SuDS Manual* (Woods-Ballard et al., 2015).

Table 17.4 Common SuDS maintenance requirements

	Maintenance needs					
Practice	Forebay cleanout and disposal	Pruning	Removing/ mowing vegetation	Inspect outlet structure	Unclog surface	Sediment removal and disposal
Retention/detention pond	X		X	X		X
Constructed wetland	X		X	X		
Bioretention	(X)	X		X	X	X
Green roof		X	X	X		
Infiltration device					X	X
Pervious pavement					X	X
Filter strip and swale			X			
Rainwater harvesting	X					

Source: Adapted from Blecken, G.T. et al. 2017. *Urban Water Journal*, 14(3), 278–290.

17.7 HEALTH AND SAFETY

It is important to appreciate the health and safety hazards involved with sewer operation. Many of the practices to be carried out are covered by the Health and Safety at Work Act 1974. In addition, the sewer environment is classified as a "confined space" under the Confined Spaces Regulations (HSE, 1997). Examples are sewers, manholes, storage tanks, and pump sumps. Specific safety advice is available for sewage workers (HSE, 2003; ICE, 1969). Clients, designers, and contractors also have specific responsibilities under the CDM, Construction (Design and Management) Regulations (www.hse.gov.uk/construction/cdm/2015/index.htm) that relate to safe operation of assets once built or refurbished.

17.7.1 Atmospheric hazards

Atmospheric hazards are probably the most dangerous: explosive or flammable gases may develop at any time. Anaerobic biological decomposition of wastewater yields methane, which is lighter than air. Petrol vapour is heavier than air and tends to form pockets at the invert of the sewer. Factories should hold trade discharge licences and report the disposal of dangerous chemicals, but they may not report accidental spills or deliberately negligent acts.

The most commonly occurring toxic gas within the sewer is hydrogen sulphide. It is a flammable and very poisonous gas that has a distinctive smell. It is particularly dangerous to workers in sewers because the ability of a person to smell the gas diminishes with exposure and as the concentration increases. The amount of breathable oxygen in a sewer can be reduced or even eliminated by displacement by a heavier gas. If there is no breathable oxygen in a sewer, the life expectancy of a person entering is approximately 3 minutes.

To provide adequate ventilation, the access manhole cover and those upstream and downstream need to be lifted well before an inspection. Those entering the sewer must use a gas detector. Because of the short life expectancy in unfavourable conditions, the current best practice is to ensure that gangs have the appropriate personal protective equipment (PPE) to self-rescue, rather than relying on fire brigade assistance. For certain activities, it can be appropriate to have rescue teams on standby. If a dangerous atmosphere exists within a man-entry sewer, the inspection can be carried out using breathing apparatus.

17.7.2 Physical injury

Physical injury is an ever-present hazard in the sewer environment. It can result from falls down manholes or in the sewer, and the dropping, throwing, or misuse of equipment. The risk of drowning must not be underestimated, whether in the residual flow within the sewer or due to a sudden rise in water level after rainfall. This can be avoided by maintaining close contact with the surface at all times.

Acids and other chemicals may find their way into the sewer system and could cause burns, so appropriate PPE should be worn at all times.

17.7.3 Infectious diseases

Infection from tetanus, hepatitis B, or leptospirosis (Weil's disease) is a potential risk that should be planned for by ensuring workers are under medical supervision and inoculated as appropriate. Discarded needles found in sewers should be avoided. Rodents (such as rats) and insects can also be a health hazard. Rodent control is covered in Section 17.7.5.

17.7.4 Safety equipment

If the sewer is to be entered for inspection or any other maintenance work, suitable PPE must be worn. This includes waterproof clothing, heavy waders, gloves, safety harnesses, eye protection, and hard hats, as well as radios, safety lamps, and, if necessary, breathing equipment (escape or full). Portable gas detectors should always be used to test for toxic gases. Two-way radios have been used to keep in contact with members of the team on the surface but may be ineffective in certain conditions; visual and vocal contact with a surface member is preferable. The surface team must have access to regular weather forecasts and knowledge of trade discharges.

17.7.5 Rodent control

The key concern over rat infestation in sewer systems is their potential link in disease transmission to humans (Keeling and Gilligan, 2000). The extent of the problem is difficult to judge, but there is objective evidence to suggest that sewer-rodent populations have declined in recent years (Channon et al., 2000). As with other operational functions described in this chapter, control is either (most commonly) reactive or proactive. The reactive approach is to respond to complaints or sightings by baiting in the locality. A proactive approach consists of planned random or semi-random baiting across the network. A protocol including elements of both approaches was developed by Ashley et al. (2003).

17.8 GAS GENERATION AND CONTROL

A side effect of conveying wastewater in pipes is the generation and release of sewer gases. The most troublesome of these are generated under anaerobic conditions and comprise a range of volatile organic compounds and reduced sulphur compound such as mercaptans. Sewerage odour is typically dominated by hydrogen sulphide. Hydrogen sulphide can cause corrosion of concrete, metal, and electrical equipment. Particularly susceptible locations are points of turbulence following long retention times (e.g., backdrop manholes, wet wells of pumping stations, and outlets of wastewater rising mains). Problems are the most serious in hot, arid climates (Ayoub et al., 2004) but will also manifest in temperate climates such as in the United Kingdom.

In addition, sewer gases, when fugitive, can cause odour nuisance (leading to customer complaint) and endanger sewer workers (as already described).

17.8.1 Mechanisms

Wastewater naturally contains sulphur as inorganic sulphate and organic sulphur compounds. The sulphate is usually derived from the mineral content of the municipal water supply or from saline groundwater infiltration. Organic sulphur compounds are present in excreta and household detergents, and in much higher concentrations in some industrial effluents such as from the leather, brewing, and paper industries (see Chapter 3).

Bacterial activity quickly depletes any dissolved oxygen that is present, and septicity can easily develop. Under anaerobic conditions, complex organic substances are reduced to form

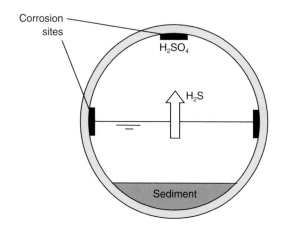

Figure 17.6 Corrosion in sewers.

volatile fatty acids resulting in a drop in pH. *Desulfovibrio* bacteria in pipe biofilms and sediment reduce organic sulphur compounds and sulphates (SO_4^{2-}) to sulphides (S^{2-}):

$$SO_4^{2-} + C,H,O,N,P,S \rightarrow S^{2-} + H_2O + CO_2 \tag{17.1}$$

Gaseous, molecular hydrogen sulphide (H_2S) results from reaction with hydrogen ions in the water and is pH dependent, with more being formed in acidic conditions:

$$S^{2-} + 2H^+ \leftrightarrow H_2S \tag{17.2}$$

In pipes flowing under gravity, H_2S escaping into the atmosphere from solution in the wastewater tends to rise and accumulate in condensation water on the soffit of the pipe (see Figure 17.6). There it is oxidised by *thiobacillus* bacteria to form sulphuric acid:

$$H_2S + 2O_2 \rightarrow H_2SO_4 \tag{17.3}$$

Sulphuric acid can cause serious damage to pipe materials through consuming alkalinity in concrete or direct attack on metal.

17.8.2 Favourable conditions

The most favourable conditions for the production of hydrogen sulphide are as follows (Newcombe et al., 1979):

- Wastewater with a large proportion of trade wastes with substantial sulphide or organic sulphur contents
- Wastewater with a relatively high sulphate concentration
- Wastewater with a low pH—the lower the pH value, the greater is the proportion of molecular hydrogen sulphide
- High rate of oxygen demand to rapidly consume available dissolved oxygen
- High wastewater temperature, which accelerates biological activity
- Retention of the wastewater flow under anaerobic conditions (e.g., long sewers with flat gradients, long rising mains, and large pump wet wells)
- Low wastewater velocity that decreases the rate of oxygen reaeration and increases sedimentation

Factors tending to increase the emission of hydrogen sulphide are as follows:

- High concentrations of molecular hydrogen sulphide in the wastewater
- High wastewater velocity or turbulence
- High relative velocity and turbulence in the head space *above* the flow
- Clean wastewater surfaces with respect to oil films, surfactants, etc.

17.8.3 Sulphide buildup

Pomeroy and Parkhurst (1977) have suggested a formula that produces an index Z that is broadly indicative of the conditions under which sulphide may be formed in gravity sewers under 600 mm in diameter. The Z formula gives the following:

$$Z = \frac{3(EBOD)}{S_o^{\frac{1}{2}} Q^{\frac{1}{3}}} \frac{P}{B} \tag{17.4}$$

where $EBOD$ is the effective biochemical oxygen demand, $BOD_5 = BOD_5 \times 1.07^{(T-20)}$ (mg/L), T is the wastewater temperature (°C), S_o is the sewer gradient (m/100 m), Q is the flow rate (L/s), P is the wetted perimeter (m), and B is the Flow width (m).

The formula contains factors representing the main influences on sulphide generation with $EBOD$ accounting for the influence of both temperature and (indirectly) sulphate content of the wastewater. Wastewater pH is assumed to be in the range 7–8. The value of Z and its interpretation are given in Table 17.5 (see also Example 17.1). The Z formula has been widely used in practice (e.g., Spooner, 2001).

Other studies (e.g., Hvitved-Jacobsen et al., 1999) indicate the particular importance of the biodegradability of the organic matter in the wastewater that is not captured in the BOD_5 test only. Vollertsen (2006) proposes a simple modification to the Z formula (Z'), such that

$$Z' = Z \left[1 + 10 \left(\frac{BOD_5}{COD} - 0.47 \right) \right] \tag{17.5}$$

where COD is the wastewater chemical oxygen demand (mg/L).

17.8.4 Control

A wide range of techniques is available to control the generation or emission of hydrogen sulphide.

Table 17.5 Likelihood of sulphide development

Z	Prevalence of sulphide
<5,000	Rarely present
~7,500	Low concentrations likely
~10,000	May cause odour and corrosion problems
~15,000	Frequent problems with odour and significant corrosion problems

17.8.4.1 Sewerage detail design

Probably the most effective way of controlling the generation and emission of sewer gases is to design the sewer system carefully in the first place. In particular, self-cleansing velocities should be designed to occur regularly. Dead spots in manholes and other structures should be avoided.

Where sulphide generation cannot be avoided, perhaps because of high ambient temperatures, sewer design should avoid excessive turbulence.

Concrete is a material susceptible to H_2SO_4 attack. Boon (1992) suggests that using calcareous aggregates rather than quartz aggregates could considerably extend the life of a concrete sewer. Alternatively, pipes protected by thin epoxy resin coatings have been used.

Vitrified clay or plastic pipes have been shown to be more resistant to corrosion by sulphuric acid. These should be specified at vulnerable locations.

17.8.4.2 Ventilation

Good ventilation was mentioned in Chapter 6. It has several benefits, including

- Helping to maintain aerobic conditions and hence preventing sulphate reduction
- Stripping any gases from the atmosphere
- Oxidising any H_2S in the atmosphere
- Reducing pipe soffit condensation

EXAMPLE 17.1

A 500 mm diameter concrete sewer ($n = 0.012$) has been laid at a gradient of 0.1%. At average flow, the pipe runs half full of wastewater. If the wastewater has a BOD_5 of 500 mg/L and the summer temperature is 30°C, estimate the likelihood that hydrogen sulphide will be generated.

Solution

Geometric properties of flow:

Area of flow, $A = \pi D^2/8 = 0.098$ m^2
Wetted perimeter, $P = \pi D/2 = 0.785$ m
Flow width, $B = D = 0.5$ m

Flow rate, Q (Equation 7.23):

$$Q = \frac{A}{n} R^{\frac{2}{3}} S_o^{\frac{1}{2}} = \frac{0.098}{0.012}\left(\frac{0.098}{0.785}\right)^{\frac{2}{3}} 0.001^{\frac{1}{2}} = 64.6 \text{ L/s}$$

$EBOD = BOD_5 \times 1.07^{(T-20)} = 500 \times 1.07^{10} = 984$ mg/L.
From Equation 17.4,

$$Z = \frac{3(EBOD)}{S_o^{\frac{1}{2}}Q^{\frac{1}{3}}}\frac{P}{B} = \frac{3 \times 984}{0.1^{\frac{1}{2}}64.6^{\frac{1}{3}}}\frac{0.785}{0.5} = 3653$$

This value of Z implies sulphide is unlikely to be present.

The latter point is particularly important—H_2S does not cause corrosion under dry conditions, as bacterial oxidation requires moisture.

17.8.4.3 Aeration

Dissolved oxygen in the form of air, or directly as molecular oxygen, has been widely used to treat rising mains. The injection point could be at the inlet end of the main with the aim being to maintain aerobic conditions to "toggle off" sulphate reduction from the obligate anaerobes. Alternatively, treatment could be at the discharge point to oxidise already formed sulphides, but this is a relatively slow reaction. Both methods will have relatively high capital and running costs. Dosing a nitrate salt, typically calcium nitrate, at the upstream location also has a similar effect of suppressing the sulphate-reducing bacteria and is considered to be best practice.

17.8.4.4 Oxidation

The addition of chlorine or hypochlorite oxidises any sulphides present and temporarily halts bio-activity, thus preventing further sulphide generation. Experience suggests that such treatment is only moderately effective, probably because wastewater has a high chlorine demand. Continuous addition of high doses of chlorine would be prohibitively expensive. A popular alternative is to use a stabilized solution of chlorine dioxide, which is an oxidant but does not produce harmful chlorinated organic residual compounds. Potassium permanganate, available as liquid or granules, can also be used and can yield large reduction in the total odour potential of the sewage.

17.8.4.5 Chemical addition

Ferric(III) salts can be dosed to precipitate existing sulphides as insoluble ferrous(II) sulphide. Typically, dose rates are kept low to ensure that only minimal, and often immeasurable, additional suspended solids are created. At rates of 1–2 mg/L, the dose is not high enough to give a coagulant effect and generate solids. Iron dosing packages are readily available with compatible pumps and vessels enabling appropriate control and storage. Iron can be dosed downstream to lock up the H_2S as an insoluble precipitate. This is of particular value if nitrate dosing cannot be used upstream to switch off the septicity. Iron is limited to eliminating dissolved sulphide and does not react with other odourants.

Another possible strategy is the addition of magnesium hydroxide as a slurry to the wastewater. The effect is to increase the pH of the wastewater, thus reducing the proportion of the sulphide present as H_2S.

PROBLEMS

17.1 Describe the main operation and maintenance functions in an urban drainage system, and the reasons for carrying them out. What are the particular challenges to be met?

17.2 Explain the differences between reactive and planned maintenance. What are the benefits of combining both approaches?

17.3 Compare and contrast manual sewer inspection with CCTV inspection. What other forms of inspection are available?

17.4 Describe the main types of sewer cleaning equipment. Discuss their main areas of application and effectiveness.

17.5 Fats, oils, and grease (FOG) can cause serious operational problems. What are these, and what are suitable methods of eliminating these problems? What are "fatbergs," and what causes them?

17.6 Explain the main differences between the maintenance needs of piped sewer systems and SuDS.

17.7 Explain the dangers to health and safety of sewer workers, and detail good working practice.

17.8 "Hydrogen sulphide causes serious corrosion of urban drainage systems." Discuss the validity of this statement in terms of how, when, and where the gas is formed.

17.9 Measurements in an existing sewer reveal the following data: diameter = 300 mm, depth of flow = 240 mm, mean velocity = 0.75 m/s, BOD = 750 mg/L, and mean temperature = 30°C. Calculate Z to assess the likelihood of H_2S generation. Assume Manning's n is 0.012.   [7,600]

17.10 List the main methods to control hydrogen sulphide problems, and discuss their relative merits.

KEY SOURCES

Boon, A. 1992. Septicity in sewers: Causes, consequences and containment. *Journal of the Institution of Water and Environmental Management*, 6, February, 79–90.

Hvitved-Jacobsen, T. 2002. *Sewer Processes—Microbial and Chemical Process Engineering of Sewer Networks*, CRC Press.

Jiang, G., Sun, J., Sharma, K.R., and Yuan, Z. 2015. Corrosion and odor management in sewer systems. *Current Opinion in Biotechnology*, 33, 192–197.

Saegrov, S. (ed.). 2006. *CARE-S. Computer Aided Rehabilitation of Sewer and Storm Water Networks*, IWA Publishing.

Sewers and Water Mains Committee. 1991. *A Guide to Sewerage Operational Practices*, Water Services Association/Foundation for Water Research.

Stuetz, R. and Frechen, F.B. 2001. *Odours in Wastewater Treatment: Measurement, Modelling and Control*, IWA Publishing.

Thomson, J. and Grada, L. 2004. *An Examination of Innovative Methods Used in the Inspection of Wastewater Collection Systems (CD)*, WERF Report 01-CTS-7.

REFERENCES

Ashley, R.M., Channon, D., Blackwood, D., and Smith, H. 2003. A new model for sewer rodent control. *International Pest Control*, 45(4), 183–188.

Ayoub, G.M., Azar, N., El Fadel, M., and Hamad, B. 2004. Assessment of hydrogen sulphide corrosion of cementitious sewer pipes: A case study. *Urban Water Journal*, 1(1), 39–53.

Bailey, J., Keedwell, E., Djordjevic, S., Kapelan, Z., Burton, C., and Harris, E. 2015. Predictive risk modelling of real-world wastewater network incidents, 13th Computer and Control for Water Industry Conference, CCWI 2015. *Procedia Engineering*, 119, 1288–1298.

Blecken, G.T., Hunt III, W.F., Al-Rubaei, A.M., Viklander, M., and Lord, W.G. 2017. Stormwater control measure (SCM) maintenance considerations to ensure designed functionality. *Urban Water Journal*, 14(3), 278–290.

BS EN 752: 2008. *Drain and Sewer Systems Outside Buildings.*

Buxton, A., Tait, S., Stovin, V., and Saul, A.J. 2001. The performance of engineered invert traps in the management of sediments in combined sewer systems. *Novatech 2001, Proceedings of the Fourth International Conference on Innovative Technologies in Urban Drainage*, Lyon, France, pp. 989–996.

Channon, D., Cole, L., and Cole, M. 2000. A long-term study of *Rattus norvegicus* in the London Borough of Enfield using baiting returns as an indicator of sewer population levels. *Epidemiology and Infection*, 125, 441–445.

Drinkwater, A., Moy, F., and Villa, R. 2015. *Fats, Oils and Greases (FOG): Where Are We and Where We Could Be: The Feasibility of Biological Dosing into Sewer Systems and the Development/Specification of a Protocol*, Report No. 15/SW/01/14, UKWIR.

Ducoste, J.J., Keener, K.M., and Groninger, J.W. 2009. *Fats, Roots, Oils, and Grease (FROG) in Centralized and Decentralized System*, WERF Report 03-CTS-16T.

Erskine, A., Watson, T., O'Hagan, A., Ledgar, S., and Redfearn, D. 2014. Using a negative binomial regression model with a Bayesian tuner to estimate failure probability for sewerage infrastructure. *Journal of Infrastructure Systems*, 20(1).

Gary, E., and Sneddon, J. 1999. Determination of the effect of enzymes in a grease trap. *Microchemical Journal*, 61, 53–57.

Gelder, P., and Grist, A. 2015. *Fats, Oils and Greases (FOG): Where Are We and Where Could We Be?* UKWIR, Report No. 15/SW/01/13.

Health and Safety Executive (HSE). 1997. *Safe Work in Confined Spaces. Approved Code of Practice, Regulations and Guidance*, HSE Books.

Health and Safety Executive (HSE). 2003. *Working with Sewage. The Health Hazards*. A Guide for Employee. Pocket Guide INDG197.

Hvitved-Jacobsen, T., Vollertsen, J., and Tanaka, N. 1999. An integrated aerobic/anaerobic approach for production of sulphide formation in sewers. *Water Science and Technology*, 41(6), 107–116.

Institution of Civil Engineers. 1969. *Safe Working in Sewers and at Sewage Works*, Water Industry Health and Safety Guideline No. 2.

Keeling, M.J. and Gilligan, C.A. 2000. Bubonic plague: A metapopulation model of zoonosis. *Proceedings of the Royal Society*, 267, 2219–2230.

Langeveld, J.G., Liefting, H.J., Schilperoot, R.P.S., Hof, A., Baars, I.B.E., Nijhof, H. et al. 2016. Stormwater management strategies: Source control versus end of pipe. *Proceedings of Ninth International Conference on Urban Water*, Novatech, Lyon, France.

Lester, J. and Gale, J.C. 1979. Sewer cleansing techniques. *The Public Health Engineer*, 7(3), July, 121–127.

Longwell, G. 2010. Inspection by robot, *Municipal Sewer and Water*, March 2010.

Mattsson, J., Hedstrom, A., Ashley, R.M., and Viklander, M. 2015. Impacts and managerial implications for sewer systems due to recent changes to inputs in domestic wastewater—A review. *Journal of Environmental Management*, 161, 188–197.

Micevski, T., Kuczera, G., and Coombes, P. 2002. Markov model for storm water pipe deterioration. *American Society of Civil Engineers, Journal of Infrastructure Systems*, 8(2), 49–56.

National Audit Office. 2004. *Out of Sight – Not Out of Mind. OFWAT and the Public Sewer Network in England and Wales*. Report by the Comptroller and Auditor General. HC 161 Session 2003–2004.

Newcombe, S., Skellett, C.F., and Boon, A.G. 1979. An appraisal of the use of oxygen to treat sewage in a rising main. *Water Pollution Control*, 78(4), 474–504.

Pomeroy, R.D. and Parkhurst, J.D. 1977. The forecasting of sulphide buildup rates in sewers. *Progress in Water Technology*, 9, 621–628.

Roberts, A. and O'Brien, D. 2016. *Turners Road Penstock Chambers Upgrade* UK Water Projects 2016, http://www.waterprojectsonline.com/case_studies/2016/Thames_Turners_Rd_2016.pdf.

Rodríguez, J.P., McIntyre, N., Díaz-Granados, M., and Maksimović, Č. 2012. A database and model to support proactive management of sediment-related sewer blockages. *Water Research*, 46(15), 4571–4586.

Rolfe, S. and Butler, D. 1990. A review of current sewer inspection techniques. *Municipal Engineer*, 7, August, 193–207.

Savic, D., Giustolisi, O., Berardi, L., Shepherd, W., Djordjevic, S., and Saul, A. 2006. Modelling sewer failure by evolutionary computing. *Proceedings of the Institution of Civil Engineers Water Management*, 159(2), 111–118.

Scott, K.M. 2012. *Investigating Sustainable Solutions for Roadside Gully Pot Management*, PhD thesis, University of Hull.

Spooner, S.B. 2001. Integrated monitoring and modelling of odours in sewerage systems. *Journal of the Chartered Institution of Water and Environmental Management*, 15, July/August, 217–222.

Ugarelli, R., Venkatesh, G., Brattebo, H., Di Federico, V., and Saegrov, S. 2010. Historical analysis of blockages in wastewater pipelines in Oslo and diagnosis of causative pipeline characteristics. *Urban Water Journal*, 7(6), 335–343.

Underground Construction. 2015. Unique Fort Worth ICAP proves effectiveness, Underground Construction Online 70, https://ucononline.com/2015/03/16/unique-fort-worth-icap-proves-effectiveness/.

Utility Engineering, Construction and Maintenance. 2016. SL-RAT sounds out a bright future, http://www.utilitymagazine.com.au/sl-rat-sounds-out-a-bright-future/.

Vollertsen, J. 2006. Description and validation of structural condition. Chapter 2 in *CARE-S. Computer Aided Rehabilitation of Sewer and Storm Water Networks* (ed. S. Saegrov), IWA Publishing.

Water UK. 2016. *Industry Facts and Figures 2016.* http://www.water.org.uk/publications/reports/industry-facts-and-figures-2016.

Williams, J.B., Clarkson, C., Mant, C., Drinkwater, A., and May, E. 2012. Fat, oil and grease deposits in sewers: Characterization of deposits and formation mechanisms. *Water Research*, 46, 6319–6328.

Winney, M. 1989. New sonar survey cracks flooded pipes problem. *Underground Magazine*, March, pp. 17–18.

Wirahadikusumah, R., Abraham, D., and Iseley, T. 2001. Challenging issues in modelling deterioration of combined sewers. *American Society of Civil Engineers, Journal of Infrastructure Systems*, 7(2), 77–84.

Woods-Ballard, B., Wilson, S., Udale-Clark, H., Illman, S., Scott, T., Ashley, R., and Kellagher, R. 2015. *The SUDS Manual*, v. 5, CIRIA C753.

Water Research Centre (WRc). 1993. *Manual of Sewer Condition Classification*, 3rd edn. Swindon.

Water Research Centre (WRc). 2005. *Sewer Jetting Code of Practice*, 2nd edn. Swindon.

Water and Wastewater Treatment (WWT). 2016. Innovation Zone: Sewerbatt, http://wwtonline.co.uk/features/innovation-zone-sewerbatt#.V7DZoY1TFwU.

Chapter 18

Rehabilitation

18.1 INTRODUCTION

18.1.1 The problem

If a sewer collapses under a busy street, causing a bus or lorry to fall down a newly opened hole, and requiring the street to be closed to traffic, something must be done. If sewers regularly flood, causing serious damage to property as well as health risks and public revulsion due to the presence of wastewater, something must again be done. But what?

Clearly the collapsed sewer must be replaced. But should we wait for another disastrous collapse before replacing the rest of that sewer length? Which other sewers may need attention? What part of the system must be improved to prevent the flooding? How do we ensure that only truly necessary work is carried out? Can the work be done without digging up the town centre, leading to public inconvenience and loss of business?

These important questions are answered by the art, science, and practice of sewer rehabilitation covered in this chapter, including an introduction (in Section 18.2) to sewerage risk management. Section 18.3 covers structural rehabilitation methods, while hydraulic rehabilitation is covered in Section 18.4. The chapter concludes with discussion in Section 18.5 on how to balance costs and risks. The emphasis throughout the chapter is on sewer pipe rehabilitation, although the broad principles apply across all parts of urban drainage including sustainable drainage systems (SuDS).

18.1.2 Sewerage Rehabilitation Manual

Sewer rehabilitation is one of the main areas of activity of drainage engineers. This has been recognised for many years, and certainly since 1983 and the publication of the first edition of the *Sewerage Rehabilitation Manual* (SRM). The current *SRM* is web based, and sewer rehabilitation is seen as part of integrated sewer system management, and, in particular, is associated with a risk management approach aligned with the Capital Maintenance Planning Common Framework (UK Water Industry Research [UKWIR], 2002). Thus, *SRM* no longer stands for *Sewer Rehabilitation Manual* but for *Sewer Risk Management* (Water Research Centre [WRc], 2017). The web-based material is presented in four sections:

1. The *SRM* approach – the basis of the approach and how it fits with other work
2. Sewerage Management Planning – a step-by-step approach to understand the drainage system, and develop and implement measures to manage the system
3. Sewer renovation – different approaches to rehabilitation

4. Surface water management – summarises the approach aligning with the Department for Environment, Food and Rural Affairs (DEFRA, 2010) surface water management planning guidance

Section 1 is freely accessible; with the remaining sections for licensed users only.

SRM is a definitive guide to good practice, and this chapter does not attempt to duplicate its contents. We deal here more with the overall philosophy, and summarise the main techniques that are available. The most important aspect of the philosophy of *SRM* is the idea that systematic definition of problems, and development of integrated solutions, can lead to savings. This is made possible by

- The acceptance of risk management approaches, and the availability of sophisticated tools for problem analysis – particularly computer models of sewer systems
- The availability of technology for in-sewer inspection and monitoring: closed-circuit television inspection systems, devices for monitoring flows (and other parameters), and technology for in-sewer rehabilitation work
- Pragmatism in the management of sewer assets: recognition of the value of what is already in the ground even if it has some problems; and systematic methods for establishing priorities in rehabilitation programmes
- The recognition that large-scale sewer reconstruction work is very expensive particularly so in urban environments, and also has a high economic cost resulting from disruption to commerce in town centres or to local residents

The result is a set of procedures that have become established and improved over time.

An important part of the strategy is the identification of priorities through application of risk management approaches. All aspects, from inspection to upgrading work, are targeted at selected parts of the system. Yet all work is planned with an awareness of its effect on the drainage area as a whole.

18.1.3 The need for rehabilitation

How does the need for rehabilitation arise? A typical sewer system in the United Kingdom consists of an older part in the centre of the town with newer sections added as the town has expanded. Some parts of the older system may now be undersized, and loadings from traffic, for example, may be far greater than anticipated when the pipe was constructed. The material of the pipe may be old. Overall, about 15% of the UK sewer system is over 100 years old, but in cities the proportion is higher. The pipe material and construction may be poor quality: a worrying characteristic of 25% of the system built between the World Wars. The system may have been poorly maintained in the past; certainly it is sometimes possible to "get away with" poor maintenance of sewers because they can still function to some extent in poor structural condition. But this may mean that eventual structural collapse is all the more catastrophic.

Seepage of water into or out of a sewer (infiltration or exfiltration) may increase the risk of sewer collapses and contribute to their seriousness when they eventually occur. Infiltration was discussed in Section 3.4 and is sometimes associated with leakage from water mains, which may also be in a poor structural condition. Any movement of water in the ground may wash soil particles with it, and over a long period of time this can lead to voids in the ground. A small pipe can cause a large cavity (Figure 18.1). A cracked sewer may be supported by the surrounding ground and not in immediate danger, but if soil is washed in through the cracks by infiltration, then subsidence or collapse becomes far more likely.

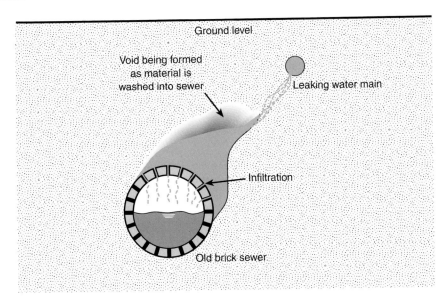

Figure 18.1 Formation of a void around an old sewer.

Symptoms of the need for rehabilitation include

- Collapse
- Flooding of roads and properties (of particular seriousness when wastewater is involved)
- Increased pollution of watercourses
- Blockages
- High levels of infiltration

In the 1980s, there were about 5,000 collapses and serious blockages per year (Finney, 1990) with the expectation that this rate would increase with time unless something was done to improve the situation. Some 5,500 collapses were recorded in 2016 (Water UK, 2016).

18.1.4 Repair, renovation, and replacement

Sewer rehabilitation can be accomplished by means of *repair, renovation*, or *replacement*. *Repair* is localised work to correct damage to the sewer fabric. *Renovation* is work to improve the performance of a length of sewer, incorporating the original sewer fabric. *Replacement* is construction of a new sewer, incorporating and possibly enhancing the functions of the old.

It is important to point out that there is nothing remarkable about the fact that sewers need rehabilitation. All elements of infrastructure have a certain design life. Within that life, they require maintenance (including repair) and, subsequently, renovation. Ultimately they require replacement.

The total length of *critical sewers* (discussed in the next section) that were renovated or replaced in the whole of England and Wales in the 11 years between 1990 and 2001 was about 2300 km, giving an average rate of 210 km per year (Battersby, 2001; based on Ofwat data). At this rate it would take 350 years to rehabilitate the total length of critical sewers (about 74,000 km), implying that this was considered to be the expected life of the asset. In two water company areas the implied asset life was over 1,000 years!

The cost of replacing sewers is high, but only about 20% is for the actual pipe, and the rest is for excavation and reinstatement. The costs of disruption to other services, people, business, and traffic in busy towns can be three times the cost of construction. This gives rise to a popular saying in sewer rehabilitation: "The greatest asset is the hole in the ground."

Sewer rehabilitation is concerned with stabilising, improving, or strengthening the surrounding of the "hole" created by the original pipe, sometimes even making the hole bigger. Access, where possible, is via the hole.

The most effective rehabilitation scheme retains as much as possible of the existing system and causes minimum disruption to the community. The actual work of inspection and rehabilitation is selective: it concentrates on core areas according to an integrated strategy.

18.2 SRM PROCEDURE

18.2.1 Steps in the procedure

The *SRM* procedure is based on the four principal activities that are identified in BS EN 752: 2008 as making up an integrated sewer system management process:

- Investigation
- Assessment
- Developing the plan
- Implementation

Figure 18.2 shows how the 11 steps of the *SRM* procedure are located within these four principal activities.

The Sewerage Management Plan (SMP) includes the objectives, performance requirements, and constraints for a particular drainage area over a defined period, and provides the business planning context for decisions. The SMP is updated when upgrading work is carried out. At all stages of the SRM procedure the focus is on the management of risk, or balancing the risk of performance failure against the cost of intervention. Assessment of risk is also used to prioritise gathering of information on the sewer system (as represented by boxes 2, 3, 4, 5, and 6 in Figure 18.2).

18.2.2 Integrated approach

The four primary elements of sewer system performance are

- Hydraulic
- Environmental
- Structural
- Operations and maintenance

The *SRM* procedure emphasises the importance of an integrated approach that considers these aspects together, with an awareness of the interactions between the elements of performance. An integrated approach also requires consideration of the features that interface with sewerage systems: the wastewater treatment plant, the receiving water, and the above-ground catchment (especially in relation to surface flooding, see Chapter 11).

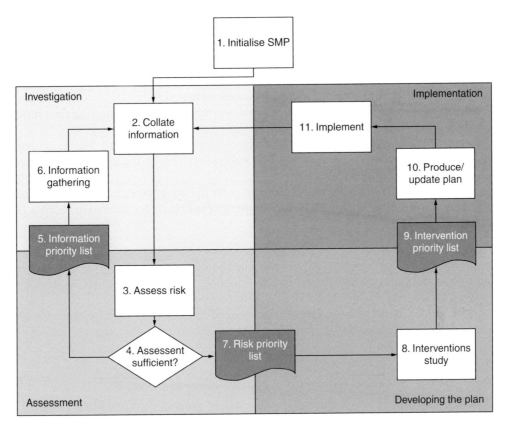

Figure 18.2 SRM procedure—flow diagram. (Reproduced with permission from WRc. 2017. Sewerage Risk Management. srm.wrcplc.co.uk.)

Step 2 in Figure 18.2, Collate information, is a review of all existing information. The focus of this is the "inventory": the dimensions, locations, and characteristics of the system components, including pipes, manholes, and ancillary structures. Existing information on hydraulic, environmental, structural, and operational performance is also assessed, with data from more than one organisation often collected (e.g., water company and environmental regulator). Of course, the basic inventory information is needed for all aspects of the later investigations, so if sewer records are missing, it is important to decide whether to undertake the necessary survey work. The collection of additional information can only be justified if sufficient assessment of risk relating to a particular aspect is not possible.

18.2.2.1 Hydraulic

Assessment of hydraulic performance is likely to involve hydraulic modelling of the system. The aim is to help to identify problem areas within the system (giving rise, e.g., to surcharging or flooding) and to investigate the possible effects of physical improvements within the system. Sewer models are discussed extensively in Chapter 19. As is pointed out there, models are heavily dependent on the accuracy of the data used to specify the physical properties of the system. An important stage (described in Chapter 19) is "model verification" in which simulations of the model are compared with actual measured performance.

Once the model has been verified, it can be used to assess the hydraulic performance of the system and investigate ways of correcting deficiencies.

The model allows an engineer to understand the impact of "do nothing" and study the overall effects of various hydraulic upgrading options. The options are considered in Section 18.4. One undersized existing pipe could cause flooding over an extensive part of the system. A proposal for overcoming the problem can be developed, and the model run again to see its benefits. The final solution can be derived from an iterative process of adjusting the proposal and seeing the consequences from the model.

The central SRM philosophy – that systematic definition of problems and development of integrated solutions can lead to savings – is strongly evident in the use of flow models to develop plans for hydraulic upgrading.

The SRM presents a detailed analysis of the potential advantages and disadvantages of typical options. The main techniques for hydraulic upgrading are considered in Section 18.4.

18.2.2.2 Environmental

The quality of receiving waters is largely the domain of the *Urban Pollution Management Manual* (Foundation for Water Research [FWR], 2012), also considered in Chapters 2 and 19. SRM guidance shows how the Urban Pollution Management approach can be coordinated with the SRM procedures. The environmental investigation entails collecting data, building water quality and impact models where necessary, and using them to identify the causes of environmental problems.

18.2.2.3 Structural

Two factors determine if pre-emptive upgrading will be necessary on structural grounds: the likelihood of failure and the consequence of failure. A third, not yet considered formally but critical to "resilience," is the ability for the "system" in its widest sense to recover.

A significant recommendation of SRM guidance has always been that priorities within the system are established by identifying *critical sewers*. These are sewers for which the costs expected to result from failure would be significantly higher than upgrading costs. Critical sewers make up 20% of a system on average. They are usually the sewers with larger diameters, predominantly made of brick or concrete (see Table 18.1). Earlier editions of the SRM procedures were associated with specific approaches to identifying critical sewers, whereas current guidance places the concept of critical sewers within a more rounded context of assessing risk generally.

Table 18.1 Sewer size and material distribution in the UK

Diameter or major dimension (mm)	All sewers (%)	Critical sewers (%)
<300	70	10
300–499	13	20
500–900	10	35
>900	7	35
Material		
Clay	75	14
Concrete	15	60
Brick	5	25
Other	5	1

Source: Adapted from Moss, G.F. 1985. *The Public Health Engineer*, 13(3), July, 157–160.

Inspection of the internal structural condition of sewers is commonly carried out by closed-circuit television (CCTV). Further details of sewer inspection techniques are given in Chapter 17. There is a standard system in the *SRM* procedure for grading the condition of a sewer. The *SRM* website provides detailed guidelines for the structural design of renovated sewers. The main techniques available for structural upgrading are considered in Section 18.3.

18.2.2.4 Operations and maintenance

There is close linkage between the hydraulic, environmental, and structural aspects of the performance of a sewer system and the issues that operation and maintenance practices address. This phase focuses on the issues such as blockages caused by solids or fats and tree roots where O&M practices and procedures are necessary to manage or reduce the impact.

18.2.2.5 Developing the plan

The results of the investigations that have been described above are used to produce an upgrading plan referred to as a SMP (or previously drainage area plan) and in some cases an operation and maintenance plan.

The solutions should be integrated where possible (solving different types of problems together), considering capital and operation interventions, be cost-effective, and consider the catchment as a whole.

Rehabilitation will be planned with the aims of overcoming hydraulic and environmental problems in the system, taking account of planned structural upgrading and allowing for known plans for future development of the area.

Increasingly, there is also greater integration and engagement between organisations responsible for different aspects of drainage in the UK, and the Drainage Strategy Framework (Gill, 2013) outlines an approach to achieve this.

18.2.2.6 Implementation

This involves detailed design and construction of upgrading works and implementation of the operation and maintenance plan (although the latter is often implemented independently).

SRM resources include detailed guidelines for structural design of renovated sewers. The main techniques available for structural upgrading are considered in Section 18.3, and for hydraulic upgrading in Section 18.4.

Monitoring of hydraulic, environmental, and structural performance, and effectiveness of the operation and maintenance plan should continue after upgrading has been carried out, and now forms an important part in demonstrating performance. Increased monitoring is taking place with the installation of permanent flow monitors (see Chapter 12).

18.3 METHODS OF STRUCTURAL REPAIR AND RENOVATION

This section is intended to provide only an introduction to this fascinating area of developing technology. Technical guidelines on the practice of sewer repair and renovation are available in the *SRM* guidance (WRc, 2017) and Read (1997), with helpful summaries on many approaches on the International Society for Trenchless Technology's website (http://www.istt.com). Naturally there are major differences between methods for man-entry sewers and those for non-man-entry sewers.

18.3.1 Man-entry sewers

Any work involving entry into confined spaces has significant health and safety issues that should be fully observed. Further details can be found in Section 17.5.

18.3.1.1 Repair

The general aims of repair are to correct defects and reduce infiltration in physically intact sewers.

Pointing is the renewal of mortar in brick man-entry sewers. If the mortar is carried to the point of application by hand, and finished off with a trowel, the process is called *hand pointing*. This is a labour-intensive and time-consuming process but produces a good finish. In longer lengths of sewer, it may be more appropriate to use equipment for delivering the mortar under pressure to the point of application (pressure pointing). In this process, excess mortar is still normally removed by hand trowelling.

Other forms of repair in man-entry brickwork sewers include replacing areas of brickwork, and rendering with high-strength mortar.

18.3.1.2 Renovation

Renovation methods generally involve providing a new lining, inside the old, to a whole length of sewer. The lining may either be constructed *in situ*, or erected from ready-made segments. There is usually a loss of area within the sewer.

Grout is a general word for a material applied under pressure as a liquid to fill a space and subsequently "cure" to a hardened state. Grouts used in sewer rehabilitation are generally made from ordinary Portland cement with added PFA (pulverised fuel ash). In some applications a "chemical grout" is used as an alternative. As well as filling the space between the new lining and the old sewer, grout can also be used to fill any voids outside the original sewer.

If the brickwork is sufficiently intact, the pipe wall can be *rendered* by hand with a lining of mortar (which may include reinforcing fibres). Alternatively, *Ferrocement* may be used consisting of a cement-rich mortar formed on layers of fine reinforcing mesh. Both techniques require relatively thin linings, resulting in limited reductions to the pipe cross-sectional area.

In situ gunite lining is formed by spraying concrete onto the old sewer wall. A reinforcing mesh is placed around the wall first and is incorporated in the sprayed lining. An *in situ* lining can also be created using *pumped concrete*. Reinforcement and specially designed steel formwork sections are put in place, and high-quality concrete is then pumped into the annular space.

Segmental linings are made up of ready-made segments erected in the sewer. Common materials are glass-reinforced cement, glass-reinforced plastic, precast gunite (sprayed concrete), and polyester resin concrete (effectively a plastic containing an aggregate in the same way as concrete). The space between the new lining and the old pipe wall is, again, filled with grout. Precast segments can be made to fit any cross-sectional sewer shape, not just a straightforward circle. Success with the technique depends on careful location of the many joints.

It should be noted in relation to all work in man-entry sewers that health and safety regulations only allow work in confined space when there is no reasonably practicable alternative.

18.3.2 Non-man-entry sewers

The most innovative developments in sewer rehabilitation have tended to be for application to sewers that are too small for direct access by a person. Some methods involve systems of remote control with high levels of technical sophistication.

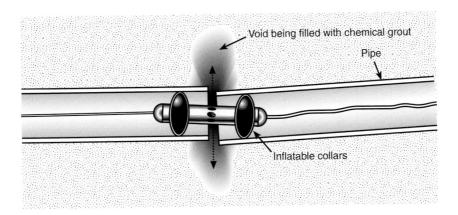

Figure 18.3 Chemical grouting.

18.3.2.1 Repair

Patch repairs are typically applied at or across a small number of joints/defect and involve remotely placing a patch of fabric containing an appropriate resin material and then curing it in place. In a *resin injection* system, an inflatable packer is used to isolate a pipe defect and to force injected resin into the defect. *Chemical grouting* uses a similar packer device to seal joints and fill any associated voids in the ground. The packer is located at a joint, and collars positioned on either side of the joint inflate to create a seal (Figure 18.3). Applying air or water pressure can first check the effectiveness of the joint. If pressure loss indicates that the joint is unsatisfactory, chemical grout is released under pressure to fill any void outside the pipe and seal the joint.

An alternative form of chemical grouting treats a whole sewer length at a time. Two chemicals are used, which react together to form a sealing gel. The sewer, laterals, and manholes for a particular length are filled with the first chemical. After a suitable period to allow penetration of the ground around the pipe, the chemical is replaced by a second, which reacts with the residue of the first to form an impermeable mass in the ground around the defects. Any surplus of the second chemical is then pumped out.

Bunting (1997), the *SRM*, and the *UK Society of Trenchless Technology* website provide further information on repair techniques for non-man-entry sewers.

18.3.2.2 Renovation

Sliplining involves forming a continuous length of plastic (such as high-density polyethylene [HDPE]) lining and pulling it through the existing pipe. One approach is for long lengths of plastic pipe (typically 5 m) to be welded/fused end-to-end on the surface. The resulting continuous length of pipe has some flexibility and is winched along the sewer via a specially excavated lead-in trench (Figure 18.4a). The winch cable is attached to a nose-cone fixed to the front end of the new lining. As an alternative to welding the lengths on the surface, pipes can be welded in an enlarged trench (Figure 18.4b). The first approach has the disadvantage that it requires space on the surface for assembly of the pipe length; the second requires more excavation (though both approaches require a significant amount). Another alternative is much shorter lengths of pipe (HDPE or polypropylene) with push-fit or screw joints. These are connected within existing manholes (Figure 18.4c). In all cases, any significant space between the new lining and the old pipe is filled with grout. The selection and use of grout can be the most difficult aspect of sliplining. The design must

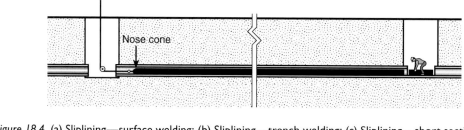

Figure 18.4 (a) Sliplining—surface welding; (b) Sliplining—trench welding; (c) Sliplining—short sections.

consider whether the grout needs to transfer load between the host and new pipe, or act as a restraint. During application of the grout, care must be taken to keep the inserted pipe fixed in position to avoid flotation, typically through filling the smaller pipe with water and applying the grout in stages.

These processes, as described here, will result in a smaller-diameter pipe. If the existing pipe includes imperfections like distorted cross-section or offset joints, the limitation on the diameter of the new lining may be even greater. It may be appropriate to carry out localised repair of imperfections before proceeding with sliplining, in order to increase the practicable diameter of the new lining. The new lining is likely to be smoother than the old, but there is still likely to be a loss of hydraulic capacity, and this is obviously a potential disadvantage.

However, the technique of *pipe-bursting* allows the new pipe to have a diameter equal to that of the old pipe, or even greater. Pipe-bursting is achieved by a pneumatically or hydraulically operated hammer that breaks the existing pipe from the inside and forces the broken pieces into the ground, to form a void larger than the former pipe interior. The mole can be drawn through the pipe ahead of a new lining. Pipe-bursting with sliplining can therefore produce a new pipe that is bigger and smoother than the old (though at the same gradient) and is a very powerful technique for renovation of non-man-entry sewers. There are circumstances in which pipe-bursting may not be practicable, for example, when the old pipe has a concrete surround, or when there is reinforcement in the pipe or surround, or when damage might be caused to neighbouring services, or in certain ground conditions. However the main problem with pipe-bursting in conjunction with sliplining is that reconnection of laterals has to be carried out externally. This creates a need for excavation at each connection and can be a significant disadvantage if the pipe is deep or there are a large number of connections. A similar but slightly different technique than bursting is pipe reaming, which removes the old pipe by grinding it away with the new pipe following behind it.

Cured-in-place linings are positioned in the sewer in a flexible state, and then cured in place to form a hard lining, usually in contact with the original sewer wall. The most common process involves a fibre/felt liner or "sock" impregnated with resin, which is inserted in the sewer by an inversion process (turning inside-out) under water pressure. When this flexible lining is in place, it is cured by circulating steam, water (Figure 18.5), or ultraviolet light. These linings can be designed to provide high structural strength, but this can lead to a significant reduction in diameter. Non-structural liners are typically thinner. These liners are less likely to require excavation for side connections.

It is possible to detect the location of laterals from inside the pipe by observing the shape of the lining, and then to cut through the lining by remote control in order to reform the connection. However, the positions of the laterals are normally confirmed prior to lining by using CCTV. Where laterals protrude into the pipe, these have to be removed, as discussed in Section 18.3.4. When curing liners, care must be taken to manage the potential of the fumes/gases entering the property as well as potential toxic discharges to the environment.

Another method using material that is not initially pipe shaped involves inserting a soft plastic lining (PVC-U) in the sewer in a folded state (so that it has a smaller area). Once in position it is heated and pushed out against the old lining by a rounding device to form its final circular shape.

Lining with spirally wound pipes is yet another alternative. A long PVC-U strip is fed into a winding machine placed at the bottom of a manhole. The machine winds the strip into a spiral of continuous pipe lining which then travels up the sewer. In this case, the annulus is normally filled with cementitious grout once the liner is in place.

18.3.3 Choice of method

If the decision is made to go ahead with rehabilitation, the next stage is to choose the most appropriate method. The following points need to be fully considered:

- Type of failure
- Pipe roundness/ovality
- Sewer type and configuration
- Change in hydraulic capacity of newly rehabilitated sewer
- Direct cost of alternative options
- Indirect costs of alternative options
- Effective life of newly rehabilitated sewer (see Table 18.2)
- Whether to renovate or replace

Differences in the properties and life expectancy of alternative materials must be considered (Table 18.2). The *SRM* website contains detailed guidance on the design of pipe linings.

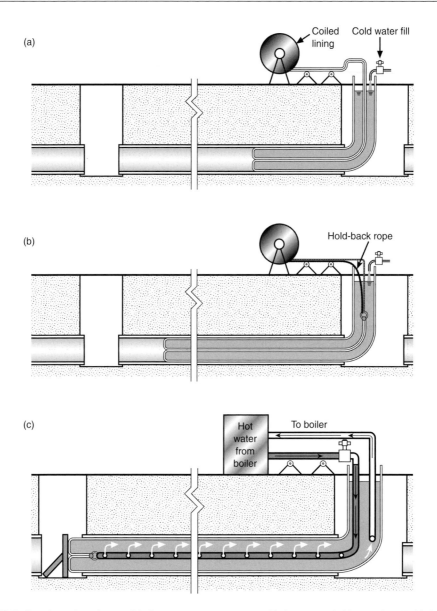

Figure 18.5 Cured-in-place lining. (a) Commencing insertion; (b) Insertion half completed; (c) Insertion completed—curing by hot water circulation.

18.3.4 Associated work

18.3.4.1 Laterals

Household drain connections, or laterals, pose one of the most awkward problems in non-man-entry renovation for two reasons:

- Badly formed laterals can protrude into the pipe, causing obstruction and potentially preventing insertion of a lining.
- Once the lining has taken place, existing connections need to be reopened.

Table 18.2 Relative properties and life expectancy of sewer materials

Pipe material	Corrosion resistance	Chemical resistance	Stiffness	Resistance to brittle failure	Abrasion resistance	Estimated life
Concrete[a]	xx	x	xxx	x	x	xx
Fibre cement[a]	xx	xx	xxx	x	x	xx
GRC[a]	xx	xx	xxx	xx	x	x(x)
GRP	xx(x)	xxx	xx	x[b]	xx	x
PVC-U	xxx	xxx	xx	x(x)	xx(x)	xxx
HDPE	xxx	xxx	x	xx(x)	xxx	xx
PP	xxx	xxx	xx	xx[b]	xxx	xxx
Resin concrete	xxx	xxx	xxx	x	xx(x)	xx
VC	xxx	xxx	xxx	x	xx	xxx

Source: Adapted and updated from Moss, G.F. 1983. *The Public Health Engineer*, 11(2), April, 31–34.

Note: Brackets indicate that current information is insufficient to differentiate between ratings.

x = low; xx = medium; xxx = high.
[a] Some of these properties may be modified by the use of a protective coating or lining.
[b] Stiffness can be specified and therefore is variable.

While it is possible to excavate down on poorly connected laterals, this is expensive and time consuming. Typically a robotically remote-controlled device cuts or grinds the lateral back to the pipe face. This is necessary to limit the liner being stretched/weakened or leaving a gap between the pipe and liner. After lining, the liner is cut at lateral connections, again typically from within the pipe using a robotic remote-controlled cutter. At times, it may be appropriate to cut from the connecting lateral.

A difficult problem is identifying which of the laterals are still live. It is important that old disused laterals are sealed.

18.3.4.2 Cleaning

Most rehabilitation operations need to be preceded by sewer cleaning. The available techniques are discussed in Section 17.4.

18.3.4.3 Overpumping

It is a common requirement that flow in the sewer needs to be diverted, often via a pump and temporary overground pipe system, especially for work in non-man-entry pipes. Overpumping can be a significant cost element in a sewer rehabilitation scheme. The overpumping system cannot be designed without estimates of flow for wastewater (including industrial) and stormwater, where appropriate, and these should be as accurate as possible. Often a sewer system model is used to help indicate the likely flows and support the planning of the lining with regards to forecast weather conditions. Nedwell and Vickridge (1997) discuss practical aspects of overpumping.

18.4 HYDRAULIC REHABILITATION

To solve problems resulting from hydraulic overloading, and achieve the performance targets set, the *SRM* guidance includes a range of upgrading options to be considered in sequence.

18.4.1 Reduce hydraulic inputs to piped system

Some adjustments to a system may be possible that reduce inflow without requiring major works. For example, it may be possible to divert certain inflows from overloaded sewers to points in the system where there is less hydraulic overloading. However, the most significant method of reducing inputs as a solution to a range of drainage problems is the use of storm-water management SuDS techniques to divert water away from entering the piped system or to limit its flow. These are described in Chapter 21.

18.4.2 Maximise capabilities of the existing system

Removal of local constrictions may allow the capacity of significant sections of the system to be increased. Sewer cleaning will also increase capacity, though if the deposition of debris had been caused by the physical nature of the sewer, then proactive maintenance will be required. There is more information on sewer cleaning in Chapter 17.

Combined sewer overflows (CSOs) are major controls on flow in combined sewer systems. Rehabilitation work in sewer systems often involves improvement, replacement, or relocation of CSOs. This may be primarily related to environmental requirements, but often has significant hydraulic effects, for example, by raising weirs and installing screens.

18.4.3 Adjust system to cause attenuation of peak flows

Most sewers flow less than full most of the time, and, by attenuating peak flows, more efficient use can be made of the capacity of the sewer. This can be achieved by

- Providing additional storage within the system in the form of detention tanks (described in Chapter 13)
- Installing throttles and controls at key points to make the best use of in-sewer storage (these devices are described in Section 8.1)
- Applying techniques for overall system management, including real-time control, as considered in Chapter 22

18.4.4 Increase capacity of the system

When the possibilities previously discussed have been exhausted, it is necessary to use what might be considered the most obvious method of overcoming hydraulic overloading – increasing capacity. Many of the structural rehabilitation techniques that have been described in Section 18.3 cause an increase in capacity, either because they reduce roughness and remove imperfections, or, in some cases where pipe-bursting techniques are being used, actually increase the size. This is why, in the planning of sewer rehabilitation schemes, structural and hydraulic effects must be considered together.

When sufficient hydraulic upgrading cannot be achieved by renovation, it may be necessary to replace existing sewers with new sewers of larger capacity. This will, of course, bring with it the high costs and associated disruption discussed in Section 18.1. Existing sewers may also be duplicated by the construction of additional sewers that flow in parallel. However, care is always needed when increasing capacity to ensure the downstream system has the ability to cope with the increased flows.

18.5 BALANCING COST AND RISK

The risk management approach of *SRM* is particularly relevant to decision-making processes associated with costs. This can be illustrated by an example.

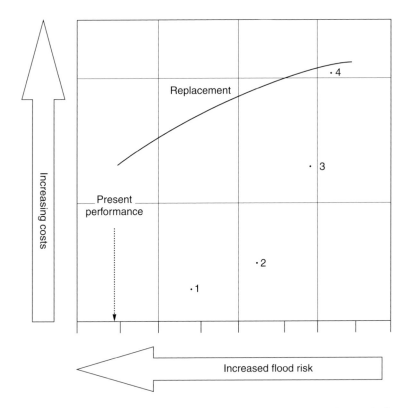

Figure 18.6 Example of balancing cost and risk. (Based on *SRM* material [Water Research Centre, 2017].)

The cost of *replacing* a sewer system, which is undersized and causing flooding, is plotted as the curve in Figure 18.6. The greater the proportion of the system replaced, the lower is the flood risk, and the higher is the cost. Options 1, 2, 3, and 4 are *upgrading* options of increasing complexity. Option 2 is a little more expensive than option 1 but gives significantly reduced flood risk. Option 3 further reduces flood risk but at significantly higher cost. Option 4 gives a further slight reduction in flood risk but is almost as expensive as the complete replacement option that would give the same flood risk. If point 4 were above the curve, upgrading would be more expensive than replacement (for the same flood risk). Overall, option 2 may be considered to give the greatest value, or a hybrid solution between 2 and 3 where the cost increases marginally but maximum flood risk reduction is gained.

Along with cost, measuring the carbon impact has become important in decision making with delivery organisations requiring low whole life embodied and operational carbon solutions. The calculation of carbon costs is explained in Section 15.8.

PROBLEMS

18.1   What does *SRM* stand for?
18.2   Why is sewer rehabilitation more common now than it was 50 years ago?
18.3   How can flow models be used in the planning of sewer rehabilitation schemes?
18.4   Describe methods of repair and renovation suited to non-man-entry sewers.
18.5   What are the merits and challenges of applying a variable performance standard to assist in decision making?

KEY SOURCES

Read, G.F. and Vickridge, I.G. (eds.). 1997. *Sewers—Rehabilitation and New Construction; Repair and Renovation*, Arnold.

Water Research Centre (WRc). 2017. *Sewerage Risk Management.* http://srm.wrcplc.co.uk.

REFERENCES

Battersby, S. 2001. Unsustainable underinvestment. *Water*, 136, 14 November.

BS EN 752: 2008. *Drain and Sewer Systems Outside Buildings*.

Bunting, N. 1997. Localised repair techniques for non-man-entry sewers, in *Sewers—Rehabilitation and New Construction; Repair and Renovation* (eds G.F. Read and I.G. Vickridge) (Chapter 11, 233–253), Arnold.

Department for Environment, Food and Rural Affairs (DEFRA). 2010. *Surface Water Management Plan Technical Guidance.* www.defra.gov.uk

Finney, A. 1990. Refurbishment of sewers. *Chemistry and Industry*, 15 October, 658–662.

Foundation for Water Research (FWR). 2012. *Urban Pollution Management Manual*, 3rd edn, www.fwr.org/UPM3.

Gill, E. 2013. *Drainage Strategy Framework*, Ofwat, www.ofwat.gov.uk/wp-content/uploads/2015/12/rpt_com201305drainagestrategy1.pdf.

Moss, G.F. 1983. Latest developments in sewer renovation. *The Public Health Engineer*, 11(2), April, 31–34.

Moss, G.F. 1985. Sewerage rehabilitation; the way forward. *The Public Health Engineer*, 13(3), July, 157–160.

Nedwell, P. and Vickridge, I.G. 1997. Ancillary works—Cleaning and overpumping, in *Sewers—Rehabilitation and New Construction; Repair and Renovation* (eds G.F. Read and I.G. Vickridge) (Chapter 9, 193–203), Arnold.

UK Water Industry Research (UKWIR). 2002. *Capital Maintenance Planning Framework*. Report No. 02/RG/05/3.

Water UK. 2016. *Industry Facts and Figures 2016*. www.water.org.uk/publications/reports/industry-facts-and-figures-2016.

Chapter 19

Modelling in practice

19.1 MODELS AND URBAN DRAINAGE ENGINEERING

There are two chapters in this text on modelling drainage systems. Here, we look at models covering hydrological, hydraulic, and water quality aspects of modelling used in practice. In Section 19.2 we consider models in context, in Section 19.3 we discuss the various elements that can be represented in such a model, and in Section 19.4 we detail the underlying equations and numerical methods that are used to model unsteady flow. The setting up and use of these types of models in practice is discussed in Section 19.5 followed by a section specifically on above-ground flood modelling (Section 19.6). Water quality system models are discussed in Section 19.7 with more detail given on pollutant transport and pollutant transformation in Sections 19.8 and 19.9, respectively. The use of water quality models in practice is described in Section 19.10 followed by Section 19.11 on how to go about planning an integrated study.

The second chapter on modelling (Chapter 20) discusses the new and rapidly changing approaches being developed in research, for future application in practice.

19.1.1 Development of flow models

The purpose of models in urban drainage engineering is to represent the system and its response to different conditions in order to answer questions about it, such as why it does or does not perform as expected now or in the future, and how to improve it.

Computer programs for drainage design and analysis emerged in the 1970s, but complex models only became standard tools of drainage engineers when appropriate computing power became available. The software package, SWMM, first appeared in the United States in the early 1970s and has continued to be developed ever since. In the United Kingdom, there was no standard software package until the early 1980s when the introduction of WASSP, based on the Wallingford Procedure (Department of the Environment/National Water Council [DoE/NWC], 1981), had a profound impact on the practice of drainage engineering. It effectively turned a branch of engineering that had relied heavily on conservative decision making based on experience, into one in which sophisticated methods of analysis were used to understand existing systems and produce better and more informed solutions.

Today, there are many modelling packages available commercially to enable the representation and replication of system performance and support design. These evolve and develop at a fast pace enabling greater functionality, the ability to model larger or more detailed networks and integrate multiple drainage system types. Furthermore, many use real-time data for purposes such as forecasting, and this is covered in Chapters 20 and 22.

There have always been two main uses for sewer system models: design (of new and retrofitted systems) and analysis (of existing systems). In design, the physical details of a proposed drainage system are determined so the system behaves satisfactorily when exposed to specific conditions. In analysis (simulation), the physical details of the system already exist, and the user is interested in how the system responds to particular conditions (in terms of flow rate and depth, the extent of surcharge and surface flooding, scale, and impact of pollution). The aim is usually to find out how the system performs, understand the causes, if the system needs to be improved, and, if so, how.

As well as hydraulic and hydrological aspects, some urban drainage flow modelling packages include the ability to replicate real-time interventions and operational rules, and receive live data streams. They have versatile data interfaces with geographic information system (GIS) compatibility and the graphical representation of the system and its response to storms.

19.1.2 Model types

Physically based computer simulation models are based on accepted mathematical relationships between physical parameters, and all involve some element of simplification. No model covers every raindrop or every variation in the catchment surface. In deterministic models, one combination of input data will always give the same output, and randomness is not accounted for. The fact that these models include simplifications and ignore random effects, combined with the uncertainty associated with input data and with field measurements, means that it would be a reckless or naive modeller who stated, "the results predict the exact behaviour of the system." That modeller would be far wiser to recommend the results as being *indicative*.

For most sewer simulations that are concerned with hydrological and hydraulic aspects, deterministic models have become the standard tools. Where necessary, sensitivity analysis is undertaken to explore the impact of uncertainty in model parameters. This is explored further in Chapter 20.

Various other approaches also exist to model urban drainage systems. These include empirical, neural network, conceptual, and stochastic methods. These are also discussed in further detail in Chapter 20.

19.2 URBAN DRAINAGE MODELS IN CONTEXT

19.2.1 The modeller

Although sewer system modelling involves application of mathematical methods, most commentators stress the fact that success is heavily reliant on the skill, experience, and judgement – even intuition – of the human beings who set up and run the models.

Osborne et al. (1996), comparing US- and UK-based runoff models and software in a large-scale application, comment that "the two approaches turn out to be similar and a good engineer can get sensible results using either approach. However an inexperienced engineer will probably get bad results using either approach." A German comparison of sewer flow and quality models (Russ, 1999) concluded that "the reliability and achievable accuracy depend more on the user's qualification, experience and care than on the performance of the model."

While there is software available that can fit variables to measured data, how the model will be used and its purpose require the modeller to fully understand the implications of their decisions that can be detrimental in its application. First, this is relevant when small storm events are used to verify or calibrate a system without realising the consequence of decisions made for when larger events are simulated and some predictions become unreliable. Second, when understanding the cause of a problem and how to address it, a model auto-calibrated without justification of the variables linked to the reality within a catchment area, limits or prevents the modeller achieving the model's purpose.

19.2.2 Confidence in the model

Models are built or updated to meet specific objectives. Whatever the purpose is, it is important to understand the level of confidence in the model to be able to replicate the system's performance. A model may not be built to the same level of detail across the whole of the catchment. This may result in differing levels of confidence across the model. Confidence can be assessed by reference to the data used to build the model, and particularly in the data used to validate it (see Section 19.5.5), such as measured from short-term flow surveys, long-term data records, and historical flooding records.

Water service providers have many drainage systems modelled to varying levels of detail and for various purposes. It is common that an existing model may be used for a purpose different to that for which it was originally built. Therefore, assessing whether it is appropriate to use the model and the level of confidence the modeller has in its use, and whether further upgrades are necessary is a critical step.

19.2.3 Model use

Models are built, maintained, and upgraded for numerous purposes. They may be used by a single stakeholder or by multiple stakeholders when working in partnership. They can be broken down into four broad categories:

- Understanding system performance now and in the future: this is typically carried out either to understand a particular problem or to predict what may occur in the future and therefore plan for it. Some models form part of wider understanding and planning, such as sewerage management plans (Water Research Centre [WRc], 2017) discussed in Chapter 18 or integrated catchment models discussed in Section 19.11.
- Developing (capital) solutions to solve a problem (based on the understanding of the system performance): here the model can be used to identify thresholds or triggers for when an intervention becomes necessary, and test alternative approaches to solving or managing a given need. It is a tool in the design development process and is best used when integrated with design and not viewed as a separate function.
- Developing and testing operational and maintenance solutions: it is becoming more common to improve drainage system performance through targeted and timely operational interventions. A model enables the testing of the solution, and importantly indicates what may happen if such interventions are not put in place.
- Forecasting and real-time "live" running to support and make dynamic system interventions: models can be run in real time and with forecast data to enable a "what if" to be determined, and dynamic decisions taken, such as when to turn on critical pumps or hold flows back in a drainage network.

The Chartered Institution of Water and Environmental Management's (CIWEM) *Code of Practice for the Hydraulic Modelling of Urban Drainage Systems* provides an overview of the use of models (CIWEM, 2017).

19.3 ELEMENTS OF URBAN DRAINAGE MODELS

19.3.1 Overview of the components

A physically based, deterministic model of sewer flow must represent the *inputs* (rainfall, infiltration, and wastewater flow) and convert them into the information that is needed: flow rate and depth within the system and at its outlets (e.g., at overflows or through flooding). These outlets can become inputs into other models (e.g., assessing the impact of overflow discharges). The model carries out this conversion by representing (mathematically) the main physical processes that take place. The model must, therefore, be reasonably comprehensive: we could not expect to leave out an important process and still produce accurate results. In order to represent processes in a physically based mathematical form, a good level of scientific knowledge about the processes is needed. Therefore, urban drainage flow models are based on a body of research information about runoff, pipe flow, and so on. However, as has already been pointed out, the model is also bound to have varying levels of detail with some elements of simplification.

At a very general level, therefore, three factors greatly influence the accuracy, confidence, and usefulness of the simulations by a particular physically based modelling package: the comprehensiveness of the model, the reliability and completeness of the scientific knowledge on which it is based, and the appropriateness of the simplifications it contains.

The word *model* tends to be used in a number of ways. We are referring throughout to mathematical, computer-based models. The mathematical representation of each process can be termed a *model*, as can the combination of all the processes (into a software package). However, these models are simply tools ready to do a job: simulation of a flow in a particular catchment. To do this job, a great deal of data are required about, for example, the surfaces of the catchment and the network of sewers. These data can improve the confidence in the simulations by reducing the uncertainty (often through physical surveys). The specific application of a software package to a particular catchment requires great effort in checking, calibration, and verification, as considered later.

Typically, a model of the minor drainage system contains *links* and *nodes*. The links are generally pipes, in which the model must represent the relationship between the main hydraulic properties: cross-sectional size, gradient, roughness, flow rate, and depth (links are also used to represent pumps and other features). The nodes are generally manholes, at which there may be additional head losses and changes of level (and are also used to represent storage). In addition to these primary building blocks, the package must also be capable of simulating conditions in more specialised ancillary structures: including tanks and combined sewer overflows (CSOs).

The emphasis throughout is on unsteady conditions: on variations of flow rate and depth with time. A crucial element in a flow-modelling package is the way in which it simulates these unsteady conditions. Common methods are presented in Section 19.4.

We now consider the main physical processes that must be represented in a software package for flow modelling.

19.3.2 Rainfall

The model will be used to find the response of the catchment and the drainage system to particular rainfall patterns. Straightforward examples would be simple constant rainfall or, more realistically, rainfall with a particular storm profile (variation of rain intensity with time). This would be generated for a specified return period using the types of relationship between intensity, duration, and frequency considered in Section 4.3. To model the operation of CSOs, or storage facilities that need to be emptied during dry periods, a typical sequence of wet and dry periods would be studied using time series rainfall stochastically generated for the specific location (see Section 4.5) or observed using rain gauge and/or radar data. Spatial variation of rainfall is an important consideration in large catchments. During model verification (described more fully in Section 19.5.5), rain gauge records or radar rainfall data are used as rainfall input to the model, and the flow simulations by the software package are compared with flows measured in the system.

19.3.3 Rainfall to runoff

The conversion of rainfall into "runoff", water destined to find its way into the sewer system, is a highly complex process. There are many reasons for rainwater not to become stormwater in the sewer. It may, for example, soak into the ground (even on "impervious" surfaces, via cracks), may form puddles and later evaporate, or may be caught in the leaves of a tree. There is an obvious distinction between water that falls on a roof or a road and that which falls in an undrained garden; but where, for example, there is a grass strip beside a pavement, some water falling on the grass may run off onto the pavement and enter the sewer, whereas some water falling on the pavement may run onto the grass and infiltrate. Once the ground is saturated though, it can become runoff again and enter the drainage system. The methods of representing these processes have been considered in Section 5.2.

19.3.4 Overland flow direct from runoff

The two main considerations here are the amount of rainwater that will enter the sewer and how much time it will take to get there. The amount is typically dealt with in the conversion to runoff. Clearly the extent to which water entering from one area will overlap with that entering from another will have a significant effect on the way the flow in the sewer builds up with time. Again, the physical processes are highly complex, with many ways in which surface irregularities can affect the flow. Overland flow is usually represented in a simplified form, and methods have been given in Section 5.3. We refer here to the initial flow of runoff. Flow over the surface in flood conditions is considered in Section 19.3.7.

19.3.5 Dry weather flow

In a combined sewer system, the stormwater joins the flow of wastewater in the sewer. Realistic simulation of wastewater generation is an important function in a flow-modelling package and can form a significant proportion of the flow in larger catchments. Methods are presented in Sections 3.2 and 3.3 outlining the wastewater components.

19.3.6 Infiltration

Another significant component of dry weather flow is infiltration or ingress of water that finds its way from the soil into the drainage system. In piped systems, this is typically through cracks and holes formed where the fabric has deteriorated (either over time or through damage). Modelling often considers different types of infiltration that may enter the drainage system in response to rainfall at different times. The largest tends to be groundwater infiltration, which occurs as a result of longer-term rainfall patterns that increase the groundwater level to a point where it can enter the drainage system. In coastal regions this may also have tidal influences and may show a tidal pattern. Non-dry weather infiltration may occur as

Table 19.1 Detail on the elements of a flow model

Topic	Chapter/section
Rainfall	4
Rainfall to runoff	5.2
Overland flow direct from runoff	5.3
Dry weather flow	3.2, 3.3, 9.3.2
Infiltration	3.4, 9.3.4
Surface flooding	11, 19.3.7

water percolates through the surface during and shortly after rainfall and finds its way into the drainage system. However, this is hard to distinguish from runoff flows.

Further details on infiltration and its representation have been given in Section 3.4. Validation is only really possible with short- and long-term monitoring, with an emphasis on long term. Care is needed when applying these models to ensure extrapolation during large events does not yield unreliable results.

19.3.7 Surface flooding

The five main processes described above must be included in a software package for flow modelling when the capacity of the (minor) sewer system is not exceeded. When the capacity is exceeded, a sixth element may become very important in modelling the hydraulic response of a drainage system: surface flooding and flood routing.

The main concepts of surface flooding are introduced in Chapter 11, and approaches to including surface flooding and routing in urban drainage models are described in Section 19.6.

The output of this modelling effort generally comes in the form of simulated variations of flow rate and depth with time at chosen points within the sewer system, and at outlets from it. These may be manholes or where greater detail is added, gullies. There is usually particular interest in the ability of the sewer system to cope with the simulated flows, and thus on the extent of possible pipe surcharge or surface flooding.

Table 19.1 summarises the parts of this text that contain more detail on each of the elements of a flow model.

19.4 MODELLING UNSTEADY FLOW

Wastewater flow varies with the time of day, and during storm conditions, inflow to the sewer system can vary dramatically with time. The representation of unsteady (time varying) flow is an important component in a sewer-flow software package.

In unsteady flow in a part-full pipe, there is a far more complex relationship between depth and flow rate than there is in steady flow (described in Section 7.4). Also, as a storm wave moves through a sewer system it *attenuates* (spreads out and the peak reduces) and it *translates* (moves along). The relationship between flow rate (or depth) and time cannot be accurately predicted without taking this effect into account. In addition, accurate simulation of unsteady flow may save on the over-design that might result from assuming that waves did not change shape. (If you are used to associating the word *wave* with an effect that lasts for a few seconds, remember that increases in flow in sewer systems resulting from rainfall, i.e., storm waves, may last many hours.)

There is a number of methods available for analysing unsteady conditions in a sewer system. Some are based on approximations, and others on attempts to give full theoretical treatment of the physics of the flow. The main methods are for free surface flows. Adjustments for surcharged pipes are considered later.

19.4.1 The Saint-Venant equations

For gradually varied unsteady flow in open channels (including part-full pipes), the full one-dimensional theoretical treatment leads to a pair of equations usually referred to as "the Saint-Venant equations" after A.J.C. Barré de Saint-Venant, who first published them in the middle of the nineteenth century. A clear derivation of these equations is available in Chow (1959).

There are two equations: a dynamic equation and a continuity equation. The dynamic equation can be written as follows:

$$S_f = S_o - \frac{\partial y}{\partial x} - \frac{v}{g} \cdot \frac{\partial v}{\partial x} - \frac{1}{g} \cdot \frac{\partial v}{\partial t} \tag{19.1}$$

$$\underrightarrow{\quad}_1$$

$$\underrightarrow{\quad}_2$$

$$\underrightarrow{\quad}_3$$

where y is the flow depth (m), v is the velocity (m/s), x is the distance (m), T is the time (s), S_o is the bed slope (–), and S_f is the friction slope (–).

In this form, the components that make up the equation can be identified. The part marked "1" above includes no variations with distance or time and applies to uniform steady conditions. Taking the terms up to "2" includes variations with distance but not with time, and applies to nonuniform steady conditions. The whole equation "3" also includes variation with time and applies to nonuniform unsteady conditions (the ones that interest us here).

Equation 19.1 is commonly presented in terms of flow rate rather than velocity and is given together with the continuity equation:

$$\frac{\partial Q}{\partial t} + \frac{\partial}{\partial x}\left(\frac{Q^2}{A}\right) + gA\frac{\partial A}{\partial x} - gA(S_o - S_f) = 0 \tag{19.2}$$

$$B\frac{\partial y}{\partial t} + \frac{\partial Q}{\partial x} = 0 \tag{19.3}$$

where Q is flow rate (m³/s), A is the area of flow cross section (m²), and B is the water surface width (m).

The Saint-Venant equations are valid in situations where the following assumptions are appropriate:

- The pressure distribution is hydrostatic.
- The sewer bed slope is so small that flow depth measured vertically is almost the same as that normal to the bed.
- The velocity distribution at a channel cross section is uniform.
- The channel is prismatic.
- Friction losses estimated by steady flow equations (see Chapter 7) are valid in unsteady flow.
- Lateral flow is negligible.

19.4.2 Simplifications of the full equations

Equation 19.1, the dynamic equation, includes terms for the bed slope, the friction slope, the variation of water depth, and the variation of flow rate with distance and time. Some of these terms may be more significant than others, giving opportunities for simplifying the equations.

The greatest simplification is to assume that most of the terms in Equation 19.1 can be ignored, and reduce it simply to

$$S_o = S_f \qquad (19.4)$$

This is the equivalent of ignoring all but part 1 of Equation 19.1, and implies that the relationship between flow rate and depth is the same as it would be in steady uniform flow (as in Section 7.4).

Combining Equations 19.3 and 19.4 gives

$$\frac{\partial Q}{\partial t} + c \frac{\partial Q}{\partial x} = 0 \qquad (19.5)$$

The wave, called a *kinematic wave*, does not attenuate but translates at the wave speed, c.

A less extreme simplification involves ignoring just the variation of flow rate with time (the "unsteady" effects), that is using Equation 19.1 up to "2". The equivalent of Equation 19.5 is now

$$\frac{\partial Q}{\partial t} + c \frac{\partial Q}{\partial x} = D \frac{\partial^2 Q}{\partial x^2} \qquad (19.6)$$

This *diffusion wave* travels at the same speed, c, as the kinematic wave but, as a result of the term on the right-hand side of the equation, is subject to diffusion. The diffusion coefficient, D, regulates the attenuation of the wave as it propagates downstream. For simplicity, c and D are usually regarded as constant although they vary.

Table 19.2 summarises the various simplified equations and their applications. Ponce et al. (1978) quantify the range of application of the simplifications as follows:

P	Diffusion
$TS_o(v_o/d_o) > 171$	$TS_o(gd_o)^{0.5} > 30$
T	Duration of flood wave (s)
v_o	Initial velocity (m/s)
d_o	Initial depth (m)

Table 19.2 gives the range of hydraulic conditions accounted for in the methods.

19.4.3 Numerical methods of solution

The main equations derived in the last two sections are partial differential equations since Q (or v) and y are functions of both distance (x) and time (t). The most common method of solution is using finite differences, involving dividing distance and time into small,

Table 19.2 Hydraulic conditions accounted for by simplifications of wave equations

Accounts for	Kinematic wave (1)	Diffusion wave (2)	Dynamic wave (3)
Wave translation	✓	✓	✓
Backwater	✗	✓	✓
Wave attenuation	✗	✓	✓
Flow acceleration	✗	✗	✓

discrete steps. This can be represented on a schematic two-dimensional grid showing x and t together, as in Figure 19.1.

In Figure 19.1, the distance step is shown as Δx and the time step as Δt. The points where the lines intersect are *calculation nodes*. The nodes marked with a circle are the distance nodes for time = 0 (say, the start of the calculation). Flow conditions for these are likely to be known (e.g., baseflow all along the pipe before the beginning of storm flow). The nodes marked with a square are the time nodes for distance = 0 (say, the upstream end of the pipe). Flow conditions for these may also be known (e.g., inflow of stormwater varying with time).

To calculate the conditions at the node marked with the arrow using a kinematic wave model, we can use the known values at neighbouring nodes. Once this calculation is complete, it can be repeated for successive distance nodes (the nodes to the right of the arrow on the same horizontal line). When we have calculated conditions at all distance nodes for that time step, we can proceed to the next time step (the horizontal line above the one we have dealt with).

We have to take into account the *boundary conditions*—the hydraulic conditions at the limits of our system, for example, the flow rate at the inlet and the water level at the pipe outlet.

There are many ways in which finite difference calculations can be carried out. In Figure 19.1 the jth distance node and the nth time node are marked.

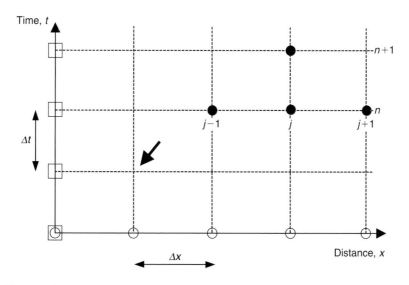

Figure 19.1 Numerical solution of flow equations: x–t grid.

Suppose we are solving Equation 19.5:

$$\frac{\partial Q}{\partial t} + c\frac{\partial Q}{\partial x} = 0$$

We write the value of Q at the jth distance node and the nth time node as Q_j^n.

Suppose we have reached the stage where we wish to calculate Q_j^{n+1}. If we assume that Q varies linearly within each distance and time step, we can write:

$$\frac{\partial Q}{\partial t} \text{ as } \frac{Q_j^{n+1} - Q_j^n}{\Delta t}$$

and

$$\frac{\partial Q}{\partial x} \text{ as } \frac{Q_{j+1}^n - Q_j^n}{\Delta x} \text{ (forward difference)}$$

$$\text{or } \frac{Q_j^n - Q_{j-1}^n}{\Delta x} \text{ (backward difference)}$$

$$\text{or } \frac{Q_{j+1}^n - Q_{j-1}^n}{2\Delta x} \text{ (central difference)}$$

If we substitute these types of expression into Equation 19.5, it can be rearranged with the unknown, Q_j^{n+1}, by itself on the left-hand side, and solved directly. In solutions where this is the case, the method is described as *explicit*.

There are many different explicit finite difference methods, some incorporating "half-steps", for example, the intermediate calculation of $Q_{j+(1/2)}^{n+(1/2)}$, to improve accuracy and stability.

More complex formulations, in which the unknown value appears on both sides of the finite difference equation, are described as *implicit*. Even though each set of equations is more difficult to solve, implicit methods can be used with longer time steps than explicit methods, and therefore often have the advantage of being more computationally efficient. More information on these methods can be found in the texts recommended at the end of this section.

These approaches to numerical solution can suffer from various anomalies, especially when the input data contain rapid changes. Two common problems are illustrated in Figure 19.2. Figure 19.2a shows *numerical diffusion*, where values are smoothed out in what should be a zone of rapid change. Figure 19.2b shows *numerical oscillation*: small fluctuations at points of change. These problems arise because the method requires variations that are actually continuous to be treated as a series of linear steps.

The problems are overcome by selection of appropriate solution methods and suitable time and distance steps. Explicit schemes usually have to satisfy the *Courant condition* to maintain stability:

$$\frac{\Delta x}{\Delta t} \geq c \tag{19.7}$$

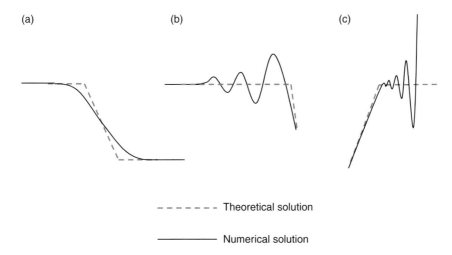

(a)   (b)   (c)

- - - - - - -   Theoretical solution

——————   Numerical solution

Figure 19.2 (a) Numerical diffusion; (b) numerical oscillation; (c) instability.

Figure 19.2c shows instability, in which errors introduced by the finite difference method are amplified as the calculation proceeds. This may cause the solution to go completely out of control. A common cause is the use of a time step that is too long.

More information on numerical methods of solving partial differential equations can be found in a number of books, some highly specialised. Accessible introductions to the subject are given by Chadwick et al. (2013), Koutitas (1983), Vreugdenhil (1989), and Yen (1986).

19.4.4 Surcharge

It is common for pipes in drainage systems to occasionally experience surcharge—to run as full pipes under pressure rather than open channels with a free surface (see Section 7.4.5) and indeed some pipes such as inverted syphons (Section 8.4) are always surcharged and can be modelled as such.

The methods described so far in this section are for free-surface flows. In a surcharged pipe, Equation 19.3, for example, would present problems; the terms B (water surface width) and y (flow depth) would be meaningless: B is equal to zero, and y is always equal to pipe diameter regardless of flow rate.

A concept that allows Equations 19.2 and 19.3 to continue to be applied in surcharge conditions is the *Preissmann slot* (Yen, 1986). An imaginary slot (Figure 19.3) is introduced above the pipe, which allows y to exceed pipe diameter and give the effect of pressurised flow.

The width of the slot b is calculated precisely to suit the conditions and must not be so wide that it has a significant effect on continuity. For a circular pipe of diameter, D, this is given by

$$b = \frac{\pi D^2 g}{4c^2} \tag{19.8}$$

where c is the speed of a wave in a pressure pipe (m/s).

In large piped systems, care is needed as the Preissmann slot can introduce significant storage that is not physically present. This can affect model accuracy and result in an under-prediction of flooding in surcharge conditions or inaccurately represent a storage calculation.

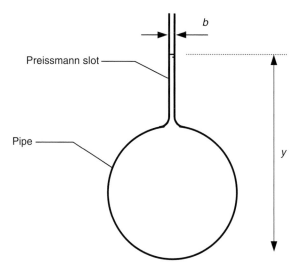

Figure 19.3 Preissmann slot.

19.5 SETTING UP AND VALIDATING A SYSTEM MODEL

19.5.1 Defining the model purpose and detail

The various purposes of a model are described in Section 19.3.3. The purpose relates to the type of detail required in the model, and ultimately to the cost of the exercise. CIWEM (2017) defines three categories of detail a model may have:

- *Limited detail/simplified*: used to provide a limited indication of performance and generate flows for other parts of the model
- *Planning/general purpose*: used to understand comparative risk across an area
- *High level of detail*: appropriate where detailed investigations and design are required

It is not only the level of detail that is important, but also the quality of the data that sit behind this, and the level of data validation and collection that help to improve the confidence in any model output.

19.5.2 Input data

A flow model requires physical definition of the drainage system (at the level of detail defined as above), information about the catchment for runoff calculations, and about specific inflows. In a conventional system in which pipes run between manholes, this generally requires the type of data listed in Table 19.3. Particular attention is given to ancillaries (CSOs, pumping stations, syphons, etc.) as they form controls in the network and require representation in the model. All manholes and pipes have reference codes (see Chapter 6).

If a package is used for design, the input data are for the proposed system, and the model simulates the response of that system to specified rainfall conditions.

If a package is used to model an existing drainage system, the input data are that which exists for the system. If the system has poor records, it may be necessary for extensive sewer survey work to be carried out. This is expensive, but good sewer records are of great value,

Table 19.3 Data requirements for flow models

Element	Data type
Manholes	Reference number
	Location (map reference)
	Ground level
	Storage volume
	Head-loss parameter
Pipes	Reference number
	Connectivity information
	Length
	Shape
	Size
	Roughness
	Invert level
Catchment	Total area contributing
	Pervious/impervious areas
	Slope of ground
	Soil data
	Details of gullies (for detailed flood analysis)
	Flooded areas
Inflows	Population-derived dry weather flow
	Infiltration
	Industrial inflows/trade effluent
	Overland flows
Ancillary structures	Data to define hydraulic performance
CSOs	Geometry
Pumping stations	Inflow/outflow arrangements
Outfalls	Screens
	Geometry
	Trigger and control levels
	Pump characteristics
	Hydraulic characteristics
	Boundary conditions, e.g., river interaction

not just for modelling (see Section 17.3). Significant effort may be needed to define the catchment area data, especially where there is incomplete information on the connection of particular properties or sub-areas to the sewer system. Data are typically held and processed within geographical information systems (GISs).

The model simulates the response of the system (represented by the previously provided data) to specified rainfall patterns. The rainfall data may take a number of forms:

- For verification or calibration, using rain gauge or rainfall radar records
- For simulation, analysis, and design, sets of synthetic storms (Section 4.4), and or time-series rainfall (Section 4.5).

19.5.3 Model testing

Initial checks on a catchment model are needed to make sure that the model is behaving satisfactorily in mathematical terms, and to eliminate obvious mistakes in the input data. It is necessary to check against instability for a variety of extreme conditions. In addition, the overall volume entering the system, or part of the system, should be compared with the overall volume leaving.

19.5.4 Flow surveys

To create a successful working model of an existing system, data are needed on how the catchment actually responds in particular rainfall conditions. Flow surveys give records of the hydraulic performance of the system, and the conditions (mainly rainfall) that produced that performance. If that rainfall is used as input to the model, a comparison between the resulting simulation and the flow survey data should tell us how much confidence we may have in our model. This is for a snapshot of time (the duration of the flow survey); hence, it is important to consider longer-term data records as well.

Appropriate rainfall measurement is a very important part of a sewer flow survey and is covered in Section 4.2.

In-sewer measurement (short-term flow surveys) is carried out at selected sites, and appropriate selection of sites is essential. There is a number of considerations. The number of sites must suit the purpose of the model to enable the required level of confidence to be achieved; their location must allow comparisons with the simulation at locations of particular interest. Critically though, they must also be appropriate from a hydraulic and a practical point of view: with safe and adequate access, and without excessive disruption to the flow, or deposited sediment.

A typical monitoring point is in a manhole, with the data-logging equipment located near the top of the manhole, and a sensor in the sewer that monitors depth and velocity. Some sensors enable depth and velocity to be measured from the same point. For example, the sensor may be positioned on the invert and is wedge shaped to minimise the buildup of solids. It measures depth using a pressure transducer, and velocity using a *Doppler shift* system. (An ultrasonic signal is transmitted against the oncoming flow and is reflected back by solid particles or air bubbles. The reflected signal changes in frequency, and the magnitude of the change is proportional to the particle velocity, thus giving water velocity.) These give reasonable results and are widely used. Accuracy is in the order of $\pm 10\%$ on flow rate under ideal conditions. An alternative method of detecting depth is a top-down ultrasonic metre, positioned above the water surface. These are often fitted along with the sensor in the invert, or alternatively may be used to monitor depth only.

Velocity may also be measured using an electromagnetic metre (where a current is induced in a conductor moving across a magnetic field that is directly proportional to the speed of movement of the conductor).

Figure 19.4 gives a plot of velocity against depth for a large number of readings at a monitoring point in a sewer system. This can be used to check the data for consistency. A similar *scattergraph* is formed by plotting log flow rate against log depth.

Long-term or permanent monitoring often takes place at ancillaries (CSOs and pumping stations) and the wastewater treatment plant (WTP). These data provide a more complete picture of the catchment response over time to rainfall across a range of conditions, seasonally and annually. Whereas short-term flow surveys may use a higher density of rain gauges, long-term monitoring typically requires good rainfall radar supported with a smaller number of rain gauges to enable validation.

19.5.5 Model calibration/verification against measured flow data

It is useful at this point to discuss some of the terminology used during the model development and building process, particularly the distinction between the terms *calibration* and *verification*. During calibration of a model, the most appropriate values of the various model parameters are sought. In verification, the model parameters and their values have now been established, and input is run through the model to produce results that are compared with

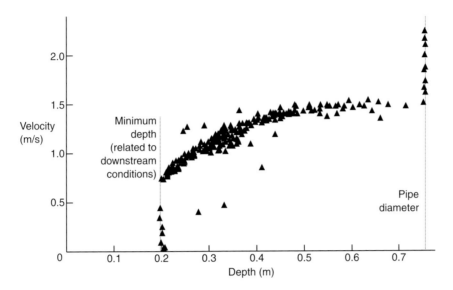

Figure 19.4 Velocity against depth "scattergraph."

known conditions (e.g., flow survey data). In this way, the agreement between computed and observed values can be verified. The verification is, strictly speaking, only valid for that particular location and only over the range of the available data.

In the past, verification often took place against three storms. This is still a focus, however it is good practice to simulate the model for the "full period" of the flow survey data and consider verification over this time frame. This provides a more representative view of performance, accounts for the antecedent conditions, and matches how the hydraulic models are run (for the full period). Comparing against longer-term or permanent monitoring also enables seasonal effects to be evaluated.

In theory, if a model is deterministic, it does not need calibration. If all the input parameter values are accurate and the physics of the processes are simulated sufficiently well, accurate results should be produced without calibration. In practice, however, many of the input parameters are not or cannot be accurately ascertained and the physics is only approximated, thereby making it necessary to resort to default values, which may not be representative of the site in question.

Price and Osborne (1986) define verification as the art of demonstrating that a model, which incorporates previously calibrated sub-models, correctly represents the reality of the particular system being studied. They see calibration as occurring only during model development, whereby the physical phenomena represented by the various sub-models are tested, under varied conditions, on many catchments to ensure approximation to observed data with sufficient accuracy. Verification is then carried out by the model user (the drainage engineer), primarily to demonstrate the physical details of the sewer system are correctly incorporated in the model. Figure 19.5 is a plot of actual and simulated hydrographs produced during verification of a flow model.

When calibration as defined above is carried out, Vojinovic and Solomatine (2006) identify a number of possible approaches. It may be a trial-and-error process in which parameters are adjusted "according to the modeller's judgement" (and should be based on data/other evidence relating to fundamental principles and mechanisms so not just to get a "better fit with observed data") in order to provide a better match with observed data. Alternatively, this adjustment process may be built into the modelling software (see also Wangwongwiraj

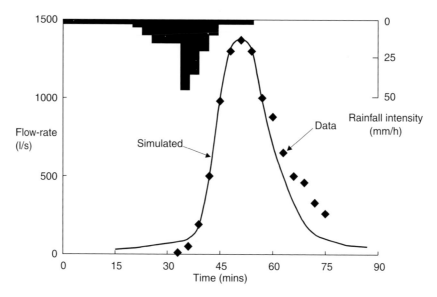

Figure 19.5 Model verification: hydrograph.

et al., 2004), or the software may be based on a number of different models whose outputs can be weighted to improve the match. Comparing these methods with each other and with what we have called verification, Vojinovic and Solomatine (2006) found that all of the approaches will "incur a degree of uncertainty." They warned that "making decisions on a single model output without the consideration of uncertainties and risk involved is very dangerous and misleading." In particular, changing values without understanding the "engineering consequences" can result in a tool that cannot be readily used to understand problems or test certain types of solutions.

19.5.6 Evaluating confidence

Evaluating model confidence provides a measure of the suitability of using the model against its intended purpose. This evaluation can be through various ways and range from simple to complex and qualitative to quantitative. CIWEM (2017) considers the confidence in various aspects of the model components, recognising their importance and contribution in defining the quality of the model. This includes

- Asset data—considers where the data come from, how much has been surveyed/confirmed, and the quality of the survey method.
- Sub-catchment—relates to the areas defined including impermeable area, method of area assessment and amount of data collected, confirmation of populations, and input data.
- Flow survey—considers the quality and extent of the flow survey and over what period. Includes both short-term, long-term, and permanent survey collection.
- Dry weather verification—considers the flow parameter predictions within an upper and lower bound tolerance.
- Storm verification—concerns the ability to represent the flow, depths, volumes, shape, and timing of flows on individual and full period events.
- Historical verification—considers the performance of the network against known flooding and longer-term data sets (such as monitors at works, pumping stations, and event duration monitors at overflows).

The CIWEM guidance provides a framework to enable model confidence to be assessed, and those who commission the building or use of a model can use it to meet their own defined needs.

19.5.7 Documentation

Setting up and developing a sewer system model is an investment in the gathering of information. Some of the benefits may be physical, in the form of improved system performance as a result of successful rehabilitation work. But many of the benefits remain as information: data on the properties of the system, understanding of its response to particular conditions, and the potential to predict its performance under new conditions. The value of this information is related to the quality and completeness of the documentation of all stages of model development. The conventions used in the documentation depend on the type of model and on the standard practice within the organisation carrying out the modelling.

19.6 MODELLING FLOODING

An important role of sewer flow models is to represent the performance of the system (both minor *and* major) during extreme rainfall events. Under those conditions the system will initially surcharge and then may well "spill" onto the urban surface: *exceedance flow* (see Chapter 11). In this section, we discuss and evaluate both the conventional approach to modelling this phenomenon and the many newer methods developed. Of course, the model will only predict flooding where it has a point to escape, and this should always be considered when comparing model predictions to reality (e.g., if the property drainage is not included in the model, this might be the lowest point of escape and hence different than that of the model).

A key component is discussed first: the site terrain model (DTM/DEM).

19.6.1 DTM/DEMs

Commercially available flow models have a GIS interface that is intimately linked to a representation of the "height characteristics" of the urban catchment surface. There is some confusion over the terms used, but in urban flood modelling a digital terrain model (DTM) is taken to be the topographic model of the ground surface only. A digital elevation model (DEM) is based on the underlying DTM but also includes elevation data of buildings and other prominent features. A DEM forms the underlying foundation to any analysis of surface flooding.

In the UK, a variety of data sources is available to derive suitable DTM/DEMs ranging from basic land survey to airborne-generated Light Detection and Ranging (LiDAR). Those now typically used are summarised in Table 19.4.

19.6.2 Virtual flood cones

Models, simulating in one dimension only use virtual flood cones or reservoir on top of each node to represent the flood flows once they have left the underground piped system in which they are temporarily stored (see Figure 19.6). Typically, the water is allowed to flow back to the underground system as long as the capacity allows, or it can be lost from the system altogether. The cone is only loosely related to the surrounding topography (by its user-defined geometry), so it can only give a rough measure of flow depth (particularly important when

Table 19.4 Available sources of DTM/DEM data

Land survey
  • The oldest and most widely used method of obtaining topographical data for a specific site
  • The most accurate form of data collection with accuracy up to ±1 cm
GIS sewer records
  • Include some form of manhole geo-referencing and a relative ground level
  • Used to produce an initial catchment DTM using thematic mapping capabilities within GIS
Light Detection and Ranging (LiDAR)
  • A process that uses an aeroplane/helicopter mounted with an optically based scanner
  • Typically available on a horizontal grid between 0.25 and 2 m
  • Vertical accuracy typically ranges from ±5 to 15 cm (although some older data sets may be less accurate)

Source: Adapted from Hale, J.N. 2004. Urban flood routing in the UK ... the next step? *Proceedings of UDM04 Sixth International Conference on Urban Drainage Modelling*, Dresden, Germany, August, 97–104.

evaluating flood damage—see Section 11.4). Although the interaction between the flood flows and pipe flows is captured to some extent, this is of limited realism. Care is needed in defining the cone size, as for example, too small a cone can create substantial head and force more flow through the system than in reality. However, this approach is quick and functional and does give an indication of surface flood volume. Where manholes cannot flood, they can be sealed in a model, and this can affect the hydraulic profile in larger return period events.

19.6.3 One-dimensional–two-dimensional (1D–2D) coupled models

As the name suggests, 1D–2D models consist of the tight coupling of 1D pipe flow models with 2D surface flow models. In 2D models, the solution domain (the whole DTM) is

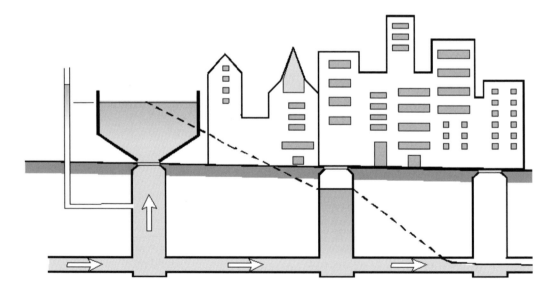

Figure 19.6 Conventional approach to sewer flood modelling using a virtual cone. (After Maksimović, C. and Prodanović, D. 2001. *Urban Drainage Modeling: Proceedings of the Speciality Symposium of the EWRI/ASCE World Water and Environmental Resources Congress*, Orlando, Florida, May.)

discretised as a coordinate system of nodes (i.e., grid points), where each point is represented by spatial coordinates (X, Y, Z). The flood routing model typically also solves a version of the Saint-Venant equations. Both models share two unknown variables, namely, velocity (or flow rate) and water level, with the difference that in the 1D case, the velocity has only one component, while in the 2D case it has two orthogonal components (X, Y).

The continuity equations for network nodes, energy equations for nodes and pipe/channel ends, and the complete Saint-Venant equations for flow in pipes and surface channels are simultaneously solved. They can represent the interaction of the major and minor systems both in terms of outfalls linked to a natural major system (e.g., watercourse, tidal) and water exiting and returning to the minor system (Figure 19.7). Results are expressed as standard sewer outflow hydrographs and exit flood volumes, but also the volume of flood flow remaining on the surface in ponds and flood levels in flooded areas.

1D–2D models, especially when coupled, are considered to be the most accurate representations of urban surface flooding currently available but achieve this accuracy at the expense of high computational burden both in term of time and data requirements (Bamford et al., 2008).

19.6.4 Rapid flood spreading models

Rapid flood spreading models take the volume of flow generated by a pipe-flow model and distribute it realistically but rapidly over the catchment surface. The earliest versions of this type of model automatically delineated the individual flooding sub-catchments. This was carried out by establishing a grid on the catchment. Then by comparing the relative elevation of each grid intersection, it was possible to compute flow direction across every cell. Knowing this for each cell helped the lowest (pit) cells to be identified and the boundary between each pit or pond to be found. It was also established whether or not each of the pits contained a sewer node (manhole) and where it was located in the pond. Individual properties were then assigned to the grid cells. Once complete, flood volume could be quickly and

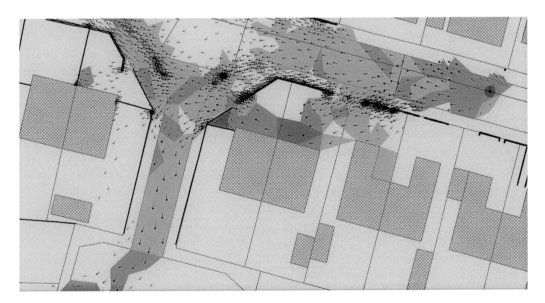

Figure 19.7 Example of a 1D–2D model with water escaping from the drainage system, flowing overland, and ponding around properties. Darker colour represents an increase in water depth, and thick lines indicate physical barriers such as walls that direct flow.

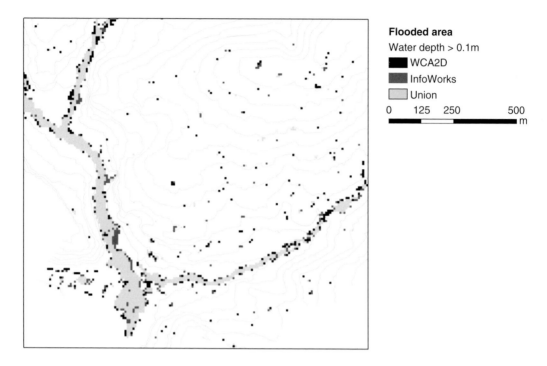

Figure 19.8 Area predicted as flooded by benchmark model only (dark grey pixel), by WCA2D only (black pixel), and by both models (light grey pixel) at 8 m resolution. (Courtesy Dr. Albert Chen.)

reliably distributed across the catchment (Ryu, 2008). The advantage of this approach is its relative ease of use and therefore speed. Flooding is assigned to local low points in the catchment and can also cross sub-catchment boundaries. However, it does not represent any flow movement and hence has no time component.

More recent approaches use 2D cellular automata (CA)–based models that employ simple transition rules and a weight-based system rather than solving the full shallow water equations. These do allow representation of flow movement. The CA technique provides a versatile method for modelling complex physical systems using simple operations (Wolfram, 1984) allowing significant increase in computational speed. Guidolin et al. (2016) report on this type of flood spreading model WCA2D that is capable of simulating water depth and velocity variables with reasonably good agreement against a benchmark 1D–2D model (Figure 19.8), but using only a fraction of the computational time and memory. In the case of a real-world example, the model run times were up to eight times faster.

19.7 WATER QUALITY MODELLING

Sewer systems can be direct contributors of pollution to waterbodies. Stormwater outfalls can discharge pollution, as discussed in Section 5.4; serious short-term pollution can arise from combined sewer overflows, as considered in Chapters 1 and 12. The effects can be counteracted by approaches such as reducing the amount of stormwater entering the system, building additional storage within the system, rehabilitating or replacing CSO structures or by treating the discharge. In order to design these measures to achieve high standards, and to evaluate them fully, it is necessary to have information on the rate at which pollutants flow into the structures from the sewer system, and how the pollutants are distributed in the dry weather and storm flow. This information is provided by models of *quality* (distribution of pollutants) in sewer flow.

Simplified models of quality have been linked to flow models of sewer systems for some time. For example, the *Sewerage Rehabilitation Manual* in 1986 gave factors that could be multiplied by pollutant concentrations in dry weather flow to give estimated average concentrations in storm flow. These could then be used, in conjunction with a flow model, to give total pollutant loads to watercourses resulting from CSO discharge. However, time-dependent effects like a first flush cannot be predicted using this method.

By the mid-1980s, deterministic *flow* models had become so popular and widespread in their use it seemed an appropriate step to develop deterministic models of sewer-flow *quality* to a comparable level of detail. In the same way that flow models gave output in the form of hydrographs (flow or depth) at specified points, quality models would give pollutographs, the variation of concentration of pollutants with time. In the same way that flow models are used to assess alternative proposals for hydraulic rehabilitation of sewer systems, quality models would be used to assess proposals for reducing pollution, particularly from CSOs.

In the United Kingdom, these plans became part of the Urban Pollution Management programme – a series of linked research projects in the 1980s and 1990s, which, as well as development of a sewer-quality model, also included extensive in-sewer monitoring, and development of models covering rainfall, wastewater treatment, and river quality. The outcome was the *Urban Pollution Management (UPM) Manual* (Foundation for Water Research [FWR], first edition: 1994, second edition: 1998, third edition: 2012), which presents a comprehensive set of procedures for controlling pollutant discharges from sewer systems, discussed in Chapter 2.

Physically based deterministic quality models are now widely available in practice, but it should be recognised that the accuracy of such models is not as high as flow models because of the complex physical, chemical, and biological processes involved. Willems (2008), for example, found that the uncertainty in the results contributed by the water quality part of a sewer model was an order of magnitude higher than that for the flow component. That said, with careful and expert calibration against measured data, these models can be a powerful tool to estimate sewer impacts on receiving waters and so develop appropriate solutions.

This section discusses some of the requirements that are common to all methods of sewer quality modelling.

19.7.1 The processes to be modelled

The main aim of a sewer-quality model is to simulate the variation of concentration of pollutants with time at chosen points in a sewer system. These simulations will be used to understand the impact on the receiving waters and where needed improve the performance of the system, for example, by aiding the design of CSOs to enhance the retention of pollutants in the sewer system. It is also common to monitor and model the receiving waterbody, as the individual discharge alone does not demonstrate its impact.

The main quality parameters, modelled separately (directly or applied as standard concentrations), will be the standard determinands described in Chapter 2, including suspended solids, oxygen demand (biochemical oxygen demand [BOD] or chemical oxygen demand [COD]), ammonia, dissolved oxygen (DO), and others depending on the model.

Pollutants find their way into combined sewers from two main sources: wastewater and the catchment surface. Once in the system, the material may move unchanged with the flow in the sewers, be transformed, or become deposited. Deposited pollutants may subsequently be re-entrained or eroded, usually in response to an increase in flow.

The main elements of the system that influence the quality of the sewer flow are indicated schematically in Figure 19.9 (for a combined sewer).

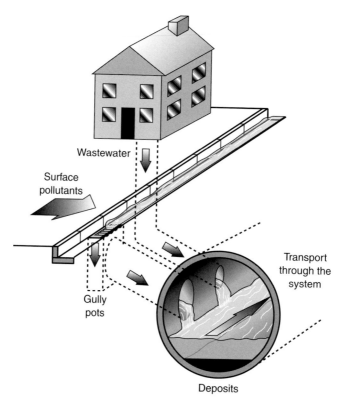

Figure 19.9 Main elements that influence sewer-flow quality (combined sewer).

19.7.2 Wastewater inflow

The flow rate and the concentration of pollutants vary with time in dry weather flow in a fairly repeatable daily (diurnal) pattern (see Chapters 3 and 9). The variation is related to patterns of human behaviour, and there tend to be differences between weekdays and weekends. Industrial and commercial flows may be present, in addition to domestic contributions, and these can lead to significant differences in pollutant load and profile. Infiltration can be a significant fraction of dry weather flow (Chapters 3 and 9), and tends to dilute the wastewater.

19.7.3 Catchment surface

In periods of dry weather, there is a buildup of pollutants on roads, roofs, and so on, and these are washed into the drainage system by the next rainfall (Chapter 5). In general terms, the amount entering the drainage system depends on the quantity of material accumulated on the catchment surface, intensity of the rain, intervention points to catch pollutants (such as gullies), level of maintenance, and nature of the overland flow.

19.7.4 Gully pots

As with catchment surfaces, pollutants tend to build up in gully pots in dry weather. Material remaining in the liquor is regularly washed into the drainage system, even by minor storms. Only high return period storms are thought to disturb previously deposited, heavier solids.

Table 19.5 Chapters relevant to quality modelling

Topic	Chapter
Wastewater inflow	3, 9
Catchment surface	5, 10
Sediment	16
Flow modelling	This chapter
First foul flush	12

19.7.5 Transport through the system

Once they have been carried into the pipe system, and provided they are not attached to solids that are deposited, pollutants are transported by the moving liquid. It is generally assumed that the pollutants move at the mean liquid velocity, though there are cases when this might be inappropriate, as considered later.

19.7.6 Pipe and tank deposits

At low flows in sewers, especially during the night, solids (and pollutants attached to them) may settle and form deposits. At higher dry weather flows or during storms, they may be re-eroded, releasing suspended solids and dissolved pollutants into the flow. The resulting increase in pollutant concentration depends on the characteristics of the system, of the catchment, of the dry weather and storm flows, and the antecedent dry period. As is clear from Chapter 16, the deposition and erosion of sediments is a complex subject, with many different types of sediment and associated pollutants. Flow patterns in tanks and large trunk sewers also lead to deposition, and the subsequent erosion can have a significant effect on quality.

The mechanisms by which pollutants are introduced to the flow, and are subsequently transported with it, are heavily dependent on hydraulic conditions. Therefore, physically based deterministic quality models need to be based on hydraulic models of the sewer system. The main deterministic quality models in current use are based on established flow models. The accuracy of quality simulations is strongly affected by the accuracy of the hydraulic simulation on which they were based.

In some systems, a significant feature in the variation of sewer-flow quality with time is the first flush in early storm flows, which may contain particularly high pollutant loads. This effect is considered in Section 12.3.2. It is important that it is represented in a model of sewer-flow quality.

Table 19.5 lists the chapters of this book that deal with the processes that should be represented in a sewer quality model.

19.8 MODELLING POLLUTANT TRANSPORT

19.8.1 Advection/dispersion

Transport of pollutants with the flow is normally represented by one of two alternative forms of equation:

$$\frac{\partial c}{\partial t} + v \frac{\partial c}{\partial x} = 0 \tag{19.9}$$

or

$$\frac{\partial c}{\partial t} + v\frac{\partial c}{\partial x} = \frac{\partial}{\partial x}\left[D\frac{\partial x}{\partial x}\right] \qquad (19.10)$$

where x is the distance (m), t is the time (s), c is the concentration of pollutant (kg/m³), v is the mean velocity of flow (m/s), and D is the longitudinal dispersion coefficient (−).

Equation 19.9 represents *advection*, movement of pollutants at the mean velocity of flow. Equation 19.10 additionally includes *dispersion*, the spreading out of pollutants relative to mean velocity. These two mechanisms are illustrated in Figure 19.10.

Figure 19.10a shows a slug of pollutant in a pipe (equally distributed horizontally and vertically), which is moved along at the mean flow velocity without spreading out: *advection only*. Figure 19.10b shows a similar slug of pollutant that is moved along the pipe at the mean flow velocity and at the same time spreads out: *advection and dispersion*.

Nearly all practical examples of sewer flow are strongly dominated by advection. For this reason not all sewer quality models include dispersion.

Numerical methods for solving these equations are similar to those outlined in Section 19.4.3.

19.8.2 Completely mixed "tank"

An alternative to Equations 19.9 and 19.10 is to treat each pipe length as a conceptual tank in which pollutants are fully mixed with the flow. The governing equation is

$$\frac{d(Sc)}{dt} = Q_1 c_1 - Q_0 c_0 \qquad (19.11)$$

where c, c_1, c_0 are concentration in pipe length, at inlet to pipe, at outlet (kg/m³); Q_1, Q_0 is flow at inlet to pipe, at outlet (m³/s); and S is the volume of liquid in the pipe length (m³).

This equation can be solved to give progressive estimates of concentration in each pipe as a whole. It does not include a distance or a velocity term, and is therefore not capable of explicitly modelling the progress of pollutants at mean velocity.

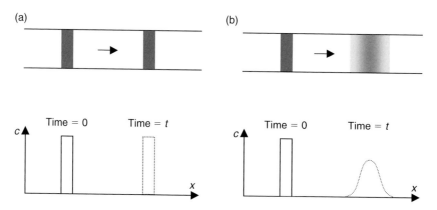

Figure 19.10 (a) Advection; (b) advection and dispersion.

19.8.3 Sediment transport

Pollutants transported in the body of the flow by advection and dispersion are either dissolved or suspended. Pollutants in solution remain in that form whatever the flow regime, although they may be transformed by biochemical processes (discussed in more detail in the next section). Pollutants in suspension may, however, be affected by the flow regime. At low flows they may concentrate in a near-bed layer or deposit to form a sediment bed, and at high flows they may re-erode. Larger, heavier solids may never (or rarely) achieve suspension, yet be transported as bed-load. The complex hydraulics of sediment movement is discussed in Chapter 16.

All sewer quality models have at least some representation of the movement of solid-associated pollutants through the system in terms of

- Mechanics: entrainment, transport, and deposition
- Sediment bed
- Solid attachment

19.8.3.1 Mechanics

A relatively straightforward way to model entrainment, transport, and deposition of pollutants is by using one of the sediment transport equations that predict the volumetric sediment carrying capacity of the flow c_v (e.g., Macke's equation for suspended solids transport, Equation 16.4). Thus, at each time step, for an incoming pollutant concentration c, if

- $c < c_v$: all the incoming pollutant is transported, and if deposited sediment is available this may be eroded up to the carrying capacity c_v.
- $c > c_v$: only c_v/c of the incoming pollutant is transported. The remainder is subject to deposition.

The most straightforward approach is to consider just one type of solid. In principle, many solid size fractions could be represented, although considerable data would be needed for calibration/verification. Representing two fractions (coarse and fine) or more can improve model performance, although would require more data to calibrate a quality model.

We saw in Chapter 16 that sediment can travel in suspension and as bed-load. Some models aim for a more realistic representation by using equations for total sediment transport. However, the mechanics of both types of transport process can be represented separately.

19.8.3.2 Sediment bed

In the appropriate hydraulic conditions, pollutants will move out of suspension and become deposited on the pipe invert. Most models allow for the development of a sediment bed. At its simplest, this can be considered to have no impact on the hydraulics of the flow. A more realistic representation will include loss of flow area and increased hydraulic roughness that can be communicated to the flow model. A further elaboration could be to include the hydraulic effects of the sediment bed forms (Chapter 16).

The simplest structure of the bed is a simple, single layer of sediment that is supplied by deposition and removed by erosion. The settling velocity of the solid may be used to fix the rate of deposition. Erosion may be controlled by the sediment transport capacity of the flow (as previously mentioned) or by a defined critical shear stress. A more refined approach is to represent two layers; one containing stored (type A) sediment, and the other, erodible type C deposits (see Section 16.5).

Time-dependent effects such as consolidation and cohesion have also been introduced in a simple form in some models. Such software packages also include the ability to transport, deposit, and erode different sediment fractions.

19.8.3.3 Solid attachment

Pollutants may be modelled in two forms: dissolved and solid-attached, for the reason mentioned at the beginning of this section. Potency factors f are specified (Huber, 1986), which simply relate concentration of solid-attached pollutant (c_s) to solid concentration c, as $c_s = fc$. An example for a pollutant n is given in Equation 19.12:

$$K_{Pn} = C1(IMKP - C2)^{C3} + C4 \tag{19.12}$$

where K_{Pn} is pollutant n potency factor, $IMKP$ is the maximum rainfall intensity over a 5-minute period (mm/h), and $C1$–$C4$ are coefficients varying for each pollutant.

19.8.4 Gross solids

A basis for predicting the behaviour of gross solids is described in Section 9.5. This information has been used by Butler et al. (2003) to develop a model of gross solid transport that was subsequently developed to link directly with commercial hydraulic modelling software to understand the drainage network and its hydraulic performance (Spence et al., 2016). The model can support the assessment of engineered solutions, including the design of storage facilities and the development of efficient screening systems at CSOs and WTPs.

The model represents two aspects of gross solid transport: advection (movement with the flow, but not necessarily at the mean water velocity) and deposition/erosion. The fact that gross solids in actual sewers possess an almost infinite variety of properties, sizes, and states of physical degradation means that simplification is needed. The model represents this range by a finite number of distinct solid *types*. For each solid type, the model requires advection and deposition properties: the values of α and β in Equation 9.8, and the critical value of velocity and depth for deposition as described in Section 9.5.1.

The computational basis of the gross solids model is to "track" the progress of individual solids or groups of solids through the system. At any point in distance and time, the mean flow velocity is known from the hydraulic model. This can be converted to solid velocity using the relationship specified for the particular solid type. The velocity of a solid in any instantaneous position is thus known, and this can be used to progressively track its movement through the system. If depth or velocity over any section decreases to below the value specified as causing deposition, the progress of solids in the section is halted until the value is again exceeded.

The model includes a simplified model of solids transport in the smaller pipes in the upstream parts of the catchment (where not specifically modelled in the drainage model), and another of solids behaviour in CSO structures, to create a comprehensive model of gross solids loadings throughout a combined sewer system (Digman et al., 2002; Spence et al., 2016). The work has also included collection of an extensive set of field data at three sites for comparison with model simulations demonstrating its ability to predict the first foul flush of solids.

19.9 MODELLING POLLUTANT TRANSFORMATION

In a gravity sewer, the main transformation processes act within or between the atmosphere, the wastewater itself, the biofilm attached to the pipe wall, and the sediment bed (see Figure 19.11). In a pressure sewer, there is no atmospheric phase and the biofilm is distributed around the pipe perimeter.

Of particular importance are those processes associated with the biodegradation of organic material. These are caused by micro-organisms occurring either on the pipe wall as a biofilm or in suspension in the wastewater. The biofilm will be more influential in smaller pipes and suspended biomass in larger sewers. These are aerobic processes, requiring the presence of adequate dissolved oxygen (DO) levels. Thus, parameters such as BOD or COD to represent the organic material and DO to represent the toxic state of the wastewater need to be modelled. Anaerobic processes are not normally modelled in detail (but see Section 17.8).

All sewer quality models have at least some representation of pollutant transformation through the system. This can range from simplistic to sophisticated, as follows:

1. Conservative pollutants
2. Simple decay expressions
3. Complex processes approach

19.9.1 Conservative pollutants

Conservative pollutants are those that are not affected by any chemical or biochemical transformation processes. Pollutant concentration may still vary due to the processes of advection and dispersion.

Some sewer quality models omit representations of pollutant degradation or biochemical interactions. The justification for this is not that all pollutants *are* conservative, but that the processes are relatively insignificant. This may be a reasonable assumption for short-retention systems but will be inaccurate in systems with, for example, long, large-diameter trunk sewers of well-aerated outfalls.

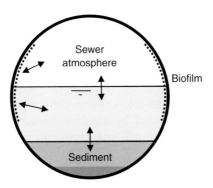

Figure 19.11 In-sewer pollutant interactions.

19.9.2 Simple decay expressions

A second simplified approach is to model the transformation of individual pollutants using a simplified, summary model of the reactions. A first-order decay model is a common example of this, where

$$\frac{dc}{dt} = -kc \tag{19.13}$$

where c is the pollutant concentration (g/m^3) and k is the rate constant (h^{-1}).

Thus, the pollutant concentration decreases with time and temperature, such that

$$k_T = k_{20}\theta^{T-20} \tag{19.14}$$

where k_T is the rate constant at T °C (h^{-1}), k_{20} is the rate constant at 20°C (h^{-1}), and θ is the Arrhenius temperature correction factor (–).

This approach ignores any interactions that may occur between the various substances (e.g., DO and BOD).

19.9.3 Complex processes approach

Current software packages offer the facility to model the complex processes discussed above in more detail by representing the oxygen balance of the flow and its various components. An example of the types and methods of representing these is given below.

19.9.3.1 Oxygen balance

DO in the flow results as a balance between oxygen supplied by aeration from the atmosphere and that consumed by the micro-organisms in the wastewater and biofilm (Almeida, 1999). Processes in the sediment bed may also exert an additional oxygen demand. This can be represented as an *oxygen balance* as follows:

$$\frac{dc_0}{dt} = K_{LA}(c_{0,S} - c_0) - (r_w + r_b + r_s) \tag{19.15}$$

where $c_{0,S}$ is the saturation dissolved oxygen concentration (g/m^3), c_0 is the actual dissolved oxygen concentration (g/m^3), K_{LA} is the volumetric reaeration coefficient (h^{-1}), r_w is the oxygen consumption rate in the bulk water (g/m^3.h), r_b is the oxygen consumption rate in the biofilm (g/m^3.h), and r_s is the oxygen consumption rate in the sediment (g/m^3.h).

19.9.3.2 Reaeration

Reaeration is a naturally occurring process of diffusion. Oxygen in the atmosphere is dissolved into the liquid up to saturation levels that depend mainly on temperature. Aeration may be increased by turbulence caused by manhole backdrops or pumps. Pomeroy and Parkhurst (1973) have derived an empirical relationship for reaeration in sewers based on a number of hydraulic parameters:

$$K_{LA} = 0.96 \left[1 + 0.17 \left(\frac{v^2}{gd_m} \right) \right] \gamma (S_f v)^{3/8} \frac{1}{d_m} \tag{19.16}$$

where v is the mean velocity (m/s), d_m is the hydraulic mean depth (m), γ is the temperature correction factor (1.00 at 20°C), and S_f is the hydraulic gradient (−).

19.9.3.3 Oxygen consumption in the bulk flow

The oxygen consumption of wastewater (also known as the oxygen uptake rate) varies with the age and temperature of the wastewater but (provided conditions are aerobic) is independent of oxygen concentration. Typical values are 1–4 mg/L.h.

19.9.3.4 Oxygen consumption in the biofilm

Oxygen consumption by biofilms is a complex phenomenon affected by substrate and oxygen availability, among other factors. Pomeroy and Parkhurst (1973) expressed consumption empirically as follows:

$$r_b = 5.3(S_f v)^{1/2} \frac{c_0}{R} \tag{19.17}$$

where R is the hydraulic radius (m).

19.9.3.5 Oxygen consumption in the sediment

Anaerobic processes in the sediment bed will produce oxygen-demanding by-products resulting in a *sediment oxygen demand* (SOD). This is exerted when the sediment bed is eroded.

19.10 USING WATER QUALITY MODELS

19.10.1 Model applications

Water quality models are needed when detailed knowledge is required on the pollutants entering, moving through, and (crucially) leaving the sewer system. The main applications for quality models are therefore studies of

- Intermittent and continuous discharges to receiving waters (river, estuary, marine)
- Loads discharged to WTPs
- Sediment transport and deposition in sewer systems

Of these three, the first application is the most common, and this is often carried out in association with a model representation of the receiving water. Therefore, it is appropriate to consider the two components together, where similar processes take place and similar methods apply.

19.10.2 Types of sewer quality models

The *UPM Manual 3* (FWR, 2012) recommends the use of two types of sewer quality model, detailed and simple, depending on the application.

Detailed models are deemed to be appropriate where the network is complex, flat, there is interaction with the WTP, or there are significant trade discharges. A verified flow *and* water quality model is used, typically with calibration of the water quality pollutant loads. Here event mean spill concentrations may be applied to the discharges or time-varying concentrations depending on the receiving waterbody modelling approach. Simple models are normally only appropriate in small catchments with little interaction, and flows are controlled, enabling representation of the system through a tank model (Section 13.4) or simplified sewer model.

19.10.3 Types and complexity of river quality models

The *UPM Manual 3* (FWR, 2012) outlines three levels of model complexity to represent the chemical and biological processes in a river depending on application. These parallel the approaches used in sewer quality models:

- Simple mass balance: no in-river processes are accounted for and simply mixes pre-dicted wastewater discharges with a volume of river water.
- Mass balance including simple BOD/DO processes: applies relatively simple decay equations to calculate the DO balance from BOD, with a number of simplifying assumptions.
- Water quality process interactions: accounts for in-river processes including disper-sion, advection, transformation, and interactions of pollutants. This approach is typi-cally applied to provide greater confidence in the results and can underpin significant investments to reduce the impact on receiving waters.

19.11 PLANNING AN INTEGRATED STUDY

19.11.1 Components to consider

An integrated study of the drainage system and receiving waters provides a greater under-standing to the causes of pollution and how to address it. Before collecting data and creating models, the following components should be considered. A river context is used as a com-mon example.

19.11.1.1 Drivers

The initial step is to understand why the study is being completed. For example, is there already a known issue and river quality target to achieve under the WFD or is the failure not understood? Is it high-level planning or a scheme to be approved by the regulator? These considerations drive the level of data collection, study approach, and model detail required.

19.11.1.2 Catchment boundary

The catchment boundary needs to be derived to establish river flows. The same is under-taken for the urban drainage catchment to understand potential interactions.

19.11.1.3 Waterbodies

To determine the level of modelling and calibration effort, it is important to understand which waterbodies require assessment. For example, it may be a whole river network or just a downstream reach. Key locations to identify for data collection and calibration purposes will be where there are permanent sampling locations and downstream of any thought to be key discharges.

19.11.1.4 River modelling approach

The approach to river flow modelling in part depends on the accuracy required but predominantly depends on the physical characteristics of the watercourses. Most studies tend to be in free-flowing reaches, and so a routing model will suffice. If there are backwater effects or slow-flowing water, then it would be appropriate to use a 1D approach. For particularly deep or wide waterbodies, then it may be necessary to use a 2D or 3D approach to capture the effects of stratification or streaming. It is possible to use a combination of approaches.

19.11.1.5 Point source discharges

Reviewing sewer models and GIS data will help CSOs, WTPs, and surface water outfalls to be identified. These can be compared to the river catchment boundary to define the extent of upstream modelling required. Typically, a sewer model is used to derive overflow pollutant discharges, whereas external distributions are applied to stormwater and final effluent discharges.

19.11.1.6 Rainfall and evaporation

As a minimum, representative historical rainfall data are needed that cover the calibration and verification periods used in the study. However, hydrological models require multi-year records due to the subsurface processes affecting catchment wetness (see Chapter 4).

During analysis, a 10-year time series is typically used, preferably historical rainfall but synthetic series can also be used. Where the catchment is large or there are changes in elevation, often more than one series are applied.

19.11.2 Input data and model calibration/verification

The concept of verifying a pre-calibrated quality model is a less realistic proposition than for a flow model (as presented earlier in the chapter). Variations and uncertainties are much larger, making it harder to transfer experience or default values from one catchment to another, even though apparently similar. Hence, a major issue is that these models require a great deal of data (either in sewer or in river). A typical deterministic quality model requires the type of input data specified in Table 19.6.

The decision to collect water quality data in the sewer and receiving waterbody depends on the nature of the study, the proposed modelling approach, and the outcomes to achieve. In some cases, despite the caveats above, the relative expense of gathering full catchment-specific data means that quality modellers may have to rely on default data.

It is common to consider dry weather flow first in this process. Typically, a starting point is to calibrate against observed WTP influent data to validate or change the default profiles. The simulations are then compared with any measured data, and default values are replaced by amending the inputs to the dry weather pollutants, in particular from trade discharges, until the model is deemed to be calibrated. Fixed calibration criteria are not

Table 19.6 Input data for a quality model

Wastewater	For a typical day in the week and at the weekend: • Variation of flow with time • Variation of pollutant concentrations with time • Trade discharges
Stormwater	For two or three separate rainfall events: • Variation of flow with time • Variation of pollutant concentrations with time
Surface pollutants	Land use characteristics Particulate characteristics
Deposits	Initial sediment depth Characteristics of sediment

typically specified, but the relative magnitude and timing of the observed response should be represented by the model (Wastewater Planning Users Group [WaPUG], 2006). Figure 19.12 shows actual and simulated dry weather flow quality data used in the verification of a model. In the waterbody, comparisons are undertaken and compared against spot samples taken monthly (with a random date and time) and any continuous monitoring by the environmental regulator. This generally provides a dry weather river calibration, typically over a 3-year period, and accounts for the background water quality, final effluent, and direct trade discharges. The process is then applied to storm flow.

Ideally at this stage a comprehensive sensitivity analysis should be carried out to capture the inherent uncertainties in the process (e.g., Schellart et al., 2008) and to explicitly estimate the statistical validity in the results. In practice, this may not be undertaken due to model run times, complexity of river models, and financial constraints. This often leads to more conservative default setups being used.

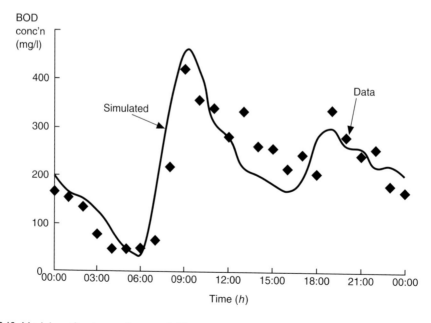

Figure 19.12 Model verification: pollutograph (dry weather flow).

19.11.3 Data collection in sewers and receiving waters

Taking samples from the flow at particular locations in the system and analysing them for pollutant concentrations is an important operation in the development of quality models for specific systems. Samples from the dry weather flow may be needed to provide catchment-specific input data, and samples from both dry weather and storm flow are needed for calibration/verification.

Other catchment-specific data collection will normally be needed. This may include taking samples of particulate material on the catchment surface, and measuring the depth and properties of the sediment layer in pipes. Most importantly, though, it is often the data collected in the receiving waterbody that enable model validation through physical sampling and continuous monitoring (e.g., sondes), such as for dissolved oxygen and temperature for dry and wet weather. A high-frequency spot sample may also be undertaken to capture the diurnal effects in dry weather and the effect of CSOs and surface water discharges in storms.

There is guidance on the collection of field data to support quality models in the *UPM Manual 3* (FWR, 2012) and the *Guide to the Quality Modelling of Sewer Systems* (WaPUG, 2006).

PROBLEMS

19.1 What is meant by "a physically based deterministic model of flow in a sewer system"? What should such a model cover? To what uses might it be put?

19.2 What has been the impact of computer models on urban drainage engineering since they were first introduced?

19.3 Substitute in Equation 19.5 using finite difference forms proposed in Section 19.4.3 (with forward differences for both terms). Rearrange to give Q_j^{n+1} on the left-hand side.

19.4 Explain why calibration and verification mean different things (in the context of sewer-flow models). How is verification carried out in practice?

19.5 "Models are always wrong." So what is the point of using them (in urban drainage)?

19.6 Explain why quality modelling of a sewer system is more difficult than flow modelling.

19.7 Describe the physical processes that should be covered in a deterministic sewer quality model.

19.8 Classify and describe the various approaches to pollutant transformation modelling. What are their relative merits?

19.9 "Attempting deterministic modelling of sewer-flow quality is a waste of time." Some experts seriously hold this view. What do you think?

19.10 Describe alternatives to physically based deterministic models for sewer-flow quality. What are their advantages and disadvantages?

19.11 What are the main steps to be carried out in an integrated study?

KEY SOURCES

Chadwick, A.J., Morfett, J.C., and Borthwick, M. 2013. *Hydraulics in Civil and Environmental Engineering*, 5th edn, Spon Press.

Chartered Institution of Water and Environmental Management (CIWEM). 2017. *Code of Practice for the Hydraulic Modelling of Urban Drainage Systems*, www.ciwem.org/groups/urban-drainage-group.

Foundation for Water Research (FWR). 2012. *Urban Pollution Management Manual*, 2nd edn.

REFERENCES

Almeida, M., do C. 1999. *Pollutant Transformation Processes in Sewers under Aerobic Dry Weather Flow Conditions*, Unpublished PhD thesis, Imperial College, University of London.

Bamford, T.B., Digman, C.J., Balmforth, D.J., Waller, S., and Hunter, N. 2008. Modelling flood risk—An evaluation of different methods. *WaPUG Autumn Conference*, Blackpool.

Butler, D., Davies, J.W., Jefferies, C., and Schütze, M. 2003. Gross solid transport in sewers. *Proceedings of the Institution of Civil Engineers, Water and Maritime Engineering*, 156(WM2), 175–183.

Chow, V.T. 1959. *Open-Channel Hydraulics*, McGraw-Hill.

Department of the Environment/National Water Council (DoE/NWC). 1981. *Design and Analysis of Urban Storm Drainage—The Wallingford Procedure*. Standing Technical Committee Report No 28.

Digman, C.J., Littlewood, K., Butler, D., Spence, K., Balmforth, D.J., Davies, J.W., and Schütze, M. 2002. A model to predict the temporal distribution of gross solids loadings in combined sewerage systems. *Global Solutions for Urban Drainage: Proceedings of the Ninth International Conference on Urban Drainage*, Portland, Oregon, on CD-ROM.

Guidolin, M., Chen, A.S., Ghimire, B., Keedwell, E.C., Djordjevic, S., and Savic D.A. 2016. A weighted cellular automata 2D inundation model for rapid flood analysis. *Environmental Modelling and Software*, 84, 378–394.

Hale, J.N. 2004. Urban flood routing in the UK ... the next step? *Proceedings of UDM04 Sixth International Conference on Urban Drainage Modelling*, Dresden, Germany, August, 97–104.

Huber, W.C. 1986. Deterministic modeling of urban runoff quality, in *Urban Runoff Pollution* (eds H.C. Torno, J. Marsalek, and M. Desbordes), NATO ASI Series G: Ecological Sciences—Vol. 10, Springer-Verlag.

Koutitas, C.G. 1983. *Elements of Computational Hydraulics*, Pentech.

Maksimović, C. and Prodanović, D. 2001. Modelling of urban flooding—Breakthrough or recycling of outdated concepts? *Urban Drainage Modeling: Proceedings of the Speciality Symposium of the EWRI/ASCE World Water and Environmental Resources Congress*, Orlando, Florida, May.

Osborne, M., Dumont, J., and Martin, N. 1996. Beckton and Crossness catchment modelling: Hydrologic modelling—Two routes to the same answer. *Proceedings of the Seventh International Conference on Urban Storm Drainage, 1*, Hannover, September, 443–448.

Pomeroy, R.D. and Parkhurst, J.D. 1973. Self purification in sewers. *Advances in Water Pollution Research. Proceedings of the Sixth International Conference*, Jerusalem, Pergamon Press, 291–308.

Ponce, V.M., Li, R.M., and Simons, D.B. 1978. Applicability of kinematic and diffusion wave models. *American Society of Civil Engineers, Journal of the Hydraulics Division*, 104(HY3), 353–360.

Price, R.K. and Osborne, M. 1986. Verification of sewer simulation models, in *Urban Drainage Modelling* (eds C. Maksimović and M. Radojkovic), Pergamon Press.

Russ, H.-J. 1999. Reliability of sewer flow quality models—Results of a North Rhine-Westphalian comparison. *Water Science and Technology*, 39(9), 73–80.

Ryu, J. 2008. *Decision Support for Sewer Flood Risk Management*, Unpublished PhD thesis, Imperial College, University of London.

Schellart, A.N.A., Tait, S.J., Ashley, R.M., Farrar, D., and Hanson, D. 2008. Uncertainty in deterministic predictions of flow quality modelling and receiving water impact. *Proceedings of the 11th International Conference on Urban Drainage*, Edinburgh, September, on CD-ROM.

Spence, K.J., Digman, C., Balmforth, D., Houldsworth, J., Saul, A., and Meadowcroft, J. 2016. Gross solids from combined sewers in dry weather and storms, elucidating production, storage and social factors. *Urban Water Journal*, 13(8), 773–789.

Vojinovic, Z. and Solomatine, D.P. 2006. Evaluation of different approaches to calibration in the context of urban drainage modelling. *Proceedings of the Seventh International Conference on Urban Drainage Modelling and Fourth International Conference on Water Sensitive Urban Design*, Melbourne, May, 2, 651–660.

Vreugdenhil, C.B. 1989. *Computation Hydraulics: An Introduction*, Springer-Verlag.

Wangwongwiraj, N., Schlütter, F., and Mark, O. 2004. Principles and practical aspects of an automatic calibration procedure for urban rainfall-runoff models. *Urban Water Journal*, 1(3), 199–208.

Wastewater Planning Users Group (WaPUG). 2006. *Guide to the Quality Modelling of Sewer Systems, v 1.0. Wastewater Planning Users Guide*, http://www.ciwem.org/wp-content/uploads/2016/05/Guide-to-the-Quality-Modelling-of-Sewer-Systems.pdf.

Willems, P. 2008. Quantification and relative comparison of different types of uncertainties in sewer water quality modeling. *Water Research*, 42(13), 3539–3551.

Wolfram, S. 1984. Cellular automata as models of complexity, *Nature*, 311, 419–424.

Water Research Centre (WRc). 2017. *Sewerage Risk Manual*, http://srm.wrcplc.co.uk/

Yen, B.C. 1986. Hydraulics of sewers, in *Advances in Hydroscience*, Vol. 14 (ed. B.C. Yen), Academic Press.

Chapter 20

Innovations in modelling

20.1 INTRODUCTION

This chapter introduces some of the more novel concepts in urban drainage modelling. First, we discuss models that are not physically based and rely more on data and less on fundamental physical equations (Section 20.2). As such, they also rely on good model calibration (discussed in Section 20.5) to make sure that these relatively simple models accurately approximate reality. We also introduce the idea that a single model may not be able to accurately (or sufficiently) describe the complete urban water system and suggest two novel ways to address this problem: integrated models (Section 20.3) and standards for coupling a series of different models (20.4). How uncertain all these models are and what we can do about it are the focus of Section 20.6.

20.2 ALTERNATIVE, NON-PHYSICALLY BASED, MODELLING APPROACHES

This section summarises approaches to modelling that can be considered alternative to more traditional, physically based models. They can be used for modelling flow as well as quality, and their basic premise is that they rely more on data or simplified conceptual descriptions of the drainage system and less on fundamental physical equations.

Every urban drainage model receives input data and produces output data. So for any given system, the modeller needs to collect data on the actual, real-world, system behaviour. These data typically consist of recorded historical events (e.g., rainfall), and consequences observed as a result from these events (e.g., runoff, possibly resulting in urban flooding). To reproduce the relationship between events driving the physical processes and their consequences within a model, the events become an *input* and the consequences an *output*. The model converts input into output using a set of mathematical procedures sometimes called *transfer functions*, representing the relationship between events and their consequences. The key difference between physically based models (discussed in previous chapters) and the alternative approaches discussed in this chapter, is that physically based models attempt to describe the system in detail, by creating a mathematical equivalent of each major, relevant, physical process: rainfall to runoff, transport of pollutants, sediment deposition in sewers, and so on. On the contrary, alternative approaches "simply" map inputs onto outputs, by finding statistical or other relationships between them. This means that their transfer function has no physical basis – we cannot point to an equation and say "This is where rainfall is converted to runoff." The formulation and potential success of the model have a mathematical explanation, but not a physical one. The model is a *black box*. It follows that a disadvantage of such models is that they cannot, for example, be used to look at options to upgrade a sewer system, since the physical properties of the system are not contained in the model. They are, however, very quick, and as such are ideal for real-time applications, such as early warning systems for urban floods (see Chapter 22). Some of the most well-known categories of "alternative" models are briefly presented in the following sections.

20.2.1 Empirical models

An empirical model is based on observation rather than theory. It usually represents the real system by simple relationships (e.g., a polynomial data-fitting equation) that rely for their accuracy on parameters that are calibrated using observed data (e.g., using a least squares optimisation method).

20.2.2 Artificial neural networks

Artificial neural networks are a product of developments in artificial intelligence. "Signals" are passed between artificial "neurons." Each neuron receives signals from a number of "upstream" neurons, applies a weighting to each input, then applies a transfer function before outputting signals to "downstream" neurons. The network is then trained (by changing the weights associated with each neuron) to reproduce the relationships in training data sets (input and output) provided by the modeller. Given enough training on good data, the network can make useful predictions for new cases, without the need for physical parameters and equations. The approach has been used successfully in sewer modelling (Loke et al., 1996) including real-time applications in sewer modelling for control (Darsono and Labadie, 2007).

20.2.3 Conceptual or meta-models

In a conceptual model, the physical system is represented by highly simplified "concepts," for example, representation of the physics of pipe flow by a simple tank system. Detailed treatment of individual processes is replaced by overall global representation. A good example of this approach is the open-source CITY DRAIN model developed by Achleitner et al. (2007).

20.2.4 Stochastic models

A stochastic model includes randomness. Unlike a deterministic model, it does not necessarily give the same output for the same input. Simple empirical models may be designed to be stochastic, and this is appropriate since the measured data on which the empirical model is based are certain to contain random elements. A complex, physically based model can also have stochastic characteristics by introducing random influences on some elements of input and showing the effect on the output. The output from a stochastic model will not be a single answer but a range of answers, possibly represented by a mean and standard deviation.

20.3 INTEGRATED MODELLING

Conventional practice has been to operate and, therefore, model the various components of the engineered urban wastewater cycle (water treatment, distribution, sewerage and storm drainage, wastewater treatment, environmental compartments) in isolation. Each component has been engineered to meet the needs of its users and the environment, but with little feedback or cross-reference to other components. Recently, the importance of a more combined management of the various urban water system components has been emerging and is now recognised by both researchers and practitioners. This move towards *integration* (1) considers all parts, or components, of the system; (2) involves water conservation and diverse fit-for-purpose water supplies; (3) works at a range of scales (both central and decentralised); and (4) allows the establishment of links with other environmental cycles (e.g., energy and nutrients) (Bach et al., 2014; Brown et al., 2009; Mitchell, 2004).

20.3.1 Why do we need integrated models?

The need for an integrated model of the system is similar to the need for a model of any of the individual elements. In addition, representing and understanding the urban system as a whole potentially allows better, more cost-effective solutions to be engineered.

The urban wastewater system as a whole contains numerous elements that can be utilised to prevent water pollution. For example, there is little point in "wasting" storage volume in tanks and pipes, or treatment capacity, on weak wastewater. If the available capacity could primarily be used to capture the most polluted flows, the potential for pollutant discharge reduction can be increased. Storage other than in the urban drainage system may also be exploited (e.g., the time lag of processes in the WTP and the receiving water). If the WTP and CSOs discharge into the same receiving water, carefully timed release of effluents could minimise overall pollutant discharges.

Calibrating and verifying such large integrated models can be challenging because of the need for large data sets from diverse parts of the system. Muschalla et al. (2008) provide advice on the procedures to be adopted, and Solvi (2007) describes in detail the calibration and verification of an integrated model case study.

BOX 20.1   VIENNA, AUSTRIA

The City of Vienna in Austria has a population of 1.8 million spread over some 260 km^2 and served by 2200 km of sewer system. The central part of the system is combined, with separate systems on the periphery. Large relief sewers have been constructed along the banks of the rivers Donau, Donaukanal, Wienfluss, and Liesing to minimise combined sewer overflow discharges after significant rainfall events. A central real-time control system has been developed that triggers control devices within the system to activate the storage within the relief sewers. The system consists of the following elements:

- Control devices to regulate water level or flow
- Measurement devices for water level, flow, and rainfall
- A supervisory control and data acquisition (SCADA) system to collect the measured data, transmit set points, and display information about the system
- A central control strategy to generate a control decision based on measured and forecast data

The system contains 25 rain gauges, 40 in-sewer flow measurement devices, and 20 water level measurement units installed at 25 sites. The data are fed into a simplified model of the sewer network. Control decisions are made online, based on "if-then" rules evaluated with the aid of fuzzy logic.

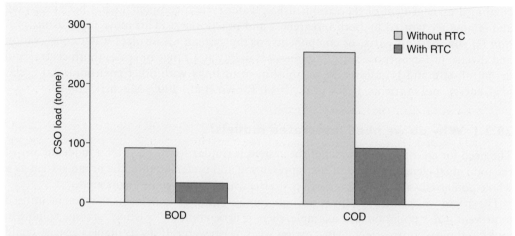

Figure 20.1 First phase CSO pollution load reduction. (Adapted from Nowak, R. 2007. *IWA Water 21*, April, 36–37.)

The scheme is being implemented in three phases. The first phase, consisting of the catchment on the left bank of the River Donau, went operational in 2005. Figure 20.1 shows predicted reductions of around 60% for CSO BOD and COD load. The second phase was integrated with the first and became operational in 2006. Evaluation of measured results between March and September 2006 showed that discharge volumes had been reduced by 40% compared with previous years. The third phase was due for completion in 2015 (Fuchs and Beeneken, 2006; Nowak, 2007).

20.3.2 Defining and classifying the level of integration

The formulation of a unique definition for integration and classification format for urban water systems modelling is rather a hard task due to the complexity of the latter and the subjectivity on what can be considered as "integrated." Rauch et al. (2005) suggested that thinking about "degrees of integration" is more useful than simply using the term *integration*. Bach et al. (2014) have suggested that to move towards integrated modelling in urban water systems, we need to consider three key points:

- Modelling of a multitude of components (biophysical, economic, and beyond) and interactions between them
- Consideration of acute, chronic, and delayed impacts of water quantity and quality processes in long simulation periods
- Ability to see both local processes and the global "big picture" to better inform decision making, policies, or scientific knowledge

Bach et al. (2014) also suggested four different integration levels to categorise urban water models according to the physical and institutional system delineation, model complexity, and differences between their development philosophies:

- *Integrated component-based models (ICBMs):* ICBMs encompass "whole-of-subsystem" or "plant-wide" integration and focus on the integration of components within the local urban water subsystem (e.g., coupling several treatment processes within a wastewater treatment plant).

- *Integrated urban drainage models (IUDMs):* IUDMs focus on integration of the sub-systems of urban drainage (wastewater and/or stormwater). They are also known as *integrated urban wastewater systems.* These are one of the most well-known forms of integrated models that are typically used to study and evaluate upgrade options for a local WTP, non-pipe decentralised technologies in stormwater management, ways of reducing CSO emissions, or showing the combined impact of different parts of the drainage stream on receiving waters.
- *Integrated urban water cycle models (IUWCMs):* At this level of integration, the IUDMs and IWSMs are combined into one modelling framework. This integration is typically used in strategic planning and conceptual assessment of sustainable solutions (e.g., combinations of central and decentralised technologies) in urban water management, rather than devising detailed control strategies. Such a model is, for example, UWOT, which is described briefly in the next section.
- *Integrated urban water system models (IUWSMs):* This is the highest level of integration in urban water systems that takes into account and draws links among environmental, social, and economic aspects of water-related issues. An inter- and multi-disciplinary approach is adopted to study and model the total urban water cycle. For an example of a complete socio-technical modelling of the urban water system, see, for example, Makropoulos (2017).

20.3.3 Methods and tools for integrated system modelling and design

Irrespective of which *level* of integration the modeller is looking to achieve, there are, typically, two different approaches to employ: The first is to use a single software package with which to build the whole integrated model, and the second is to integrate existing models (built using different software packages) into one model "chain." Extensive reviews of these modelling approaches can be found in Elliott and Trowsdale (2007) and in Bach et al. (2014).

Typical examples of the first approach include *SIMBA* (IFAK, 2007) and *WEST* (Vanhooren et al., 2003), while combined modelling of water and wastewater flows can also be performed with *CITY DRAIN* (Achleitner et al., 2007; Burger et al., 2010), *MIKE URBAN* (DHI, 2009), and *UrbanCycle* (Hardy et al., 2005). Finally, *UrbanBEATS* (Bach, 2012), *DAnCE4Water* (Rauch et al., 2012), and *UWOT* (Rozos and Makropoulos, 2013) also support a higher integration level by allowing the modelling of additional aspects of the urban water cycle (i.e., social, economics, energy, etc.). UWOT is presented in Box 20.2 as an example.

20.3.4 Integrated control

With the advent of integrated modelling tools that can represent the dynamic hydraulic and water quality processes within the entire system, opportunities are created for controlling the performance of the system as a whole (see also Chapter 22). Therefore, *integrated control*, with the objective of minimising detrimental impacts on the receiving water, becomes possible and is characterised by two aspects (Schütze et al., 1999):

- *Integration of objectives:* control objectives within one sub-system may be based on criteria measured in other sub-systems (e.g., operation of pumps in the drainage system directed at minimising oxygen depletion in the receiving water body).
- *Integration of information:* control decisions taken in one sub-system may be based on information about the state of other sub-systems (e.g., operation of pumps in the drainage system based on WTP effluent data).

BOX 20.2   THE URBAN WATER OPTIONEERING TOOL (UWOT)

The UWOT model (Makropoulos et al., 2008; Rozos and Makropoulos, 2013) enables simulation of the three main components of the urban water cycle (i.e., water supply, wastewater disposal, and stormwater drainage) in an integrated framework. Unlike other typical urban water models that simulate actual water flows, UWOT adopts an alternative approach based on the generation, aggregation, and transmission of demand signals, starting from household water appliances and moving towards the source. This demand-oriented conceptualisation approach has the advantage of directly representing the principal purpose of infrastructure, which is to serve the need for water supply and wastewater disposal. Figure 20.2 presents a simplified example of the water flows from the abstraction, through the transmission network to the treatment plant, and then to the distribution network (left panel), along with the equivalent representation in UWOT (right panel). The part inside the ellipse describes the external water system, including abstractions, transmission, and treatment of raw water, whereas the lower part of the panel shows the generation of the demand and the disposal of wastewater and stormwater to water bodies (the internal water system).

UWOT distinguishes between two signal types: *push* and *pull* signals. The push signals express a need to dispose of a specific volume of water (e.g., the output of a washing machine). The pull signals express a demand for a specific volume of water (e.g., the water required for the operation of a washing machine). Pull signals do not bear a qualitative characterisation because water that covers a demand is assumed to meet the quality standards imposed by regulations. Push signals are characterised by a qualitative value that can express any preselected water quality parameter.

The information required by UWOT to simulate the operation of water and wastewater components (water consumption, required energy, BOD values, capital and operational cost, etc.) is stored in a database that is called a *technology library*. The library is populated with information obtained from surveys, practitioner manuals, scientific publications, laboratory and pilot tests, and so on, comprising both central and distributed management technologies. The user

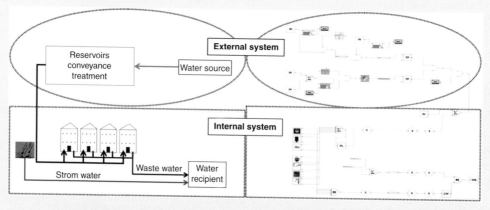

Figure 20.2 Left panel: representation of the urban water cycle. Right panel: source-to-tap modelling of the complete urban water cycle with UWOT in Matlab Simulink. The external water supply system is presented inside ellipses, while the water consumption and the drainage system are shown inside rectangles. (After Rozos, E. and Makropoulos, C. 2013. *Journal of Environmental Modelling and Software*, 41, 139–150.)

is able to define system components (from appliances to reservoirs and from WTPs to CSOs) and its topology (connections between components) via an advanced (drag and drop) graphical user interface. UWOT has been successfully applied to simulate the urban water cycle at a wide range of spatial and temporal scales, from a single household, all the way to city-wide scales.

Real-time control of the components of the integrated system, therefore, takes into account the state of the whole system when utilising the control devices available in order to reach operational objectives defined for any location in the system. Table 20.1 summarises some of the measures and objectives of control, as well as methods applied to determine control strategies.

Figure 20.3 shows diagrammatically how individual sub-systems can be placed in an integrated framework to include control and optimisation. Here, the integrated model has been applied to a catchment, and various degrees of control were simulated. These include a base case with local control only, an optimised variation of this with fixed set points, and

Table 20.1 Components of control of the urban wastewater system

Sub-system	Devices	Objectives	Decision-finding methods
Sewer system	• Pumps • Weirs • Gates	• Prevention of flooding • CSO reduction (frequency, volumes, loads) • Equalisation of flows	• Heuristics, intuition • Self-learning expert system • Offline optimisation • Online optimisation
Treatment plant	• Weirs, gates • Return sludge rate • Waste sludge rate • Aeration	• Maintenance of effluent standards • Process maintenance • Improved water quality	• Model-based control • Application of control theory
Receiving river	• Weirs • Gates	• Flood protection	

Source: Schütze, M., Butler, D., and Beck, M.B. 1999. *Water Science and Technology*, 39(9), 209–216.

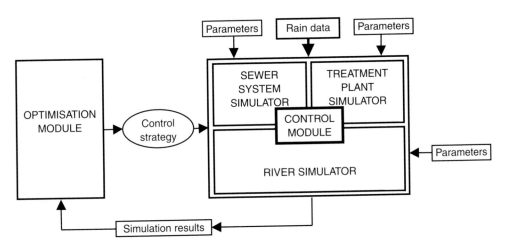

Figure 20.3 Overview of the "Integrated Simulation and Optimisation Tool." (Reproduced from Schütze, M., Butler, D., and Beck, M.B. 1999. *Water Science and Technology*, 39[9], 209–216, with permission of publishers Pergamon Press and copyright holders IAWQ.)

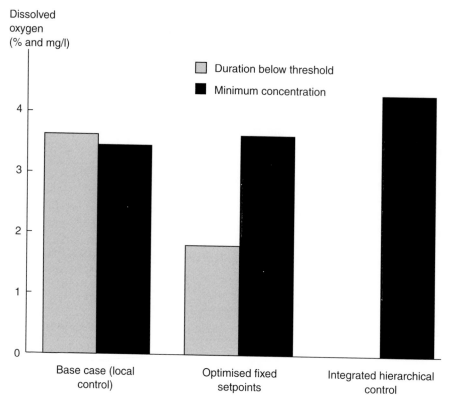

Figure 20.4 Effect of control scenario on river DO. (Adapted from Schütze, M., Butler, D., and Beck, M.B. 1999. *Water Science and Technology*, 39[9], 209–216, with permission of publishers Pergamon Press and copyright holders IAWQ.)

an example of simple integrated control. The performance of each of the control scenarios is optimised to allow fair comparison.

Figure 20.4 illustrates the effects of these scenarios in terms of reaching a single objective, namely, (1) the duration (% of run time) of the oxygen concentration below a 4 mg/L threshold value at any location in the receiving water (a river in this case), or (2) the minimum DO concentration in the river during the simulated time period. It can be seen that control with optimised fixed set points leads to improved performance over the base case. Further improvement is achieved for the integrated control scenario.

20.3.4.1 Multi-objective optimisation

Within this optimisation context, more recent work (Fu et al., 2008; Muschalla et al., 2005) has focussed on multi-objective optimisation of the integrated urban wastewater system to establish possible trade-offs between key parameters such as water quality determinants, energy consumption, and cost. Figure 20.5 shows the output of a study on an integrated system where both river water dissolved oxygen (DO) concentration and ammonium concentration are used as optimisation objectives for many different control strategies. A clear trade-off is evident between the two parameters; an increase in one implies a decrease in the other (a *Pareto front*). It is not possible to both fully maximise river DO concentration and at the same time minimise ammonium concentration.

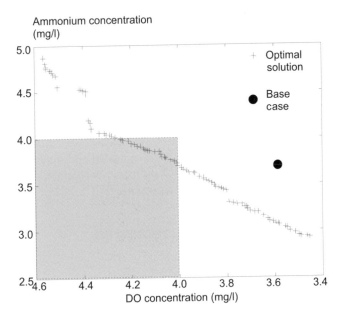

Figure 20.5 Pareto optimal solutions with respect to DO and ammonium concentration. (Adapted from Fu, G., Butler, D., and Khu, S.T. 2008. *Environmental Modelling and Software*, 23[2], 225–234.)

In practice, feasible solutions on the Pareto front can be reduced to a great extent by only considering those that comply with river water quality standards. For example, as discussed in Chapter 2, the *UPM Manual* (FWR, 2012) defines DO and ammonium concentration thresholds that should not be breached beyond a specified frequency. So, based on a 4 mg/L constraint for DO and ammonia concentration (the area shaded grey on Figure 20.5), a much smaller group of promising solutions (control strategies) emerges. Comparison should also be made with the non-optimised base case indicating the improvement that can be achieved in terms of both DO and ammonium concentrations.

20.4 COUPLING MODELS TO OTHER MODELS: EMERGING STANDARDS

As discussed earlier, an alternative approach to building the integrated model within one software package is to link up existing models, built using different software packages into a model chain. To be able to do this, one needs to make sure that the outputs of one model (e.g., the rainfall-runoff urban hydrology model) become an input for the next "downstream" model in the chain (e.g., the hydraulic model of the drainage system), and so on, all the way down to the recipient water body. The typical way of accomplishing this model link is to write bespoke code (sometime termed *gluing routines*). This approach, however, has a serious disadvantage: one needs to change these routines every time a new model is linked in the chain, or indeed a new variable is being passed from model to model.

A significant innovation addressing this issue has been the development of a common model coupling standard, the *Open Modelling Interface (OpenMI) Standard*, which allows for consistent and efficient integration of different models at runtime (Gregersen et al., 2007). OpenMI-compliant models can be easily combined into integrated modelling systems without the need for bespoke gluing routines. The OpenMI Association is dedicated to supporting the approach (www.openmi.org), and independent studies are already emerging following this standard (e.g., Castronova et al., 2013; Smolders et al., 2008).

More integrated models, require, however, more raw data, and the need for standardisation not only of the data exchanged between models, but also of water observations that serve as inputs to these models is emerging as an important issue. A noteworthy development in standardisation, aiming to provide a common data exchange format for hydrological time series, and therefore facilitate the exchange of such data sets across different models and information systems is *WaterML*. WaterML is an open standard and information model for the encoding, representation, and communication of hydro-meteorological observations and water measurements. It has been designed and developed by many national and international organisations from public and private sectors, within the OGC (Open Geospatial Consortium) Hydrology Domain Working group (Open Geospatial Consortium [OGC]. 2014).

20.5 CALIBRATION ASPECTS FOR INTEGRATED URBAN DRAINAGE SYSTEMS

As discussed in Chapter 19, the majority of urban drainage applications require the calibration of the parameters of the model. In the calibration problem, the internal properties of the model, either physical or conceptual, are unknown and have to be fine-tuned by optimising (in this case minimising) the departures of the simulated responses of the model at hand against the observed, real-life system responses. So, calibration is a particular case of optimisation.

The calibration of urban drainage models is, however, characterised by several difficulties, the most significant of which are

- The high computational burden of drainage models, which makes calibration a highly time-consuming process
- The vast amount of data and measurements that are required to capture the nature of each component of the urban water cycle
- The fact that calibration procedures may lead to equally good combinations of parameters, reducing the confidence in model predictions. This concept is associated with calibration uncertainty and is also known as *equifinality*.

These difficulties are further exacerbated in the calibration of integrated models (Beck, 1997; Rauch et al., 2002; Vojinovic and Seyoum, 2008). In this especially challenging case, three different calibration approaches are followed (Bach et al., 2014):

- Calibration/optimisation of the whole integrated model at once (often a very difficult and sometimes impossible task).
- Initial calibration/optimisation of the upstream models, gradually adding the next immediate downstream component.
- Calibration/optimisation of individual sub-models separately before integrating them. In this case, of course it is not possible to assess the effect of remaining errors in the sub-models after integration.

20.5.1 Optimisation methods

To address these and other challenges in the calibration of complex models, such as urban drainage models, several different optimisation approaches have been developed. These range from simple Monte Carlo methods (Freni et al., 2008), controlled random search

(Butler and Schütze, 2005), and evolutionary algorithms (Muschalla, 2008), all the way to multi-objective algorithms such as the Pareto preference ordering (Khu and Madsen, 2005), genetic algorithms (Schütze et al., 2001), and surrogate-enhanced evolutionary annealing simplex algorithms (Tsoukalas et al., 2016).

In the computational procedure of calibration, the simulation runtime required by the drainage model is by far the most time-consuming element, and the one that imposes a practical barrier to optimisation: typically, several thousand runs of the model, each driven by very long (often stochastically generated) time series, are required by the most powerful optimisation algorithms (of the evolutionary kind, such as genetic algorithms) to reach a good (calibrated) parameter set. If, however, each model run needs several minutes to complete, then the whole calibration procedure may well cost days or even weeks of simulation time. This is clearly impractical.

Typical approaches to alleviate this high computational burden imposed by simulation models can be classified into four main categories (Razavi et al., 2010): (1) parallel computing (e.g., Cheng et al., 2005; Dias et al., 2013); (2) computationally efficient optimisation algorithms (e.g., Tolson et al., 2009; Tolson and Shoemaker, 2007); (3) strategies to opportunistically avoid (expensive) model evaluations (e.g., Razavi et al., 2010); and (4) surrogate modelling techniques (e.g., Forrester and Keane, 2009; Razavi et al., 2012; Tsoukalas and Makropoulos, 2014).

The latter category, surrogate models (also termed *meta-models* or *response surface models*) are particularly useful as they can be applied to virtually any optimisation problem, including the urban drainage calibration variety. The key concept of this technique is to generate (inexpensive) meta-models that are accurate in a certain region (i.e., in a potentially optimal area) of the search space and use these meta-models in combination with the more expensive complete models to intelligently guide the optimisation algorithm (Couckuyt et al., 2013). The basic steps of a typical surrogate-based optimisation algorithm are as follows (Tsoukalas and Makropoulos, 2015): (1) optimal selection of some points, usually via a random sampling procedure, which are going to be evaluated with the expensive model; (2) construction of the surrogate model to approximate the expensive simulator (typical surrogate models are radial basis functions [RBFs], polynomials, artificial neural networks, and support vector machines); (3) searching of the surrogate model (optimised/explored) with an internal optimisation algorithm; (4) identification and evaluation of the optimised samples with the expensive model and updating of the surrogate model to improve its accuracy. This iterative process continues until a termination condition is satisfied. Typical surrogate-based optimisation algorithms are the Multistart Local Metric Stochastic RBF (Regis and Shoemaker, 2007), DYnamic COordinate Search-Multistart Local Metric Stochastic RBF (Regis and Shoemaker, 2013), and Surrogate-Enhanced Evolutionary Annealing Simplex Algorithm (Tsoukalas et al., 2016).

The key promise these algorithms hold is that they can be used to calibrate complex (even integrated) models, using very long time series, also driven by stochastic inputs (such as rainfall), allowing us to perform otherwise impractical, formal uncertainty analysis of (calibrated) model parameters and model results. A discussion of the process of uncertainty analysis in urban drainage is the subject of the following section.

20.6 UNCERTAINTY ANALYSIS

Urban drainage models, as any other mathematical models, provide a simplified representation of reality and subsequently are characterised by an inherent uncertainty. The total elimination of uncertainty is not possible to achieve, and hence, an uncertainty analysis is of high importance to quantify the level of reliability of the models, providing a robust basis for their application in practice (e.g., to provide a level of confidence for a model used for risk analysis). Extensive reviews on uncertainty analysis in urban drainage models can be found in Deletic et al. (2012), Dotto et al. (2012), and Freni et al. (2009).

20.6.1 Types of uncertainty in urban drainage modelling

The total uncertainty in model results is typically attributed to three main sources:

1. Model input uncertainty: derives from the imperfect or infrequent measurement of data used in model calibration and validation, and as initial or boundary conditions (e.g., systematic and/or random errors)
2. Parameter uncertainty: defined as the uncertainty in the model parameter vector and is related with the processes and data used in model calibration. This uncertainty is influenced by the
   a. Accuracy and availability of observations (e.g., measurement errors in both inputs and outputs)
   b. Appropriateness of calibration data sets (e.g., use of non-representative calibration data)
   c. Efficiency and appropriateness of calibration/optimisation algorithms
   d. Appropriateness of objective functions
3. Model structural uncertainty: also known as "model error," is related to the ability of the model to represent the real mechanism. This error can be further discerned into conceptualisation error (i.e., scale issues, omission or simplification of key processes), ill-posed equations, and poor identification of boundary conditions and numerical methods.

The above sources of uncertainty are interlinked, and their decomposition into individual components is rather a hard task requiring assumptions on the statistical properties of the individual errors. Figure 20.6, graphically presents the main sources of uncertainty along with their interconnections.

20.6.2 Uncertainty analysis in urban drainage modelling

The complexity and non-linearity of drainage models do not allow for the analysis, estimation, and propagation of uncertainties via analytical methods. This process typically involves probability theory, Monte Carlo (MC) sampling techniques, and formal or informal Bayesian statistical methods.

The most popular, Monte Carlo–based uncertainty assessment method is the Generalised Likelihood Uncertainty Estimator (GLUE), developed by Beven and Binley (1992). This approach rejects the idea of an *optimum model* or parameter set in favour of the *equifinality concept* that assumes that many parameter sets or model structures may have an equal likelihood to produce acceptable simulations.

In GLUE, different candidate model structures are identified along with the parameters that most affect the output. A large number of parameters are generated by prior probability distributions using a MC-type sampling process. The models are run for each generated parameter set, and the model outputs are compared to the available calibration data. A quantitative measure is then used to assess the performance of each model, based on the model residuals. Different formal or informal likelihood measures or combinations of them can be used. The models and their corresponding parameters sets that provide a likelihood

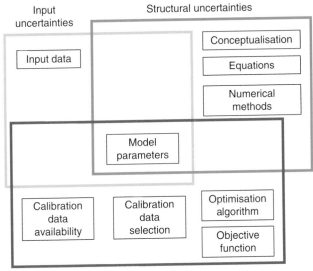

Figure 20.6 The main sources of uncertainties in urban drainage models and interlinks between them. (Adapted from Deletic, A. et al. 2012. *Physics and Chemistry of the Earth*, 42–44, 3–10.)

measure that reaches a minimum threshold are retained and rescaled so that their cumulative probability sum is equal to 1. The uncertainty bands of model predictions are then calculated.

The implementation of the above general methodology requires the modeller to make subjective decisions at each step of the process (Beven, 2012):

- The model or models to include in the analysis
- The feasible range for each parameter value
- The sampling strategy for the parameter sets
- An appropriate likelihood measure or set of measures
- The conditions for rejecting models

The main advantage of the GLUE methodology is the flexibility to account for different sources of uncertainty along with the fact that there is no need for an assumption of the error distribution functions. For instance, by considering different parameter sets or model structures, one can explicitly assess parameter uncertainty or structural uncertainty, respectively. By using different input and output data, the effect of measurement errors can be estimated. Further to the high computational burden imposed by the great number of required simulations, the method has been mainly criticised for the use of informal likelihood measures for statistical inference and for the choice of the threshold of some likelihood measure above which models are considered as acceptable.

A comparative analysis of the use of GLUE and other methods for parameter uncertainty analysis in urban drainage models is provided by Dotto et al. (2012).

PROBLEMS

20.1   List the main alternative modelling approaches, and explain their main characteristics.

20.2   Define integration of urban models, and discuss main principles.

20.3 List the main parts of the urban wastewater system, and consider ways in which they interact with each other. What are the potential benefits of integrated system control?

20.4 What is the difference between single and multi-objective modelling? What benefits can be gained from the latter?

20.5 Describe the main standards for model and data integration.

20.6 What is calibration of a model? What are the main difficulties in the calibration of integrated urban models?

20.7 List and describe the typical approaches to address the high computational burden of urban water models.

20.8 What are the main sources of uncertainty in urban drainage modelling? Explain a suitable method to assess them.

KEY SOURCES

Bach, P.M., Rauch, W., Mikkelsen, P.S., McCarthy, D.T., and Deletic, A. 2014. A critical review of integrated urban water modelling—Urban drainage and beyond. *Environmental Modelling and Software*, 54, 88–107.

Deletic, A., Dotto, C.B.S., Mccarthy, D.T., Kleidorfer, M., Freni, G., Mannina, G. et al. 2012. Assessing uncertainties in urban drainage models. *Physics and Chemistry of the Earth*, 42–44, 3–10.

Dotto, C.B.S., Mannina, G., Kleidorfer, M., Vezzaro, L., Henrichs, M., Mccarthy, D.T. et al. 2012. Comparison of different uncertainty techniques in urban stormwater quantity and quality modelling. *Water Research*, 46(8), 2545–2558.

Freni, G., Mannina, G., and Viviani, G. 2009. Uncertainty assessment of an integrated urban drainage model. *Journal of Hydrology*, 373(3–4), 392–404.

Makropoulos, C.K., Natsis, K., Liu, S., Mittas, K., and Butler, D. 2008. Decision support for sustainable option selection in integrated urban water management. *Environmental Modelling and Software*, 23(12), 1448–1460.

Razavi, S., Tolson, B.A., and Burn, D.H. 2012. Review of surrogate modeling in water resources. *Water Resources Research*, 48(7), W07401.

Schütze, M., Butler, D., and Beck, M.B. 2001. Parameter optimisation of real-time control strategies for urban wastewater systems. *Water Science Technology*, 43(7), 139–146.

REFERENCES

Achleitner, S., Möderl, M., and Rauch, W. 2007. CITY DRAIN © An open source approach for simulation of integrated urban drainage systems. *Environmental Modelling and Software*, 22(8), 1184–1195.

Bach, P.M. 2012. *UrbanBEATS: An Exploratory Model for Strategic Planning of Urban Water Infrastructure* (Online). Melbourne. www.urbanbeatsmodel.com

Beck, M.B. 1997. Applying systems analysis in managing the water environment: Towards a new agenda. *Water Science Technology*, 36(5), 1–17.

Beven, K. 2012. *Rainfall-Runoff Modelling: The Primer*, 2nd edn, John Wiley and Sons, Ltd, Chichester, UK. doi:10.1002/9781119951001.

Beven, K. and Binley, A. 1992. The future of distributed models: Model calibration and uncertainty prediction. *Hydrological Processes*, 6, 279–298.

Brown, R.R., Keath, N., and Wong, T.H.F. 2009. Urban water management in cities: Historical, current and future regimes. *Water Science Technology*, 59(5), 847–855.

Burger, G., Fach, S., Kinzel, H., and Rauch, W. 2010. Parallel computing in conceptual sewer simulations. *Water Science Technology*, 61(2), 283–291.

Butler, D. and Schütze, M. 2005. Integrating simulation models with a view to optimal control of urban wastewater systems. *Environmental Modelling and Software*, 20, 415–426.

Castronova, A.M., Goodall, J.L., and Ercan, M.B. 2013. Integrated modeling within a hydrologic information system: An OpenMI based approach. *Environmental Modelling and Software*, 39, 263–273.

Cheng, C.-T., Wu, X.-Y., and Chau, K.W. 2005. Multiple criteria rainfall–runoff model calibration using a parallel genetic algorithm in a cluster of computers. *Hydrological Sciences Journal*, 50(6), 1069–1087.

Couckuyt, I., Deschrijver, D., and Dhaene, T. 2013. Fast calculation of multiobjective probability of improvement and expected improvement criteria for Pareto optimization. *Journal of Global Optimization*, 60(3), 575–594.

Darsono, S. and Labadie, J.W. 2007. Neural-optimal control algorithm for real-time regulation of in-line storage in combined sewer systems. *Environmental Modelling and Software*, 22(9), 1349–1361.

DHI 2009. *MIKE URBAN Model Manager*. Denmark.

Dias, B.H., Tomim, M.A., Marcato, A.L.M., Ramos, T.P., Brandi, R.B.S., Junior, I.C.d.S., and Filho, J.A.P. 2013. Parallel computing applied to the stochastic dynamic programming for long term operation planning of hydrothermal power systems. *European Journal of Operational Research*, 229(1), 212–222.

Elliott, A.H. and Trowsdale, S.A. 2007. A review of models for low impact urban stormwater drainage. *Environmental Modelling and Software*, 22, 394–405.

Forrester, A. and Keane, A. 2009. Recent advances in surrogate-based optimization. *Progress in Aerospace Sciences*, 45(1–3): 50–79.

Foundation for Water Research (FWR). 2012. *Urban Pollution Management Manual*, 3rd edn, London.

Freni, G., Maglionico, M., Mannina, G., and Viviani, G. 2008. Comparison between a detailed and a simplified integrated model for the assessment of urban drainage environmental impact on an ephemeral river. *Urban Water Journal*, 5(2), 87–96.

Fu, G., Butler, D., and Khu, S.T. 2008. Multiple objective optimal control of integrated urban wastewater systems. *Environmental Modelling and Software*, 23(2), 225–234.

Fuchs, L. and Beeneken, T. 2006. Comparison of measured and simulated real-time control. *Seventh International Conference on Developments in Urban Drainage Modelling and Fourth International Conference on Water Sensitive Design*, Melbourne, Australia, V2.687–V2.695.

Gregersen, J.B., Gijsbers, P.J.A., and Westen, S.J.P. 2007. OpenMI: Open modelling interface. *Journal of Hydroinformatics*, 9(3), 175–191.

Hardy, M.J., Kuczera, G., and Coombes, P.J. 2005. Integrated urban water cycle management: The UrbanCycle model. *Water Science Technology*, 52(9), 1–9.

Institut für Automation und Kommunikation (IFAK). 2007. *SIMBA (Simulation of Biological Wastewater Systems): Manual and Reference*. Magdeburg, Germany.

Khu, S.T. and Madsen, H. 2005. Multiobjective calibration with Pareto preference ordering: An application to rainfall-runoff model calibration. *Water Resources Research*, 41, 1–14.

Loke, E., Warnaars, E.A., Jacobsen, P., Nelen, F., and Almeida, M. 1996. Problems in urban storm drainage addressed by artificial neural networks. *Proceedings of the Seventh International Conference on Urban Storm Drainage*, 3, Hannover, September, 1581–1586.

Makropoulos, C. 2017. Thinking platforms for smarter urban water systems: Fusing technical and socio-economic models and tools. *Geological Society, London, Special Publications*, 408(1), 201–219.

Mitchell, V.G. 2004. Integrated urban water management, a review of current Australian. Technical Report of the Australian Water Conservation and Reuse Research Program, a joint initiative of CSIRO and Australian Water Association, Melbourne, Australia.

Muschalla, D. 2008. Optimization of integrated urban wastewater systems using multi-objective evolution strategies. *Urban Water Journal*, 5(1), 57–65.

Muschalla, D., Ostowski, M., Schröter, K., and Wörsching, S. 2005. Integrated modelling and multi-objective evolution strategy as a method for water quality oriented optimisation of urban drainage systems. *Proceedings of the 10th International Conference on Urban Drainage*, Copenhagen, August, on CD-ROM.

Muschalla, D., Schütze, M., Schroeder, K., Bach, M., Blumensaat, K., Klepiszewski, K., et al. 2008. The HSG guideline document for modelling integrated urban wastewater systems. *Proceedings of the 11th International Conference on Urban Drainage*, Edinburgh, September, on CD-ROM.

Nowak, R. 2007. Real-time control of Vienna's sewer system. *IWA Water 21*, April, 36–37.

Open Geospatial Consortium (OGC). 2014. SensorML: Model and XML Encoding Standard, Mike Botts.

Rauch, W., Bertrand-Krajewski, J., Krebs, P., Mark, O., Schilling, W., Schütze, M., and Vanrolleghem, P.A. 2002. Deterministic modelling of integrated urban drainage systems. *Water Science Technology*, 45(3), 81–94.

Rauch, W., Seggelke, K., Brown, R., and Krebs, P. 2005. Integrated approaches in urban storm drainage: Where do we stand? *Environmental Management*, 35(4), 396–409.

Rauch, W., Bach, P.M., Brown, R., Deletic, A., Ferguson, B., De Haan, J. et al. 2012. Modelling transition in urban drainage management. *Ninth International Conference on Urban Drainage Modelling*. Belgrade, Serbia.

Razavi, S., Tolson, B.A., Matott, L.S., Thomson, N.R., MacLean, A., and Seglenieks, F.R. 2010. Reducing the computational cost of automatic calibration through model preemption. *Water Resources Research*, 46(11), W11523. doi:10.1029/2009WR008957.

Regis, R.G. and Shoemaker, C.A. 2007. A stochastic radial basis function method for the global optimization of expensive functions. *Informs Journal on Computing*, 19(4), 497–509.

Regis, R.G. and Shoemaker, C.A. 2013. Combining radial basis function surrogates and dynamic coordinate search in high-dimensional expensive black-box optimization. *Engineering Optimization*, 45(5), 529–555.

Rozos, E. and Makropoulos, C. 2013. Source to tap urban water cycle modelling. *Journal of Environmental Modelling and Software*, 41, 139–150.

Schütze, M., Butler, D., and Beck, M.B. 1999. Optimisation of control strategies for the urban wastewater system – An integrated approach. *Water Science and Technology*, 39(9), 209–216.

Smolders, S., Neyskens, I., Willems, P., Vaes, G., and Van Assel, J. 2008. Bidirectional sewer-river linking through the OpenMI software. *Proceedings of the 11th International Conference on Urban Drainage*, Edinburgh, September, on CD-ROM.

Solvi, A.M. 2007. *Modelling the Sewer-Treatment-Urban River System in View of the EU Water Framework Directive, PhD thesis*, Ghent University, Belgium.

Tolson, B.A. and Shoemaker, C.A. 2007. Dynamically dimensioned search algorithm for computationally efficient watershed model calibration. *Water Resources Research*, 43(1), WR004723.

Tolson, B.A., Asadzadeh, M., Maier, H.R., and Zecchin, A. 2009. Hybrid discrete dynamically dimensioned search (HD-DDS) algorithm for water distribution system design optimization. *Water Resources Research*, 45(12), W12416.

Tsoukalas, I. and Makropoulos, C. 2014. A surrogate based optimization approach for the development of uncertainty-aware reservoir operational rules: The case of nestos hydrosystem. *Water Resources Management*, 29, 4719–4734.

Tsoukalas, I. and Makropoulos, C. 2015. Multiobjective optimisation on a budget: Exploring surrogate modelling for robust multi-reservoir rules generation under hydrological uncertainty. *Environmental Modelling and Software*, 69, 396–413.

Tsoukalas, I., Kossieris, P., Efstratiadis, A., and Makropoulos, C. 2016. Surrogate-enhanced evolutionary annealing simplex algorithm for effective and efficient optimization of water resources problems on a budget. *Environmental Modelling and Software*, 77: 122–142.

Vanhooren, H., Meirlaen, J., Amerlinck, Y., Claeys, F., Vangheluwe, H., and Vanrolleghem, P.A. 2003. WEST: Modelling biological wastewater treatment. *Journal of Hydroinformatics*, 5(1), 27–50.

Vojinovic, Z. and Seyoum, S.D. 2008. Integrated urban water systems modelling with a simplified surrogate modular approach. In: *11th International Conference on Urban Drainage*, Edinburgh, Scotland, UK.

Chapter 21

Stormwater management (SuDS)

21.1 INTRODUCTION

Stormwater management (using non-pipes systems) has already been referred to a number of times in this text where we look to drain developed areas in a more natural way at the source, using the infiltration and storage capacities of devices such as soakaways, swales, and ponds. This approach has been given different names in different countries. In the United States, the techniques are now called *stormwater control measures* (SCMs) but were formerly known as *best management practices* (BMPs). In Australia, the general expression *water sensitive urban design* (WSUD) communicates a philosophy for water engineering in which water use, reuse, and drainage, and their impacts on the natural and urban environments, are considered holistically. China is developing a concept of the *Sponge City*. In the United Kingdom, since the mid-1990s, the label has been SuDS (sustainable drainage systems). Several other terms are in use conveying similar meanings (see Fletcher et al., 2015, for more detail and historical context).

Earlier chapters, especially Chapter 10, have dealt with piped systems for stormwater, so in this chapter we consider the alternatives to pipes including the wider benefits of SuDS above and beyond their water quantity and quality function. Section 21.2 describes the common types of SuDS devices in turn, Section 21.3 gives guidance on aspects of their design, and Section 21.4 considers how they can be linked together. A range of important issues still surround SuDS, and these are considered in Section 21.5. Depending on one's point of view, these can be seen either as disadvantages of SuDS or as barriers to wider acceptance that need to be removed. Finally, a range of other stormwater management measures, including community engagement, is discussed in Section 21.6. The place of stormwater management within wider water management initiatives such as water-sensitive urban design and green infrastructure is discussed in Chapter 24.

21.2 DEVICES

Details of many individual SuDS devices and approaches to their selection to maximise benefits based on land use, site and catchment characteristics, and required performance, are set out in the comprehensive Construction Industry Research and Information Association (CIRIA) document, *The SuDS Manual* (Woods-Ballard et al., 2016). Another resource on SuDS is available online at: www.susdrain.org. This section summarises the most common types of individual SuDS devices:

- Inlet controls
- Infiltration devices
- Vegetated surfaces

- Pervious pavements
- Filter drains
- Infiltration basins
- Detention basins
- Ponds
- Constructed wetlands

21.2.1 Inlet controls

Runoff can be managed at or near its source by a variety of means including blue roofs, green roofs, rainwater harvesting and water butts, and paved area ponding.

21.2.1.1 Blue roofs

Stormwater can be retained on flat roofs, thus exploiting their reservoir storage potential by using flow restrictors on the roof drains. Obviously, this will induce an additional live load to be taken into account in the structural design, and this will depend on the maximum depth of the retained water. A 75 mm depth system will induce a loading of 0.75 kN/m² (equivalent to a typical snow load). Particular care is needed in specifying and maintaining the roof waterproof membranes. In practice, flow restrictors can become blocked, leading either to overtopping or prolonged ponding, again pointing to the need for regular maintenance and exceedance routes. Blue roofs can also be specified in combination with green roofs.

21.2.1.2 Green roofs

A green roof is a planted area that has rainfall storage potential, encourages evapo-transpiration, and provides the added benefit of water quality improvement as stormwater travels through the soil layer (Figure 21.1). Extensive green roofs typically take the form of a "carpet" of plants, supported by lightweight growing media and overlying a drainage layer, whereas intensive green roofs incorporate more deeply rooted vegetation. During a storm event, rainfall is intercepted by the plant layer, infiltrated and stored in the substrate, and then stored and transported downstream in the drainage layer.

Green roofs have been shown to retain between 50% and 100% of precipitation over the long term, but this depends on the intensity and duration of the rainfall events and is influenced by the roof slope, substrate depth, substrate moisture content, evapo-transpiration rates, and plant morphology (Whittinghill et al., 2015). In a UK context, Stovin et al. (2012) found that over a 29-month period, cumulative retention of a trial green roof was 50%. However, this varied widely between storm events, ranging from 0 to 100%, and the roof provided only 13% retention for the largest event (16-year return period). For storms with rainfall depth > 2 mm, mean peak flow attenuation was 60%. In their UK study, Nawaz et al. (2015) found the overall retention to be 66%.

Green roofs also have numerous wider benefits, including reduced air and noise pollution, increased habitat and biodiversity, improved aesthetics, increased roof lifespan, energy savings, and mitigation of the urban heat island effect (Berardi et al., 2014; Early et al., 2007).

21.2.1.3 Rainwater harvesting and water butts

Rainwater harvesting (RWH) consists of a system to collect, filter, and store rainfall runoff from roofs with the intention to use the stored water for non-potable uses around the house

Figure 21.1 Green roof, Portland, Oregon.

and garden. Particularly in the drier months, tanks may be fully or partly empty for some of the time and so some storage volume will be available to retain flows and provide attenuation providing water demand is greater than runoff yield. It is difficult to predict the performance of existing systems due to the wide variety of factors in play, but evidence suggests this can be significant (Campisano et al., 2014). BS 8515: 2009+A1: 2013 provides a methodology for sizing the volume of new tanks to provide additional stormwater management benefits based on water use behaviour, rainfall statistics, and roof size. The method has been applied to a UK case study by Kellagher and Gutierrez-Andres (2015) using number of bedrooms and average property occupancy as a proxy for water use.

An alternative approach partitions the rainwater storage tank as shown in Figure 21.2. The retention and throttle concept includes dedicated storage for retaining runoff and limiting outflow, while protecting the volume required for non-potable water supply. The throttle can be either passive (e.g., an orifice or vortex regulator) or active (e.g., powered and controlled). Evidence is beginning to emerge on the benefits of this type of system, at least in modelling studies. For example, Melville-Shreeve et al. (2016a) compared a traditional RWH system with a dual-purpose RWH system when both were combined with that of a downstream storage tank for runoff control from a new development. While the total storage volume for the latter was found to be greater, its overall cost was lower due to cheaper

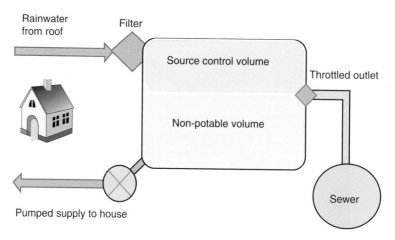

Figure 21.2 Dual-purpose rainwater harvesting system.

RWH tanks. Many different RWH system configurations are under test or coming onto the market (Melville-Shreeve et al., 2016b.)

As an alternative to a full RWH system, a much simpler store can be provided at the foot of the downpipe, either above or below ground. Small volumes (a water butt will have a capacity of about 200–350 L) used in large numbers, can have effects comparable with rooftop ponding or RWH systems. A water butt is still fundamentally a means of harvesting rainwater for garden use, but to be most effective in providing attenuation for stormwater management, it must allow some overflow to the drainage system so there is some capacity for the next rainfall.

A further simple alternative is to discharge runoff away from the building and over stable pervious areas (such as lawns, swales, porous pavements) rather than directly to the pipe system. Therefore, the surface runoff is delayed, infiltration is increased, and pollutants are removed to a certain extent. This can be carried out in existing urban areas as well as new-build.

21.2.1.4 Paved area ponding

In principle, similar benefits to rooftop ponding can be gained by exploiting the storage available on paved areas. Potential sites include car parks, paved storage yards, and other large impervious surfaces (also see Chapter 11 on flooding). The advantage over roof storage is that much larger surfaces are available and ponding depth can be greater. Disadvantages relate particularly to the nuisance value of ponded water and, in extreme cases, possible damage and safety issues. In addition, unless the system is properly maintained, it will not function. Methods to mobilise ground-level storage usually involve restricting flow into the sewer system via gullies with orifices or vortex regulators (see Chapter 8).

21.2.2 Infiltration devices

The two most common infiltration devices are soakaways and infiltration trenches. A soakaway is an underground structure that can be stone filled, formed with plastic mesh boxes, dry wall lined, or built with precast concrete ring units (see Figure 21.3). It is recommended that any filling has a void ratio, e (defined as the ratio of interstitial volume to soil volume), of at least 30%. An infiltration trench is a linear excavation lined with a filter fabric, backfilled with stone, and possibly covered with grass. Runoff is diverted to the soakaway or trench and either infiltrates the soil or evaporates (see Figure 21.4). The device provides

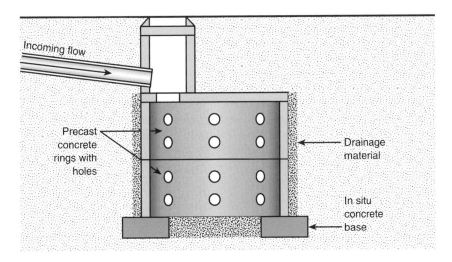

Figure 21.3 Soakaway.

storage and enhances the ability of the soil to accept water by creating a surface area of contact. Soakaways and infiltration trenches should not be located within 5 m of the foundations of buildings, or under roads. As a result of their shape characteristics, trenches are usually more efficient than soakaways at controlling runoff.

Soakaways and trenches can be used in any area that has pervious subsoils such as gravel, sand, chalk, and fissured rock. However, trenches installed on land gradients greater than about 4% need "flow checks" at regular intervals. These systems are only suitable in areas where the water table is low enough to allow a free flow of the stormwater into the subsoil at all times of the year. The base of the soakaway or trench should therefore be at least 1 m above the groundwater level.

As well as disposing of the stormwater, soakaways and trenches can reduce the concentration of some of the pollutants it contains, by processes of physical filtration, absorption, and biochemical activity. Mean annual removal efficiencies for suspended solids, metals, polyaromatic hydrocarbon (PAH), oil, and chemical oxygen demand (COD) of 60%–85% have been recorded for infiltration trenches draining highway runoff (Colwill et al., 1984).

A potential problem is clogging by sediments carried with the stormwater. Siriwardene et al. (2007) have studied the processes involved and have shown that even very fine sediments can cause clogging.

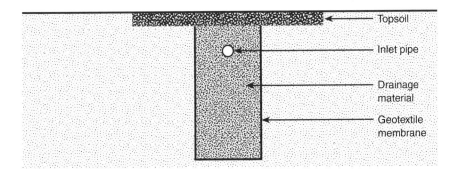

Figure 21.4 Infiltration trench.

There are some restrictions on infiltration where there is a risk of pollution to groundwater resources (see Section 21.5).

21.2.3 Vegetated surfaces

The main types of vegetated surfaces used in stormwater management are filter strips, grassed swales (see Figure 21.5), and bioretention systems. *Swales* are grass-lined channels used for the conveyance, storage, infiltration, and treatment of stormwater. Runoff enters directly from adjoining buildings or other impermeable surfaces. The runoff is stored either until infiltration takes place, or until the filtered runoff is conveyed elsewhere, to the sewer system, for example. *Filter strips*, also known as *vegetative buffer strips*, are gently sloping areas of ground designed to promote sheet flow of stormwater runoff.

To function well, swales require shallow slopes (< 5%) or the introduction of check dams and soils that drain well. Typically, they have side slopes of no greater than one in three, allowing them to be easily maintained by grass-cutting machinery. The bottom width is usually between 0.5 m and 2 m, they are 0.25–2 m deep, and can be readily incorporated into landscape features. Filter strips should allow a minimum flow distance of about 6 m. Swales and strips delay stormwater runoff peaks and provide a reduction in runoff volume due to infiltration and evapo-transpiration. Typical velocities should be below 0.3 m/s to encourage settlement.

They are often used as a pre-treatment in combination with other control measures. Pollutants are removed by sedimentation, filtration through grass and adsorption onto it, and infiltration into the soil. Runoff quality can be considerably improved, and Ellis (1992) found that a swale of length 30–60 m could retain 60%–70% of solids and 30%–40% of metals, hydrocarbons, and bacteria. Controlled experiments on grass swales in Australia (Fletcher et al., 2002) have demonstrated reductions in total suspended solids *concentrations* of between 73% and 94% and in total suspended solids *loads* of between 57% and 88%, together with significant reductions in total nitrogen and total phosphorus. Revitt et al. (2017) found that infiltration from swales had negligible impact on groundwater quality if properly maintained.

Figure 21.5 Swale. (Courtesy of Prof. Chris Jefferies.)

Figure 21.6 Car park with pervious pavement. (Courtesy of Formpave Limited.)

Bioretention systems (and rain gardens) are shallow landscaped depressions that typically consist of a filter medium (usually sandy), underlain by a gravel drainage layer, and may be lined. Their role is to intercept runoff from frequent storm events and treat stormwater by mechanical filtration, sedimentation, adsorption, and plant and microbial uptake (Hatt et al., 2009).

21.2.4 Pervious pavements

Pervious pavements are used mostly for car parks (Figure 21.6) and can also be used for other surfaces where there is no traffic or very light traffic. A typical arrangement for the pavement structure is illustrated in vertical section in Figure 21.7. There are a number of alternatives for the surface layer. It could consist of one of a variety of types of block, or could be a layer of porous asphalt. Blocks may be *porous*, allowing water to seep through them via pores in the material, or *permeable*, where the material is not porous but the blocks are laid in such a way that water can pass *between* them. Permeable blocks may fit tightly,

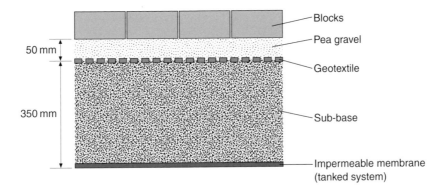

Figure 21.7 Typical pervious pavement structure, vertical section.

Figure 21.8 Permeable surface blocks. (Courtesy of Formpave Limited.)

with water passing through narrow slots between blocks (Figure 21.8), or may be laid with a pattern of larger voids, which are filled with soil and grass, or gravel.

Below the surface layer of blocks is a bedding layer of sand or small-size gravel, separated from the sub-base below by a layer of geotextile. (The bedding layer is not necessary with porous asphalt.) The sub-base consists of crushed rock or other sufficiently hard material, or a plastic mesh structure.

At the lower surface of the sub-base is a pervious geotextile, if infiltration to the ground is intended, or an impervious geomembrane if it is not. If there is no infiltration, the pavement structure is a *tanked system*, providing considerable attenuation to the storm flow but still requiring arrangements for outflow.

The storage potential of a pervious car park structure can also be exploited by receiving further runoff from roof surfaces or other impermeable surfaces. There is also the potential for runoff stored in the sub-base to be used for applications such as toilet flushing and garden/landscape watering. Antunes et al. (2016) confirm the great potential for reuse but provide caution on the quality of the water produced.

Infiltration rates through permeable pavement surfaces, especially new ones, are generally high. For new surfaces, rates are commonly in excess of 1000 mm/h, and sometimes considerably higher, depending on the type of surface. Studies from a number of countries reported by Pratt et al. (2002) indicate that the infiltration rate of a pavement surface may reduce over its life to 10% of its original (new) value. To limit blockage at the surface, it is recommended (Woods-Ballard et al., 2016) that the surface should be cleaned three times a year. However, Rommel et al. (2001) showed that other parts of the system, especially the geotextile, were more likely than the surface to limit infiltration through the structure as a whole.

Monitored sites have consistently demonstrated substantial reduction and delay in storm peaks and reduction in overall volumes. The sites have included tanked systems (Abbott et al., 2003; Schlüter and Jefferies, 2001) and systems with infiltration to the ground (Macdonald and Jefferies, 2001).

Pervious pavements may also have a positive effect on water quality by providing mechanisms that encourage filtration, sedimentation, adsorption, and chemical/biological treatment, as well as storage (Pratt et al., 2002). Laboratory and field tests have demonstrated excellent performance in retention and degradation of oils (Pratt et al., 1999). Effluent pollutant levels have been shown to be significantly correlated with the type of sub-base used (Sañudo-Fontaneda et al., 2014).

21.2.5 Filter drains

Filter drains are linear devices consisting of a perforated or porous pipe in a trench of filter material. They have traditionally been constructed beside roads to intercept and convey runoff, but they can be used simply as conveyance devices. They may or may not allow infiltration to the ground, in the same way as pervious pavements.

21.2.6 Infiltration basins

Infiltration basins are depressions with vegetative cover that store runoff for infiltration into the ground. They are used where there is sufficient capacity for infiltration and where infiltration is appropriate. The bottom of the basin is flat to provide uniform infiltration. The side slopes should be no steeper than one in four. They are generally used for relatively small catchments.

21.2.7 Detention basins

Detention basins provide storage for stormwater with controlled outflow to the next stage of stormwater management or to a watercourse. They are effectively storage facilities formed out of the landscape. They are not intended to encourage infiltration to the ground and may be lined if infiltration is to be prevented completely. After controlled outflow has taken place the basin is commonly left dry until the next rainfall. In many cases detention basins are hardly noticed by the public as a consequence of thoughtful landscaping and choice of vegetation. They are often multifunctional, also operating as recreation areas and only filling during storms. Detention basins may be online or offline. They have relatively low pollutant removal efficiencies because of the re-erosion of previously deposited solids during filling.

21.2.8 Ponds

Ponds provide storage and treatment for stormwater within a permanent volume of water (Figure 21.9). They have aesthetic, recreational value (e.g., sailing, fishing) and environmental benefits such as returning wildlife habitats into urban areas, in addition to their flood control function. They can also play a significant role in pollution control since sedimentation and biological processes may enhance the water quality of the outflow. This is because many pollutants are attached to suspended solids, which are themselves captured by sedimentation. The depth of the pool is usually limited to 1.5–3 m to avoid thermal stratification. Shallow side slopes and well-chosen marginal vegetation help ensure safety. McLean et al. (2005) give practical guidance on ponds and detention basins.

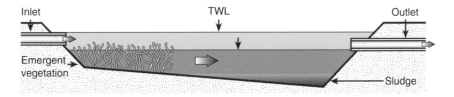

Figure 21.9 Simplified wet detention pond.

21.2.9 Constructed wetlands

Constructed or *artificial wetlands* (including reed beds, reed marshes, and vegetative systems) are shallow areas of excavated land filled with earth, rock, or gravel, saturated with water or covered by shallow flowing water at some time during the growing season, and planted with selected aquatic plants. The key roles of the plants are to transmit oxygen from the atmosphere to the root system (thus the soil) and to encourage microbial growth. Wetlands require relatively large areas of flat to gently sloping (less than about 5%) land.

It is generally agreed that wetlands are simple and inexpensive to build, but there is some disagreement on their ease of operation. In order to remain effective, wetlands need long-term programmes of maintenance, involving the planting and extraction of the aquatic plants. It has been estimated that a wetland has a life of about 15–20 years.

In addition to substantially attenuating and reducing runoff flows, wetlands can significantly improve water quality by removing large quantities of particulate and soluble contaminants (including suspended solids, metals, excess nutrients, and bacteria) through biological action and sedimentation. Wetlands also trap silt, and promote recovery of DO in the effluent. Results show that wetlands can, on an annual basis, remove almost all bacteria and suspended solids, and just under half the total phosphorus and nitrogen load (Ellis, 1992). They also have valuable secondary uses, in providing wildlife habitats and recreational/educational areas. However, wetlands are sensitive and must be carefully managed to avoid vegetation die-off.

There should be a mixture of water depths. Typically depths over 1 m should cover no more than 20% of the area, and depths of under 0.5 m should cover more than 50%. There should be continuous baseflow, but velocities should be low (less than 0.1 m/s). A sediment removal forebay is normally included. Further details are given in Scholz (2006).

21.3 ELEMENTS OF DESIGN

The ultimate aim in design is to find the right tools for the job, used in the best combination. The result may be a system of drainage based entirely on SuDS and involving no conventional piped drainage. Alternatively, it may contain elements of source control in combination with oversized sewers or underground storage tanks, discharging heavily attenuated storm flows to a conventional drainage system. In a densely developed area, specifying and installing SuDS may well require far greater effort and imagination than relaying solely on a standard piped solution.

We consider here elements in the design of SuDS devices. The detailed design of a whole SUDS system is beyond the scope of this book, but sources of further guidance are given throughout this chapter.

21.3.1 Rainfall

Clearly all SuDS design work involves some prediction of rainfall. Appropriate methods have been presented in Section 4.3. Predicting the impact of climate change (Section 4.6) is as relevant to SuDS as it is to other methods of drainage.

21.3.2 Runoff

Methods for predicting runoff have been presented in Sections 5.2 and 5.3. As well as requiring predictions of runoff from developed areas, SuDS design often calls for values of greenfield runoff (before development). Further guidance is given in Section 10.7.

21.3.3 Conveyance

Clearly any pipes that form part of a SuDS system can be designed using the principles set out in Sections 7.3 and 7.4. Flow in swales and over filter strips can be treated as open channel flow, and calculations can be based on the Manning equation (Section 7.5). It is recommended that Manning's n is taken as 0.25 when depth of flow is below the top of the grass and 0.1 when it is above. When estimating the roughness value to be used, vegetative growth should be considered, as at times, it may be short and hence increased conveyance rates and at other times considerably longer, and therefore reduced conveyance rates (and the potential to improve water quality).

21.3.4 Inlets and outlets

Inlet and outlet arrangements are important details in SuDS design. Elements of outlet design that relate to the control of outflow are covered in Sections 8.1 and 8.2. Inlet arrangements may include sediment detention, and also arrangements for flow distribution, for example, kerb details to guide runoff from a road onto a filter strip, or overflow arrangements for different levels of runoff. Details are given in *The SuDS Manual* (Woods-Ballard et al., 2016).

21.3.5 Storage volume related to inflow and outflow

A simple method for determining storage volume by considering storms of different durations is given in Section 13.3. More detailed methods of considering the interaction between inflow, outflow, and storage are demonstrated in Sections 13.4 and 13.5.

21.3.6 Infiltration from a pervious pavement sub-base

CIRIA Report 156 (Bettess, 1996) gives the following method for *plane infiltration systems*. The formula below gives the maximum depth of water in the sub-base:

$$h_{max} = \frac{D}{n}(Ri - f) \tag{21.1}$$

h_{max} is the maximum depth (m), D is the duration of rainfall (h), n is the porosity of sub-base (volume of voids ÷ total volume), R is the ratio of drained area to infiltration area, i is the rainfall intensity (m/h), and f is the infiltration rate (m/h).

Accurate estimation of soil infiltration rate is difficult because rates depend on numerous factors, such as

- Soil particle size and grading
- The presence of organic material
- Plant, animal, and construction activity
- Soil history

In particular, considerable differences in value may occur at different times of the year and with different antecedent conditions. The best, but still not ideal, information can be obtained from undertaking on-site trials. General soil data, for use in preliminary calculations, is given in Table 21.1. Infiltration systems are not suitable in soils with infiltration rates less than 0.001 mm/h. Porosity ranges from 0.2 for graded sand to as high as 0.9 for some proprietary geocellular products.

Table 21.1 Typical soil infiltration rates

Soil type	Rate (mm/h)
Gravel	10–1000
Sand	0.1–100
Loam	0.01–1
Chalk	0.001–100
Clay	<0.0001

Source: After Bettess, R. 1996. *Infiltration Drainage—Manual of Good Practice*, CIRIA 156.

Table 21.2 Factor of safety applied to measured infiltration rate

Area drained (m²)	Factor of safety, related to consequence of failure		
	No damage or inconvenience	Minor inconvenience	Major consequences
<100	1.5	2	10
100–1000	1.5	3	10
>1000	1.5	5	10

The recommendation in CIRIA 156 is that the infiltration rate is determined from on-site tests, and that a factor of safety is then applied in calculations to account for progressive siltation (Table 21.2).

EXAMPLE 21.1

Determine the maximum depth of water in the sub-base of a pervious car park for the following storm event: rain intensity 50 mm/h, duration 1 hour. Infiltration rate to soil has been measured as 15 mm/h. Area drained is 500 m², area of pavement is 250 m². Surface flooding would cause minor inconvenience. Porosity of sub-base material is 0.4.
    Repeat for a tanked system.

Solution

Safety factor is 3. Therefore, $f = 15/3 = 5$ mm/h. From Equation 21.1:

$$h_{max} = \frac{1}{0.4}(2 \times 50 \times 10^{-3} - 5 \times 10^{-3}) = 0.238 \text{ m}$$

For a tanked system,

$$h_{max} = \frac{1 \times 2 \times 50 \times 10^{-3}}{0.4} \, 0.250 \text{ m}$$

If there is no infiltration from the bottom of the sub-base (a tanked system), and outflow by other means can be ignored, the expression is even simpler:

$$h_{max} = \frac{DRi}{n}$$

21.3.7 Infiltration from a soakaway or infiltration trench

The main design methods available for design of non-plane infiltration systems are BRE Digest 365 (BRE, 1991) and CIRIA Report 156 (Bettess, 1996). These have some differences, which include procedures for the use of test pits to measure infiltration rates. Also, as described above, CIRIA 156 recommends factors of safety to be applied to the infiltration rate in calculations, whereas BRE Digest 365 assumes that infiltration takes place through the sides but not through the base (which gives an implicit factor of safety by assuming that the base will become clogged with fine particles). BRE Digest 365 assumes that infiltration from the system into the soil is constant and corresponds to that when the system is half full of water, whereas CIRIA 156 includes infiltration through the base and uses a more complex representation of infiltration overall. The method in BRE Digest 365 (which leads to more straightforward calculations than CIRIA 156) is presented in most detail here.

21.3.7.1 BRE Digest 365 method

Storage for runoff from the critical storm, at a given return period (10 years is suggested), is given by

$$\text{Storage} = \text{Runoff volume} - \text{Infiltration during storm} \tag{21.2}$$

So, if the critical storm has duration, D (h), storage, S (m³), is given by

$$S = iA_iD - fa_{50}D \tag{21.3}$$

I is the rainfall intensity (m/h), A_i is the impervious area (m²), f is the soil infiltration rate (m/h), and a_{50} is the effective surface area for infiltration (m²).

This can be applied to an infiltration trench as follows. A trench has a width of B_d, an effective depth (i.e., depth below invert of the incoming pipe) y, and a length L. Assuming discharge *from* the trench *to* the soil is through the sides (length and ends) only, and that the average water level is half the effective trench depth, the effective infiltration surface area is

$$a_{50} = y(B_d + L) \tag{21.4}$$

Active storage volume S in the trench is

$$S = yB_dLn \tag{21.5}$$

n is the porosity of the trench material (free volume ÷ total volume).

After the rainfall event, it is recommended that the infiltration rate should be sufficient to empty at least half the stored volume within 24 hours.

The suggested design procedure is

1. Determine the soil infiltration rate
2. Adopt a device cross section
3. Determine the required storage volume by considering a range of durations of 10-year design storms
4. Review the suitability of the design and check that the device will half empty within 24 hours

A calculation using this method is presented as Example 21.2.

EXAMPLE 21.2

Design an infiltration trench to serve an individual property draining an impermeable area of 100 m² if it is filled with an aggregate, which gives a free volume that is 0.3 of total volume. The soil infiltration rate is estimated from an on-site trial pit at 10 mm/h. Rainfall statistics for the 10-year storm are as derived in Example 4.2.

Solution

1. $f = 10$ mm/h (no factor of safety is used, but it is assumed that there is no infiltration through the base).
2. Say: Width $(B_d) = 1$ m
   Effective depth $= 1$ m
   ... determine L
3. Equation 21.3:

$$S = iA_iD - fa_{50}D$$

Substitute using Equations 21.4 and 21.5:

$$yB_dLn = iA_iD - fy(B_d + L)D$$

Sample calculation for $D = 1$ hour: $i = 24.8$ mm/h

$$1 \times 1 \times L \times 0.3 = 24.8 \times 10^{-3} \times 100 \times 1 - 10 \times 10^{-3} \times 1 \times (1 + L) \times 1$$

giving $L = 7.97$ m.

Considering the range of durations of 10-year design storms from Example 4.2:

Storm duration, D (h)	(min)	Intensity, i (mm/h)	Length, L (m)
	5	112.8	3.13
	10	80.4	4.44
	15	62	5.12
	30	38.2	6.25
1		24.8	7.97
2		14.9	9.23
4		8.6	9.96
6		6.1	9.93
10		4	9.64
24		2	8.27

The critical case is at 4 hours: length 9.96 m—say 10 m.

4. Time for emptying is evaluated assuming half the stored volume discharges through the effective surface area for infiltration:

$$t = \frac{0.5B_dL_n}{y(B_d + L)f} = \frac{0.5 \times 1 \times 10 \times 0.3}{1 \times (1 + 10) \times 10 \times 10^{-3}} = 13.6 \text{ hours}$$

which is acceptable (< 24 hours).

21.3.7.2 CIRIA 156 method

For a particular rainfall event discharging to an infiltration device, maximum depth of water in the device is given by

$$h_{max} = a(\exp(-bD) - 1) \tag{21.6}$$

in which

$$a = \frac{A_b}{P} - \frac{iA_D}{Pf}$$

and

$$b = \frac{Pf}{nA_b}$$

D is the storm duration (h), A_b is the area of base (m²), P is the perimeter of infiltration device (m), i is the rainfall intensity (m/h), A_D is the impermeable area from which runoff is received (m²), and f is the infiltration rate (m/h) = measured rate ÷ factor of safety.

Time for half emptying (t in hours) is given by

$$t = \frac{nA_b}{fP} \ln \left[\frac{h_{max} + \dfrac{A_b}{P}}{\dfrac{h_{max}}{2} + \dfrac{A_b}{P}} \right] \tag{21.7}$$

21.3.8 Water quality

21.3.8.1 Mass balance

The basis of water quality design for local disposal techniques is essentially a matter of mass balance of pollutants, in many ways similar to hydraulic design with an additional term:

$$\text{Accumulated mass} = \text{Mass inflow} - \text{Mass outflow} \pm \text{Mass change} \tag{21.8}$$

The mass inflow for a given time period is the product of the flow volume and the mean pollutant concentration. The mass change is a function of the chemical, biological, or physical changes that take place within the device. Change can be both positive and negative, and it is possible that, for a given storm event, there is a net *export* of pollutants due to release of pollutants in some way bound within the device (e.g., in sediments or biomass).

The difficulties in the design become apparent when the complexities of urban rainfall and runoff are compounded with the great number of pollutant types and concentrations, and with the many reactions that can take place. In addition, the pollutant removal rates of devices vary considerably.

For practical purposes, it can be assumed that the percent of pollutant mass captured is proportional to the percent of runoff flow volume captured (i.e., diverted away from the piped system). Capture of about 10–15 mm of effective rainfall (i.e., runoff) should capture in the range of 80%–90% of the annual effective rainfall and thus that percentage of the pollution. The reason is that the majority of the storms in any given year only produce a small amount of runoff. So, unlike stormwater quantity management, where infrequent major storms are of most concern, in water quality control it is total *volumes* that are of most significance.

21.3.8.2 Treatment volume

A feature of storage is that it encourages the settlement of some of the suspended sediment and associated pollutants contained in the incoming stormwater. In practice, removal efficiencies vary considerably, but well-designed ponds have the capacity to make reductions in many pollutants.

For significant treatment to take place, the captured runoff must be retained for at least 24 hours. CIRIA (2000) suggest the following expression to estimate required treatment volume, based on data obtained from the Wallingford Procedure:

$$V_t = 9(M5 - 60\,\text{min})(SOIL/2 + PIMP(1 - SOIL/2)/100) \tag{21.9}$$

V_t is the basic treatment volume (m^3/ha), *SOIL* is the soil index for the UK based on the Winter rain acceptance potential (WRAP) classification, *M5—60 min* is a 5-year return period, 60-minute duration (standard) rainfall depth, and *PIMP* is the percentage imperviousness of the catchment served.

Dry detention ponds are designed to fully contain the treatment volume V_t, and to drain down within 24 hours. Wet retention ponds require a volume of 4 V_t to allow time for biological treatment in addition to physical processes.

Other important factors for optimum pollutant reduction (Ellis, 1992) are that

- The ratio of pond volume to mean storm runoff volume is between 4 and 6.
- The storage volume exceeds 100 m^3 per hectare of effective contributing area.
- The ratio of pond surface area to effective contributing area is 3%–5%.

21.3.9 Amenity and biodiversity

Amenity is defined in *The SuDS manual* (Woods-Ballard et al., 2016) as "a useful or pleasant facility or service." Many SuDS facilities can, with careful design, detailing, and ongoing maintenance contribute towards providing amenity benefits in many ways, including environmental quality, health and well-being, education opportunities, recreation, visual appeal, and more. Each of these can be expressed as design criteria. More detailed advice and examples are given in *The Manual*.

Biodiversity, while linked to amenity, is a separate concept and refers to "the number and variety of different organisms found within a particular region." This is expressed locally as the character of animals and plants that share the same living space as humans. Design objectives include supporting and enhancing local habitats, contributing to habitat connectivity, and providing diversity of local ecosystems (Woods-Ballard et al., 2016).

21.3.10 Modelling

The main applications for models of drainage systems are in *design* and *simulation*, as described in Chapter 19 (in the context of conventional sewered catchments). This is equally true for drainage systems based on SuDS, and also to sewered systems that include SuDS.

SuDS devices are typically included in the main commercial sewer modelling packages and a wide range of research models and tools has also been developed. All provide a representation of device hydrologic and hydraulic performance, but other issues such as water quality and economic impacts are less comprehensively covered (Elliott and Trowsdale, 2007; Lerer et al., 2015). Caution is advised in evaluating model results as already complex models now have the added burden of representing (perhaps poorly characterised) two- or three-dimensional devices and flow patterns, depending on the level of effort and detail undertaken during modelling.

Compared with linear pipe systems, SuDS provision has a spatial element to it, and greater knowledge is required about the local urban context in order to specify appropriate devices. This has given rise to the development of mapping tools (typically geographic information system [GIS] based), which can support option choice and decision making. For example, Makropoulos et al. (2001), Ellis and Viavattene (2014), and Voskamp and Van de Ven (2014) describe tools that seek to prioritise the placement of SuDS devices and quantify the impact on flood risk.

Digman et al. (2016) have developed technical guidance and a spreadsheet tool *BeST— Benefits of SuDS Tool* to evaluate, quantify, and compare the multiple benefits of SuDS, accounting for uncertainty in their valuation.

21.4 SuDS APPLICATIONS

21.4.1 Management train

The use of SuDS devices in combination is important to maximise their overall performance, termed a *management train*, or also referred to as a *treatment train*. The recommended sequence of possibilities, with devices appropriate for each stage, is given in Figure 21.10. It is preferable to find a drainage solution as close to the top of this diagram as possible, but if all drainage needs cannot be achieved at a particular stage, the designer must move further down the list. A good example of the management train principle in practice is described by Bray (2001), where runoff from different sections of a motorway services site passes through separate trains of linked devices including trenches, swales, ponds, and wetlands.

Jefferies et al. (2008) describe a tool and scoring system for providing guidance on the appropriate level and combination of SuDS depending on the type and scale of a development and the nature of the receiving water. Intended to provide greater consistency in meeting regulatory requirements for SuDS, the tool balances the risks of pollution to the receiving water with the treatment provided in the management train. Bastien et al. (2010) concluded that the use of a treatment train approach provides a more flexible solution than individual SuDs or end-of-pipe solutions taking into account cost and performance objectives.

21.4.1.1 Retrofit

In existing sewered catchments, there may also be benefits in *retrofitting* SuDS devices. This is slowly becoming a mainstream approach to managing stormwater (Stovin and Swan, 2003). Heal et al. (2005) describe a retrofit SuDS system consisting of a pond and wetland to remediate pollution of the Caw Burn in Livingston, Scotland, resulting from runoff from an industrial estate. Field monitoring established that the system was effective in reducing concentrations of sediment, oil, and other pollutants in the runoff, and had resulted in an increase in the diversity of life in the watercourse. Stovin et al. (2013) present a specific example of the viability of SuDS retrofit in the context of addressing combined sewer

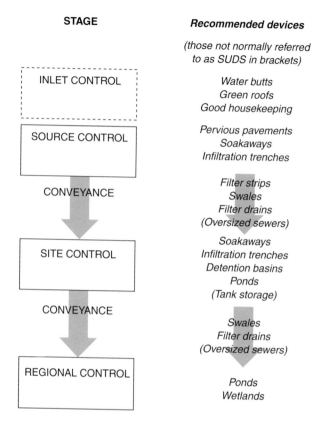

Figure 21.10 SuDS management train.

overflow discharges in London. *Susdrain* (www.susdrain.org) contains numerous examples of retrofit, and CIRIA's guidance (Digman et al., 2012) *Retrofitting to Manage Surface Water* provides practical steps to support the retrofitting of SuDS in urban areas.

21.5 ISSUES

We consider here some of the wider issues that are involved when SuDS are considered as drainage options. Some of these issues are described as "barriers," and the growth in the use of SuDS has been associated with their progressive removal. Most commentators point to the importance of partnerships between developers, local planning authorities, and environmental agencies, potentially also including consulting engineers, water companies, highway authorities, third-sector organisations, residents, and the wider community.

21.5.1 Long-term performance

Concerns have been expressed about the performance of SuDs devices over the long term and reported experiences have been mixed. Jefferies (2004), for example, collated the results of monitoring of a wide range of SuDS sites in Scotland including ponds, pervious pavements, detention basins, filter drains, and swales. All systems were found to operate effectively in terms of flow attenuation, with systems closest to the source of runoff having the most impact. No evidence was found to suggest that the systems studied would not continue to operate as designed, provided they continued to be maintained properly. However, Bergman et al. (2011) examined the performance of two stormwater infiltration trenches installed in the late 1990s in Copenhagen and found that the life span of the infiltration trenches was much shorter than expected due to siltation and clogging. Similar concerns were expressed by Achleitner et al. (2011) on the hydraulic permeability of an infiltration and swale system.

This evidence points to the conclusion that SuDS are not "fit and forget" systems, and there is an ongoing need for appropriate maintenance as discussed next.

21.5.1.1 Maintenance

Accepting responsibility to maintain SuDS devices in perpetuity is seen by some as a risk because of a lack of data on long-term performance. Possible deterioration in performance, perhaps due to blockage by silt, makes others question the appropriateness of the word *sustainable* for such devices. (For more thoughts on this controversial word, see Chapter 24.) However, all infrastructures require maintenance. The maintenance needs of piped systems are described in Chapter 17, and the obvious fact that SuDS also need to be maintained does not necessarily put them at a disadvantage.

It is recommended that owners of developments with SuDS should be provided with an owner's manual and model agreements for SuDS—legal documents that can be used as the basis of arrangements for maintenance—and these have been published by CIRIA (Shaffer et al., 2004).

Many SuDS devices promote sedimentation, and therefore sediment *removal* can be an important element in maintenance. Sediment removal operations and subsequent disposal of waste are considered in *The SUDS Manual* (Woods-Ballard et al., 2016).

21.5.1.2 Adoption

As described in Section 6.4.2, if a developer constructs a piped drainage system, it is normal for the system to be "adopted" by the sewerage undertaker who will then include the system as a revenue-earning asset and accept responsibility for operation and maintenance.

It is not clear that the same procedure is suitable for all SuDS devices because many could be considered as public space landscape features (rather than drainage features) and therefore more appropriately the responsibility of the local authority. Adoption of SuDS therefore remains a complex and as yet unresolved issue (Grant et al., 2017).

21.5.1.3 Costs

Despite some concerns over cost, evidence suggests that in many cases for new or redevelopment, it should be cost beneficial to install a SuDS system in preference to a traditional drainage system. The larger and less dense the new development to be served is, the more likely it is that this will be the case (Royal Haskoning DHV, 2012). Retrofit costs are often higher than conventional solutions, but this is predominantly because they will often provide wider drainage benefits and other benefits that traditional solutions tend not to provide. It is becoming more common for these benefits along with the costs to be considered in making investment decisions, particularly by water service providers. Operation and maintenance costs must be taken into account in addition to construction costs. *The SUDS Manual* recommends and presents a whole life cost approach, and provides some typical costs.

21.5.1.4 Planning

Following significant flooding in the UK in 2007 (and reinforced by subsequent major flooding events), an overhaul of flood risk management procedures was carried out with the aim of simplification, avoiding duplication, and clarifying lines of communication. Most notably, this included designating Lead Local Flood Authorities operating at a regional level. In parallel, changes to the national planning policy framework in England gave priority to the use of SuDS in new major developments. However, while these initiatives refer to the preferential use of SuDs, these must be *economically proportionate*, meaning SuDS still only remain desirable rather than mandatory obligations within development control (Ellis and Lundy, 2016; Shaffer, 2010).

21.5.1.5 Groundwater pollution

The risk of groundwater pollution from infiltrated runoff is rightly a potential concern. The groundwater protection policy of the Environment Agency (2007) gives detailed guidance in terms of the required aquifer protection. There is generally no objection to infiltration of roof drainage, but, depending on groundwater vulnerability, an oil interceptor may be required for other impermeable areas, or infiltration simply may not be allowed. However, Ellis (2006) cautions that there is contradictory evidence as to whether infiltration systems pose a long-term threat to groundwater quality in the context of the EU Water Framework Directive.

21.6 OTHER STORMWATER MANAGEMENT MEASURES

21.6.1 Oil separators

Oil separators are underground chambers with compartments designed to separate oil and water, and retain the oil fraction. They are usually provided for small catchments, particularly where heavy oil or petrol spills are expected (e.g., petrol stations). Outflow is routed to the drainage system.

In most types of interceptor, the flow rate of the effluent is reduced, and the lighter oil fraction separated by gravitational means. Interceptors suitable for treating stormwater should also contain a storage section for silt and suspended matter. The captured oil and sediment must be regularly removed as, otherwise, a heavy storm event may cause oil accumulated from previous storms to be discharged to the receiving water. Current designs include more efficient tilting plate separators and coalescing filters.

Information on applications of oil separators, on types and classes of separator, and on design, sizing, and operation, are given in PPG 3, one of the Pollution Prevention Guidelines by EHSNI/SEPA/EA (2015).

21.6.1.1 Non-structural measures

Stormwater quality can be affected not only by the use of engineered control devices, but also by management practices. This can include altering current maintenance programmes (e.g., street sweeping, gully pot cleaning—see Section 17.4) or discouraging or preventing inappropriate practices (e.g., in pesticide management, chemical storage).

21.6.1.2 Public attitudes and community engagement

SuDS are far more obviously part of the urban landscape than underground pipe systems, and therefore, to be successful, they must be welcomed by the public. Amenity benefits are only genuine if they are perceived as such by the public. A survey in Scotland of public attitudes to SuDS (Apostolaki et al., 2001) found that public opinion varied according to the type of device, with more positive attitudes towards ponds than swales, for example. The only real concerns over ponds were over safety aspects. Little understanding of the purpose of swales was thought to contribute to the negative attitudes.

Public education has the potential to increase the popularity and acceptance of SuDS. This can be facilitated pre- and post-installation through a variety of means including clear and visually attractive information about local SuDS systems in the form of leaflets, information boards, and signs. Schools and their teachers have an important role to play here. This can help local people understand the purpose and function of SuDS, appreciate the services they can provide, and so engage in good behaviour around installations. However, for even greater benefit, consideration should be given to involving the community in the co-design of installations so they can offer informed opinion and local experience, and options can be tailored, where possible, towards local preferences (Everett and Lamond, 2016).

PROBLEMS

21.1 What are SuDS, and why is their use being promoted?

21.2 Outline the range of SuDS devices, and describe how the main types of device operate.

21.3 What is meant by the term *surface water management train*?

21.4 A 0.9 m wide, 1 m effective depth infiltration trench is required to drain an impermeable area of 120 m². The trench fill material is known to have a porosity of 0.33, but the soil infiltration rate can only be estimated as 25–50 mm/h. Calculate the required length of trench. Rainfall statistics are the same as those used in Example 21.1.   [10.1 m]

21.5 Explain the function and operation of constructed wetlands in stormwater management.

21.6 What non-structural methods are available to improve the quality of stormwater from residential developments? What are their benefits compared with structural measures?

21.7 Describe and discuss the multiple benefits provided by SuDS.

21.8 How might the local community be more actively engaged in SuDS design and operation?

KEY SOURCES

Ashley, R., Lundy, L., Ward, S., Shaffer, P., Walker, L., Morgan, C. et al. 2013. Water-sensitive urban design: Opportunities for the UK. *Proceedings of the Institution of Civil Engineers, Municipal Engineer*, 166(ME2), 65–76.

Ashley, R., Walker, L., D'Arcy, B., Wilson, S., Illman, S., Shaffer, P. et al. 2015. UK sustainable drainage systems: Past, present and future. *Proceedings of the Institution of Civil Engineers, Civil Engineering*, 168(CE3), 125–130.

BS EN 8582: 2013. *Code of Practice for Surface Water Management for Development Sites*.

Charlesworth, S.M., and Booth, C.A. Eds. 2016. *Sustainable Surface Water Management. A Handbook for SUDS*, Wiley Blackwell.

National Research Council. 2008. *Urban Stormwater in the United States*, National Academies Press, Washington, DC.

Woods-Ballard, B., Wilson, S., Udale-Clark, H., Illman, S., Scott, T., Ashley, R., and Kellagher, R. 2016. *The SUDS Manual*, 5th edn, CIRIA C753.

REFERENCES

Abbott, C.L., Weisgerber, A., and Woods-Ballard, B. 2003. Observed hydraulic benefits of two UK permeable pavement systems. *Proceedings of the Second National Conference on Sustainable Drainage*, Coventry University, June, 101–111.

Achleitner, S., Engelhard, C., Stegner, U., and Rauch, W. 2011. Local infiltration devices at parking sites—Experimental assessment of temporal changes in hydraulic and contaminant removal capacity. *Water Science and Technology*, 55, 193–200.

Antunes, L.N., Thives, L.P., and Ghisi, E. 2016. Potential for potable water savings in buildings by using stormwater harvested from porous pavements. *Water*, 8, 11.

Apostolaki, S., Jefferies, C., and Souter, N. 2001. Assessing the public perception of SUDS in two locations in Eastern Scotland. *Proceedings of the First National Conference on Sustainable Drainage*, Coventry University, June, 28–37.

Bastien, N., Arthur, S., Wallis, S., and Scholz, M. 2010. The best management of SuDS treatment trains: A holistic approach. *Water Science and Technology—WST*, 61(1), 263–272.

Berardi, U., GhaffarianHoseini, A., and GhaffarianHoseini, A. 2014. State-of-the-art analysis of the environmental benefits of green roofs. *Applied Energy*, 115, 411–428.

Bergman, M., Hedegaard, M.R., Petersen, M.F., Binning, P., Mark, O., and Mikkelsen, P.S. 2011. Evaluation of two stormwater infiltration trenches in central Copenhagen after 15 years of operation. *Water Science and Technology*, 63, 2279–2286.

Bettess, R. 1996. *Infiltration Drainage—Manual of Good Practice*, CIRIA 156.

Bray, R. 2001. Environmental monitoring of sustainable drainage at Hopwood Park motorway service area M42 junction 2. *Proceedings of the First National Conference on Sustainable Drainage*, Coventry University, June, 58–70.

BS 8515 2009+A1: 2013. *Rainwater Harvesting Systems—Code of Practice*.

Building Research Establishment. 1991. *Soakaway Design*, Digest 365.

Campisano, A., Di Libertoa, D., Modicaa, C., and Reitanob, S. 2014. Potential for peak flow reduction by rainwater harvesting tanks. *Procedia Engineering*, 89, 1507–1514.

Construction Industry Research and Information Association (CIRIA). 2000. *Sustainable Urban Drainage Systems—Design Manual for Scotland and Northern Ireland*, CIRIA C521 and *Sustainable Urban Drainage Systems—Design Manual for England and Wales*, CIRIA C522.

Colwill, D.M., Peters, C.J., and Perry, R. 1984. *Water Quality in Motorway Run-off*, TRRL Supplementary Report 823, Transport and Road Research Laboratory.

Digman, C.J., Ashley, R.M., Balmforth, D.J., Balmforth, D., Stovin, V., and Glerum, J. 2012. *Retrofitting to Manage Surface Water*, CIRIA C713.

Digman, C.J., Horton, B., Ashley, R.M., and Gill, E. 2016. *BeST (Benefits of SuDS Tool) User Manual, Version 2*, CIRIA W045d.

Early, P., Gedge, D., Newton, J., and Wilson, S. 2007. *Building Greener—Guidance on the Use of Green Roofs, Green Walls and Complementary Features on Buildings*, CIRIA C644.

EHSNI/SEPA/EA. 2015. *Use and Design of Oil Separators in Surface Water Drainage Systems, PPG 3 (Pollution Prevention Guidelines)*. Environment and Heritage Service (Northern Ireland), Scottish Environment Protection Agency, Environment Agency.

Elliott, A.H. and Trowsdale, S.A. 2007. A review of models for low impact urban stormwater drainage. *Environmental Modelling and Software*, 22, 394–405.

Ellis, J.B. 1992. Quality issues of source control. *Proceedings of CONFLO 92: Integrated Catchment Planning and Source Control*, Oxford.

Ellis, J.B. 2006. A UK and European perspective of Sustainable Urban Drainage (SUDS) with particular reference to infiltration systems and groundwater pollution. *It Doesn't Just Go Away, You Know..., SUDS and Groundwater Monitoring, Proceedings of the IAH, 26th Annual Groundwater Conference*, Co. Offaly.

Ellis, J.B. and Lundy, L. 2016. Implementing sustainable drainage systems for urban surface water management within the regulatory framework in England and Wales. *Journal of Environmental Management*, 183, 630–636.

Ellis, J.B. and Viavattene, C. 2014. Sustainable urban drainage system modeling for managing urban surface water flood risk. *CLEAN-Soil, Air, Water*, 42(2), 153–159.

Environment Agency. 2007. *Groundwater Protection: Policy and Practice.*

Everett, G. and Lamond, J. 2016. SuDS and human behaviour: Co-developing solutions to encourage sustainable behaviour. *FLOODrisk 2016, Third European Conference on Flood Risk Management, E3S Web of Conferences*, doi:10.1051/e3sconf/2016.

Fletcher, T.D., Peljo, L., Fielding, J., Wong, H.F., and Weber, T. 2002. The performance of vegetated swales for urban stormwater pollution control. *Global Solutions for Urban Drainage: Proceedings of the Ninth International Conference on Urban Drainage*, Portland, Oregon, September, on CD-ROM.

Fletcher, T.D., Shuster, W., Hunt, W.F., Ashley, R.A., Butler, D., Arthur, S. et al. 2015. SUDS, LID, BMPs, WSUD and more—The evolution and application of terminology surrounding urban drainage. *Urban Water Journal*, 12(7), 525–542.

Grant, L., Chisholm, A., and Benwell, R. 2017. *A Place for SuDS? Assessing the Effectiveness of Delivering Multifunctional Sustainable Drainage through the Planning System*, CIWEM Report.

Hatt, B.E., Fletcher, T.D., & Deletic, A. 2009. Hydrologic and pollutant removal performance of stormwater biofiltration systems at the field scale. *Journal of Hydrology*, 365(3), 310–321.

Heal, K.V., Scholz, M., Willby, N., and Homer, B. 2005. The Caw Burn SUDS: Performance of a settlement pond/wetland SUDS retrofit. *Proceedings of the Third National Conference on Sustainable Drainage*, Coventry University, June, 19–29.

Jefferies, C. 2004. *SUDS in Scotland—The Monitoring Programme of the Scottish Universities SUDS Monitoring Group*. Scotland and Northern Ireland Forum for Environmental Research (SNIFFER) Report SR(02)51.

Jefferies, C., Duffy, A., Berwick, N., McLean, N., and Hemmingway, A. 2008. SUDS treatment train assessment tool. *Proceeding of the 11th International Conference on Urban Drainage*, Edinburgh, September, on CD-ROM.

Kellagher, R. and Gutierrez-Andres, J. 2015. Rainwater harvesting for domestic water demand and stormwater management, in *Alternative Water Supply Systems* (eds. F.A. Memon and S. Ward), IWA Publishing, 62–83.

Lerer, S.M., Arnbjerg-Nielsen, K., and Mikkelsen, P.S. 2015. A mapping of tools for informing water sensitive urban design planning decisions—Questions, aspects and context sensitivity. *Water*, 7, 993–1012.

Macdonald, K. and Jefferies, C. 2001. Performance comparison of porous paved and traditional car parks. *Proceedings of the First National Conference on Sustainable Drainage*, Coventry University, June, 170–181.

Makropoulos, C., Butler, D., and Maksimovic, C. 2001. GIS-supported stormwater source control implementation and urban flood risk mitigation, in *Advances in Urban Stormwater and Agricultural Runoff Source Controls* (ed. J. Marsalek), Kluwer, 95–105.

McLean, N., Campbell, N., Bray, R., and D'Arcy, B.J. 2005. New approaches to detention basins and ponds. *Proceedings of the Third National Conference on Sustainable Drainage*, Coventry University, June, 211–221.

Melville-Shreeve, P., Ward, S., and Butler, D. 2016a. Dual-purpose rainwater harvesting system design, in *Sustainable Surface Water Management. A Handbook for SUDS* (eds. S.M. Charlesworth and C.A. Booth), Wiley Blackwell, 255–268.

Melville-Shreeve, P., Ward, S., and Butler, D. 2016b. Rainwater harvesting typologies for UK houses: A multi criteria analysis of system configurations. *Water*, 8, 129.

Nawaz, R., McDonald, A., and Postoyko, S. 2015. Hydrological performance of a full-scale extensive green roof located in a temperate climate, *Ecological Engineering*, 82, 66–80.

Pratt, C.J., Newman, A.P., and Bond, P.C. 1999. Mineral oil bio-degradation within a permeable pavement: Long term observations. *Water Science and Technology*, 39(2), 103–109.

Pratt, C.J., Wilson, S., and Cooper, P. 2002. *Source Control Using Constructed Pervious Surfaces—Hydraulic, Structural and Water Quality Performance Issues*. CIRIA C582.

Revitt, D.M., Ellis, J.B., and Lundy, L. 2017. Assessing the impact of swales on receiving water quality. *Urban Water Journal*, 14(8), 839–845.

Rommel, M., Rus, M., Argue, J., Johnston, L., and Pezzaniti, D. 2001. Carpark with '1 to 1' (impervious/permeable) paving: Performance of 'Formpave' blocks. *Novatech 2001 Proceedings of the Fourth International Conference on Innovative Technologies in Urban Drainage*, Lyon, France, June, 807–814.

Royal Haskoning DHV. 2012. *Costs and Benefits of Sustainable Drainage Systems*, Committee on Climate Change, Final Report, 9X1055.

Sañudo-Fontaneda, L.A., Charlesworth, S.M., Castro-Fresno, D., Andrés-Valeri, V.C.A., and Rodriguez-Hernandez, J. 2014. Water quality and quantity assessment of pervious pavements performance in experimental car park areas. *Water Science and Technology*, 69(7), 1526–1533.

Schlüter, W. and Jefferies, C. 2001. Monitoring the outflow from a porous car park. *Proceedings of the First National Conference on Sustainable Drainage*, Coventry University, June, 182–191.

Scholz, M. 2006. *Wetland Systems to Control Urban Runoff*, Elsevier.

Shaffer, P. 2010. *Planning for SuDS—Making It Happen*, CIRIA C687. London.

Shaffer, P., Elliott, C., Reed, J., Holmes, J., and Ward, M. 2004. *Model Agreements for Sustainable Water Management Systems, Model Agreements for SUDS*, CIRIA C625.

Siriwardene, N.R., Deletic, A., and Fletcher, T.D. 2007. Clogging of stormwater gravel infiltration systems and filters: Insights from a laboratory study. *Water Research*, 41, 1433–1440.

Stovin, V., Vesuviano, G., and Kasmin, H. 2012. The hydrological performance of a green roof test bed under UK climatic conditions. *Journal of Hydrology*, 414–415, 148–161.

Stovin, V.R. and Swan, A.D. 2003. Application of a retrofit SUDS decision-support framework to a UK catchment. *Proceedings of the Second National Conference on Sustainable Drainage*, Coventry University, June, 171–180.

Stovin, V.R., Moore, S.L., Wall, M., and Ashley, R.M. 2013. The potential to retrofit sustainable drainage systems to address combined sewer overflow discharges in the Thames Tideway catchment. *Water and Environment Journal*, 27, 216–228.

Voskamp, I. and Van de Ven, F. 2014. Planning support system for climate adaptation: Composing effective sets of blue-green measures to reduce urban vulnerability to extreme weather events. *Building and Environment*, 83, 159–167.

Whittinghill, L.J., Rowe, D.B., Andresen, J.A., and Cregg, B.M. 2015. Comparison of stormwater runoff from sedum, native prairie, and vegetable producing green roofs. *Urban Ecosystems*, 18, 13–29.

Chapter 22

Smart systems

22.1 INTRODUCTION

Recent technological innovations and developments in information and communication technologies (ICTs), provide new perspectives towards a more dynamic, real-time and "smart" management of urban drainage infrastructure. This chapter starts off with an account of real-time control for urban drainage systems (Section 22.2) – well understood, but not yet widely practised. Here we suggest that technological and methodological developments now make it possible to model and optimise the effect of control on the urban wastewater system as a whole. Then the chapter describes the emergence of early warning systems (EWSs) as a key tool in the management of water-related hazards (Section 22.3), allowing us to prevent loss of life and limit property and infrastructure damage by predicting storm events in advance and warning vulnerable communities. Especially in the urban setting, EWSs deal with (pluvial) flash floods and (fluvial) surface water/river floods, as well as coastal floods and storm-induced landslides, mud flows, and debris flows. Last, we present recent developments towards *citizen-observatories*, getting the public involved in a more hands-on manner, both as sensors detecting problems in the functioning of the urban drainage infrastructure or the aftermath of floods and as recipients of real-time observations on (imminent) flood events (Section 22.4).

22.2 REAL-TIME CONTROL

Most urban drainage systems are managed in an essentially passive mode. That is, fixed elements are provided to operate in one way only without the opportunity for intervention of any kind. Systems are conventionally designed (see Chapter 10) to handle flows from storms of high return period (low frequency) and, therefore, contain large amounts of storage volume (even without dedicated stormwater detention). For most storm events, this storage is not fully utilised. So there should be significant scope to improve the performance of the system by actively managing this storage in response to changing input (e.g., rainfall) and output (e.g., flooding, overflows, interaction with treatment plant).

22.2.1 Definition

Schilling (1989) considers an urban drainage system to be operated under real-time control (RTC) when "process data, which is currently monitored in the system, is used to operate flow regulators during the actual process". Thus, system information (e.g., rainfall, water level) is continuously collected and processed and is used to make decisions about operation of devices (e.g., pumps, weirs) in the system in real time to limit the occurrence of adverse effects. This is also sometimes called *active control* or *active system control*.

The ultimate target in the application of RTC in urban drainage systems is the reduction of risk of flood events and sewer overflows, without expanding the capacity of the existing system. More recently, RTC systems have been studied in conjunction with water quality and pollution load issues (García et al., 2015).

22.2.2 Equipment

The main hardware elements to be found in RTC systems are

- *Sensors* that monitor the ongoing process
- *Regulators* that manipulate the process
- *Controllers* that activate the regulators
- *Data transmission systems* that carry measured data from sensors to controllers and signals from controllers to regulators

Together, these four elements form the *control loop* that is common to all RTC systems (considered in Section 22.2.3). More details on recent developments in RTC components can be found in Campisano et al. (2013).

22.2.2.1 Sensors

Although there is a large number of sensor types, only a few are suitable for RTC of drainage systems. These include rain gauges, water level gauges, flow rate gauges, and water quality sensors. In general, requirements include long-term resistance to water and chemicals, the suitability for continuous recording, remote data transmission, long life or continuous energy source, robustness, and reliability.

Water level measuring devices are the most commonly used sensors. They are used to determine the state of storage facilities, to evaluate flow depths into pipes, and to monitor the discharges through the gate orifice. Four different types of water level sensors are available (Campisano et al., 2013): capacity probes, pressure sensors, ultrasonic probes, and microwave sensors.

Flow rate estimates can be obtained from weirs and gates via level-flow converting relationships. More accurate measurements can be provided via electromagnetic and ultrasonic flow rate metres.

Recently, technological improvements (Campisano et al., 2013) make sensors more durable in the difficult environment of drainage systems, enabling the measurement of various quality parameters, such as temperature, pH, turbidity, suspended solids, organic, carbon, chemical oxygen demand, conductivity, ammonia, nitrate, and total nitrogen. The use of water quality sensors, although showing considerable potential, remains limited in the RTC framework.

In modern RTC systems, rain gauges along with radar stations are used to deliver information on the spatio-temporal distribution of rainfall over large catchments. Rainfall and flow forecasts from radar measurements are used to configure the control strategy (Fuchs and Beeneken, 2005).

22.2.2.2 Regulators

Sewer flow regulators, also known as actuators, are the components of the RTC system that are used to control water flows and levels. They include a variety of pumps (constant or variable speed), moveable weirs, penstocks, inflatable dams, valves, and flow splitters (see Chapters 8 and 14). Recently, chemical dosing devices and aeration devices have been used to adjust the quality parameters of the flow. Some of the more specialised devices are described by Schilling et al. (1989).

22.2.2.3 Controllers

A controller is required for every regulator. It accepts an input signal and adjusts the regulator accordingly. The most common types of digital controller are programmable logic controllers (PLCs) or remote terminal units (RTUs). Controllers can be broadly classified into two categories: continuous and discrete. The most common method of discrete control is the two-point controller that has only two positions: on/off or open/closed. An example of two-point control is shown in Figure 22.1 for a simple pumping station. Here, the pump switches off at a low level and on at a high level as shown.

The disadvantage of two-point control is that the frequency of switching can become excessive. Three-point controllers have been developed to overcome this problem, and they are typically used for regulators such as automatic penstocks and moveable weirs. The most commonly used controller for continuously variable regulator settings is the *PID* (*Proportional-Integral-Derivative*) *controller*. Simplified versions (P, PI, and PD) are also available.

To prevent adverse effects in the case of controller breakdown or data transmission failures, emergency strategies should be planned and implemented on-site.

22.2.2.4 Data transmission systems

The RTC components described above need to be well distributed over the entire drainage system, and hence a data transmission (telemetry) system is required to ensure communication among them. Data transmission is either wired (e.g., public telephone, radio communication) or wireless and should be fast and safe.

22.2.3 Control

22.2.3.1 Classification

RTC systems can be broadly classified as local, global, or integrated. Under local control, regulators use process measurements taken directly at the regulator site (e.g., by a float) but are not remotely manipulated from a control center, even if operational data are centrally

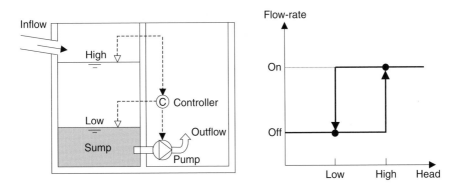

Figure 22.1 Two-point control at a pumping station.

acquired. An example is an automatic penstock being operated in relation to water level monitored by an ultrasonic level monitor.

In global RTC systems, regulators are operated in a coordinated way by a central computer with respect to process measurements throughout the entire system. This central computer is often part of the control system, also known as SCADA system (supervisory control and data acquisition) that monitors and controls the main components of the drainage system. For example, mobilisation of upstream and downstream storage volumes is linked in order to avoid emptying the upper tank into the already-full lower tank. In large and complex drainage systems, both local and global RTC systems co-exist.

The newest class (and least proven in practise) is that of integrated control where control can be influenced by process measurements derived from systems outside the sewer system (e.g., wastewater treatment plant [WTP], river).

22.2.3.2 Control loop

The *control loop* is the basic element of any RTC system. In this loop, measured values of the controlled variable are compared with a set of reference values, known as *set-points*. The outcome of this comparison then determines how the variable will be adjusted. Two main types of control loop can be distinguished (see Figure 22.2): feedback and feed forward.

- Feedback control: Control commands are actuated depending on the measured deviation of the controlled process from the set-points. Unless there is a deviation, the feedback controller is not actuated.

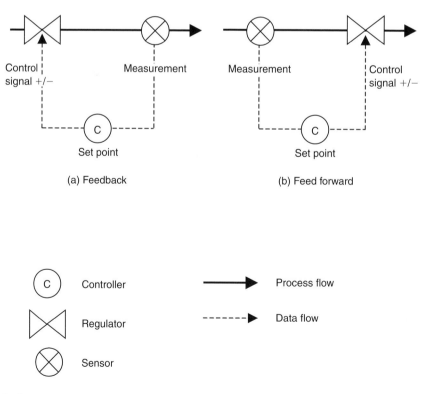

Figure 22.2 Control loop.

- Feed forward control: This anticipates the immediate future values of these deviations using a model of the process controlled, and activates the controls ahead of time to compensate. Its accuracy therefore depends on the effectiveness of the model.

22.2.3.3 Control strategy

A *control strategy*, also known as *control algorithm*, can be defined as either the time sequence of all regulator set-points or the set of control rules in a RTC system (Schilling et al., 1989). Control can either be performed *offline*, via static rules, or *online* via rules that are continuously updated depending on fast computer forecasts of the system state. Clearly, the simplest strategy is to keep the set-points constant, but time-varying set-points will probably give better performance, allowing the system to react to the non-regular transient storm events to which it is subjected.

A great variety of methods has been proposed for the development of control strategies (García et al., 2015; Ocampo-Martinez, 2010; Schilling et al., 1996; Schütze et al., 2004). In general, control strategies can be a product of either a heuristic approach or a more rigorous optimisation procedure. Heuristic control algorithms are based on the experience of the operations personnel to derive the initial set-points. Then, multiple simulation runs can be performed to improve the performance of the controllers (trial-and-error approach).

General control strategies are

- Preferential upstream storage: Wastewater/stormwater is retained first in the upper reaches of the network to reduce downstream flooding impacts.
- Preferential downstream storage: Wastewater/stormwater is retained first in the lower reaches of the network to minimise upstream combined sewer overflow (CSO) impacts.
- Balanced storage: The various storage elements are evenly filled throughout the catchment.

Through this process, *decision matrices* containing control actions for all possible combinations of the systems inflow and state variables can be produced *a priori* (*offline optimisation*). The control actions are presented in the form of "if … then … else" rules, where the first part of the statement describes the current state of the system (e.g., Almeida and Schilling, 1993). Relatively simple rules can be amended heuristically to improve system performance further. Although this algorithm is simple to implement and understand, it requires high levels of expertise from the staff and good knowledge of the system's behaviour. Fuzzy logic has also been used to obtain control values according to fuzzy rules (Klepiszewski and Schmitt, 2002; Nikoli et al., 2010).

Further to the trial-and-error approach, mathematical optimisation algorithms have also been used to derive optimum control rules on the basis specific criteria. Operational objectives are translated into the minimisation or maximisation of an *objective function* subject to (physical) constraints. For example, a simple RTC optimisation procedure would be to minimise the sum of all CSO discharge volumes V_i over a time horizon t_i to t_f:

$$\sum_{i=t_i}^{t_f} V_i \rightarrow \min$$

The aim is to identify the optimum solution and, therefore, the "best" performance of the system, given all constraints and assumptions. In the case of *multi-objective control*, more than one operational objective is considered. Some commonly used criteria are as follows (Joseph-Duran et al., 2014):

- Minimisation of flooding in streets
- Maximisation of the treated sewage in the system
- Minimisation of operation costs (pump stations and treatment plants)
- Minimisation of the water pollution released to the environment

This multi-objective problem is solved either by the scalarization of objective functions or via the typical Pareto-based optimisation techniques (Regneri et al., 2010).

Further to the *a priori* specified control strategies, periodical determination of the control rules depending on the current state of the system can also be implemented. In *online optimisation*, a number of possible control actions are evaluated, at pre-defined time steps (e.g., 5 minutes), and the most beneficial (on the basis of certain criteria) is applied. In the model-based predictive control (MPC) method, the mathematical model of the process is frequently updated with current measurements, generating a sequence of future control actions within a pre-defined, finite-time, horizon (García et al., 2015). The sequence of actions derives via an optimisation procedure that uses the predicted outputs of previous steps and the set of system constraints (operational, technical, etc.). The MPC approach has the advantage of anticipating the system's response to forthcoming rainfall events and the capacity to capture the complex non-linear system dynamics. However, the high computational effort required to evaluate the impact of different control actions in drainage system remains a critical issue, especially in the case of complex systems. To alleviate this burden, control-oriented drainage models, compromising accuracy and complexity, are usually used (see Chapter 20). Features of these models are discussed in García et al. (2015) and Antonio and Gutierrez (2014). The application of MPC methods in drainage systems is discussed in Marinaki and Papageorgiou (2005) and Ocampo-Martinez (2010).

Information collection and interpretation is a vital part in the processing of any control strategy. Historical data are very useful, but forecast information can be even more valuable to allow the system to become ready for the expected load (see Section 22.3). Thus, possible sources of information are as follows (Smart and Herbertson, 1992):

- Flow, level, and quality measurements in upstream sewers: System reaction must be within the time of flow.
- Rainfall measurements and results from rainfall/runoff models: Available reaction time is extended to the time of concentration of the catchment.
- Rainfall forecasts: These gain additional time depending on the forecast lead time.

22.2.4 Applicability

The implementation of an RTC system may be regarded as too complex and too expensive, depending on the special characteristics of the drainage system (Beeneken et al., 2013). For this reason, a comprehensive investigation should be conducted to identify the benefits and costs of deploying RTC solutions to a particular drainage system. There are numerous indicators and criteria to help make such a decision (García et al., 2015). Perhaps the greatest single indicator of the benefit to be derived from RTC of a particular catchment is the magnitude of the useful storage. In under-designed systems, with low storage volumes, real-time control will be of little benefit. But, in an over-designed system with large storage volumes, flooding and CSO discharges will be infrequent anyway, and little additional benefit will accrue. Properly designed systems with sufficient but not excessive storage that can be "activated" should produce the best results. Appropriate distribution of storage around the catchment is also important.

A second major factor is the hydraulic load to which the network is subjected. Assuming storage volume is fixed, RTC will provide no benefit for minor storm events that are handled effectively by the passive system. For large storm events, the benefits are more limited because, once all the storage capacity is fully utilised, the only remaining option is to allow discharges to the receiving water. However, even for large events, RTC offers some potential to more evenly distributed CSO discharges and to better control the first foul flush (see Chapter 12). RTC is, however, most effective at reducing the frequency of CSO spills from smaller but still significant storms. These minor spills are common occurrences in conventional passive systems where controls are pre-set at the design stage (often many years earlier).

Other network characteristics that favour RTC application are

- Spatially distributed inputs (rainfall)
- Spatially distributed storage
- Larger, flatter, more looped sewer networks
- Many controllable elements (e.g., storage tanks, pumps, overflows)

Guidelines for the practical assessment of RTC potential have been published by the German Association for Water, Wastewater and Waste (Erbe and Haas, 2007). The document includes valuable advice on the various stages from planning to implementation.

22.2.5 Benefits and drawbacks

In general, RTC has the following major benefits (Schilling et al., 1996):

- Reduction in the risk of flooding by utilising the full storage of the system
- Reduction in pollution spills by detaining more wastewater within the system
- Reduction in capital costs by minimising the storage and flow-carrying capacity requirements of the system
- Reduction in operating costs by optimising pumping and maintenance costs
- Enhancement of WTP performance by balancing inflow loads and allowing the plant to operate closer to its design capacity

Indirect benefits (Kellagher, 1996) concern flexibility to respond to changes in the catchment or local failure in the system, and better understanding of the performance of the network (see Section 17.7).

Typical RTC benefits include large reductions in CSO spills at the most sensitive locations, reduced frequencies of overflow operation (by about 50%), and reduced annual CSO volumes (by 10%–20%). Secondary benefits include lower energy costs (from less pumping), improved wastewater treatment, control of sewer sediments, and better system supervision, understanding, and record keeping (Nelen, 1993). However, operational experience and verified benefits have not been widely reported.

Drawbacks are relatively few, and reluctance to use more active systems in general and RTC in particular seems to rest mainly on lack of operational experience. However, there are also legitimate concerns regarding the increased maintenance commitment needed and issues surrounding the risk of failure. Currently, in the United Kingdom, local control is very common, but globally controlled systems are rare. As more RTC systems go online worldwide, the experience gained should improve confidence and accelerate uptake in appropriate circumstances. This will be further propelled by the rising prominence of the whole smart systems agenda and industrial capacity building.

22.3 EARLY WARNING SYSTEMS

Among natural hazards, storms and floods are the most lethal, affecting millions and causing heavy economic losses. According to the Emergency Event Database (EM-DAT), in Europe between 2000 and 2016, severe events affected more than 1 million people, caused around 1,500 human deaths, and resulted in EUR 20 billion direct economic losses (Centre for Research on the Epidemiology of Disasters [CRED], 2016). At the same time, climate change projections show a future increase in the magnitude and frequency of extreme events (see Chapter 4). To mitigate and reduce these impacts, various risk management frameworks and plans have been issued, both in European and at national/regional levels (European Commission, 2007), aiming to improve the prevention, protection, preparedness, and emergency response to such events.

22.3.1 Definition and elements

An early warning system is defined as "The set of capacities needed to generate and disseminate timely and meaningful warning information to enable individuals, communities and organizations threatened by a hazard to prepare and to act appropriately and in sufficient time to reduce the possibility of harm or loss" (UNISDR, 2009). According to the UN Office for Disaster Risk Reduction (World Meteorological Organization [WMO], 2015), an effective and people-centred EWS is compiled by the integration of four key elements: (1) *risks knowledge*, (2) *monitoring and warning service*, (3) *dissemination*, and (4) *emergency response capacity* (Figure 22.3).

The implementation of a complete EWS requires the synergy of these four components, based on an appropriate legislative and legal framework, along with the close collaboration of many agencies and actors (e.g., international bodies, national and local governments, regional institutions and organisations, communities, non-governmental organisations, science community, media, private sector). Failure of any part of the system may imply failure of the whole system. For instance, accurate warnings will have no impact if the population is not prepared or if the alerts are received but not disseminated by the agencies receiving the messages.

22.3.1.1 Risk knowledge

Risk knowledge on the one hand supports the development of an efficient EWS, and on the other, enables the development of effective preparedness measures and management plans in advance of an event. The assessment of risks is based on the combined study of hazards along with the vulnerabilities of the population and assets (e.g., agricultural production, infrastructure, and homes). In this regard, hazard characteristics (e.g., magnitude, duration, frequency, location) are analysed on the basis of key socio-economic factors (e.g., casualties, construction, and crop damages). Information of the patterns and trends of hazards and vulnerabilities can be used by the decision makers at different levels (local, national, regional, and global) to plan risk reduction strategies, such as (WMO, 2015) (1) casualties reduction, (2) medium and long-term planning to reduce economic losses and develop resilience, and (3) risk financing and transfer (e.g., insurance) to transfer and/or redistribute the financial impacts of disasters.

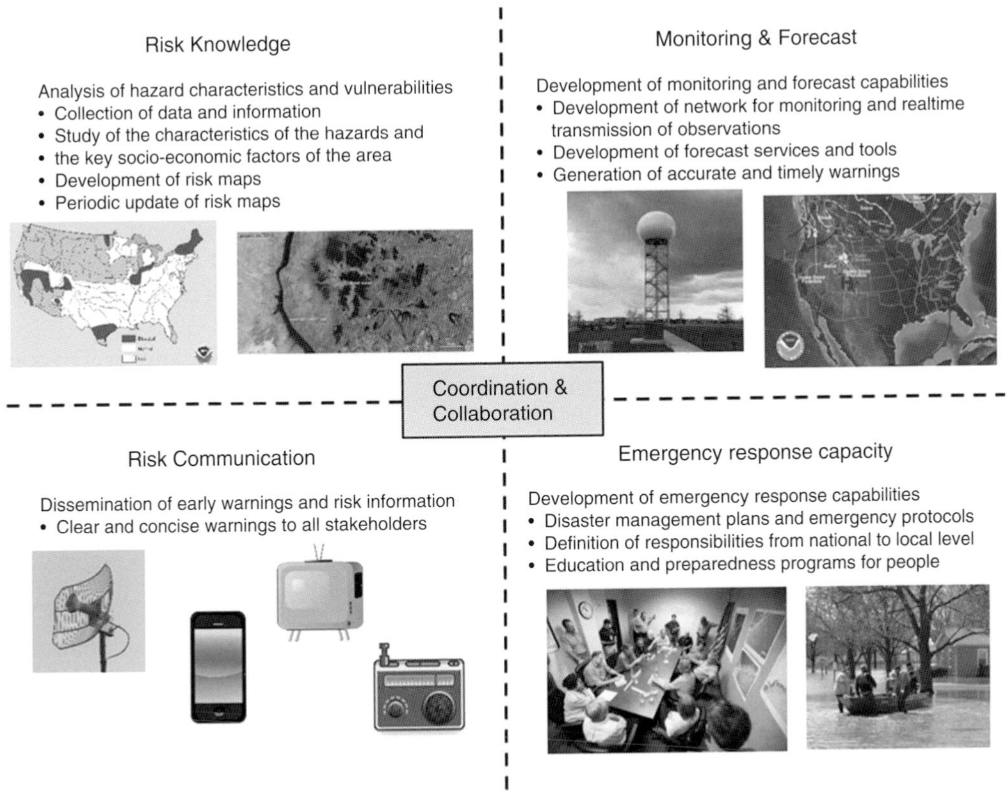

Risk Knowledge

Analysis of hazard characteristics and vulnerabilities
- Collection of data and information
- Study of the characteristics of the hazards and
- the key socio-economic factors of the area
- Development of risk maps
- Periodic update of risk maps

Monitoring & Forecast

Development of monitoring and forecast capabilities
- Development of network for monitoring and realtime transmission of observations
- Development of forecast services and tools
- Generation of accurate and timely warnings

Coordination & Collaboration

Risk Communication

Dissemination of early warnings and risk information
- Clear and concise warnings to all stakeholders

Emergency response capacity

Development of emergency response capabilities
- Disaster management plans and emergency protocols
- Definition of responsibilities from national to local level
- Education and preparedness programs for people

Figure 22.3 An efficient and people-centered EWS.

22.3.1.2 *Monitoring and warning service*

The two core components of an EWS are the sensor network system for continuous monitoring of hazard variables and the warning service, able to detect, forecast, and communicate the occurrence of a hazard. Both systems must reliably operate on a 24/7 basis, while their systematic maintenance and update are required to remain operational.

The monitoring network includes the sensors and the appropriate telecommunication infrastructures, for the real-time transmission of observations to the forecast system. For flood-oriented EWS, the network is typically composed of rainfall gauges, streamflow gauges, weather radars, and satellites (Hill et al., 2010). The last two are the key sources of hydrometeorological information used as input in weather prediction models, due to their ability to detect and track the evolution of clouds, providing precipitation estimates over large areas. Ground gauges provide accurate rainfall observations that are used to correct the error in radar and satellite estimates (see Chapter 4). Historic rainfall data along with discharge measurements from streamflow gauges are also used to calibrate the hydrological and flood models of the EWS.

Communication of data from the monitoring network to the forecast centre is conducted by various means. Data from local rain and runoff gauges are usually transmitted via wired or wireless Internet links, public telecommunication lines, and radio links (VHF, HF, and UHF). Observations from radar and satellite networks, operated mainly via international and national meteorological services, are typically available via Internet links and satellite communication links (VSAT). Towards the efficient implementation of a multi-hazard early warning system, the World Meteorological Organization has established the Global Telecommunications System to enable the exchange of data at local, national, and international levels (Hill et al., 2010).

22.3.1.3 Dissemination

An efficient EWS should ensure that the warnings and forecast information are communicated to the decision makers and/or directly to the general public in a timely fashion. This is achieved via a multi-channel communication system that is operated 24 hours a day, every day of the year. The warning information must be clear, concise, and in a familiar format to enable the immediate response of those at risk.

A typical typology to communicate information on the severity, timing, and level of confidence of a forecast hazard is the three-level "Ready, Set, Go" concept, adopted by many EWS in the United States (Hill et al., 2010).

The dissemination system must be reliable and redundant, while it is important to provide update and cancellation capabilities.

22.3.1.4 Emergency response capacity

Once a warning is issued, the line of actions must be clear and understood by all EWS stakeholders. In this regard, well-prepared and well-tested disaster management plans and emergency protocols should be developed by the appropriate authorities, defining roles and responsibilities from the national level down to the neighbourhood level. Equally crucial is the systematic education and preparedness programmes that inform the general public about the nature of the hazards, the potential risks, and options for safe behaviour. Towards this, there are many community-based disaster risk reduction programmes, such as the U.S. Storm Ready Program (see www.stormready.noaa.gov/), which aims to increase resilience to floods by establishing community-based plans, raising public awareness, and training people.

22.3.2 Forecast services

The forecast service is the core of any EWS, able to collect, process, and analyse various types of data in order to produce forecasts on the characteristics of a forthcoming flood event. Many EWS of different complexity and sophistication levels have been developed during the last decades, tailored to the special characteristics of the hazard and the focal area (Alfieri et al., 2012), as well as the interplay between different hazards (such as floods and fires (Kochilakis et al., 2016). The spatial and temporal extent of the different water-related hazards is summarised in Figure 22.4 (top panel). The corresponding meteorological inputs that are used in the forecasting process are shown in the bottom panel, with a qualitative assessment of their skill towards the forecast lead time.

The simplest forecast systems, usually employed for long-lasting river floods, use water level data from the upstream regions. If the measurements exceed pre-defined critical thresholds (derive from past flood observations), then the warning system is activated for downstream areas. More sophisticated EWS systems can provide greater lead times by coupling meteorological predictions with hydrological simulation models to provide runoff forecasts, which are then compared with warning levels.

In general, a complete flood forecasting system requires (Borga et al., 2011; Collier, 2007) (1) a precipitation detection system, typically consisting of remote sensing (radar, satellites) and ground rainfall and water-level gauges; (2) a regional and global weather prediction model, able to provide short-range meteorological forecasts, typically 24–48 hours ahead;

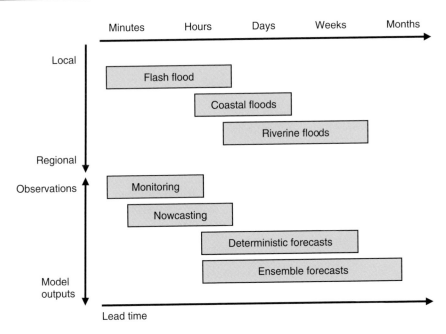

Figure 22.4 Spatio-temporal extent of water-related hazards and the corresponding meteorological input for forecasting. (Adapted from Alfieri, L. et al. 2012. *Environmental Science and Policy*, 21, 35–49.)

and (3) hydrological and/or hydraulic models to simulate the response of the catchment, at a wide range of time scales, over predictions. These components along with the data involved in the hydrological forecasting chain are presented schematically in Figure 22.5.

Especially in the case of urban areas, the timely forecasting of flash flood events is of vital importance for the protection and awareness of the people. Such events are usually associated with heavy storms of short duration and high intensity, while the main difference between flash floods and other water-related hazards is their rapid onset (from minutes to few hours) that restricts the opportunity for effective warning and preparation. Flash floods

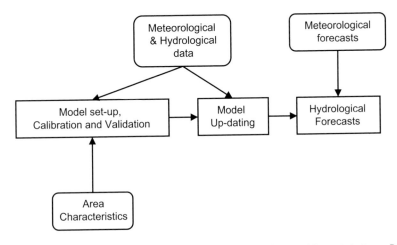

Figure 22.5 Main components, data, and tasks of forecast service. (Adapted from Arheimer, B., Lindström, G., and Olsson, J. 2011. *Atmospheric Research*, 100, 275–284.)

typically occur in streams and small catchment areas of a few hundred square kilometres, whose characteristics (e.g., steep slopes, saturated soils, extensive impervious surfaces) favour quick response to intense rainfall rates (Hapuarachchi et al., 2011).

22.3.2.1 Flow forecasting models

"The heart of any flow forecasting system is a hydrological model" (Serban and Askew, 1991). Hydrological and hydraulic models support the transformation of meteorological predictions (e.g., precipitation, air temperature, wind speed, humidity, etc.) into runoff forecasts (see Chapters 5, 7, and 10). These are generally classified into data-driven (black-box) and conceptual or physical-process models. The main representatives of the first category are the transfer function models (Sene and Tilford, 2004) and neural network models (Sahoo et al., 2006). Forecast frameworks that couple neural network models with multiple sources of information (i.e., gauged and satellite data as well as numerical weather prediction model forecasts) can be found in Chiang et al. (2007).

Unlike data-driven models, conceptual and physical-process hydrological models try to mathematically represent the various components of hydrological cycles. In flood forecasting, both lumped and (semi-) distributed models have been widely used (Hapuarachchi et al., 2011). Representative of the first category are the Sacramento soil moisture accounting model (SAC-SMA model) (Burnash et al., 1973), used by the U.S. National Weather Services. Typical spatially distributed models applied in a forecasting framework include the TOP-MODEL model (Beven and Binley, 1992) and the LISFLOOD model (De Roo et al., 2000). Distributed models support simulations at finer temporal and spatial scales, and hence, they can be considered as more suitable in flash flood applications. However, the high computational burden along with the high demands for observed data may restrict their operational applicability.

The application of hydrological models in a real-time basis requires their continuous updating in order to constrain the various sources of uncertainty (i.e., input error, parameter, and structural uncertainty) at each time step of the forecast procedure. The update concerns (Serban and Askew, 1991) (1) the input variables (e.g., meteorological and hydrological predictions); (2) the state variables of the modelled system (e.g., water depths, extent of saturated and unsaturated areas, water level in reservoir and the other flood protection structures, etc.); and (3) the model parameters or structure (i.e., recalibration of the model with updated data). Typically, formal (Vrugt et al., 2003) or informal (i.e., GLUE; Beven and Binley, 1992) Bayesian methodologies are used in this process.

22.3.2.2 Meteorological predictions

The efficiency of any EWS depends on lead time, where accurate weather forecasts, especially on precipitation, are available. In recent years, many different techniques have been developed to provide precipitation forecasts at different spatial extents, temporal analysis, and forecast range. These techniques are mostly based on numerical weather predictions (NWPs) and weather radar and satellite estimates (see Chapter 4), as well as on their blending.

Short-range precipitation forecasts (up to 6 hours ahead), also known as nowcasts (Golding, 2000), have until recent years been based on extrapolation techniques that use radar and satellite observations in order to capture the spatio-temporal dynamic of precipitation field and forecast the motion of a discrete set of rain objects (Alfieri et al., 2012; Borga et al., 2011; Collier, 2007). Real-time measurements from ground rain gauges are typically used to adjust the error in precipitation estimation (see Chapter 4). Forecasting extrapolation techniques can be grouped into those that use correlation techniques, track the centroid of an object, and use NWP wind advection techniques (Bowler et al., 2004). Hybrid approaches, artificial neural network applications, and Bayesian statistical techniques have also been applied (Hapuarachchi et al., 2011).

Although these types of methods have shown some success in simulating and forecasting the precipitation field, there are weaknesses in developing convection of rainfall fields in new areas and capturing cell splitting and decay that usually control heavy rain dynamics and, hence, flash floods (Borga et al., 2011). The new generation of high-resolution NWP models, with finer temporal resolution (lower than 1 hour) and spatial resolution of a few kilometres, offer the prospect of producing useful forecasts of convective storms on scales applicable for flood prediction. Further prediction accuracy can be achieved by combining NWP forecasts with radar rainfall estimation. A characteristic operational forecast system that follows this data assimilation approach is the Nimrod precipitation nowcasting system (Golding, 1998) of the UK Met Office that provides precipitation forecasts from 1 up to 6 hours lead time (Collier, 2007).

To take into account the uncertain and complex nature of the atmosphere ensembles of many NWPs, known as ensemble prediction systems (EPSs), rather than single deterministic forecasts, can be used. EPSs are typically derived by perturbating the initial conditions of the model. The EPS is then imported into a hydrological model to produce a set of different possible flood predictions (Cloke and Pappenberger, 2008). With the use of ensembles, the accuracy of general weather forecasts can be improved at medium-range (Tracton and Kalnay, 1993) and at short-range (Du et al., 1997) lead times (Collier, 2007). A typical operational EPS system is the short-term ensemble prediction system (STEPS), which merges an extrapolated nowcast with different downscaled NWP model forecasts in order to provide the probability of precipitation (Bowler et al., 2004).

22.3.2.3 Flood occurrence criteria

To infer whether a rainfall (and subsequent runoff) forecast may lead to flooding, criteria and reference thresholds need to be established.

The *flow comparison method* compares the forecasted runoff with a flooding flow. In the simplest case, flooding flow is defined by an observed flooding threshold. However, the estimation of a flooding flow is a rather hard task, especially in urban areas where flood modelling is complex due to human-made interventions and constructions. As such, bank-full discharge, calculated using the uniform flow resistance formula, is typically used as a flooding threshold (Hapuarachchi et al., 2011). Despite its apparent simplicity, this method requires extensive and detailed information on the characteristics of the river or urban drainage system (i.e., Manning's roughness coefficient (see Section 7.5), channel slope, and cross-section geometry) as well as their spatial variation. Otherwise, this method is highly uncertain. Bank-flow discharges can also be derived via the statistical analysis of historical discharge data. According to Hapuarachchi et al. (2011), there is a good statistical relationship between the bank-full flow and a flow with a return period of between 1 and 2 years. Simulated runoff series from models can also be used in the case of inadequate observed data (Reed et al., 2007). Further to observed flooding thresholds, a more probabilistic-distributed approach can also be used, employing long-term simulated records to develop a flooding flow frequency curve for each cell of the grid (Reed et al., 2007).

Similarly, the *rainfall comparison method* compares the rainfall forecasts against the rainfall required to produce flooding flow in a given area. The main representative of this approach is the flash flood guidance (FFG) that is defined as the volume of rainfall of a given duration distributed uniformly over a small catchment (<300 km^2) that is just enough to cause minor flooding at the catchment outlet. If the forecasted rainfall volume is greater than the characteristic rainfall volume, then flooding in the catchment is likely to occur. FFG was developed in the United States (Georgakakos, 2006) and is fully operational within the U.S. National Weather Service (NWS) River Forecast Centers that issue FFG throughout the day for every county in their area. The determination of the FFG is conducted via a hydrological model that is run with increasing amounts of accumulated rainfall for a particular duration (e.g., 1, 3, 6 hours). The simulated runoff is plotted against the required rainfall, and the FFG is retrieved for a known threshold runoff value. U.S. national estimates of threshold runoff have been derived by Carpenter et al. (1999). The hydrological model is updated periodically with the current soil moisture state. In the UK, the Flood Forecasting Centre (FFC) issues a Flood Guidance Statement (FGS) that provides a daily flood risk assessment for all types of natural flooding: coastal/tidal, river, groundwater, and surface water. The FGS presents risk assessment by county and unitary authority across England and Wales over 5 days.

22.4 CITIZEN OBSERVATORIES

Today's wide use of smart mobile communication devices (such as smartphones) along with the advent of new ICT services (i.e., greater capabilities of sharing, storing, processing, and displaying vast amounts of data and information on the fly) allows for an active engagement of citizens, beyond scientists and water professionals, in water management. In the "human observatories" framework, citizens act as "sensors" that collect and share valuable information and observations, providing up-to-date awareness on water infrastructures and environmental problems, such as urban flooding (Wehn et al., 2015).

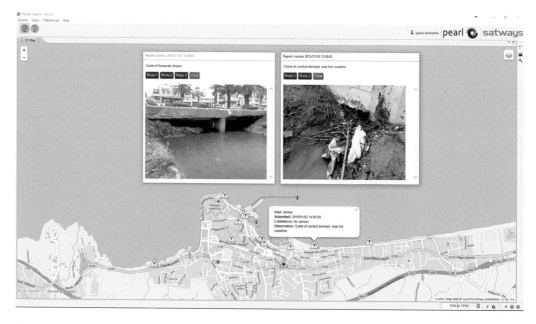

Figure 22.6 The PEARL Detective app applied for flood protection in the city of Rethymno. (Reproduced from http://www.pearl-fp7.eu/, courtesy of Christos Makropoulos.)

In an urban setting, citizens-as-sensor technologies find wide applicability, especially in flood risk management. A characteristic example of such a technology is the Pearl Detective smartphone application (Figure 22.6) that enables users to share information about flood incidents with the authorities.

The user is able to submit, through the smartphone application, text reports and other information (including photos) on a flood-related event (e.g., increase in river water level, overtopping) or on situations that may cause problems during a future event (e.g., damaged pipes, obstructed inlets, etc.). The system supports a two-way communication, also informing the users about flood events that are forecasted in their area. It is expected that such approaches will increasingly play a more important role in shaping smart urban drainage systems of the future, creating a more direct link between infrastructure, authorities, and citizens.

PROBLEMS

22.1 Define real-time control, and describe the main hardware components needed to implement it in an urban drainage system.

22.2 What are the advantages and disadvantages of urban drainage RTC? In what situations is RTC likely to be beneficial?

22.3 Explain how you would go about developing an RTC strategy, and give an example of a possible practical approach. What sources of data would you need?

22.4 Define early warning system, and describe the main key elements needed for its efficient implementation.

22.5 Describe the main components, data, and tasks that are involved in the forecast service of a typical early warning system.

22.6 Explain the concept of the citizen-observatories framework. What sources of data would it use?

KEY SOURCES

García, L., Barreiro-Gomez, J., Escobar, E., Téllez, D., Quijano, N., and Ocampo-Martinez, C. 2015. Modeling and real-time control of urban drainage systems: A review. *Advances in Water Resources*, 85, 120–132.

Hapuarachchi, H.A.P., Wang, Q.J., and Pagano, T.C. 2011. A review of advances in flash flood forecasting. *Hydrological Processes*, 25, 2771–2784.

Marinaki, M. and Papageorgiou, M. 2005. *Optimal Real-Time Control of Sewer Networks. Advances in Industrial Control*, Springer.

Ocampo-Martinez, C. 2010. *Model Predictive Control of Wastewater Systems, Advances in Industrial Control*, Springer, London.

Schilling, W. 1989. *Real-Time Control of Urban Drainage Systems: The State-of-the-Art*, Scientific and Technical Report No. 2.

World Meteorological Organization (WMO). 2015. Synthesis of the status and trends with the development of early warning systems. *Background Paper prepared for the Global Assessment Report on Disaster Risk Reduction 2015*. Geneva.

REFERENCES

Alfieri, L., Salamon, P., Pappenberger, F., Wetterhall, F., and Thielen, J. 2012. Operational early warning systems for water-related hazards in Europe. *Environmental Science and Policy*, 21, 35–49.

Almeida, C. and Schilling, W. 1993. Derivation of If ... Then ... Else rules from optimized strategies for sewer systems under real-time control. *Proceedings of Sixth International Conference on Urban Storm Drainage*, Canada, pp. 1525–1530.

Antonio, L. and Gutierrez, G. 2014. On the modeling and real-time control of urban drainage systems: A survey. *11th International Conference of Hydroinformatics.*

Arheimer, B., Lindström, G., and Olsson, J. 2011. A systematic review of sensitivities in the Swedish flood-forecasting system. *Atmospheric Research*, 100, 275–284.

Beeneken, T., Erbe, V., Messmer, A., Reder, C., Rohlfing, R., Scheer, M. et al. 2013. Real time control (RTC) of urban drainage systems—A discussion of the additional efforts compared to conventionally operated systems. *Urban Water Journal*, 10, 293–299.

Beven, K. and Binley, A. 1992. The future of distributed models: Model calibration and uncertainty prediction. *Hydrological Processes*, 6, 279–298.

Borga, M., Anagnostou, E.N., Blöschl, G., and Creutin, J.-D. 2011. Flash flood forecasting, warning and risk management: The HYDRATE project. *Environmental Science and Policy*, 14, 834–844.

Bowler, N.E.H., Pierce, C.E., and Seed, A. 2004. Development of a precipitation nowcasting algorithm based upon optical flow techniques. *Journal of Hydrology*, 288, 74–91.

Burnash, R.J.C., Ferral, R.L., McGuire, R.A., McGuire, R.A., and Center, U.S.J.F.-S.R.F. 1973. *A Generalized Streamflow Simulation System: Conceptual Modeling for Digital Computers*. U.S. Department of Commerce, National Weather Service, State of California, Department of Water Resources.

Campisano, A., Cabot Ple, J., Muschalla, D., Pleau, M., and Vanrolleghem, P.A. 2013. Potential and limitations of modern equipment for real time control of urban wastewater systems. *Urban Water Journal*, 10, 300–311.

Carpenter, T.M., Sperfslage, J.A., Georgakakos, K.P., Sweeney, T., and Fread, D.L. 1999. National threshold runoff estimation utilizing GIS in support of operational flash flood warning systems. *Journal of Hydrology*, 224, 21–44.

Chiang, Y.-M., Hsu, K.-L., Chang, F.-J., Hong, Y., and Sorooshian, S. 2007. Merging multiple precipitation sources for flash flood forecasting. *Journal of Hydrology*, 340, 183–196.

Cloke, H.L. and Pappenberger, F. 2008. Evaluating forecasts of extreme events for hydrological applications: An approach for screening unfamiliar performance measures. *Meteorological Applications*, 15, 181–197.

Collier, C.G. 2007. Flash flood forecasting: What are the limits of predictability? *Quarterly Journal of the Royal Meteorological Society*, 133, 3–23.

Centre for Research on the Epidemiology of Disasters (CRED). 2016. *OFDA/CRED International Disaster Database*, Université Catholique de Louvain, Brussels, Belgium.

De Roo, A.P.J., Wesseling, C.G., and Van Deursen, W.P.A. 2000. Physically based river basin modelling within a GIS: The LISFLOOD model. *Hydrological Processes*, 14, 1981–1992.

Du, J., Mullen, S.L., and Sanders, F. 1997. Short-range ensemble forecasting of quantitative precipitation. *Monthly Weather Review*, 125, 2427–2459.

Erbe, V. and Haas, U. 2007. *Application of a Guideline Document for Sewer System Real Time Control 761–768*, NOVATECH.

European Commission. 2007. *Directive 2007/60/EC of the European Parliament and of the Council of 23 October 2007 on the Assessment and Management of Flood Risks*, 27–34.

Fuchs, L. and Beeneken, T. 2005. Development and implementation of a real-time control strategy for the sewer system of the city of Vienna. *Water Science and Technology*, 52, 187–194.

Georgakakos, K.P. 2006. Analytical results for operational flash flood guidance. *Journal of Hydrology*, 317, 81–103.

Golding, B. 2000. Quantitative precipitation forecasting in the UK. *Journal of Hydrology*, 239, 286–305.

Golding, B.W. 1998. Nimrod: A system for generating automated very short range forecasts. *Meteorological Applications*, 5. doi:10.1017/S1350482798000577.

Hill, C., Verjee, F., and Barrett, C. 2010. Flash flood early warning system reference guide. University Corporation for Atmospheric Research, NCAR/UCAR.

Joseph-Duran, B., Jung, M.N., Ocampo-Martinez, C., Sager, S., and Cembrano, G. 2014. Minimization of sewage network overflow. *Water Resources Management*, 28, 41–63. doi:10.1007/s11269-013-0468-z.

Kellagher, R. 1996. *Evaluating Real Time Control for Wastewater Systems: A Managerial and Technical Guide*. Report SR479, HR Wallingford.

Klepiszewski, K. and Schmitt, T.G. 2002. Comparison of conventional rule based flow control with control processes based on fuzzy logic in a combined sewer system. *Water Science and Technology*, 46, 77–84.

Kochilakis, G., Poursanidis, D., Chrysoulakis, N., Varella, V., Kotroni, V., Eftychidis, G., et al. 2016. A web based DSS for the management of floods and wildfires (FLIRE) in urban and periurban areas. *Environmental Modelling and Software*, 86, 111–115.

Nelen, F. 1993. On the potential of real time control of urban drainage systems. *Water Science and Technology*, 27, 111 LP-122.

Nikoli, V., Žarko, Ć., Ivan, Ć., and Petrovi, E. 2010. Intelligent decision making in wastewater treatment plant SCADA system. *Automatic Control and Robotics*, 9, 69–77.

Reed, S., Schaake, J., and Zhang, Z. 2007. A distributed hydrologic model and threshold frequency-based method for flash flood forecasting at ungauged locations. *Journal of Hydrology*, 337, 402–420.

Regneri, M., Klepiszewski, K., Ostrowski, M., and Vanrolleghem, P. 2010. Fuzzy decision making for multi-criteria optimization in integrated wastewater system management. *Proceedings of the Sixth International Conference on Sewer Processes and Networks*.

Sahoo, G.B., Ray, C., and De Carlo, E.H. 2006. Use of neural network to predict flash flood and attendant water qualities of a mountainous stream on Oahu, Hawaii. *Journal of Hydrology*, 327, 525–538.

Schilling, W., Andersson, B., Nyberg, U., Aspegren, H., Rauch, W., and Harremoës, P. 1996. Real time control of wastewater systems. *Journal of Hydraulic Research*, 34, 785–797.

Schütze, M., Campisano, A., Colas, H., Schilling, W., and Vanrolleghem, P.A. 2004. Real time control of urban wastewater systems—Where do we stand today? *Journal of Hydrology*, 299, 335–348.

Sene, K. and Tilford, K. 2004. Review of transfer function modelling for fluvial flood forecasting. *R&D Technical Report W5C-013/6/TR, Defra/Environment Agency*, Flood and Coastal Defence R&D Program.

Serban, P. and Askew, A.J. 1991. Hydrological forecasting and updating procedures. *Hydrology for the Water Management of Large River Basins*, 201, 357–369.

Smart, P. and Herbertson, J.G. (Eds.) 1992. *Drainage Design*, Springer Science + Business Media.

Tracton, M.S. and Kalnay, E. 1993. Operational ensemble prediction at the national meteorological center: Practical aspects. *Weather and Forecasting*, 8, 379–398.

UNISDR. 2009. *UNISDR Terminology on Disaster Risk Reduction*, International Strategy for Disaster Reduction, Geneva.

Vrugt, J.A., Gupta, H.V., Bouten, W., and Sorooshian, S. 2003. A shuffled complex evolution metropolis algorithm for optimization and uncertainty assessment of hydrologic model parameters. *Water Resources Research*, 39.

Wehn, U., Rusca, M., Evers, J., and Lanfranchi, V. 2015. Participation in flood risk management and the potential of citizen observatories: A governance analysis. *Environmental Science and Policy*, 48, 225–236.

Chapter 23

Global issues

23.1 INTRODUCTION

Providing water, sanitation, and drainage for all people regardless of socio-economic status, age, gender, and mobility is one of the most important global issues of our time. Despite this, relatively limited progress has been made in the last 40 years. Governments and international agencies have made repeated commitments to achieve universal provision of water and sanitation (Satterthwaite et al., 2015):

- 1976: UN Conference on Human Settlements (Habitat I)
- 1977: UN Water Conference
- 1980–1990: UN International Drinking Water Supply and Sanitation Decade
- 2000–2015: UN Millennium Development Goals (MDGs)

The MDG target called for halving the proportion of the population without sustainable access to basic sanitation between 1990 and 2015. Although during the MDG period, use of improved sanitation facilities rose from 54% to 68% globally, the target set of 77% was clearly missed. By 2015, 2.4 billion people still lacked access to improved sanitation facilities, and many of the lowest income countries actually had urban populations with reduced provision of improved sanitation as compared with 1990 (UNICEF and World Health Organization, 2015). The number of deaths that can be attributed to water and excreta-related diseases is still around 2 million per year, of which children under 5 years old are the most affected group. The deep irony is that although these diseases account for such a high proportion of illness, debility, and death among the poor, and are a substantial cause of high infant mortality rates, they are largely preventable. Notable by its absence from the MDGs was a specific goal relating specifically to urban drainage.

The UN's latest initiative consists of 17 sustainable development goals (SDGs). These aim to build on lessons learned from the MDGs and end poverty, protect the planet, and ensure prosperity for all. One of these (SDG 6) is focussed on "Clean water and sanitation," and aims by 2030 to "achieve access to adequate and equitable sanitation and hygiene for all and end open defecation, paying special attention to the needs of women and girls and those in vulnerable situations." SDG 11 "Make cities inclusive, safe, resilient and sustainable" includes the target to "significantly reduce the number of deaths and the number of people affected and substantially decrease the direct economic losses relative to global gross domestic product caused by disasters, including water-related disasters, with a focus on protecting the poor and people in vulnerable situations." We can only hope that better progress towards these water-related goals is achieved than hitherto.

The greatest water and sanitation needs can often be found in the informal peri-urban areas where slums and shantytowns proliferate in many Third World cities. These are rarely

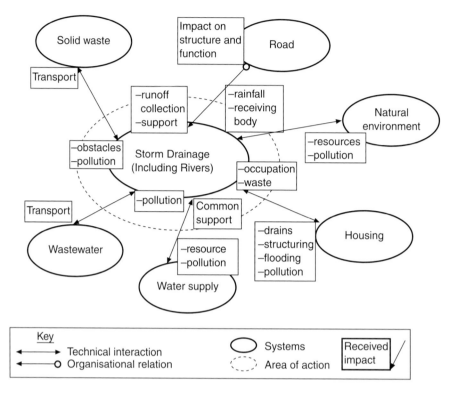

Figure 23.1 Integrated municipal services. (Reproduced from Wondimu, A. and Alfakih, E. 1998. Urban drainage in Addis Ababa (Ethiopia): Existing situation and improvement ideas. *Fourth International Conference on Developments in Urban Drainage Modelling—UDM '98*, London, 823–830, with permission of the authors.)

given services prior to their establishment and subsequent provision places severe financial strain on already over-stretched government resources. Yet, if these basic services can be provided, the intention is that better public health will allow an upward spiral of social and economic development, leading to increased productivity, higher standards of living, and improved quality of life (Parkinson et al., 2007).

An important part of the "upward spiral" of development is provision of other services, such as storm drainage and solid waste collection, and organisation of community hygiene education. If these can be deployed, both water supply and sanitation services will, in turn, function more effectively, and additional health benefits will accrue. In fact, an integrated whole of municipal services (see Figure 23.1) will bring maximum benefits. The significant role (and deficiencies) of local authorities and municipalities in supplying these services to a growing population that cannot afford to pay has been highlighted (Ashipala and Armitage, 2011; Kobel and Del Mistro, 2012).

In addition to health benefits, sanitation is also valuable in giving dignity and privacy to people, and in providing a cleaner, more pleasant living environment. Similarly, good drainage is also valuable in reducing nuisance and economic loss due to flooding.

In this chapter, the emphasis is placed on urban drainage services in the context of the social and financial constraints of low-income communities. Section 23.2 concentrates on health implications, Section 23.3 gives an overview of options available, and Sections 23.3 through 23.7 cover, in turn, the major approaches available: on-site sanitation, off-site sanitation, storm drainage, and grey water management.

23.2 HEALTH

First in public health importance are the many faeco-oral infections where the contaminated faeces of one person are transmitted via water, hands, insects, soil, or plants to other individuals. These diseases include the well-known "waterborne" diseases such as cholera and typhoid, but also the many common diarrhoeal diseases that particularly affect young children, contributing to malnutrition and death. In fact, these diarrhoeal diseases are often the greatest cause of child mortality (Cairncross and Ouano, 1991).

The larvae of helminths like roundworms and hookworms may be transmitted from person to person when infected faeces are left on the ground. Children are particularly at risk when playing or bathing in faecally contaminated stormwater (see Figure 23.2). Some roundworm eggs (e.g., *Ascaris*) remain viable for long periods.

A further group of diseases may be classified as "water-related" rather than waterborne. For example, mosquitoes spread several different diseases. Their connection with water is that each species selects different types of water body in which to breed. Malaria is the best known of these diseases and is transmitted by the female *Anopheles* mosquito. These do not usually breed in heavily polluted water but can multiply in swamps, pools, puddles, and other poorly drained areas. *Culex* mosquitoes favour polluted water (e.g., pit latrines, blocked storm drains) and are a vector for filariasis, which can lead to elephantiasis. Other mosquito species spread diseases such as dengue and yellow fever.

Another important disease is schistosomiasis (bilharzia), which can be transmitted in poorly drained urban areas (Figure 23.2). If standing stormwater becomes contaminated with egg-infected excreta, the microscopic larvae (miracidia) that hatch out can multiply in the bodies of small aquatic snails. From every infected snail, thousands of infective cercariae emerge and swim in the water. Local residents become infected through their skin when they stand or play in the water. Table 23.1 summarises the major diseases and their link to urban drainage.

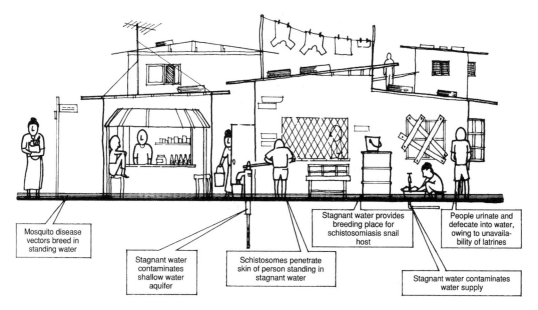

Figure 23.2 Health implications of poor drainage. (Reproduced from Cairncross, S. and Ouano, E.A.R. 1991. *Surface Water Drainage for Low-income Communities*, World Health Organization, with permission of WHO, Geneva.)

Table 23.1 Classification of diseases linked to lack or precariousness of urban drainage

Group		Disease
I	Diseases transmitted by flying vectors that can multiply in pools and wetlands	Urban yellow fever Dengue Filariasis Malaria
II	Diseases in which the etiological agent uses an intermediate aquatic host that can multiply in wetlands	Schistosomiasis
III	Diseases transmitted by direct contact of water or soil (with the presence of the hosts)—contamination is favoured by floods and wetlands	Leptospirosis
IV	Diseases transmitted by ingestion of water contaminated by etiological agents present in wetlands and floods that enter into water distribution systems	Typhoid fever (water) Cholera and other diarrhea (water)
	Diseases transmitted by direct contact to contaminated soil—contamination is favoured by floods and wetlands	Hepatitis A (water) Ascaridiasis (water) Trichuriasis (water) Hookworm (water and soil)

Source: After Souza, C. et al. 2002. *Water21*, October, 40–41.

Improved water supply, sanitation, drainage, and hygiene education are important components in obstructing the transmission route of these diseases. Of course, in providing engineering works for this purpose, great care is needed to ensure that new breeding sites are not created inadvertently.

23.3 OPTION SELECTION

23.3.1 Sanitation

Alternative types of sanitation can be conveniently classified as on-site or off-site. In on-site methods, the excreta storage/treatment is in or near the individual dwelling. Off-site systems remove the excreta from the dwelling for disposal. Additionally, systems can be designated as dry or wet, with wet systems using water to transport the excreta (see Table 23.2). Whatever type is chosen, the basic functions are the same, and they are to

- Collect the excreta
- Transport it to a suitable location and/or store it for treatment
- Treat it
- Reuse it and/or discharge it to the environment

Table 23.2 Classification of sanitation options

	Dry	Wet
On-site	Pit latrines	Pour-flush latrines
	VIP latrines	Septic tank systems
	Composting latrines	Aqua privies
	Communal latrines	
	Urine-diverting dry toilet	
Off-site	Bucket latrines	Conventional sewerage
	Vault latrines	Unconventional sewerage

A good sanitation system also minimises or removes health risks and negative impacts on the environment.

The final choice between on- and off-site systems will normally be financial. In low-density, low-income settlements, on-site systems will almost certainly be the most cost-effective option. In higher-density areas, the feasibility of using on-site methods becomes less, and some form of sewerage may be appropriate. In northeast Brazil, for example, Sinnatamby (1986) demonstrated that unconventional (shallow) sewerage is cost-effective for population densities exceeding about 160 hd/ha. Population densities in slums and shantytowns can be much greater than this (2000 hd/ha is not uncommon).

Conventional sewerage, the most expensive option, will only be appropriate where property values are high and occupiers can pay for the full costs involved. Suitable upgrade paths also need to be considered. However, each case is different, and the merits of all the options need to be thoroughly evaluated in technical terms and by considering social, cultural, financial, and institutional factors. This may involve considerably more consultation with community members than in conventional planning and design practise. Indeed, it has been argued that co-designed and co-produced schemes involving the urban poor working with local government without external funding are the most likely to succeed in the long term (Nance and Ortolano, 2007; Satterthwaite et al., 2015).

This approach broadly matches the IWA analytical framework *Sanitation 21*, which argues that one size does not fit all, but that a properly functioning system may well have elements of both on- and off-site systems (Parkinson et al., 2014). Scott et al. (2015) have developed a citywide sanitation planning framework that combines existing sanitation planning approaches (such as Sanitation 21) with householders' tenure security and status. Their "Sanitation Cityscape" maps the fluxes of faecal material through the city providing a rationale for targeted interventions.

23.3.2 Storm drainage

Storm drainage options are more limited. The main classification is between "closed" systems, relying on underground pipes, "open" systems requiring open channels, and on-site options (see Table 23.3).

In most situations, conventional piped drainage will not be an option, unless it is part of a simplified system. Open channels are widely used but need to be carefully designed, constructed, and maintained. On-site options include many of the approaches discussed in Chapter 21 (Stormwater management), which may have application in this context.

23.4 ON-SITE SANITATION

On-site disposal is widely practised in low-income communities. This can be a perfectly satisfactory urban solution (even in crowded areas) provided the plot-size is large enough and sub-soil conditions are suitable for disposal of the effluent to the ground without danger of contamination of nearby water sources (wells, for example).

Table 23.3 Classification of drainage options

Open	Open channels
	Road-as-drain
Closed	Conventional piped drainage
	Dual drainage
On-site	Stormwater management

The following sections contain an outline of the on-site options available. Design methods and typical details can be found in Pickford (1995) and Mara (1996b).

23.4.1 Latrines

23.4.1.1 Pit latrine

The simplest form is the *pit latrine* (Figure 23.3a). This consists of a squatting hole or plate directly above a pit in the ground into which the excreta falls. The conditions within the pit are anaerobic, promoting gaseous products (mainly CO_2 and CH_4 but also some malodourous gases) that escape into the atmosphere. The excreta gradually decompose and a solid residue accumulates in the pit bottom. Water, urine, and other liquids infiltrate to the ground through the pit walls and base. When the residue reaches about 0.5 m from the top, the latrine needs to be either cleaned out or abandoned. Pits are usually approximately 1 m in diameter (or 1 m square) and up to 3 m deep.

As the pit latrine works by allowing infiltration of liquids into the surrounding soil, it follows that there must be sufficient open space available on the individual plot, the actual size depending on the conductivity of the soil. However, pit latrines require only 1–2 m² of space, making them suitable sanitation options even in high-density areas (Reed, 1995).

23.4.1.2 VIP latrines

Ventilating pipes can be used to overcome the common complaints of bad smells and insect nuisance. The ventilated improved pit (VIP) latrine consists of a slightly offset pit with a vertical vent having fly-proof netting at its exit (see Figure 23.3b). The latrine is kept dark so any flies hatched in the pit are attracted to the light at the top of the vent and are trapped and die. The wind across the top of the vent causes low pressure and, therefore, an updraught extracting any foul odours. The pipe can also be painted black, which helps heat up the air

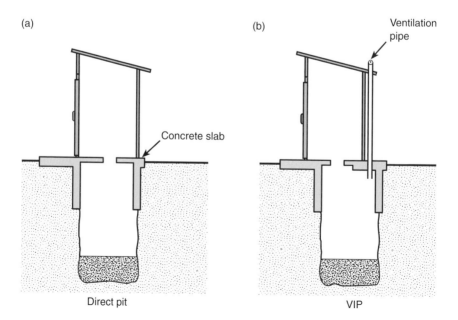

Figure 23.3 Types of latrine.

inside, causing it to rise and ventilate the pit. The shelter needs to be well-ventilated to allow a through-flow of air.

Permanent VIP latrines can also be built with two chambers (alternating twin pit VIPs) where pits are filled and emptied alternatively. This allows safe manual emptying of "old" sludge, but does, however, require more of the householder than other options (Tayler, 1996). Thus, VIP latrines constitute a very useful option, even in highly built-up, low-income areas.

23.4.1.3 Pour-flush latrines

Another solution to the odour and insect problem is to provide a pan and trap with a water seal above the pit. This has the additional benefit of removing the direct line-of-sight between the user and the faeces below. Well-designed pans can be washed down with 1–3 L of water poured from a handheld vessel.

Pour-flush latrines are widely and successfully used in low-income communities where there is a nearby water source, such as a well or standpipe.

23.4.1.4 Composting latrines

Composting faeces with vegetable wastes in an enlarged pit latrine or other chamber offers another method of on-site treatment with reuse potential. These arrangements require considerable care and continuous attention, with 5 months or so needed to produce pathogen-free compost. The compost can be either sold or used locally to replace chemical fertilisers. Odours can be reduced by separating urine. Some users have shown a reluctance to handle the by-products, and there is concern over affordability. Further work is still required before this ecological sanitation (*ecosan*) option can be considered a fully viable option in urban or peri-urban areas.

23.4.1.5 Communal latrines

Communal latrines are acceptable in some situations (e.g., city-centre areas), although this depends on the attitudes and habits of the local people. In some places, privacy is considered so important that local people refuse to use them. Where they are acceptable, and where arrangements can be made for frequent and regular cleaning, their cost is much lower than that of individual household latrines

23.4.2 Septic tank systems

Septic tank systems consist of a tank and, ideally, a drainage field (see Figure 23.4). The tank is a watertight, underground vessel that provides conditions suitable for the settlement, storage, and (temperature-related) anaerobic decomposition of excreta. Sludge accumulates

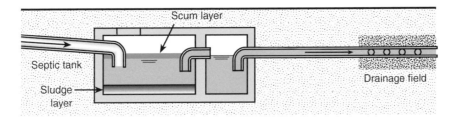

Figure 23.4 Septic tank system.

at the bottom of the tank and has to be emptied periodically. A hard crust of solidified grease and oil forms on the surface. Wastewater is fed directly to the tank, through which it flows, and then on to the drainage field. Direct discharge to a ditch, stream, or open drain is not recommended but is common practise in many Third World cities. The *drainage field* consists of a soakaway or sub-surface irrigation pipe system, which drains the effluent into the surrounding soil and provides additional treatment.

Tanks should have a minimum volume of 1 m³. The desludging interval should be short enough to ensure the tank does not become blocked but long enough to allow the benefits of anaerobic reduction in sludge volume. This will be typically 2–5 years.

Septic tanks require significantly more space than pit latrines. Depending on water use and soil conditions, this can range from 10 to 100 m² (Reed, 1995). They are also expensive to construct and operate but have proved satisfactory in low- and high-income countries alike, especially in low-density housing areas (Butler and Payne, 1995).

23.4.3 Aqua privies

An *aqua privy* consists of a latrine set over a septic tank. The squatting plate is connected to a pipe that dips below the surface of the liquor in the chamber below. Overflowing liquor infiltrates into the soil through a drainage field, and the water level in the tank is made up with small amounts of cleaning water. Regular desludging is required.

Aqua privies were popular, but problems have always been encountered in maintaining the necessary water level in the tank and with faecal fouling of the drop-pipe. Both can cause odour nuisance and health hazard.

23.4.4 Urine diverting dry toilets

A urine-diverting dry toilet operates without water and has a pan that diverts the urine away from the faeces. The urine is collected and drained from the front area of the toilet, while faeces fall through a large chute in the back. Drying material such as lime, ash, or earth is added into the same hole after defecating. These toilets facilitate the on-site reuse of urine in gardens and other small plots. The concept of this *ecosan* option is to exploit the nutrient value of the excreta for food production (see Chapter 24).

23.4.5 New technologies

Due in part to the lack of commercial drivers, there has been relatively little development in sanitation technologies over the years. This deficiency has been noticed and is being addressed by the Bill and Melinda Gates Foundation. They began funding research in this area in 2011 as part of their "Re-invent the Toilet Challenge." This aims to create a toilet that

- Removes "germs" from human waste and recovers valuable resources such as energy, clean water, and nutrients
- Operates "off the grid" without connections to water, sewer, or electrical lines
- Costs less than US$0.05 per user per day
- Promotes sustainable and financially profitable sanitation services and businesses that operate in poor, urban settings
- Is a truly aspirational next-generation product that everyone will want to use, in developed as well as developing nations

(www.gatesfoundation.org/What-We-Do/Global-Development/Reinvent-the-Toilet-Challenge)

System configuration

A waterless self-contained toilet
for private household of 10
people

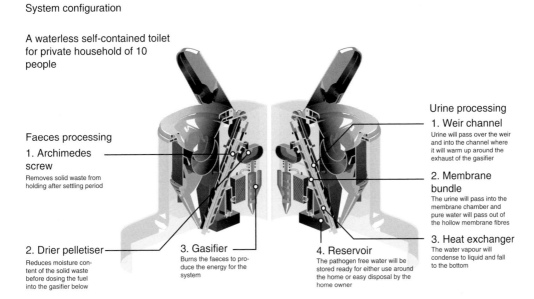

Faeces processing

1. Archimedes screw
Removes solid waste from
holding after settling period

2. Drier pelletiser
Reduces moisture content of the solid waste
before dosing the fuel
into the gasifier below

3. Gasifier
Burns the faeces to produce the energy for the
system

Urine processing

1. Weir channel
Urine will pass over the weir
and into the channel where
it will warm up around the
exhaust of the gasifier

2. Membrane bundle
The urine will pass into the
membrane chamber and
pure water will pass out of
the hollow membrane fibres

3. Heat exchanger
The water vapour will
condense to liquid and fall
to the bottom

4. Reservoir
The pathogen free water will be
stored ready for either use around
the home or easy disposal by the
home owner

Figure 23.5 Cranfield nanomembrane toilet system. (Courtesy of Ewan McAdam.)

One of the new prototype toilets is shown in Figure 23.5. Below the toilet rim, there is a drum that is mechanically rotated by 180° to empty the contents into a settling tank below using the opening/closing action of the toilet seat. This is supported by a "swipe" to ensure complete urine/faeces separation. Once the toilet seat is re-opened, the rotating drum is reset for the next use. Faeces are then separated in the tank below under gravity. At the base of the tank is a mechanical screw with an even pitch and 60° incline giving further solid/liquid separation and partial dewatering. Solids then enter into a convection drier sited above the combustor, which operates on a continuous basis. Residual liquid is passed through a thermally driven membrane process that produces pathogen-free water that can be collected locally through a thermal recovery process (Hanak et al., 2016). Power is provided through novel harvesting processes using thermal or chemical gradients.

23.5 OFF-SITE SANITATION

23.5.1 Bucket latrines

Bucket systems are a traditional form of excreta removal. Waste is simply deposited into a container, and the "nightsoil" is collected on a daily basis. This method is still widely practised in urban areas, although it is objectionable from most points of view (including being hazardous to health), but at least it does remove excreta from the household.

23.5.2 Vault latrines

In principle, these are similar to bucket latrines except the squatting plate/seat is joined directly to a closed, watertight chamber where the excreta are deposited. Vaults need to be periodically emptied often by scoops or ladles. Objections similar to those for bucket latrines can be raised. The vault system is widely and successfully used in Japan where vacuum trucks empty the waste and transport it to be treated centrally. Similar systems connected to conventional WCs (*cesspools*) are also used in some remote properties in high-income countries. However, the need for very regular emptying makes this option expensive.

23.5.3 Conventional sewerage

As mentioned, the most expensive sanitation option is conventional sewerage. The advantages of using conventional sewerage (as described in the rest of this book) should by now be clear. However, the disadvantages for low-income communities are manifest: high cost; the need for an ample water supply; the difficulty of construction, operation, and maintenance; and the potential for serious pollution at the outfall (unless expensive wastewater treatment is proposed). In addition, there are two other practical causes for concern.

23.5.3.1 Septicity

High temperatures accelerate decomposition and limit the amount of oxygen that can be dissolved in water, leading to the rapid development of anaerobic conditions. Such conditions can give rise to H_2S production, resulting in corrosion of cementitious materials (see Chapter 17).

23.5.3.2 Blockage

Blockage can be caused or exacerbated by

- Abuse of the system through ignorance
- Use of traditional methods for anal cleansing (e.g., leaves, rags, stones, newspaper)
- Use of traditional methods for pot cleansing (e.g., sand, ash)
- Low water use
- Too few sewer connections

Also see Chapter 17.

23.5.4 Unconventional sewerage

23.5.4.1 Simplified sewerage

Simplified sewerage (also known as shallow or condominial sewerage) is similar to conventional separate foul sewerage except it is reduced to the basics and less-conservative assumptions are used in its design. Thus, sewer diameters, depths, and gradients are reduced compared with conventional systems, and locally available materials are utilised. Hydraulic design is also similar to conventional foul sewers as described in Chapter 9 (see Table 23.4).

In addition to relaxation in the hydraulic design criteria, the following practical details are changed (Reed, 1995):

- Conventional access points are replaced by ones of smaller diameter or rodding eyes.
- Access point spacing is increased.
- More maintenance responsibility is taken on by residents.
- Layout is amended, in particular, back-of-property collectors are used to minimise sewer length.

Table 23.4 Typical unconventional sewerage design criteria

Criteria	Simplified	Settled
Per capita flow (L/d)	100	100
Peak factor (× DWF)	2	1.5
Minimum velocity (m/s)	0.5	0.3
Maximum proposed depth of flow	0.75	1.0
Minimum pipe size (mm)	100	75
Minimum slope	1/200	—
Minimum cover (mm)[a]	350–500	350–500

[a] Depending on location.

The latter point is characteristic of the variant known as condominial sewerage. Stormwater drainage is needed to exclude runoff from such systems.

Simplified sewerage systems have most obvious application in high-density, low-income areas where space is at a premium and on-site solutions are inappropriate. However, their main advantage is their increased affordability with costs found to be approximately one-third of that of conventional sewerage in an Indian context (Mara and Broome, 2008). Some of the disadvantages of conventional systems also apply to simplified systems, with the main concern being the problem of blockages; however, both laboratory work (Gormley et al., 2013) and feedback from the field indicate blockage propensity to be low (Mara, 1996c).

23.5.4.2 Settled sewerage

Settled sewerage (also termed *small-bore*, *solids-free*, or *interceptor tank sewerage*) consists of small-diameter sewers connected to small tanks that collect individual household waste-water and capture much of the solid material. Solids accumulate in the tank and require periodic removal. The tank is provided to

- Settle out heavier material that normally requires relatively high self-cleansing velocities.
- Settle out larger solids, grease, and scum that might potentially block smaller pipes.
- Attenuate individual flow inputs to reduce the peak flow.

This allows small pipes to be provided at nominal fall. Some designs (Otis and Mara, 1985) allow sections of adverse pipe gradient and pipes flowing surcharged, provided there is an overall positive hydraulic gradient.

As with simplified sewerage, hydraulic design is similar to that for conventional foul sewers but using the criteria summarised in Table 23.4. However, because surcharged pipes are allowed in some systems, extra care is needed in designing the system to ensure all tanks can empty under gravity. Mara (1996c) gives a worked example and further explanation. The tanks can be designed as single-chamber septic tanks (see Section 23.4.2), although there are reports of much smaller tanks being used in certain areas (Tayler, 1996).

Settled sewerage may be the best option and the most economic choice where septic tanks are already in existence. Probably the main concern about these systems is their long-term performance. How well and for how long will they operate if tank desludging is neglected? What provision, if any, has been made for proper collection and safe disposal of the sludge? At the time of writing this edition, there has still been relatively little uptake and reported positive experience of these systems.

23.6 STORM DRAINAGE

Provision of drainage for low-income communities often lags behind water supply and sanitation. This is a pity because, if flooding is frequent, water supply and sanitation will be difficult to install and ineffective in operation. As has already been argued, storm drainage also fulfils a disease control function.

23.6.1 Flooding

When using conventional storm drainage in developed countries, systems are designed to run full at relatively modest storm return periods (1 or 2 years) with the knowledge and expectation that flooding will occur even less frequently (e.g., once every 5–10 years).

This situation can be contrasted with many low-income communities living with the monsoon. During the rainy season, high-intensity storms take place frequently, and drainage systems can fail almost immediately from lack of capacity and blockage, and widespread flooding occurs. As storm drains are routinely contaminated with sullage and faeces, the flooding is with dilute wastewater. However, detailed study of some low-income householders' attitudes towards flooding is very revealing (see Box 23.1).

BOX 23.1   COMMUNITY ATTITUDES IN INDORE, INDIA

Flooding was ranked low in comparison to other risks and problems, such as improvements in job opportunities, provision of housing, mosquitoes, and smelly back lanes.

A major concern mentioned by residents relates to the predictability of the flooding event—even extensive inundation is bearable if expected. Interventions aimed at ameliorating the effects of flooding should try to take account of these needs of the community to understand and adapt their coping strategies, if necessary.

Residents in flat areas give equal, if not more, weight to the after-effects of rains, as compared to immediate effects. Water may stand for long periods in very flat areas. Respondents feel that this makes walking difficult and allows the breeding of mosquitoes. Faeces-contaminated mud caused by stormwater is seen as most problematic as it is also a perceived source of mosquitoes and noxious smells.

In areas of the slum improvement programme, residents had high expectations of drainage improvement when the projects were initiated and appear to have expected that flooding and inundations would cease or be reduced substantially. It is likely that the feeling that their expectations have not been met is in part an immediate sense of disappointment that flooding has not ceased altogether.

Drainage interventions would gain favour if residents understood and were clearly informed about the effects (good and bad) on the environmental risk, which they perceive as inherently a natural event. This, of course, necessitates technical personnel being able to predict the consequences of technical interventions. Such a strategy might reduce the scale of expectation (Stephens et al., 1994).

So, in low-income communities subject to frequent flooding, even in improved areas, local people want to know most of the very things the drainage engineer finds difficult to tell them even with detailed modelling and forecasting technology. When will it flood and how long will it last, where will it flood, and how high will the water rise (Kolsky et al., 1996)?

23.6.2 Open drainage

23.6.2.1 Open channels

Open drains have a number of advantages when compared to closed pipes because

- They are cheaper to build as they are simpler and shallower than closed pipes.
- Blockage with refuse and washed-in sediment can be more easily monitored and safely cleaned out.
- They use available head more efficiently.
- Mosquito breeding is easier to control than in closed drains.

The simplest and cheapest drains are unlined channels along the roadside with water shedding from the road to the drain by positive fall. The sides of an unlined drain should not slope by more than 1:2 to ensure that they will be stable. If the gradient along the drain exceeds about 1:100, lining will usually be required to protect the channel from damage by scouring (Figure 23.6a).

One of the keys to making open drains work in practise is to maintain them properly. Drains require cleaning at regular intervals (whatever the self-cleansing velocity specified), since material from the street inevitably gets washed, blown, or dumped in. Marais et al. (2004) estimated litter loads as high as 6000 kg/ha.yr in some South African informal settlements lacking storm drainage systems and organised refuse collection. Kolsky et al. (1996) have demonstrated how the performance of the drainage system is substantially reduced by even small amounts of sediment in the system. Cleaning technique is also important. "Sweepings" from drains are traditionally left to dry in roadside piles prior to collection. Unfortunately, this material can quickly find its way back to the drain from whence it came.

A compromise solution is to build channels (under the footpath, for example) with removable covers (see Figure 23.6b) along their length. The covers discourage entry of rubbish and sediment but can be removed for cleaning, if necessary.

Design can be undertaken in a similar way to that of storm sewers described in Chapter 10. The major differences lie in the roughness of the channel being used and in ensuring velocities are not so high as to damage unlined channels (see Example 23.1).

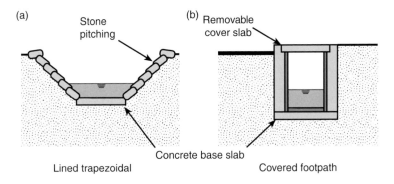

Figure 23.6 Typical open drainage channels.

Figure 23.7 Road-as-drain in Indore, India.

23.6.2.2 Road-as-drain

Experience suggests that blockage of drains is likely to be a recurrent problem, and that frequent stormwater flow along the streets of low-income communities during the monsoon season is inevitable. Therefore, logic dictates that engineers should consider the deliberate and controlled routing of surface runoff exceedance flows over the streets of the slum (see Figure 23.7). The street surface should be depressed below the level of housing sites and not *vice versa*, as is usually the case. This may involve more extensive site grading but will assure more reliable drainage than any system based on conduits (open or closed) that are subject to solids deposition. Roads as drains are also easier to maintain; street sweeping is easier than drain cleaning, and residents have an interest in keeping roads reasonably clear for access (Kolsky, 1998).

This is by no means a universal panacea, however. It is most appropriate in narrow streets where heavy vehicles do not pass at all and traffic is light. It will not be appropriate on steep streets (say >5%), and in flatter areas the micro-topography may well prevent surface routing of all flows. Road-surfacing material needs to be carefully selected (e.g., concrete,

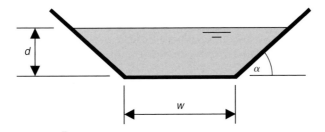

Figure 23.8 Trapezoidal channel geometry (Example 23.1).

compacted gravel, or stone) to provide erosion protection. Many Western nations are now considering similar options (see Chapter 11).

23.6.3 Closed drainage

23.6.3.1 Conventional drainage

The main advantage of closed drains is that they do not take up surface space. They also reduce the risk of children playing in or falling into polluted water, and the possibility of vehicles damaging the drains or falling into them. Also, in principle, sediment entry to the system can be minimised with catchpits and gully pots. The main problem is blockage due to refuse or sediment that does enter, and then the difficulty of cleaning out such material. Satisfactory simplified versions of storm drainage systems have not yet been developed.

EXAMPLE 23.1

An unlined (but vegetation-free) trapezoidal open channel (geometry shown in Figure 23.8) is to be built alongside a road of gradient of 1:100. If the design stormwater flow is estimated at 300 L/s, calculate the depth of flow in the channel, and check its velocity. Use Manning's equation (Equation 7.23), and assume a roughness of $n = 0.025$.

Solution

A trapezoidal channel of bottom width $w = d/2$ and side slope 1:2 has
Cross-sectional area, A:

$$A = d(w + d \cot \alpha) = 2.5d^2$$

Hydraulic radius, R:

$$R = \frac{d(w + d \cot \alpha)}{w + 2d / \sin \alpha} = 0.5d$$

Thus,

$$Q = \frac{1}{n} 2.5d^2 (0.5\,d)^{2/3} S_o^{1/2}$$

which for $Q = 0.3$ m³/s and $S_o = 0.01$ gives $d = 0.32$ m.
Velocity, $v = Q/A = 0.3 \times 2/(5 \times 0.32^2) = 1.2$ m/s.
This velocity is unlikely to be high enough to self-cleanse, and may cause erosion of the unlined channel.

As with conventional sewerage, conventional drainage using closed drains is the most expensive option available. Even though it is the most highly engineered approach, uncritical transference of European or North American standards has littered developing countries with expensive infrastructure that does not work. Hence, piped drains should be built in low-income tropical areas only after very careful consideration of the other options.

23.6.3.2 Dual drainage

Dual major/minor drainage, as described in Chapter 11, also shows promise for low-income communities.

23.6.4 Stormwater management

The topic of stormwater management has been dealt with comprehensively in Chapter 21. However, the question to be posed here is, will such an approach be viable in a low–income community context? The answer is a qualified "yes," and Reed (2004) found that sustainable urban drainage systems (SuDS) are already being used, albeit informally. The problems encountered by systems in informal settlements are essentially similar to those found in an industrialised country context, but to a greater degree compounded by their "contentious, turbulent and legally uncertain nature" (Jiusto and Kenney, 2016). Strengths, weaknesses, opportunities, and threats are summarised in Table 23.5.

23.7 GREY WATER MANAGEMENT

Grey water (or sullage) is defined as all the wastewater produced by a dwelling except that used in association with excreta disposal and arises from personal washing, laundry, food preparation, and the cleaning of kitchen utensils. Grey water can have a significant quantity

Table 23.5 Potential for use of SuDS in Sub-Saharan cities

Strengths	Weaknesses	Opportunities	Threats
Viable flood risk management option	Difficulty in quantifying long-term performance	Augmentation of water supply	Lower prioritisation on urban agenda (true of drainage in general)
Addresses triple goals of controlling quantity and quality of runoff and supports diversity	Operation and maintenance may be costly and require new skills	May be part of infrastructure upgrading programmes	Difficulty in convincing decision makers of new approaches
Predicted to be cheaper in the long run and can use local materials	Requires multi-stakeholder decision making	Opportunity for more inclusive decision making	May become health and safety threats if not maintained
More adaptive option	Relatively untested approach with lack of system data	Support for urban agriculture	Gentrification based on SuDS may lead to displacement of poor
Contributes to urban water cycle management	May require large open spaces of land	Jobs for relatively unskilled workers	Poor solid waste management

Source: Adapted from Mguni, P. et al. 2016. *Natural Hazards*, 82, 241–257.

of chemical pollutants and pathogenic microorganisms. Chapter 3 gives further details on the quantity and quality of wastewater, including this component. However, the whole issue of its management in an informal settlement context is often neglected.

It is commonplace that grey water (especially cooking and laundry sullage) is simply thrown outside the dwelling and inevitably finds its way into stagnant pools of water. These pools will ultimately overflow into open-channel stormwater drains. Even when grey water pipes are provided, they can still lead to the same stormwater drains. The resulting pollution can create a host of health and environmental impacts.

23.7.1 Options

The key guiding principles for management options are that grey water must not be allowed to (Carden et al., 2007)

- Pond on the surface
- Enter stormwater systems (including any SuDS if provided)
- Build up in the soil to such an extent that it damages it or significantly pollutes the groundwater

On- or off-site options for grey water include

- Beneficial use (e.g., irrigation)
- Treatment
- Disposal

Further discussion on these options is given by Carden et al. (2007) in the context of in South Africa, Al-Mamun et al. (2009) in Malaysia, and Sikder et al. (2016) in Bangladesh. Grey water recycling is further discussed in Chapter 24.

PROBLEMS

23.1 Explain the importance of taking an integrated approach to municipal service provision.

23.2 What are the main health-related issues concerned with poor sanitation and storm drainage?

23.3 Classify and compare the main sanitation options available to low-income communities.

23.4 Classify and compare the main storm drainage options available to low-income communities.

23.5 Compare and contrast the various types of pit latrine.

23.6 Discuss what is meant by "simplified" and "settled" sewerage, and explain how they differ from conventional sewerage.

23.7 Assess the relative merits of open and closed storm drains.

23.8 For the channel cross section described in Example 23.1, what gradient is needed to reduce the flow velocity to 1 m/s to avoid erosion?   [1:140]

23.9 Discuss the pros and cons of using stormwater management in low-income communities.

23.10 What is grey water, and why does it needs to be managed in an informal urban setting?

KEY SOURCES

Ludwig, H.F., Fennerty, H., Sow, K.L., and Mohit, K. 2005. *Textbook of Appropriate Sewerage Technology for Developing Countries*, South Asian Publishing.

Mara, D.D. (ed.). 1996a. *Low-Cost Sewerage*, John Wiley and Sons.

Mara, D.D. 1996b. *Low-Cost Urban Sanitation*, John Wiley and Sons.

Mihelcic, J.R., Fry, L.M., Myre, E.A., Phillips, L.D., and Barkdoll, B.D. 2009. *Field Guide to Environmental Engineering for Development Workers: Water, Sanitation and Indoor Air*, ASCE Press.

Parkinson, J. and Mark, O. 2005. *Urban Stormwater Management in Developing Countries*, IWA Publishing.

Tilley, E., Lüthi, C., Morel, A., Zurbrügg, C., and Schertenleib, R. 2008. *Compendium of Sanitation Systems and Technologies*, Swiss Federal Institute of Aquatic Science and Technology (Eawag), Dübendorf, Switzerland.

Vojinovic, Z. and Price, R.K. (eds.). 2007. Urban drainage in developing countries, Special Issue. *Urban Water Journal*, 4(3), 135–231.

REFERENCES

Al-Mamun, A., Zahangir Alam, M., Idris, A., and Sulaiman, W.N.A. 2009. Untreated sullage from residential areas—A challenge against inland water policy in Malaysia. *Pollution Research*, 28(2), 279–285.

Ashipala, N. and Armitage, N.P. 2011. Impediments to the adoption of alternative sewerage in South African urban informal settlements. *Water Science and Technology*, 64(9), 1781–1789.

Butler, D. and Payne, J.A. 1995. Septic tanks: Problems and practice. *Building and Environment*, 30(3), 419–425.

Cairncross, S. and Ouano, E.A.R. 1991. *Surface Water Drainage for Low-Income Communities*, World Health Organization.

Carden, K., Armitage, N., Sichone, O., and Winter, K. 2007. The use and disposal of greywater in the non-sewered areas of South Africa: Part 2—Greywater management options. *Water SA*, 33(4), 433–442.

Gormley, M., Mara, D.D., Jean, N., and McDougall, I. 2013. Pro-poor sewerage: Solids modelling for design optimization. *Municipal Engineer*, 166(ME1), 24–34.

Hanak, D.P., Onabanjo, T., Wagland, S.T., Patchigolla, K., Fidalgo, B., Manovic, V. et al., 2016. Conceptual energy and water recovery system for self-sustained nano membrane toilet. *Energy Conversion and Management*, 126, 352–361.

Jiusto, S. and Kenney, M. 2016. Hard rain gonna fall: Strategies for sustainable urban drainage in informal settlements. *Urban Water Journal*, 13(3), 253–269.

Kobel, D. and Del Mistro R. 2012. Evaluation of non-user benefits towards improvement of water and sanitation services in informal settlement. *Urban Water Journal*, 9(5), 347–359.

Kolsky, P. 1998. *Storm Drainage. An Engineering Guide to the Low-Cost Evaluation of System Performance*, Intermediate Technology Publications.

Kolsky, P.J., Parkinson, J., and Butler, D. 1996. Third world surface water drainage: The effect of solids on performance. Chapter 14, in *Low-cost Sewerage* (ed. D.D. Mara), John Wiley and Sons.

Mara, D.D. 1996c. Unconventional sewerage systems: Their role in low-cost urban sanitation. Chapter 2, in *Low-Cost Sewerage* (ed. D.D. Mara), John Wiley and Sons.

Mara, D.D. and Broome, J. 2008. Sewerage: A return to basics to benefit the poor. *Proceedings of the Institution of Civil Engineers—Municipal Engineer*, 161(4), 231–237.

Marais, M., Armitage, N., and Wise, C. 2004. The measurement and reduction of urban litter entering stormwater drainage systems: Paper 1—Quantifying the problem using the City of Cape Town as a case study. *Water SA*, 30(4), 469–482.

Mguni, P., Herslund, L., and Jensen, M.B. 2016. Sustainable urban drainage systems: Examining the potential for green infrastructure-based stormwater management for Sub-Saharan cities. *Natural Hazards*, 82, 241–257.

Nance, E. and Ortolano, L. 2007. Community participation in urban sanitation: Experiences in northeastern Brazil. *Journal of Planning Education and Research*, 26, 284–300.

Otis, R.J. and Mara, D.D. 1985. *The Design of Small Bore Sewer Systems*, TAG Technical Note. 14, The World Bank.

Parkinson, J., Tayler, K., and Mark, O. 2007. Planning and design of urban drainage systems in informal settlements in developing countries. *Urban Water Journal*, 4(3), 137–149.

Parkinson, J., Lüthi, C., and Walther, D. 2014. *Sanitation 21: A Planning Framework for Improving City-wide Sanitation Services*, IWA Publishing.

Pickford, J. 1995. *Low-cost Sanitation. A Survey of Practical Experience*, Intermediate Technology Publications.

Reed, R.A. 1995. *Sustainable Sewerage. Guidelines for Community Schemes*, WEDC, Intermediate Technology Publications.

Reed, B. 2004. *Sustainable Urban Drainage in Low-Income Countries—A Scoping Study*, WEDC, Loughborough University.

Satterthwaite, D., Mitlin, D., and Bartlett, S. 2015. Editorial: Is it possible to reach low-income urban dwellers with good-quality sanitation? *Environment and Urbanization*, 27(1), 3–18.

Scott, P., Cotton, A.P., and Sohail, M. 2015. Using tenure to build a "sanitation cityscape": Narrowing decisions for targeted sanitation interventions. *Environment and Urbanization*, 27(2), 1–18.

Sikder, T., Hossain, Z., Pingk, P. B., Biswas, J.D., Rahman, M., Hossain, M. et al. 2016. Development of low-cost indigenous filtration system for urban sullage: Assessment of reusability. *Future Cities and Environment*, 5(2), 1–8.

Sinnatamby, G.S. 1986. *The Design of Shallow Sewer Systems*, United Nations Centre for Human Settlements.

Souza, C., Bernardes, R., and Moraes, L. 2002. Brazil's modelling hopes. The public health perspective. *Water21*, October, 40–41.

Stephens, C., Pathnaik, R., and Lewin, S. 1994. *This Is My Beautiful Home: Risk Perceptions Towards Flooding and Environment in Low Income Communities*, London School of Hygiene and Tropical Medicine.

Tayler, K. 1996. Low-cost sewerage systems in South Asia, Chapter 4, in *Low-Cost Sewerage* (ed. D.D. Mara), John Wiley and Sons.

UNICEF and World Health Organization. 2015. *Progress on Sanitation and Drinking Water—2015 Update and MDG Assessment*.

Wondimu, A. and Alfakih, E. 1998. Urban drainage in Addis Ababa (Ethiopia): Existing situation and improvement ideas. *Fourth International Conference on Developments in Urban Drainage Modelling—UDM '98*, London, 823–830.

Chapter 24

Towards sustainable urban water management

24.1 INTRODUCTION

This chapter looks to the future and asks whether the urban drainage systems we now have are sustainable, and how they might need to change. Sustainable development is defined, and is interpreted in the context of sustainable water management (this section). Aspects of what makes an urban drainage system sustainable are presented (Section 24.2) together with example approaches that are "steps in the right direction" (Section 24.3). Approaches to and methods of assessing sustainability are explained in Section 24.4. Further to sustainability, this chapter discusses the planning, design, and adaptation of urban drainage systems in the context of resilience (Section 24.5). The key definitions, theoretical approaches, and available tools are briefly presented. The chapter concludes with a discussion on urban futures and possible urban water systems that will evolve to support them (Section 24.6).

24.1.1 Sustainable development

Sustainable development has been on the world agenda since the 1987 World Commission on Environment and Development. The outcome (the *Brundtland Report* [World Commission on Environment and Development, 1987]) offered a viable alternative to the commonly held view of environmentalists in developed countries: that pursuit of economic growth is incompatible with a responsible policy towards the environment. Such an attitude was unacceptable in developing countries, where increasing national wealth had to be a primary aim, and where any global environmental policy that threatened growth was seen as the rich countries "pulling up the ladder after them." In the context in which the phrase "sustainable development" was coined, it was the inclusion of the word *development* that was particularly significant: it affirmed the right of a country to seek to develop – but in a way that does not compromise opportunities for the future.

Brundtland defined sustainable development as the development "which meets the needs and aspirations of the present generation without compromising the ability of future generations to meet their own needs."

Further reflection has shown that sustainable development must satisfy environmental, economic, and social criteria (Bruce, 1992). Specifically,

- It must not damage or destroy the basic life support system of our planet: the air, water, and soil, and the biological systems.
- It must be economically viable to provide a continuous flow of goods and services derived from the earth's natural resources.
- It requires developed social systems, at international, national, local, and family levels, to ensure the equitable distribution of the benefits of the goods and services produced.

24.1.2 Sustainability

Even though the key components of sustainability – society, environment, and economy – are widely agreed on in principle, what they mean in practice is open to interpretation. So, the answer to the question, "what is sustainable?" depends on who is asking the question and why. So although sustainable development, when considered as an outcome, is difficult to fully define, when considered as a process of learning and communication, a journey that is constantly reviewed and adapted in the light of new information and knowledge, *more* sustainable development becomes a realistic target (see Box 24.1). Furthermore, a multi-disciplinary, multi-actor approach is particularly beneficial in arriving at agreed-upon solutions, especially in long-term collaborations. This is because the development of a common

BOX 24.1 A JOURNEY INTO THE UNKNOWN

The year 1927 saw the first flight across the Atlantic Ocean from New York to Paris. The most striking aspect of Charles Lindbergh's pioneering plane – *The Spirit of St. Louis* – was the fact that it had no front window (Figure 24.1). Lindbergh could not see what lay ahead; he could only see to the sides. He did have a compass to give him a heading, but there were few other navigational aids. This meant he only knew where he had been (the past) and where he was currently (the present), but he did not know or see exactly where he was going (the future).

We can draw a parallel with the journey towards sustainable development – it is a journey into the unknown. Arguably, we have a compass of sustainability principles (society, economy, environment), but we are not certain of our final destination. History records that Lindbergh was successful in navigating purely by the past and present and extrapolating intelligently into the future. What we can learn is twofold: that the journey of discovery along the route to sustainable development is as important as the final destination, and that the most important step to take is the first one, accepting that some wrong turns may be made along the way (Butler et al., 2010).

Figure 24.1 The Spirit of St Louis.

language, which will allow an informed debate about sustainability to take place, is a long and difficult process (Sharp, 2006). Lack of dialogue can lead to a continuation of "tried and tested" practices and an over-reliance on monetary evaluation to determine appropriate solutions (Brown et al., 2009).

24.1.3 Sustainable urban water management

According to the American Society of Civil Engineers/UN Educational, Scientific, and Cultural Organization (ASCE/UNESCO, 1998), sustainable water systems are "those systems designed and managed to fully contribute to the objectives of society, now and in the future, while maintaining their ecological, environmental and hydrological integrity." Butler and Maksimović (1999) pointed out how the notion of sustainable development has contributed to a rethinking of traditional approaches to urban water systems. They emphasised the importance of public participation, individual responsibility, and of water systems that are environmentally friendly, socially acceptable, and financially viable. Generic priority areas were identified:

- *Integration*: acknowledging and exploiting the reality of the urban water system as a unified whole
- *Interaction*: identifying synergistic effects and guarding against antagonism
- *Interfacing*: prioritising the need for engagement with the public and developing a greater urgency concerning the part urban water systems play in protection of the environment
- *Instrumentation*: acknowledging the key role of sensor technology and the need for greater active control of the system
- *Intelligence:* wider use of decision support tools and information management
- *Interpretation*: wisdom to bring about ordered change and the willingness to think laterally
- *Implementation*: the need to make progress on the ground, including "learning by doing"

Malmqvist et al. (2006) present a strategic approach to the planning of sustainable urban water systems. Technical, economic, and environmental aspects are covered together with the challenges of institutional capacity development and public participation in the planning process.

24.1.4 Transition states

Brown et al. (2009) propose a "transitions framework" with a representation of different temporal, ideological, and technological states that cities transition through when moving between different management paradigms, towards more sustainable futures. As shown in Figure 24.1, the framework includes six city states, namely the Water Supply City, the Sewered City, the Drained City, the Waterways City, the Water Cycle City, and the Water Sensitive City. Each of the six city states is characterised by a distinct shift in the following pillars of institutional practise:

- Cognitive: dominant knowledge, thinking, and skills
- Normative: values and leadership
- Regulative: administration, rules, and systems

In Figure 24.2, the "cumulative socio-political drivers" reflect shifts over time in the agreements between communities, businesses, and governments on how water should be

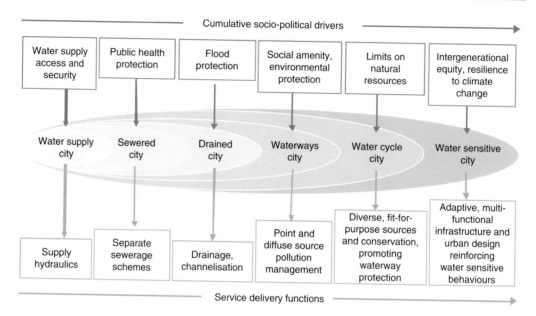

Figure 24.2 Urban water management transition states. (Reproduced from Brown, R. et al. 2009. *Water Science and Technology*, 59[5], 847–855, with permission from the copyright owners, IWA Publishing.)

managed (also termed a *hydro-social contract*; Lundqvist et al., 2001). The "service delivery functions" represent the key practical concerns and priorities being addressed in those paradigms.

24.2 SUSTAINABILITY IN URBAN DRAINAGE

24.2.1 Objectives

The difficulty with the Brundtland definition of sustainable development is that, although its principles are widely accepted, it is not immediately clear how to put them into practice. The following more specific objectives have been suggested (Butler and Parkinson, 1997):

- Maintain an effective public health barrier.
- Avoid local or distant flooding.
- Avoid local or distant degradation/pollution of the environment (water, soil, air).
- Minimise the utilisation of natural resources (water, nutrients, energy/carbon, materials).
- Be reliable in the long term and adaptable to future (as yet unknown) requirements.
- Be affordable to the community it serves.
- Be socially and culturally acceptable.

24.2.2 Strategies

The search for economically viable solutions with low water, energy/carbon, and maintenance requirements, which are both flexible to change and hygienically acceptable under changing conditions, is a great challenge. Butler and Parkinson (1997) suggest three fundamental strategies that should be pursued:

- Reduce the reliance on water as transport medium for waste.
- Avoid mixing industrial wastes with domestic wastewater.
- Avoid mixing storm runoff with wastewater.

They argue that if priority were given to these strategies, many benefits would be immediately realised, even if they are only introduced singularly or incrementally. Wholesale replacement of existing systems, on the other hand, is unlikely to be the more sustainable approach (but see Section 24.1.4). The strategies and their potential advantages and disadvantages are summarised in Table 24.1.

24.2.2.1 Water transport

Centralised urban drainage systems arguably utilise water inefficiently. Large quantities of water are abstracted and treated using expensive treatment technology to drinking standard, yet are subsequently used to flush faeces and urine from the WC (see Chapter 3). Not only is this wasteful of a precious resource, but it also promotes unnecessary mixing and

Table 24.1 Strategies towards sustainable urban drainage

Component of wastewater	Problems	Proposed strategies	Potential advantages	Potential disadvantages
Carriage water	• Unnecessary water consumption • Dilution of wastes • Requires expensive end-of-pipe treatment	• Introduce water conservation/efficiency techniques • Reuse water • Seek alternative means of waste conveyance	• Conserves water resources • Improves efficiency of treatment processes	• Increases possibility of sedimentation in sewers • Health hazards associated with water reuse
Industrial waste	• Disrupts conventional biological treatment • Increases cost of wastewater treatment • Causes accumulation of toxic chemicals in the environment • Renders organic wastes unsuitable for agricultural reuse	• Remove from domestic waste streams • Pre-treat, reduce concentration of problematic chemicals • Promote alternative industrial processes using biodegradable substances	• Improves treatability of wastewaters • Improves quality of effluents and sludges • Reduces environmental damage • Saves costs associated with reuse of recovered chemicals	• Costs associated with implementing new practice • Lack of monitoring facilities • May promote illicit waste disposal
Stormwater	• Requires large and expensive sewerage systems • Transient flows disrupt treatment processes • Discharge from overflows causes environmental damage • Causes floods	• Utilise overland drainage patterns • Store and use stormwater as a water resource • Provide infiltration ponds, percolation basins, and permeable pavements • Promote ecologically sensitive engineering, e.g., constructed wetlands	• Reduces pollution from overflows • Improves efficiency of treatment • Recharges groundwater • Reduces demand for potable water • Reduces hydraulic capacity requirements of conduits	• Decentralised facilities harder to monitor • Increases space requirements • Risk of groundwater contamination

dilution of wastes and contamination of previously unpolluted water. End-of-pipe treatment technology is then needed to extract the solid and dissolved pollutants from the liquid component of the waste flow to avoid receiving water pollution.

Household water consumption can be reduced substantially by a number of means (Butler and Memon, 2006), and comprehensive performance evidence is now available (Omambala et al., 2011). Box 24.2 gives a good example of an innovative, ultra-low-flush toilet with the potential to significantly reduce domestic water consumption. A key challenge is to ensure that existing systems will operate effectively with much lower dry weather flows (see Box 24.3). Alternative no-flow technologies for the conveyance of sanitary waste, such as dry sanitation systems (Chapter 23) or some non-gravity systems (Chapter 14), should be investigated as feasible technical options for the future.

BOX 24.2   PROTOTYPE ULTRA-LOW-FLUSH TOILET

The *Propelair* toilet is a prototype, patented ultra-low-flush toilet (ULFT). As can be seen (Figure 24.3), in appearance it is similar to a standard toilet. However, in function and performance it is very different. It works by locally pressurising air, which is then expelled through the toilet under pressure during the flushing cycle instead of water. Just 1.5 L of water per flushing cycle is used to cleanse the bowl and refill the water seal. Cleansing of the pan is excellent, exceeding the requirements of BS EN 997: 2003. The ULFT can only be flushed when its lid is fully closed. It can be connected to a standard gravity drainage system.

The prototype was thoroughly evaluated in the WaND (Water Cycle Management for New Developments; Butler et al., 2010) project to determine the following key aspects: downstream performance, user acceptance, and water and energy savings.

DOWNSTREAM PERFORMANCE

The ULFT's flushing performance was tested in a full-scale test rig and compared with that of a 6/4 L dual flush toilet. Results showed that the toilet works best with a 50 mm diameter downstream drain (which can be flexible and does not need to have a constant gradient) and that its limiting solid transport distance (LSTD) performance (see Section 9.5.2) exceeds that of a 6 L toilet (Littlewood et al., 2007).

USER ACCEPTANCE

Figure 24.3 shows one of two toilets installed in an office building, which was monitored over a period of 8 months to gauge functionality, water and energy savings, and user acceptance. During the trial, no incidents of blockage were reported, and there was no interference with the water traps of adjacent toilets. Some 58% of respondents to a questionnaire thought the ULFT was easy to use, and 93% described the flushing performance as being good. However, some users viewed the need to close the toilet as unhygienic and were concerned about not knowing when the lid was closed properly (Millán et al., 2007).

WATER AND ENERGY CONSUMPTION

When compared with the adjacent conventional toilets (9 L), water saving per flush was 84%. Although the ULFT uses a small amount of energy for each flush, taking into account the energy displaced by not using water, 80% energy savings were also made (Millán et al., 2007).

Figure 24.3 Prototype ultra-low-flush toilet. (Courtesy of Garry Moore.)

24.2.2.2 *Mixing of industrial and domestic wastes*

The mixing of industrial effluents with domestic wastewaters can cause problems for conventional wastewater treatment resulting in the creation of water pollution problems. The reuse potential of nutrients and minerals contained in sludge is diminished by small concentrations of industrial contaminants. In many cases, this results in the disposal of sludge to landfill sites or by incineration, and the waste of a valuable resource. Synthetic organic chemicals and heavy metals accumulate in the environment causing problems to ecosystems as they move through the food chain.

Removing the industrial waste component from municipal wastewater is, therefore, critically important to sustainable drainage strategies. Dealing with the waste as close to the location of production as possible reduces the problems associated with treating a highly complex mixture of substances and increases the reuse potential of the more valuable components of waste (Hvitved-Jacobsen et al., 1995). Where the complete isolation of an industrial waste stream is impractical, pre-treatment using best available technology to reduce concentrations of the undesirable wastes (be they refractory chemicals or excessively high concentrations of biodegradable waste) is required prior to discharge into the sewer.

BOX 24.3 DROYLSDEN, UNITED KINGDOM

Droylsden is a largely residential suburb of Manchester. It has a population of about 20,000 and an area of 420 ha. The majority of the catchment is urbanised, although there are substantial undeveloped areas adjacent to the River Medlock and its tributaries. The existing sewerage system is combined, with just a few newer developments having separate sewers. The system drains to a single outfall crossing the river before discharging into a large interceptor sewer.

A detailed theoretical exercise (Parkinson et al., 2005) was carried out to assess the implications on the existing sewers of introducing water-conserving devices throughout the catchment. To do this, a sewer simulation model was developed and verified using *in situ* measured data. Figure 24.4 shows the relative frequency of velocities during dry weather in all the modelled sewers for the base-case of 9 L flush toilets. As illustrated, in this system, very few sewers reach velocities required for self-cleansing (a minimum of 0.6 m/s).

According to model predictions, application of lower-flush toilets only has a small effect. However, the effect is interesting in that it increases the proportion of pipes that have lower velocities, and therefore decreases those with higher velocities. Thus, fewer sewers will reach the self-cleansing threshold.

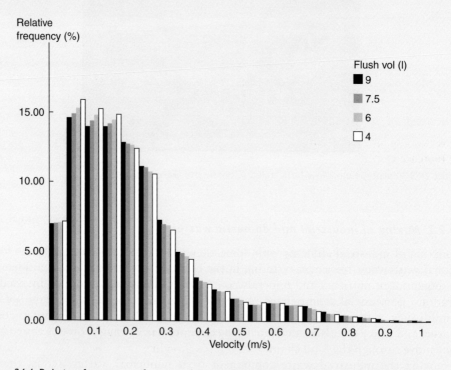

Figure 24.4 Relative frequency of sewer dry weather flow velocities with and without water-conserving toilets.

24.2.2.3 Mixing of stormwater and wastewater

By disconnecting stormwater from overloaded combined systems, a number of benefits are achieved: flooding frequency is reduced, energy costs and carbon emissions are decreased, and combined sewer overflow (CSO) discharges are reduced, thus reducing the extent of environmental damage, and the problems caused at treatment works by peak and transient flows are reduced; the potential for use of rainwater for reuse or recreation is also increased.

Isolating storm runoff from wastewater is not a new idea. As discussed in Chapter 1, separate systems were regarded in the 1950s as being the drainage systems of the future and, indeed, are still the current practice in most developed countries. Generally, these systems are more costly to construct, generate higher emissions during manufacture (Gigerl and Rosenwinkel, 1998), and yet have not delivered the expected environmental improvements. However, it seems that the overall strategy of separating the two components is not at fault, rather the method through which the strategy has been implemented.

Current thinking advocates separation of storm runoff from other urban wastewaters but, instead of routing stormwater directly into a separate piped system, recommends utilising the natural drainage patterns of the catchment, or using stormwater management techniques as described in Chapter 21. Many of these methods utilise on-site, small-scale infiltration. Where the soil and the quality of the runoff permits, direct infiltration into the ground is preferable in order to recharge groundwater reserves. Where infiltration is not possible, the development of natural drainage patterns (see also Chapter 11) offers a range of opportunities for conservation, recreation, and amenity, as well as providing basic flood and pollution control.

24.2.3 Integration

Dixon et al. (2014) argue that in order to meet the needs of a sustainable future, current water infrastructure systems and technologies must be comprehensively reconfigured or retrofitted not as individual entities, but by integrating the urban water cycle as a whole and promoting multiple benefits. This should aim to

- Reduce water demand
- Restore urban ecosystems
- Increase resource use efficiency
- Develop new and sustainable resources
- Change water cultures and practices
- Use end-use water qualities
- Increase the flexibility, resilience, and adaptability of water, sanitation, and drainage systems

Not only should this retrofit be technologically based, but it also needs to be in terms of societal values and practices associated with water, its quality, and our intended uses. Figure 24.5 presents a range of retrofit interventions and how they are associated with the main water challenges (drought, flooding, and water pollution), the emphasis being on multi-functional interventions. For example, rainwater harvesting systems can now be successfully designed to both save water and mitigate flooding (Melville-Shreeve et al., 2016). Green infrastructure can reduce the impacts of diffuse pollution, surface runoff, and combined sewer overflows on receiving water bodies (Rozos et al., 2013).

24.2.4 Does size matter?

Over the last two to three decades there has been a great deal of interest and activity in researching and indeed promoting small-scale, decentralised approaches as more appropriate solutions than traditional, large-scale, centralised systems (Otterpohl et al., 1997; Tjandraatmadja et al., 2005). To date, there have been no definitive studies of whether size

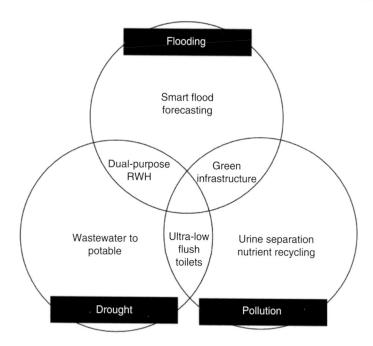

Figure 24.5 Future retrofit interventions to water challenges.

does indeed matter, or of the circumstance in which it might. However, the prevailing if not fully substantiated view of many is that small-scale systems are more sustainable than large-scale ones (Makropoulos and Butler, 2010). Additionally, they are likely to contribute towards increased system resilience (see Section 24.5).

24.3 STEPS IN THE RIGHT DIRECTION

24.3.1 Domestic grey water recycling

The reuse of grey water (all the wastewater produced in a house except by the WC) potentially reduces the need to use potable water for non-potable applications, with the water effectively being used twice before discharge to the sewer. The major reuse potential is for WC flushing and garden watering, relieving demand on public water supplies and wastewater collection and treatment facilities (Nolde, 2014). The main elements of a grey water reuse system are a collection and distribution pipe network, sufficient storage volume to balance inflows and outflows, and appropriate treatment to render the recycled water fit for the purpose (Gross et al., 2015).

If WC flushing and garden watering demand can be fully satisfied using recycled water, comparison with Table 3.1 indicates that a 36% reduction in demand will be realised. In theory, this demand could easily be met by reusing water first used for personal washing (26%) and clothes washing (12%). System modelling shows that 90% WC water saving efficiency can be achieved using a storage capacity of approximately 200 L (Dixon et al., 1999a).

Widespread adoption of such systems will inevitably depend on their cost. Retrofitting systems to individual houses is unlikely to be financially attractive, although systems built into new houses may well be. Larger-scale application (e.g., hotels, office blocks) may prove to be more cost effective (Butler and Dixon, 2002; Memon et al., 2005) or when used synergistically between buildings (Zadeh et al., 2013). There are still some nagging concerns about health issues (Dixon et al., 1999b) and whether systems are carbon positive or negative (Memon et al., 2007).

24.3.2 Nutrient recycling

It is an important fact that, in principle, each person produces in their excreta enough nutrients to grow 250 kg of cereal per year, which is about enough for one person to live on (Drangert, 1998). Of course, some of these nutrients are currently used on agricultural land as sludge derived from wastewater treatment plants (WTPs) (about 50% of sludge goes to land in the UK). However, significant quantities of nutrients are discharged through the WTP to the receiving water and so are effectively lost to agriculture. Also, industrial discharges add heavy metals to wastewater, rendering it less usable.

One way to make better (more sustainable) use of wastewater nutrients is to isolate and handle urine separately and use it as a fertiliser. One important reason for this strategy is the fact that urine is the source of around 30% of phosphorus and around 70% of nitrogen in wastewater (Figure 24.6). Additional benefits include the fact that urine is relatively low in heavy metal concentration and also relatively easy to handle, and much less water is needed to flush it. The concept of managing and exploiting excreta and wastewater in this way has come to be known as *ecological sanitation* (or *ecosan*) (Winblad and Simpson-Hébert, 2004).

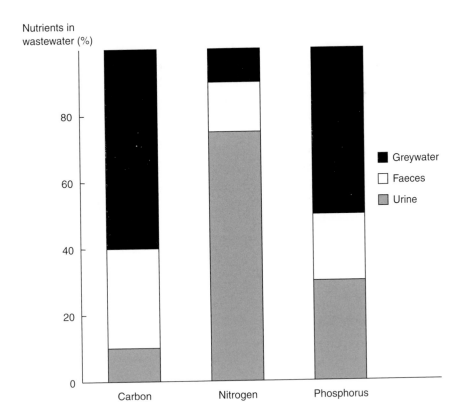

Figure 24.6 Percentage source of nutrients in wastewater.

To isolate and reuse urine, special no-mix toilets are required, which can range from the simple to the sophisticated, with choice depending on the socio-economic culture. Box 24.4 describes a case-study application of urine-separation technology.

24.3.3 Heat recovery

A large amount of thermal energy is wasted as warm wastewater is transported in sewers to the wastewater treatment plant. Recovering that heat has been practiced in some Scandinavian cities but is set to be applied more widely as the technology develops. Sewer-heat recovery systems use exchangers set in the invert of the pipes to extract the heat from the wastewater. Heat pumps then boost temperatures to a range useful for residential space heating and domestic hot water. Compared with other traditional solutions, the energy is

- Renewable
- Locally available
- Associated with significantly reduced greenhouse gas emissions

A case study of this approach has been presented in Box 1.4.

BOX 24.4   BJÖRSBYN, SWEDEN

Björsbyn is an "ecological" village built in 1994 in northern Sweden. It was designated as such to emphasise that the inhabitants' way of living is deliberately different from traditional urban dwellers, moving towards a more sustainable lifestyle. Interestingly, however, ecological concern was a major reason for moving into the village for only two of the families. (The main attraction was an appealing rural area within a reasonable distance from the city centre.) The village has urine-separating toilets, compost bins for biodegradable organic wastes at each house, and a small outhouse for the collection of paper and glass separated from the solid wastes (Hanaeus et al., 1997).

- Excreta are separated at source using no-mix toilets (Figure 24.7). The toilets used were equipped with a small collection unit for urine with its own flush-water system (0.1 L per flush). The urine is collected separately and led through a sewer system to a collection tank with 8 months' storage capacity. When the tanks are full, or urine is needed, they are emptied onto nearby farmland.
- The faeces and other grey water are collected and treated by septic tanks followed by infiltration beds. The sludge from the septic tanks is treated locally by a combined sludge drying, freezing, and composting unit. After treatment, the sludge is used as a fertiliser and soil conditioner on agricultural land.
- Stormwater flows via ditches to the surrounding countryside where it is discharged and infiltrates to groundwater.

When studying this system, Hanaeus et al. (1997) concluded that the successful operation of a urine separation system is particularly dependent on complete separation (to avoid dilution or contamination of the urine). This can be promoted by well-designed toilets, careful design of the collection system, and education of users.

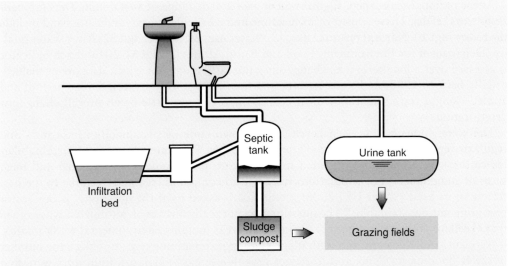

Figure 24.7 The Björsbyn system. (Reproduced from Hanaeus, J. et al. 1997. *Water Science and Technology*, 35[9], 153–160, with permission of publishers Pergamon Press and copyright holders IAWQ.)

24.3.4 Disposal of domestic sanitary waste

In some countries, including the UK, it has been common for the WC to be used as a disposal system for a variety of solid waste items (listed in Chapter 3). Each individual selects the method of disposal, in the privacy of the bathroom: the item is either dropped in the WC and flushed away, or placed in a bin for subsequent removal via the solid waste system.

Disposal of sanitary waste by the waterborne route is known to cause operational and environmental problems of small pipe blockage, screen ragging and blinding, and offence to the public if discharged to the environment. Common sense would suggest that this is an inefficient method of disposal – solid waste mixed and transported with wastewater requires an increase in effort and expense to re-separate.

A non-technical approach to reducing the problems caused by these wastes emphasises cultural rather than technical aspects. It is to create a change in public disposal practices: to reduce disposal by the waterborne route and increase disposal by the solid waste route. There have been many public awareness campaigns drawing attention to the operational and environmental problems caused by waterborne disposal as described in Chapter 1.

Research by Ashley et al. (1999) investigated the advantages of encouraging a change in public disposal practice, and has shown that such a change improves the sustainability of the system as a whole. For public campaigns to be successful, there needs to be an emphasis on activity at a local level. Sharp (2017) emphasises the great need for sympathetic and sustained engagement with the public in urban water management.

24.4 ASSESSING SUSTAINABILITY

How can we assess whether one urban drainage option is more sustainable than another? The most common approach is to use *indicators* at various scales.

At the national scale, the UK government uses a wide range of *Sustainable Development Indicators* (SDIs). These consist of 12 headline and 23 supplementary indicators and include the following: 10. Natural resource use, 12. Water use, 29. Waste, and 32. River water quality (Department for Environment, Food and Rural Affairs [DEFRA], 2013). Each indicator is assessed over a long-term (i.e., change since the earliest date for which data are available – usually back to 1990) and short-term period (i.e., change for the latest 5-year period) and graded using a traffic light system to denote improvement, little or no overall change, or deterioration.

The water industry has been developing its own suite of sustainability indicators and regularly reports "direction of travel" (see Table 24.2). These are a mixture of sustainability indicators (e.g., average energy used for treatment, greenhouse gas emissions) and measures of compliance (e.g., population connected to compliant treatment plants). In the first review of its kind (Water UK, 2008), the industry asked itself the question, "is our sector becoming more sustainable?" The question was not really answered, although it was argued that excellent progress had been made in many areas (notably environmental water quality, investment, and efficiency) but with many challenges remaining (e.g., to reduce the number of sewer flooding incidents and to reduce greenhouse gas emissions). Indicators were first revised and updated for 2007/2008, while for the assessment of period 2010/2011 (Water UK, 2012) the number of indicators was reduced to 17 and organised according to the five industry priorities agreed by Water UK:

- Protect and promote public health.
- Provide a high-quality, reliable, and value-for-money service.
- Respond to climate change.
- Reduce our impacts on the environment.
- Earn the trust of the community.

At the city level, the KWR Watercycle Research Institute has developed a standardised indicator framework to support strategic planning of integrated water management called the City Blueprint (Van Leeuwen et al., 2012, 2016). The methodology was further updated and applied in the framework of the EU TRUST project (Van Leeuwen and Marques, 2013). The main objective of the blueprint is to measure and assess a city's (sustainability) performance in order to envision, develop, prioritise, and implement interventions and measures towards a water-sensitive city (see Figure 24.7). City Blueprint consists of 26 indicators

Table 24.2 Selected sustainability indicators

Indicator	2002/2003	2003/2004	2004/2005	2005/2006	2006/2007
Proportion of rivers with good biological water quality (%)	74.6	75.7	75.5	76.3	76.5
Total number of pollution incidents	2426	2820	2235	2352	2384
Properties at risk of sewer flooding (%)	0.05	0.05	0.04	0.03	0.03
Properties flooded (%)	0.02	0.02	0.02	0.02	0.02
Population connected to a compliant WTP (million)	86.6	93.7	97.0	88.0	86.1
Average energy used to treat 1 ML sewage (kWh)	814	645	663	634	756
Total CO_2 emissions (Mt)	NA	NA	4.14	4.15	5.03

Source: Water UK. 2008. *Sustainable Water*. State of the Water Sector Report. www.water.org.uk.

classified into 7 categories: (1) water quality, (2) solid waste treatment, (3) basic water services, (4) wastewater treatment, (5) infrastructure, (6) climate robustness, and (7) governance (Koop et al., 2015). A city's sustainability assessment is provided in the form of a spider-graph as depicted in Figure 24.8, while the overall performance is derived from the geometric mean of all indicators, namely the Blue City Index. The City Blueprint also provides a set of indicators to assess the trends and pressures on a city (e.g., urbanisation rate, education rate, urban drainage flood, river peak discharges) to facilitate the determination of limitations and opportunities for urban sustainable water management. The framework has been applied to assess the performance of more than 45 cities to date (Van Leeuwen et al., 2016).

At the level of assessing sustainability in relation to a choice between urban drainage options, a more specifically focussed approach is needed. One approach has been developed by a project named SWARD, Sustainable Water Industry Asset Resource Decisions (Ashley et al., 2004). This project investigated the way asset investment decisions are made in the water industry, and how issues of sustainable development should be included. The decision support processes recommended are set out in seven stages:

1. Define objectives
2. Generate options
3. Select criteria
4. Collect data
5. Analyze options
6. Select and implement preferred option
7. Monitor outcome

The options, considered in this way, can be quite general (e.g., alternative strategies for managing domestic sanitary waste) or quite specific (alternative locations for a wastewater treatment facility).

Candidate criteria are proposed under four headings: environmental, social, economic, and technical. Recommended "primary criteria" are listed in Table 24.3. Each primary criterion leads to a number of secondary criteria, each with a recommended indicator to be

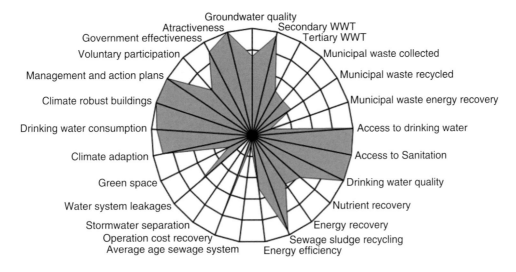

Figure 24.8 Example city blueprint spider-graph.

Table 24.3 SWARD primary criteria

Type	Primary criterion
Environmental	Resource utilisation
	Service provision
	Environmental impact
Social	Impact on risks to human health
	Acceptability to stakeholders
	Participation and responsibility
	Public awareness and understanding
	Social inclusion
Economic	Life-cycle costs
	Willingness to pay
	Affordability
	Financial risk exposure
Technical	Performance of the system
	Reliability
	Durability
	Flexibility and adaptability

used in the assessment. The proposed secondary criteria and indicators for one of the primary environmental criteria, *resource utilisation*, are given in Table 24.4.

Analysis and comparison of the options using the selected criteria and their indicators require some form of multi-criterion analysis approach. Case studies using this general approach are described by Ashley et al. (2003).

A formal multi-criterion method was subsequently developed in the WaND project – the Project Assessment Tool, based on SWARD indicators (Butler et al., 2010). This was used with a wide variety of stakeholders to promote discourse, to illustrate innovations, and ultimately to provide a means to operationalise sustainability (Ashley et al., 2004).

Table 24.4 SWARD: Secondary criteria and indicators under the primary criterion resource utilisation

Secondary criteria	Indicator
Water resource use	
– Withdrawal	Annual freshwater withdrawal ÷ annual available volume (%)
– River water quality	% of rivers of good or fair quality
– Nutrients in water	% of river length with greater than guideline nutrient concentrations
Land use	Land area used in km^2
Energy use	
– Energy for water supply	Energy use ($kW.h/m^3$)
– Energy for wastewater treatment	Energy use ($kW.h/m^3$)
Chemical use	
– WTP or on-site (herbicides)	Litres/year
Material use	
– Aggregates, plastics, metals	Total material requirement (tonnes/year)

24.5 RESILIENCE

The concept of resilience was introduced in Chapter 11 in the context of flooding. This section contains a more detailed consideration of the topic and its importance to sustainable water management.

24.5.1 Definitions and objectives

Water services, and especially urban drainage systems, face significant internal and external challenges and pressures (Makropoulos et al., 2016). For instance, flooding can be caused not only by *external* threats such as extreme rainfall events and increasing urbanisation (Mugume and Butler, 2016), but also by *internal* system threats such as equipment malfunction, sewer collapse, and blockages (Mugume et al., 2015). These threats may result in system or component failures whose impacts may lead to adverse consequences for the users of the system.

To tackle these pressures and challenges, we need to build more *resilient* water systems to enhance their ability to both maintain their level of service and to minimise the consequences during unexpected or exceptional loading conditions (Butler et al., 2016; Mugume et al., 2015). Although the concept of *resilience* has only recently been applied to infrastructure systems, there is an extensive literature on its definitions and interpretations, which are varied across different scientific fields and types of systems (Mugume et al., 2015; Pizzol, 2015). The common ground between them points towards the characterisation of resilience as a given performance objective delivered by the property of the system as a whole, rather than attributed to individual elements or units. According to Morecroft et al. (2012), all definitions revolve around two key system performance objectives:

- The amount of disturbance that a system can withstand without changing self-organised processes and structures (Holling, 1973)
- The return time to a stable state following a perturbation (Brede and de Vries, 2009)

Furthermore, Pizzol (2015) highlights three common ideas across different interpretations and applications of resilience:

- Resilience depends on both system elements and how they are structured. Specific designs of this connectivity lead to increased resilience (e.g., by increasing the number of connections between elements or their strength – ideas also loosely connected with redundancy or overdesign).
- There is a trade-off between resilience and efficiency, with some natural systems favouring resilience while most human systems favour efficiency
- Resilience is associated with the sustainability of a system, intended as the capacity of systems to maintain their functions in a context of continuous change. From this perspective, resilience is a key attribute or prerequisite of a sustainable system. In many ways, differences between these two terms could be understood as varying in the boundaries of the system in question and the number and scale of interactions of relevance.

Focussing on engineered water systems, such as urban drainage, work on "Safe and SuRe" Water Management defines resilience as the "the degree to which the system minimises level of service failure magnitude and duration over its design life when subject to exceptional conditions" (Butler et al., 2016). These exceptional conditions can be considered the threats or pressures that lead to system failure, such as extreme rainfall events, sewer collapse, or blockage. According to this definition, the goal of resilience is the maintenance of acceptable functionality levels (by withstanding service failure) and the rapid recovery from failure once it occurs (Butler et al., 2016; Lansey, 2012; Park et al., 2013). According to Butler et al. (2014), resilience can be further classified into three broad categories:

1. *General* (attribute-based) *resilience* that refers to the state of the system that enables it to limit failure duration and magnitude to any threat (i.e., all hazards)
2. *Specified* (performance-based) *resilience* that refers to the agreed performance of the system in limiting failure magnitude and duration to a given (known) threat
3. *Technology-based resilience* that refers to the equipment that can improve the preparedness of the user for extreme events

24.5.2 Resilience and sustainability

As discussed in the previous section, the key idea behind sustainability is the maintenance or even enhancement of socially derived objectives (e.g., social, economic, environmental), not only for now but also for future generations. In the framework of the Safe & SuRe work, sustainability is defined as "the degree to which the system maintains levels of service in the long-term whilst maximising social, economic and environmental goals" (Butler et al., 2014). So, sustainability targets long-term horizon planning and hence requires the design of resilient systems that will be able to cope with any possible future threat or pressure (Scholz et al., 2012). Although it is clear that resilience lies at the core of sustainability in an operational context (Pickett et al., 2014), the details of the exact relationship remain unknown; hence, currently it is not clear how to maximise sustainability by building resilience (Butler et al., 2014). The conceptual relationship of three notions, as defined in the context of the Safe & SuRe programme, is presented schematically in Figure 24.9, where a safe/reliable system should be built upon by resilience and topped off with sustainability.

24.5.3 Resilience in strategic planning

Water service providers, tasked to manage urban water and the infrastructure that delivers it to customers (or protects customers from it, as is the case in urban drainage), come face to face with the concept of resilience during the process of strategic planning. To decide

Figure 24.9 The Safe & SuRe pyramid.

between alternative system configurations and large-scale interventions that will change the function of the system in the longer term, providers need to understand the future behaviour/performance of their systems in the longer term, under accidents/incidents and/or extreme events, and under changing external conditions (e.g., societal, economic, or climatic trends) that involve significant uncertainties. In this setting, *resilience* has come to dominate the discussion around "future proofing" water systems.

An operational methodological framework to assess the resilience of the complete urban water cycle was developed for the Dutch Water Sector (Makropoulos et al., 2016). Here, resilience is defined as "the degree to which a water system continues to perform under progressively increasing disturbance." The degree of performance in this definition is quantified through *reliability*, which is in itself defined, in a somewhat more expansive way than we have seen in earlier sections, as "the ability of the system to consistently deliver its objectives, considered over a timespan." This extension of the typical definition of the reliability term (Mays, 1989) enables the assessment and operationalisation of the effect of different "failure modes" and pressures on the water system. Progressively increasing disturbance comes from progressively more "stressful" scenarios, not unlike the approach discussed in Mugume et al. (2015). The result is a form of *stress-testing of water systems*, whereby the system is exposed to progressively increasing disturbance and the change in its behaviour (i.e., performance) is assessed and traced in a form of a graph, termed the *Resilience Profile* graph.

Each point of a typical *Resilience Profile Graph* (Figure 24.10) gives the reliability of a given objective or group of objectives (*y*-axis) under the conditions of a particular stress scenario (*x*-axis). The *x*-axis of the resilience profile graph is constructed as a series of progressively more extreme disturbances in the form of scenarios and is therefore by definition an ordinal scale.

It is useful to observe that this graph also provides an assessment of the system's *robustness*, which is here defined as "the extent to which a system can keep performing within design specifications under increasing stress." In other words, robustness is defined as the level of pressure that the system can take without failing, while resilience refers to the ability of the system to cope (well or otherwise) with failure. A robust system is also resilient, but the opposite does not necessarily hold: a system can be resilient without being robust.

To scale resilience and robustness to maximum of 1, the area under the curve is divided by the area of a "completely robust" system, and robustness is divided with the number of points in the resilience profile diagram (i.e., the number of stress scenarios).

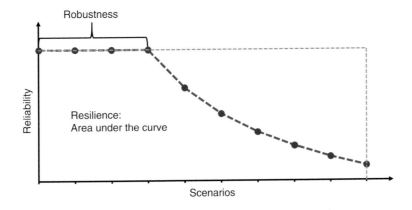

Figure 24.10 Resilience profile graph.

Resilience assessment for strategic planning, resulting in the above *Resilience Profile Graph*, comprises the following steps (Makropoulos et al., 2016):

1. Identify and evaluate the main properties of the water system under study to create a baseline scenario. Typical aspects include geography, geomorphology, climate, socio-cultural-economic background, water sources, urban typology-topology, infrastructure, demand types, etc.
2. Identify and define alternative design philosophies and set of interventions, including technical and non-technical measures, that are available for the system under study. These alternative configurations are also termed *levels of ambition* (see Makropoulos and Butler, 2010).
3. Set up one or several models (e.g., using the Urban Water Optioneering Tool [UWOT] – see Chapter 20) to simulate the urban water system.
4. Identify the external pressures/threats, which are expected to influence the system, and compose a set of scenarios of increased disturbance.
5. Run each model to the *same scenario*, and evaluate the performance of the system.
6. For each set of interventions, the *resilience profile graph* is composed.
7. To explore total system resilience and receive answers to different questions, one may
   a. Test different interventions to explore which improve more the system resilience.
   b. Test the same interventions under different scenarios to explore system resilience under a specific setting.

An example of using this approach is given in Box 24.5.

BOX 24.5   RESILIENT PLANNING OF WATERCITY

The Resilience Assessment for the strategic planning, discussed previously, was demonstrated using the semi-hypothetical "WaterCity," a typical, but anonymised Dutch City shown in Figure 24.11 (Makropoulos et al., 2018). The scope of this analysis was to provide evidence-based support to investment decisions of future water system configurations. The following three alternative sets of interventions (levels of ambition) were assessed:

1. The current state, *business as usual* (BAU), model of the system (following standard practices of the Dutch Water Sector).
2. *Next step* (NS) interventions that could be implemented tomorrow (or be included in the next 5-year plan and are to a large extent dependent on internal system variables). These interventions include local treatment of grey water at the household level, drinking water production from seawater, drinking water transport bundles, and aquifer storage recovery (ASR) stormwater collection technology for horticulture uses.
3. *Further ahead* (FA) interventions that need more time and possibly more investment (which would typically need 10 or more years to be implemented).

Each of the above system configurations was tested against a series of increased disturbance scenarios. Each scenario is composed of a set of pressures (model parameters) with varying degrees of magnitude and rates of change. The schematic overview of both scenarios and parameters is presented in Figure 24.12.

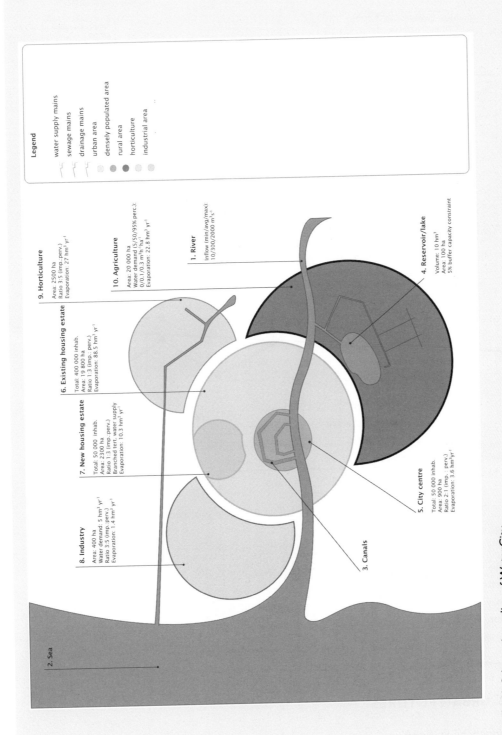

Legend

water supply mains
sewage mains
drainage mains
urban area
densely populated area
rural area
horticulture
industrial area

2. Sea

8. Industry
Area: 400 ha
Water demand: 5 hm³ yr⁻¹
Ratio 3:5 (imp.:perv.)
Evaporation: 1.4 hm³ yr⁻¹

7. New housing estate
Total: 50 000 inhab.
Area: 2300 ha
Ratio 1:3 (imp.:perv.)
Branched tert. water supply
Evaporation: 10.3 hm³ yr⁻¹

6. Existing housing estate
Total: 400 000 inhab.
Area: 19 800 ha
Ratio 1:3 (imp.:perv.)
Evaporation: 88.5 hm³ yr⁻¹

9. Horticulture
Area: 2500 ha
Ratio 3:5 (imp.:perv.)
Evaporation: 27 hm³ yr⁻¹

10. Agriculture
Area: 20 000 ha
Water demand (5/50/95% perc.):
0/0.1/0.3 m³h⁻¹ha⁻¹
Evaporation: 22.8 hm³ yr⁻¹

3. Canals

5. City centre
Total: 50 000 inhab.
Area: 900 ha
Ratio 2:1 (imp. : perv.)
Evaporation: 3.6 hm³yr⁻¹

1. River
Inflow (min/avg/max):
10/300/2000 m³s⁻¹

4. Reservoir/lake
Volume: 10 hm³
Area: 100 ha
5% buffer capacity constraint

Figure 24.11 Schematic outline of WaterCity.

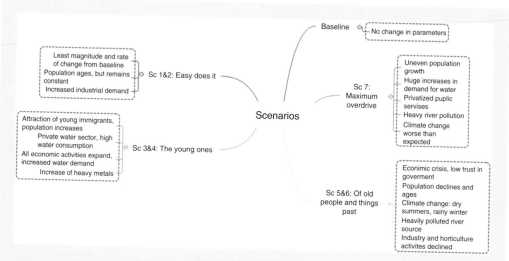

Figure 24.12 Schematic overview of scenarios and parameters.

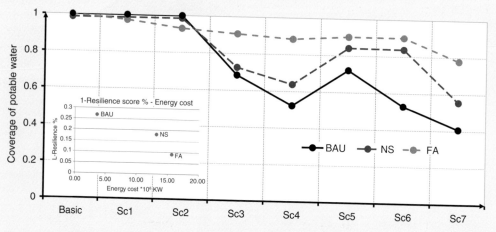

Figure 24.13 Resilience plot graph for BAU, NS, and FA configurations and trade-off between resilience and energy costs.

The simulation of the urban water cycle for the three system configurations, under alternative scenarios, was performed using UWOT (see Chapter 20). As a performance measure, the system's reliability towards *potable water coverage* was used, although other reliability metrics may also be used (including metrics directly relevant to urban drainage, such as CSO spills or flooding extent). The *Resilience Profile Graph* along with the trade-off between energy costs and resilience for the three system configurations are depicted in Figure 24.13.

24.6 URBAN FUTURES

Although Niels Bohr was surely correct when he said, "Prediction is very difficult, especially about the future," there is definite merit in assessing what might happen in the future to understand what options are feasible and what strategies are required to achieve them. It is sometimes argued that it is impossible to know *anything* about the future so it can be ignored, only to be responded to when it actually happens. Of course, only a little can be known about the future, but that little is important because *some* knowledge of the future is vital in making good decisions. In addition, learning what is knowable about the future enables action in the present that will contribute to achieving a desirable goal. Three main approaches to thinking about the future are commonly adopted: trend, precursor, and scenario analysis, and the latter is discussed here.

Scenario analysis consists of constructing explanatory "futures." such as those used to stress test the urban water system as the *x*-axis of the Resilience Profile described in the previous section. These futures describe broad social, economic, and political changes that *may* rather than *will* occur in the future. To that extent they are coherent, internally consistent, and plausible, typically including a narrative element and some quantitative indicators (Berkhout et al., 2002). Of the many scenarios produced, typically no attempt is made to predict which, if any, is most likely to occur. The most well-known are the socio-economic futures first developed for and used by the UK government: the Foresight programme (OST, 1998), the Intergovernmental Panel on Climate Change in the context of emission scenarios (IPCC, 2014), and the UK Climate Impacts Programme (UKCIP, 2001).

The Foresight Futures approach can be thought of as a generic framework based on two key drivers of future change: social values (*x*-axis) and systems of governance (*y*-axis), which are illustrated in Figure 24.14. Social values range from individualistic to more community-orientated ideals, including the political and economic implications that arise from them. Governance ranges from autonomy where power remains at local level, through regional and national levels, to interdependence where globalisation is the norm. Four futures can

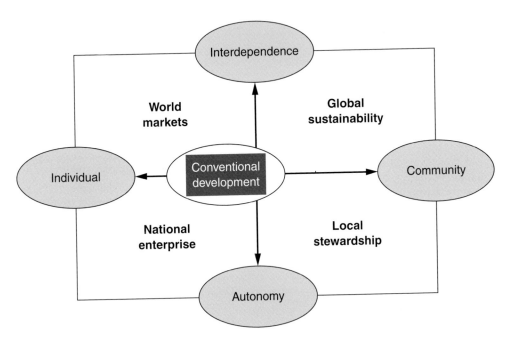

Figure 24.14 Socio-economic futures. (Adapted from OST. 1998. *Environmental Futures*, Department of Trade and Industry/Office of Science and Technology.)

Table 24.5 Proposed futures and their urban drainage implications

Future	Characteristics	Urban drainage components
Green	Ecologists are in charge. They have imposed a *balanced integrated approach* attempting to combine both "soft" and "hard" technologies in achieving sustainability of urban systems, including drainage. This scenario may be hard to imagine, because in most developed societies, only the engineers are licensed with the responsibility to build infrastructure. Soft measures include source controls (both legislated and voluntary).	On-site: • Minimisation. • SuDS. • Reuse and recycling. • Existing sewerage has very low flows and is being progressively abandoned or used for other purposes. • Engages users (who may subcontract responsibility). Problems occur due to: • On-site failures/misuse/abuse. • Transition from existing to new systems is costly and slow.
Technocratic	Publicly employed engineers are in charge of the water cycle and decide which technical approaches should be used and the ways the systems are managed. This scenario represents the classical approach, before water privatisation became the international norm. Application of largely proven technology, coupled with minimum risk, with duplication, and with large safety factors, engineers make sure that the systems do not fail. Long-term planning (Grand Master Plan).	End-of-pipe systems maintained: • Centralisation with large treatment plants. • Large fully sewered networks. • Continues to utilise valuable existing infrastructure assets. • Innovation limited although incremental changes in technology will solve new problems. Problems due to: • Lack of sustainability and high maintenance costs. • Inappropriate for large-scale new developments in developing world. • Costs rising. • Discourages user engagement and responsibility.
Privatisation	Investors and economists are in charge, and the mission is cost efficiency and shareholder dividends. This is achieved by control of labor costs and limited investments in both new and existing infrastructure. Main idea of the privatisation scenario is not to find the most effective solution to a technical problem, but to make, or at least minimise loss of, money. In the worst situation, the water service provider might define the level of service.	As above, but motives are profits. • Likely to continue to utilise serviceability of existing assets as long as possible (operating licence) and minimise investment in new systems. • A lot of subcontracting out of service provisions. Problems due to: • Service monopolies. • Lack of incentive to reduce demand (customer). • Reluctance to innovate due to risk responsibility framework. • Poor performance of subcontractors. • Mergers/acquisitions/lack of loyalty to staff. • Remoteness of end users (customers). • Unlikely that the true costs of service provision can be passed to users.
Business as usual	Continuation of what has been done for the past 30 years. Nobody is in charge. This scenario is poorly defined, because the current levels of drainage services and approaches vary widely between countries and between jurisdictions. Practically everywhere, it is recognised as (or called) not sustainable, yet it is the most common. Generally because of the large inertia of existing infrastructure in the developed countries, radical change is not feasible.	A confusing mixture of all of the above. • Recognises the need to continue to utilise the valuable existing infrastructure base, but introduction of innovation possible. • A perceived low-risk approach (always done this way). Unlikely to have many catastrophic failures. Problems: • Possibly all of those from above. • Unlikely to lead to sustainable systems in time. • Passes much of the financial burden to future generations. • Will not attract the best people into a career in this area.

Source: Schilling, W. 2003. Urban drainage—quo vadis? *Proceedings of XXX IAHR Congress*, Thessaloniki, Greece, 1–17.

BOX 24.6   URBAN FLOODING

The remit of the UK government's 2002 Foresight exercise on flooding was to produce a long-term vision for the future of flood and coastal defence. As part of that process, the Foresight futures were married with the UKCIP02 climate scenarios. Four emission levels—low, medium-low, medium-high, and high—were matched (approximately) with global responsibility, local stewardship, national enterprise, and world markets, respectively.

In their analysis of the urban flooding and urban drainage components, the authors ranked the importance of a wide range of drivers within a framework of source, pathway, and receptor. The driver with the greatest influence was precipitation and its spatial and temporal changes. Model studies revealed large (>50%) rises in flooding volume for sewer systems subjected to simple representations of climate-affected rainfall. Other important physical drivers were urban creep, increase in groundwater infiltration, and increase in receiving water pollution. Also thought to be influential in the future was the degree and type of regulation, public attitudes towards flooding, and the ability or willingness to pay for improvements (Evans et al., 2003).

then be developed that are conveniently placed in the four different quadrants of the axes: world markets, global sustainability, national enterprise, and local stewardship. The "position" of current development is also shown on Figure 24.14. A relevant application of this framework is given in Box 24.6.

A similar approach, but using different futures, was taken by the International Water Association/International Association on Hydraulic Engineering and Research (IWA/IAHR) Joint Committee on Urban Drainage (www.jcud.org). Their four futures are summarised in Table 24.5. In the first case, the green future, political ecology prevails and the environment is put first. The second future is technology based and describes a world where engineers are in charge and presumes that "technological fixes" will prevail. In the third future, the key issue is finance, and widespread privatisation is assumed. The final future is business as usual, which presumes that financial, technical, and political driving forces act together, as today, but not in a particularly coordinated way. Table 24.5 also includes notes on what type of urban drainage system might be prevalent in each future.

While the use of futures does give insights into future possible states and can help guide us to a more sustainable future, it does have its weaknesses (Butler, 2004):

- Technical solutions do not naturally or necessarily map onto any particular socio-economic future.
- Technical solutions need not be *exclusive* to any particular future but may be used in combination.
- The high sunk asset value, inertia, and conservatism of the water industry mean that any future that is not closely related to the present is unlikely.

That said, more detailed work under the WaND project has been attempted to logically and consistently map the state of urban water systems to six socio-political futures (Makropoulos et al., 2008).

PROBLEMS

24.1   Define *sustainable* development, and interpret its implications in terms of society, economics, and the environment. Illustrate your answer with examples from urban drainage.

24.2   A sustainable city is a contradiction in terms. Discuss.

24.3   Describe the six urban water management transition states, the socio-political drivers, and their delivery functionalities. Pick a city you know, and explain which state it is currently in.

24.4   Explain what you understand by the term sustainable urban drainage.

24.5   In what ways is centralised urban drainage unsustainable?

24.6   Describe three techniques or technologies that claim to be more sustainable, and cite evidence to support the claims.

24.7   What sustainability indicators would you use to evaluate urban drainage systems and why?

24.8   Explain what you understand by the term resilience and how it can be increased in urban drainage systems.

24.9   Explain the relationship between resilience, sustainability, and reliability. Discuss possible differences in working definitions and what each one highlights.

24.10 Explain the main steps to create a resilience profile graph for strategic planning.

24.11 Explain what you understand by urban futures, and explain how they can be used. Should an assessment be made of the future that is most likely to occur in the real world?

KEY SOURCES

Ashley, R., Blackwood, D., Butler, D., and Jowitt, P. 2004. *Sustainable Water Services: A Procedural Guide*, IWA Publishing.

Butler, D. and Memon, F.A. eds. 2006. *Water Demand Management*, IWA Publishing.

Butler, D., Memon, F.A., Makropoulos, C., Southall, A., and Clarke, L. 2010. *Guidance on Water Cycle Management for New Developments*, CIRIA C690.

Butler, D., Ward, S., Sweetapple, C., Astaraie-Imani, M., Diao, K., Farmani, R., and Fu, G. 2016. Reliable, resilient and sustainable water management: The Safe & SuRe approach. *Global Challenges*, 1(1), 63–77.

Hunt, D.V.L., Lombardi, D.R., Atkinson, S., Barber, A.R.G., Barnes, M., Boyko, C.T. et al., 2012. Scenario archetypes: Converging rather than diverging themes. *Sustainability*, 4(4), 740–772.

Makropoulos C., and Butler, D. 2010. Distributed water infrastructure for sustainable communities. *Water Resources Management*, 24(11), 2795–2816.

Maksimović, Č. and Tejada-Guibert, J.A. eds. 2001. *Frontiers in Urban Water Management. Deadlock or Hope?*, IWA Publishing.

Malmqvist, P.-A., Heinicke, G., Korrman, E., Stenstrom, T., and Svensson, G. 2006. *Strategic Planning of Sustainable Urban Water Management*, IWA Publishing.

Mugume, S.N., Gomez, D.E., Fu, G., Farmani, R., and Butler, D. 2015. A global analysis approach for investigating structural resilience in urban drainage systems. *Water Research*, 81, 15–26.

Novotny, V. and Brown, P. 2007. *Cities of the Future. Towards Integrated Sustainable Water and Landscape Management*, IWA Publishing.

REFERENCES

American Society of Civil Engineers/UN Educational, Scientific, and Cultural Organization (ASCE/UNESCO). 1998. *Sustainability Criteria for Water Resource Systems*, Cambridge University Press.

Ashley, R.M., Souter, N., Butler, D., Davies, J., Dunkerley, J., and Hendry, S. 1999. Assessment of the sustainability of alternatives for the disposal of domestic sanitary waste. *Water Science and Technology*, 39(5), 251–258.

Ashley, R.M., Blackwood, D., Butler, D., Davies, J.W., Jowitt, P., and Smith, H. 2003. Sustainable decision making for the UK water industry. *Proceedings of the Institution of Civil Engineers, Engineering Sustainability*, 156(ES1), 41–49.

Berkhout, F., Hertin, J., and Jordan, A. 2002. Socio-economic futures in climate change impact assessment: Using scenarios as "learning machines." *Global Environmental Change*, 12, 83–95.

Brede, M. and B. J. M. de Vries. 2009. Networks that optimize a tradeoff between efficiency and dynamical resilience. *Physics Letters A*, 373(43), 3910–3914.

Brown, R., Keath, N., and Wong, T. 2009. Urban water management in cities: Historical, current and future regimes. *Water Science and Technology*, 59(5), 847–855.

Bruce, J.P. 1992. *Meteorology and Hydrology for Sustainable Development*, World Meteorological Organization Report No. 769.

BS EN 997: 2003. *WC Pans and WC Suites with Integral Trap*.

Butler, D. 2004. Urban water—Future trends and issues, in *Hydrology—Science and Practice for the 21st Century* (eds B. Webb, M. Acreman, Č. Maksimović, H. Smithers, and C. Kirby), British Hydrological Society, Vol II, 233–242, IAHS Press.

Butler, D. and Dixon, A. 2002. Financial viability of in-building grey water recycling. *Proceedings of International Conference on Wastewater Management abd Technologies for Highly Urbanized Cities*, Hong Kong, June, 57–63.

Butler, D. and Maksimović, Č. 1999. Urban water management—Challenges for the third millennium. *Progress in Environmental Science*, 1(3), 213–235.

Butler, D. and Parkinson, J. 1997. Towards sustainable urban drainage. *Water Science and Technology*, 35(9), 53–63.

Butler, D., Farmani, R., Fu, G., Ward, S., Diao, K., and Astaraie-Imani, M. 2014. A new approach to urban water management: Safe and SuRe. *Procedia Engineering*, 89, 347–354.

Department for Environment, Food and Rural Affairs (DEFRA). 2013. *Sustainable Development Indicators*. https://www.gov.uk/government/statistics/sustainable-development-indicators-sdis.

Dixon, A., Butler, D., and Fewkes, A.F. 1999a. Water saving potential of domestic water reuse systems using greywater and rainwater in combination. *Water Science and Technology*, 39(5), 25–32.

Dixon, A., Butler, D., and Fewkes, A.F. 1999b. Guidelines for greywater reuse—Health issues. *CIWEM Water and Environmental Management Journal*, 13(5), 322–326.

Dixon, T., Eames, M., Hunt, M., and Lannon, S. (eds.). 2014. *Urban Retrofitting for Sustainability: Mapping the Transition to 2050*, Routledge.

Drangert, J.-O. 1998. Fighting the urine blindness to provide more sanitation options. *Water SA*, 24(2), 157–164.

Evans, E.P., Thorne, C.R., Saul, A., Ashley, R., Sayers, P.N., Watkinson, A. et al., 2003. *An Analysis of Future Risks of FLooding and Coastal Erosion for the UK between 2030–2100. Overview of Phase 2*. Foresight Flood and Coastal Defence Project, Office of Science and Technology.

Gigerl, T. and Rosenwinkel, K.-H. 1998. Life cycle assessment of sewer systems. Conceptual idea. *Fourth International Conference on Developments in Urban Drainage Modelling*, UDM '98, London, September, 673–680.

Gross, A., Maimon, A., Alfiya, Y., and Friedler, E. 2015. *Greywater Reuse*, CRC Press.

Hanaeus, J., Hellstorm, D., and Johansson, E. 1997. A study of a urine separation system in an ecological village in northern Sweden. *Water Science and Technology*, 35(9), 153–160.

Holling, C.S. 1973. Resilience and stability of ecological systems. *Annual Review of Ecology and Systematics*, 4, 1–23.

Hvitved-Jacobsen, T., Nielsen, P.H., Larsen, T., and Jensen, N.A. (eds.). 1995. The sewer as a physical, chemical and biological reactor. Water Science and Technology series, Book 31, International Water Association, London.

Intergovernmental Panel on Climate Change (IPCC). 2014. *Climate Change 2014 Synthesis Report Summary Chapter for Policymakers*. IPCC 31. doi:10.1017/CBO9781107415324.

Koop, S.H.A. and van Leeuwen, C.J. 2015. Assessment of the sustainability of water resources management: A critical review of the city blueprint approach. *Journal of Water Resources Management*, 29, 5649–5670.

Lansey, K. 2012. Sustainable, robust, resilient, water distribution systems. *14th Water Distribution Systems Analysis Conference, WDSA 2012*, Adelaide, Australia, 1–18.

Littlewood, K., Memon, F.A., and Butler, D. 2007. Downstream implications of ultra-low flush WCs. *Water Practice and Technology*, 2(2), doi:10.2166/wpt.2007.037.

Lundqvist, J., Turton, A., and Narain, S. 2001. Social, institutional and regulatory issues, in *Frontiers in Urban Water Management: Deadlock or Hope* (eds C. Maksimovic and J. A. Tejada-Guilbert), IWA Publishing, Cornwall, 344–398.

Makropoulos, C., Memon, F.A., Shirley-Smith, C., and Butler, D. 2008. Futures: An exploration of scenarios for sustainable urban water management. *Water Policy*, 10(4), 345–373.

Makropoulos, C., Nikolopoulos, D., Palmen, L., Kools, S., Segrave, A., Vries, D., Koop, S. et al. 2018. A resilience assessment method for Urban Water Systems, *Urban Water Journal*, 15.

Makropoulos, C., Palmen, L., Kools, S., Segrave, A., Vries, D., Koop, S., van Alphen, H.J., Vonk, E., and van Thienen, P. 2016. Developing water wise cities: A methodological proposition. KWR Watercycle Research Institute, BTO Report 400695/044.

Mays, L.W. (ed.). 1989. *Reliability Analysis of Water Distribution Systems*. ASCE.

Melville-Shreeve, P., Ward, S., and Butler, D. 2016. Dual-purpose rainwater harvesting system design, in *Sustainable Surface Water Management. A Handbook for SUDS* (eds S.M. Charlesworth and C.A. Booth), Wiley Blackwell, 255–268.

Memon, F.A., Butler, D., Han, W., Liu, S., Makropoulos, C., Avery, L., and Pidou, M. 2005. Economic assessment tool for greywater recycling systems. *Institution of Civil Engineers Engineering Sustainability*, 158(ES3), 155–161.

Memon, F.A., Zheng, Z., Butler, D., Shirley-Smith, C., Liu, S., Makropoulos, C., and Avery, L. 2007. Life cycle impact assessment of greywater treatment technologies for new developments. *Journal of Environmental Monitoring and Assessment*, 129, 27–35.

Millán, A.M., Memon, F.A., Butler, D., and Littlewood, K. 2007. User perceptions and basic performance of an innovative WC, in *Water Management Challenges in Global Change* (eds B. Ulanicki, K. Vairavamoorthy, D. Butler, P.L.M. Bounds, and F.A. Memon), Taylor and Francis, 629–633.

Morecroft, M.D., Crick, H.Q.P., Duffield, S.J., and Macgregor, N.A. 2012. Resilience to climate change: Translating principles into practice. *Journal of Applied Ecology* 49(3), 547–551.

Mugume, S.N. and Butler, D. 2016. Evaluation of functional resilience in urban drainage and flood management systems using a global analysis approach. *Urban Water Journal*. doi:10.1080/1573062X.2016.1253754.

Nelson, D.R., Adger, W.N., and Brown, K. 2007. Adaptation to environmental change: Contributions of a resilience framework. *Annual Review of Environment and Resources* 32, 395–419.

Nolde, E. 2014. Greywater recycling in buildings, in *Water Efficiency in Buildings: Theory and Practice* (ed K. Adeyeye), Chapter 10, 169–189.

Omambala, I., Russell, N., Tompkins, J., Bremner, S., Millar, R., Rathouse, K., and Cooper, M. 2011. *Evidence Base for Large-Scale Water Efficiency. Phase II Final Report*. Waterwise. www.waterwise.org.uk/data/resources/12/evidence-base-report_april-2011_final.pdf.

Office of Science and Technology (OST). 1998. *Environmental Futures*, Department of Trade and Industry/Office of Science and Technology.

Otterpohl, R., Grottker, M., and Lange J. 1997. Sustainable water and waste management in urban areas. *Water Science and Technology*, 35(9), 121–133.

Park, J., Seager, T.P., Rao, P.S.C., Convertino, M., and Linkov, I., 2013. Integrating risk and resilience approaches to catastrophe management in engineering systems. *Risk Analysis*, 33(3), 356–367.

Parkinson, J., Schütze, M., and Butler, D. 2005. Modelling the impacts of domestic water conservation on the sustainability of the urban wastewater system. *Chartered Institution of Water and Environmental Management, Water and Environment Journal*, 19(1), 49–56.

Pickett, S.T.A., McGrath, B., Cadenasso, M.L., and Felson, A.J. 2014. Ecological resilience and resilient cities. *Building Research and Information*, 42, 143–157.

Pizzol, M. 2015. Life cycle assessment and the resilience of product systems. *Journal of Industrial Ecology*, 19(2), 296–306.

Rozos, E., Makropoulos, C., and Maksimović, Č. 2013. Rethinking urban areas: An example of an integrated blue-green approach. *Water Science and Technology: Water Supply*, 13(6), 1534–1542.

Schilling, W. 2003. Urban drainage—Quo vadis?. *Proceedings of XXX IAHR Congress*, Thessaloniki, Greece, 1–17.

Scholz, R.W., Blumer, Y.B., and Brand, F.S. 2012. Risk, vulnerability, robustness, and resilience from a decision-theoretic perspective. *Journal of Risk Research*, 15(3), 313–330.

Sharp, L. 2006. Water demand management in England and Wales: Constructions of the domestic water user. *Journal of Environmental Planning and Management*, 49(6), 869–889.

Sharp, L. 2017. *Reconnecting People and Water. Public Engagement and Sustainable Urban Water Management*, Earthscan.

Tjandraatmadja, G., Burn, S., McLaughlin, M., and Biswas, T. 2005. Rethinking urban water systems—Revisiting concepts in urban wastewater collection and treatment to ensure infrastructure sustainability. *Water Science and Technology*, 5(2), 145–154.

UK Climate Impacts Programme (UKCIP). 2001. *Socio-economic Scenarios for Climate Change Impact Assessment: A Guide to Their Use in the UK Climate Impacts Programme*, United Kingdom Climate impacts Programme, Oxford.

Van Leeuwen, K. and Marques, R.C. 2013. *Current State of Sustainability of Urban Water Cycle Services. Transition to the Urban Water Services of Tomorrow (TRUST)*. Report D11.1. http://www.trust-i.net/downloads/index.php?iddesc = 68.

Van Leeuwen, C.J., Frijns, J., Van Wezel, A., and Van De Ven, F.H.M. 2012. City blueprints: 24 indicators to assess the sustainability of the urban water cycle. *Journal of Water Resources Management*, 26, 2177–2197.

Van Leeuwen, C.J., Koop, S.H.A., and Sjerps, R.M.A. 2016. City Blueprints: Baseline assessments of water management and climate change in 45 cities. *Environment, Development and Sustainability*, 18(4), 1113–1128.

Water UK. 2008. *Sustainable Water. State of the Water Sector Report*. www.water.org.uk.

Water UK. 2012. *Sustainability Indicators 2009/10*. www.water.org.uk.

Winblad, U. and Simpson-Hébert, M. eds. 2004. *Ecological Sanitation, revised and enlarged edition*, Stockholm Environment Institute.

World Commission on Environment and Development. 1987. *Our Common Future*, Oxford University Press.

Zadeh, S.M., Hunt, D.V.L., Lombardi, D.R., and Rogers, C.D.F. 2013. Shared urban greywater recycling systems: Water resource savings and economic investment. *Sustainability*, 5(7), 2887–2912.

Index